Neurotransmitters and Epilepsy

Contemporary Neuroscience

Neurotransmitters and Epilepsy

Edited by
Phillip C. Jobe and Hugh E. Laird II

Humana Press • Clifton, New Jersey

Library of Congress Cataloging-in-Publication Data

Neurotransmitters and epilepsy.

    (Contemporary neuroscience)
    Includes bibliographies and index.
    1. Epilepsy—Pathophysiology. 2. Epilepsy—Animal models. 3. Neuro transmitters. I. Jobe, Phillip C. II. Laird, Hugh E. III. Series.
    [DNLM: 1. Epilepsy—physiopathology. 2. Neuroregulators—physiology. WL 385 N4935]

RC372.5.N48   1987   616.8'53071   87-27579
ISBN 0-89603-101-2

Preface

The idea for this book has evolved from our desire to present a conceptual approach to the study of neurotransmitters in epilepsy. Such an approach requires an understanding of the function of neurotransmitter systems in various experimental models of epilepsy. Toward this goal we have included in this book chapters on neurotransmitter systems in nine different epilepsy models. To complete the coverage of this topic, there is a chapter on the role of selected neurotransmitters in the various types of human epilepsies. In the final chapter the editors integrated the neurotransmitter data from the various epilepsy models into a matrix from which a better understanding of the function of these neurotransmitters in modulating epileptogenesis may be obtained.

The information found in this book is not the result of a symposium on this topic, but rather a review of available information on neurotransmitters in each of the experimental models. The evidence is presented by knowledgeable researchers using these models. This approach gives a current, broader, and more thorough presentation of each of the topics. We therefore feel that this is not just a glimpse at the subject matter, but a panoramic view of the topic.

The selection of the experimental models of epilepsy for inclusion in this book was determined by the desire of the editors to give a broad exposure to neurotransmitter information available from both genetic and nongenetic epilepsy models. Both of these types of experimental models have given important information regarding neurotransmitter function in modulating the epileptic state. The types of experimental questions addressed in each model must be carefully considered, however, so that the model system most appropriate for the question can be selected.

The goal of the editors is to provide the reader with a source of information on the roles of neurotransmitter systems in currently used and biologically diverse epilepsy models. In addition, an attempt is made to provide an updated view of those neurotransmitter systems thought to play a role in the human epileptic state. Finally, in the last chapter the editors have sought to construct an integrated conceptual framework for viewing the neurotransmitter data obtained in the various models. Such a chapter was included to give the editors an opportunity to express their interpretation of the information presented on the various model systems and to develop a broader perspective on the role of neurotransmitter systems in regulating the susceptibility to and severity of the epileptic state.

Phillip C. Jobe
Hugh E. Laird II

Dedication

We are deeply indebted to Drs. Albert L. Picchioni and Lincoln Chin, our mentors. They ignited our interest in epilepsy research through their dedication, diligence, integrity, and creativity. *Neurotransmitters and Epilepsy* is inscribed to these men as a mark of our respect and affection.

Acknowledgment

The editors would like to thank each of the chapter authors for their contributions since without their work this book would not have been possible. We are grateful to Mr. Thomas Lanigan for giving us the opportunity to edit a book on this topic. In addition, we are deeply indebted to Wendee Higa and Annette Sherman for their unstinting efforts during the editorial process. The participation of Ms. Donna Birkhahn in preparing the index is greatly appreciated. Finally, we thank our wives, Susie and Marilyn, for their encouragement and understanding during the preparation of this book.

Contents

Neurotransmitter Systems and Epilepsy: *An Overview*
John W. Dailey and **Phillip C. Jobe**

Epilepsy-Prone Mice: *Genetically Determined Sound-Induced Seizures*
A. G. Chapman and **B. S. Meldrum**

The Spontaneously Epileptic Mongolian Gerbil
Peter Lomax, Randall J. Lee, and Richard W. Olsen

The Genetically Epilepsy-Prone Rat
Hugh E. Laird II and Phillip C. Jobe

The Epileptic Chickens
D. D. Johnson and **J. M. Tuchek**

Biochemical and Pharmacologic Studies of Neurotransmitters in the Kindling Model
James O. McNamara, Douglas W. Bonhaus, Barbara J. Crain, Randy L. Gellman, and **Cheolsu Shin**

In Vitro Models of Epilepsy
Roger D. Traub, Robert K. S. Wong, and Richard Miles

Experimental Epilepsy Induced by Direct Topical Placement of Chemical Agents on the Cerebral Cortex
Charles R. Craig and Brenda K. Colasanti

Seizures Induced by Convulsant Drugs
Carl L. Faingold

The Role of Neurotransmitters in Electroshock Seizure Models
Ronald A. Browning

Neurotransmitters in Human Epilepsy
Mitchell J. Kresch, Bennett A. Shaywitz, Sally E. Shaywitz, George M. Anderson, James L. Leckman, and Donald J. Cohen

Neurotransmitter Systems and the Epilepsy Models: *Distinguishing Features and Unifying Principles*
Phillip C. Jobe and Hugh E. Laird II

Contributors

GEORGE M. ANDERSON · *Laboratory Medicine and Yale Child Study Center, Yale University, New Haven, Connecticut*

DOUGLAS W. BONHAUS · *Department of Neurology, Duke University Medical Center and Epilepsy Research Laboratory, Veterans Administration Medical Center, Durham, North Carolina*

ROBERT A. BROWNING · *Department of Medical Physiology and Pharmacology, School of Medicine, Southern Illinois University, Carbondale, Illinois*

A. G. CHAPMAN · *Institute of Psychiatry, Department of Neurology, De Crespigny Park, London*

DONALD COHEN · *Department of Pediatrics and Yale Child Study Center, Yale University, New Haven, Connecticut*

BRENDA K. COLASANTI · *Department of Pharmacology and Toxicology, West Virginia University Medical Center, Morgantown, West Virginia*

CHARLES R. CRAIG · *Department of Pharmacology and Toxicology, West Virginia University Medical Center, Morgantown, West Virginia*

BARBARA J. CRAIN · *Departments of Anatomy and Pathology, Duke University Medical Center and Epilepsy Research Laboratory, Veterans Administration Medical Center, Durham, North Carolina*

JOHN W. DAILEY · *Department of Basic Sciences, College of Medicine at Peoria, University of Illinois, Peoria, Illinois*

CARL L. FAINGOLD · *Department of Pharmacology, School of Medicine, Southern Illinois University, Springfield, Illinois*

RANDY L. GELLMAN · *Department of Pharmacology, Duke University Medical Center and Epilepsy Research Laboratory, Veterans Administration Medical Center, Durham, North Carolina*

PHILLIP C. JOBE · *Department of Basic Sciences, College of Medicine at Peoria, University of Illinois, Peoria, Illinois*

D. D. JOHNSON · *Department of Pharmacology, College of Medicine, University of Saskatchewan, Saskatoon, Saskatchewan, Canada*

MITCHELL J. KRESCH · *Department of Pediatrics, Yale University, New Haven, Connecticut*

HUGH E. LAIRD II · *Department of Pharmacology and Toxicology, College of Pharmacy, University of Arizona, Tucson, Arizona*

JAMES L. LECKMAN · *Department of Pediatrics and Yale Child Study Center, Yale University, New Haven, Connecticut*

RANDALL J. LEE · *Department of Pharmacology, School of Medicine and the Brain Research Institute, University of California, Los Angeles, California*

PETER LOMAX · *Department of Pharmacology, School of Medicine and the Brain Research Institute, University of California, Los Angeles, California*

B. S. MELDRUM · *Institute of Psychiatry, Department of Neurology, De Crespigny Park, London*

JAMES O. MCNAMARA · *Departments of Neurology and Pharmacology, Duke University Medical Center and Epilepsy Center, Veterans Administration Medical Center, Durham, North Carolina*

RICHARD MILES · *University of Texas Medical Branch, University of Texas, Galveston, Texas*

RICHARD W. OLSEN · *Department of Pharmacology, School of Medicine and the Brain Research Institute, University of California, Los Angeles, California*

BENNETT A. SHAYWITZ · *Departments of Pediatrics and Neurology, Yale Child Study Center, Yale University, New Haven, Connecticut*

SALLY B. SHAYWITZ · *Department of Neurology and Yale Child Study Center, Yale University, New Haven, Connecticut*

CHEOLSU SHIN · *Department of Neurology, Duke University Medical Center and Epilepsy Research Laboratory, Veterans Administration Medical Center, Durham, North Carolina*

ROGER D. TRAUB · *IBM T. J. Watson Research Center, Yorktown Heights, New York and Neurological Institute, New York, New York*

J. M. TUCHEK · *Department of Pharmacology, College of Medicine, University of Saskatchewan, Saskatoon, Saskatchewan, Canada*

ROBERT K. S. WONG · *University of Texas Medical Branch, University of Texas, Galveston, Texas*

Neurotransmitter Systems and Epilepsy

An Overview

John W. Dailey and Phillip C. Jobe

1. Overview

A conceptual matrix encompassing the role of neurotransmitter systems in the etiology of epilepsy is emerging. Interesting and provocative data are developing both from animal and human investigations. Experimentally, data have been obtained from four potentially divergent research approaches. First, normal nervous systems have been exposed to seizure-provoking stimuli. The intent has been to determine the neurochemical consequences of the stimulus alone or of the stimulus plus the resulting seizure activity. The assumption has been that such consequences in normal nervous systems may provide clues to an understanding of epilepsy.

Second, neurotransmitters in normal nervous systems have been modified experimentally so that the consequences of these imposed neurochemical alterations on seizure activity could be determined. The assumption in these studies has been that induced alterations in otherwise normal nervous systems may yield clues to the causes of epilepsy.

Third, the effects of clinically useful antiepileptic drugs on neurotransmitter systems in normal nervous systems have been extensively investigated. Attempts have been made to determine whether the neurochemical alterations coincide with the capacity of the same treatment to alter experimentally induced seizure activity. In some instances, precautions have been made to determine whether the induced neurochemical changes are produced by amounts of the antiepileptic drug that are within the range of serum concentrations associated with seizure control in human epileptics. The assumption has been that the neurochemical alterations produced under these conditions reflect the abnormalities responsible for epilepsy.

Fourth, genetically epileptic animals have been examined to identify abnormalities in neurotransmitter systems that are responsible for seizure predisposition. The assumption has been that the factors responsible for epilepsy in these abnormal subjects are neurochemical analogs of the dysfunctional conditions that underlie human epilepsy.

At this point in time, experimental protocols have been designed so that epilepsy models produced by kindling, electroshock, chemoshock, topical convulsants, and the in vitro slice preparations have been primarily restricted to the first three research approaches (McNamara et al., 1986; Browning, 1986; Faingold, 1986; Craig and Colasanti, 1986; Traub and Wong, 1986). Thus, neurochemical data from these models have been obtained from animals that are not epileptic when the seizure-inducing stimulus is applied and that fail to exhibit any type of seizure predisposition, except perhaps in response to the seizure-evoking stimulus.

The epilepsy-prone mice, genetically epilepsy-prone rats, epileptic gerbils, epileptic chickens, and some epileptic humans fall into the fourth category of epilepsy models. Through the use of appropriate experimental protocols, neurochemical data from these models may reveal the neurobiological abnormalities that exist before the seizure-inducing stimulus is applied (Chapman and Meldrum, 1986; Laird and Jobe, 1986; Lomax et al., 1986; Johnson and Tuchek, 1986; Kresch et al., 1986). Consequently, we may have some confidence that these abnormalities represent the underlying causes of seizure predisposition.

Observations that neurochemical abnormalities may be caused by seizure-evoking stimuli or the resulting seizure episodes provide a basis for speculation. Perhaps these abnormalities represent nothing more than the neurobiological consequences of seizure episodes. Are these changes also the causes of seizure predisposition inherent in the epileptic subject before a first seizure-evoking stimulus is experienced? Are these the determinants of seizure pattern or severity in epileptic subjects? If the neurochemical consequences contribute to an increased probability of future seizure activity, do they cause kindling?

Neurochemical alterations caused by seizure episodes may be anticonvulsant in nature. Therefore, they may be responsible for terminating seizure episodes and rendering the subject temporarily refractory to future seizures. Alternatively, neurochemical consequences of seizure activity might be devoid of a seizure-regulating role. Perhaps they influence other types of behavior or psychic activity that are not manifested as seizure episodes.

Definitive resolutions of these issues are not available at this time. However, provocative clues are emerging. Some paradigms such as kindling have been used to determine the neurochemical abnormalities that result during the process of seizure development (McNamara et al., 1986). Which changes occur early in the process before seizures occur? Which ones precede the onset of spontaneous convulsions? Genetic models of epilepsy have been used to examine neurochemical abnormalities that may be inherited with seizure predisposition and predetermined levels of seizure severity (Chapman and Meldrum, 1986; Laird and Jobe, 1986; Lomax et al., 1986; Johnson and Tuchek, 1986). Because of their genetic background, these animals are known to have a high degree of seizure susceptibility, as well as abnormally low seizure thresholds, even though they have not each been exposed to a seizure-provoking stimulus. Thus, the neurochemistry of these animals can be examined in the seizure-naive state. Such studies provide assurance that the detectable abnormalities have not occurred as a consequence of seizure activity. Are these abnormalities similar to those that develop during the process of kindling? Are they related to the abnormalities that occur in nonepileptic animals subjected to multiple electroshock or chemoshock seizure episodes? Are they similar to the neurochemical alterations associated with indices of seizure generation in the in vitro slice preparation?

From a more comprehensive point of view, the data from all of these diverse models may eventually provide a single unifying hypothesis of the epilepsies. Is a single neurochemical defect responsible for all of the epilepsies? Will we eventually realize that the epilepsies are not a group of different diseases? Are the varied clinical or overt manifestations caused merely by variations in the interaction between the complex of environmental and endogenous factors with the single underlying cause?

At the present stage of understanding, it is probably reasonable to conclude that seizures in the various animal models of epilepsy are not regulated by the same neurotransmitter systems (Jobe and

Laird, 1981). For example, noradrenergic terminal fields play an important role in regulating audiogenic seizure intensity in the genetically epilepsy-prone rat (Ko et al., 1982). Similar experiments from the same laboratory strongly suggest that dopamine plays a much more important role in regulating audiogenic seizures in the genetically epileptic DBA/2J mouse (Dailey and Jobe, 1984). Does this dicotomy mean that audiogenic seizures are regulated by different neuronal tracts or different brain areas in these two models? Are the relevant neurotransmitter receptors in the genetically epileptic mouse activated by dopamine, whereas the analogus receptors in the genetically epilepsy-prone rat brain require norepinephrine?

These observations do not prove the absence of a single unifying abnormality responsible for all types of epilepsy. However, they do emphasize that seizure-regulating neurochemical differences exist among the various models. This divergence of neurochemical seizure regulation is compatible with the concept that the epilepsies are a group of disorders. In our view, the existence of a common neurotransmitter defect underlying all seizure episodes and disorders in man or other animals is doubtful. Indeed, the different epileptic conditions may occur because different underlying abnormalities are responsible for their emergence (Jobe and Laird, 1981). As the search for the neurochemical basis of epileptogenesis continues, it is becoming increasingly clear that different neurochemical abnormalities cause different types of epilepsy.

The possibility of a singular neurotransmitter defect responsible for all of the epilepsies is also challenged by the enormous complexity of central nervous system circuitry and functional neurochemical anatomy. For several years, evidence has implicated the inhibitory neurotransmitters in seizure mechanisms. More recently attention has been focused on the excitatory and potentially neurotoxic amino acid neurotransmitters. Assumptions have sometimes been made that a deficit of neurotransmission at inhibitory synapses, or an increment at excitatory synapses, may form the basis of seizure disorders. However, in a given neuronal circuit or cascade, either an excess or a deficit of excitatory or inhibitory transmission could produce the same effect on the level of neuronal activity in other parts of the brain. For example, a deficit of an excitatory neurotransmitter at synapses with inhibitory neurons could cause an excessive level of downstream neuronal activity. Failure of excitatory transmitters to activate inhibitory neurons could conceptually lead to widespread inhibitory deficits within the central nervous system and consequently to seizure episodes. In contrast, an increment of an inhibitory neurotransmitter at synapses with other inhibitory neurons could also cause an excessive level of neuronal activity throughout the brain. Excessive inhibition of inhibitory neurons would lead to seizure episodes. These considerations suggest that neurotransmitters that are inhibitory to cell firing for individual neurons may actually participate in seizure genesis. It is also conceivable that excitatory neurotransmitters may participate in seizure suppression.

Such complexities in the circuitry and functioning of the central nervous system provide additional possibilities for the existence of a multiplicity of neurochemical causes of epilepsy. Conceptually, one type of epilepsy could be caused by a deficit in a particular excitatory neurotransmitter, whereas another type of epilepsy could result from an increment in this same excitatory substance, albeit the deficit would probably not occur in the same synapses as the increment.

Two caveats are especially relevant to the possibility that several defects rather than one underlie the epilepsies. First, experimental conditions used to test for the roles of the neurotransmitters in seizures may produce a multiplicity of effects (Faingold, 1986). Under such circumstances it may be unclear whether the convulsant effects of the drug are caused by a reduction in GABAergic transmission or by alterations in ionic movements and distributions. As another example, electroconvulsive stimuli alter numerous neuronal elements, including the multiplicity of neurotransmitter systems (Browning, 1986). Moreover, most attempts to genetically select for seizure predisposition have been associated with several neurochemical alterations (Chapman and Meldrum, 1986; Laird and Jobe, 1986; Lomax et al., 1986; Johnson and Tuchek, 1986). Which of these is etiologically significant?

The second caveat derives from the observation that in studies of disease mechanisms biological systems can respond to an experimental condition without that response being of pathophysiological significance to the disease under study. This would be true even if the experimental manipulation were capable of producing only one neurochemical effect. In epilepsy studies, we might assume that a completely selective drug was given to alter only one neurochemical factor and that as a result seizures occurred. Such an observation would suggest that the manipulated system has the capacity to determine the presence or absence of seizures. Whether the system actually regulates seizures in epilepsy is not definitively answered by this type of experiment. As an analogy, acetylcholine occurs within the cardiovascular system. Exogenously administered acetylcholine causes vasodilation. Drugs that release acetylcholine cause vasodilation indirectly by releasing acetylcholine into the blood stream. Drugs that block the effects of acetylcholine prevent the vasodilation caused by acetylcholine. These circumstances show that acetylcholine has the capacity to regulate vascular diameter. They do not prove that acetylcholine participates in the physiological control of blood pressure. Neither do they prove that lack of acetylcholine participates in the pathophysiology of hypertension. Indeed, most cardiovascular authorities have concluded that factors other than acetylcholine are responsible for blood pressure regulation and/or the appearance of hypertension.

2. Conclusion

In summary, the roles of the neurotransmitter systems in the epilepsies have been revealed through studies of a multiplicity of models. Most investigations have employed experimental protocols in which the central nervous systems were not epileptic at the time the seizure-inducing stimulus was applied and that were not characterized by seizure predisposition. However, a few protocols have utilized animals that are exquisitely susceptible to seizures and do not require exposure to a seizure-provoking stimulus to establish the presence of the seizure predisposition. Interpretation of data derived from both types of experimental approaches are subject to ambiguities. Unequivocal conclusions regarding the pathophysiology of the epilepsies are rarely possible. The likelihood that a single neurochemical defect is responsible for all of the epilepsies seems remote. Indeed, it is becoming increasingly apparent that different neurochemical abnormalities underlie different types of epilepsy. One of the pressing tasks of epilepsy research is to identify animal models of epilepsy that can serve as neurochemical analogs of correspondingly determined human epileptic disorders.

References

Browning, R. A.: The Role of Neurotransmitters in Electroshock Seizure Models. In: *Neurotransmitters and Epilepsy* (P. C. Jobe and H. E. Laird, II, eds.) Humana, New Jersey, 1986.

Chapman, A. G. and Meldrum, B. S.: Epilepsy Prone Mice: Genetically Sound-Induced Seizures. In: *Neurotransmitters and Epilepsy* (P. C. Jobe and H. E. Laird, II, eds.) Humana, New Jersey, 1986.

Craig, C. R. and Colasanti, B. K.: Experimental Epilepsy Induced by Direct Topical Placement of Chemical Agents on the Cerebral Cortex. In: *Neurotransmitters and Epilepsy* (P. C. Jobe and H. E. Laird, II, eds.) Humana, New Jersey, 1986.

Dailey, J. W. and Jobe, P. C.: Effect of increments in the concentration of dopamine in the central nervous system on audiogenic seizures in DBA/2J mice. *Neuropharmacology* **23**(9): 1019–1024, 1984.

Faingold, C. L.: Seizures Induced by Convulsant Drugs. In: *Neurotransmitters and Epilepsy* (P. C. Jobe and H. E. Laird, II, eds.) Humana, New Jersey, 1986.

Jobe, P. C. and Laird, H. E.: Neurotransmitter abnormalities as determinants of seizure susceptibility and intensity in the genetic models of epilepsy. *Biochem. Parmacol.* **30**(23): 3137–3144, 1981.

Johnson, D. D. and Tuchek, J. M.: The Epileptic Chickens. In: *Neurotransmitters and Epilepsy* (P. C. Jobe and H. E. Laird, II, eds.) Humana, New Jersey, 1986.

Ko, K. H., Dailey, J. W., and Jobe, P. C.: Effect of increments in norepinephrine concentrations on seizure intensity in the genetically epilepsy-prone rat. *J. Pharmacol. Exp. Ther.* **222**(3): 662–669, 1982.

Kresch, M. J., Shaywitz, B. A., Shaywitz, S. B., Anderson, G. M., Leckman, J. L., and Cohen, D.: Neurotransmitters in Human Epilepsy. In: *Neurotransmitters and Epilepsy* (P. C. Jobe and H. E. Laird, II, eds.) Humana, New Jersey, 1986.

Laird, H. E. and Jobe, P. C.: The Genetically Epilepsy-Prone Rat. In: *Neurotransmitters and Epilepsy* (P. C. Jobe and H. E. Laird, II, eds.) Humana, New Jersey, 1986.

Lomax, P., Lee, R. J., and Olsen, R. W.: The Spontaneously Epileptic Mongolian Gerbil. In: *Neurotransmitters and Epilepsy* (P. C. Jobe and H. E. Laird, II, eds.) Humana, New Jersey, 1986.

McNamara, J. O., Bonhaus, D. W., Crain, B. J., Gellman, R. L., and Shin, C.: Biochemical and Pharmacologic Studies of Neurotransmitters in The Kindling Model. In: *Neurotransmitters and Epilepsy* (P. C. Jobe and H. E. Laird, II, eds.) Humana, New Jersey, 1986.

Traub, R. D., Wong, R. K. S., and Miles, R.: In Vitro Models of Epilepsy. In: *Neurotransmitters and Epilepsy* (P. C. Jobe and H. E. Laird, II, eds.) Humana, New Jersey, 1986.

Epilepsy-Prone Mice

Genetically Determined Sound-Induced Seizures

A. G. Chapman and B. S. Meldrum

1. Epilepsy-Prone Mice

Many different strains of mice are prone to epilepsy, manifest as either spontaneous seizures or as seizures occurring in response to specific sensory inputs, such as a loud sound (*see* Table 1). At least 12 single locus mutations are known that produce neurological syndromes associated with spontaneous seizures (Noebels, 1979; Seyfried, 1982). The syndrome in the mutant mouse tottering (tg/tg) has been proposed as a model of "absence attacks" (Noebels and Sidman, 1979). Distinctive cerebellar abnormalities (loss of granule cells and abnormal synaptic morphology) are found in staggerer mice (Sax et al., 1968). In a mutant (sps recessive) in C57BL/6 Bg mice, spontaneous seizures, perhaps of limbic origin, are frequently fatal (Maxson et al., 1983).

Of the convulsions evoked by sensory stimulation, the best studied (in terms of genetics, biochemistry, and pharmacology) are audiogenic seizures in DBA/2 mice. Some limited biochemical data are available on the El (epilepsy-like, also called EP) mouse in which postural stimulation is the sensory trigger (Kurokawa et al., 1966; Suzuki and Nakamoto, 1977). This chapter presents a summary of the information available concerning the abnormality in the DBA/2 mouse and other strains showing sound-induced seizures. The approaches adopted for studying this syndrome could be usefully applied to the other genetic seizure-prone mice strains.

TABLE 1
Mouse Strains Showing Reflexly Induced or Spontaneous Seizures

Strain	Seizure type	Refs.
Reflexly induced seizures		
DBA/2J		
DBA/2Bg, DBA/1 Bg		
Frings	Audiogenic seizure,	Frings et al., 1956
Swiss albino Rb	wild running +	Simler et al., 1973
O'Grady (Swiss-Webster albino)	clonus, tonus (sub-cortical), age-dependent	Alexander and Gray, 1972
A2G		Pasquini et al., 1968
C57 BL/6 Bg *Gad* la		Maxson, 1980
BXD-13	Audiogenic and spontaneous seizures	Seyfried et al., 1985
Epileptiform (*epf* recessive)		Hare and Hare, 1979
El (or ep) "epilepsy like"	Vestibular stimulation tonic-clonic seizure	Kurokawa et al., 1963; 1966; Suzuki et al., 1983
Quaking mouse (qk/qk) (mutants of C57 B1/6J)	Handling gen. clonus—extension of head and limbs	Chermat et al., 1979; Chauvel et al., 1980
Spontaneous seizures		
Tottering (tg) (mutant of C57 BL/6J)	Spike and wave and focal motor seizures	Noebels and Sidman, 1979; Heller et al., 1983
Spontaneous seizure (sps) (mutant of C57 BL/10Bg)	Absence or arrest, generalized con-vulsions	Maxson et al., 1983
Lurcher (*Lc*, semi-dominant)	Tonic seizures	Noebels, 1979; Seyfried et al., 1985
Staggerer (sg/sg)		Sax et al., 1968
Jimpy (jp)	Tonic seizures	Seyfried et al., 1985

2. Sound-Induced Seizures in Mice

Sound-induced seizures in the DBA (dilute brown agouti) strain of the house mouse were described by Hall (1947). Subsequently numerous reviews have considered the nature of the syndrome, its genetics, biochemistry, and pharmacology (Vicari, 1951; Bevan, 1955; Hamburgh and Vicari, 1960; Lehmann, 1970; Fuller and Collins, 1970; Boggan et al., 1971; Collins, 1972; Schlesinger and Sharpless, 1975; Kellogg, 1976; Seyfried, 1979; 1982; Seyfried et al., 1985; Jobe, 1981; 1984; Jobe and Laird, 1981; Chapman et al., 1984a; Laird et al., 1984).

The characteristic seizure response induced by exposure to a loud sound in DBA/2 mice is age-dependent and consists of a sequence of convulsive phenomena, commencing with an explosive burst of ill-coordinated locomotion ("wild running"), followed by rhythmic clonic jerking, with the animal lying on one side, followed by tonic flexion and extension of trunk, limbs, and tail. The latter phase may terminate with respiratory arrest and death. These sequential seizure phases provide a very simple and consistent method of scoring seizure severity. Schlesinger and Uphouse (1972) and Horton et al. (1980) score 1 for wild running, 2 for clonus, 3 for tonic extension, and 4 for respiratory arrest. Other authors variously omit the score for respiratory arrest and/or amalgamate the scores for the clonic and tonic phases.

Seizure susceptibility in DBA/2 mice is shown in the age range of 16–30 d. A restricted peak susceptibility (at 16–21 d) is described by Schlesinger et al., 1965; Schreiber, 1981; and Seyfried, 1982) (*see* Fig. 1). The details of the age dependence of the response depend on the genetic composition of the strain and the conditions of rearing. A longer period of peak susceptibility in DBA/2 mice is described by Vicari (1951) (20–39 d) and Suter et al. (1958a) (peak at 30 d). In other strains (e.g., O'Grady and Frings), more sustained seizure susceptibility is shown.

Crossing with a non-seizure-susceptible strain can change the developmental pattern of susceptibility (as shown in Fig. 1). It is important to be aware of this when designing genetic studies. Many researchers have attempted to relate developmental studies of neurotransmitter mechanisms to the ontogeny of seizure susceptibility. An abnormality of neurotransmitter function, however, might contribute importantly to seizure susceptibility and yet not show a developmental time course matching that of the seizure susceptibility, if other factors associated with development play a permissive or protective role.

3. Genetic Studies: Recombinant Inbred Strains

Experiments involving cross-breeding DBA/2 mice with non-seizure-prone strains have established that at least three separate genes determine susceptibility to sound-induced seizures.

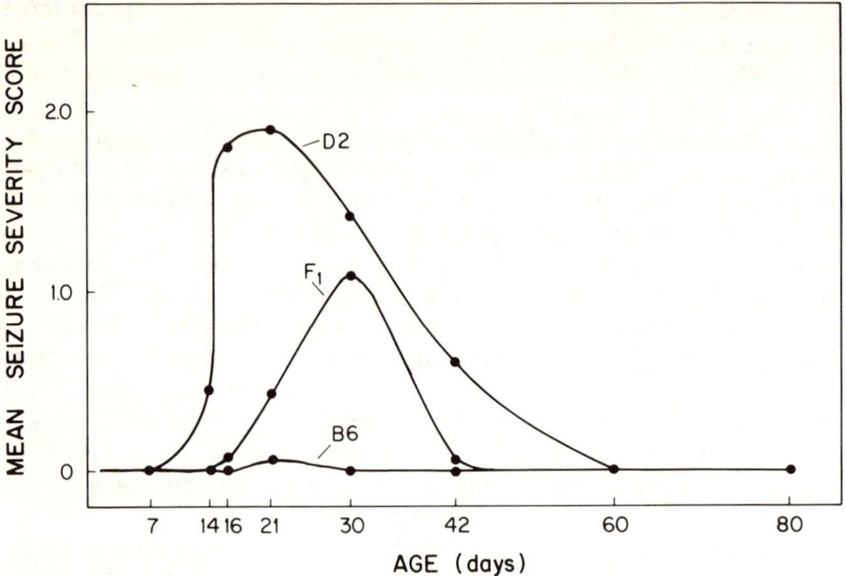

Fig. 1. Developmental profile of audiogenic seizure susceptibility in the C57 BL/6 and DBA/2 inbred strains and their F_1 hybrids, reproduced with permission from Seyfried (1982).

Collins (1970) designated the autosomal recessive genetic locus determining high seizure risk as *asp* (audiogenic seizure prone) and showed that it resided in chromosome 4 (linkage group VIII, as indicated by its linkage to other behavioral variants). This genetic factor is, however, different from two potent genes identified by cross-breeding with C57BL/6J mice and deriving recombinant inbred strains from the F_2 generation (Seyfried et al., 1985). This procedure can separate factors producing biochemical differences that are merely characteristic of the strain from those that are specifically associated with the high seizure susceptibility. In this way it has been shown that Ca^{2+} ATPase activity is low in brainstem homogenates of recombinant strains that show high seizure susceptibility, but not in others. This deficiency appears to be produced by a major gene associated with the *Ah* locus (and designated *Caa*) (Palayoor and Seyfried, 1984b). A separate gene (designated *Ias*) confers resistance to seizure spread (Seyfried et al., 1985). The biochemical correlate of this has not been identified.

Other strains that have high audiogenic seizure susceptibility include Frings and O'Grady mice (Frings et al., 1956; Alexander

and Gray, 1972). The developmental profile of seizure susceptibility
is somewhat different in these strains; the decline in susceptibility
above 30 d of age is less marked. The range of biochemical and
pharmacological data is limited for these strains compared with the
DBA/2 strain; the latter therefore forms the major concern of the
following presentation.

4. Biochemical Studies General

We shall consider first biochemical differences of a general
nature observed between DBA/2 mice and other non-seizure-prone
strains. These differences may directly or indirectly influence synaptic function and neurotransmitter metabolism. Many of the studies
have involved comparison with a single non-seizure-prone strain,
most often C57BL/6J. It might appear more appropriate to compare all available seizure-prone strains with all available non-seizure-
prone strains (*see* Ingram and Corfman, 1980). This does not provide
an ideal solution. The division into seizure-prone and non-seizure-
prone strains is not absolute either phenotypically or genotypically.
Careful testing often reveals wild-running responses (but not clonic
or tonic components) in a proportion of animals in non-seizure-
prone strains. The possibility that some non-seizure-prone strains
carry a gene conferring seizure susceptibility and additionally a
separate gene conferring some measure of protection has already
been documented in recombinant inbred studies (see below). Furthermore different seizure-prone strains may have quite different
biochemical abnormalities underlying the seizure susceptibility. The
recent studies of Seyfried and colleagues (Seyfried et al., 1985) have
unequivocally demonstrated the advantages of using recombinant
inbred strains to establish the role of genetically determined biochemical differences in seizure susceptibility.

A wide range of early and recent studies has concerned general
cellular metabolism, particularly oxygen consumption, glucose and
glycogen metabolism, various ATPases, and the effects of thyroid
hormone. Others have concerned enzyme cofactors and heavy
metals that influence the activity of a very wide range of enzymes
including several that directly control neurotransmitter synthesis.
These topics will be discussed prior to consideration of changes
in specific neurotransmitter systems.

4.1. Glucose Metabolism

Schreiber (1981) has proposed that the onset of audiogenic seizure susceptibility in DBA/2 mice at the time of weaning may be related to the shift from ketone bodies to glucose as the metabolic precursor. During suckling, plasma β-hydroxybutyrate levels are 30–40% higher in DBA/2 mice than in C57 mice (Schreiber and Ungar, 1984). At the time of AGS susceptibility the levels of glycogen (16–30 d of age) and glucose (24–40 d) are lower in the brains of DBA/2 mice than in those of C57 mice (Schreiber, 1981). Administration of D-glucose, 10 g/kg, ip to 21-d-old DBA/2 mice provides (after 4–6 h) complete protection against AGS (Schreiber and Ungar, 1984). On this evidence a defect in the uptake or utilization of glucose at the critical age has been proposed as the primary cause of the seizure susceptibility (Schreiber and Ungar, 1984).

Regional rates of glucose utilization (autoradiographic determination of cerebral 2-deoxyglucose incorporation) in the epileptic El mouse are similar to those in a resistant DDY mouse under control conditions (Nonaka et al., 1980; Suzuki et al., 1983). The pattern of metabolic activation during stimulus-induced seizures in the El mouse (preferential activation of hippocampal and cortical glucose utilization) (Suzuki et al., 1983) is typical of many experimentally induced seizures (Chapman, 1985), and differs from the pattern observed during evoked focal seizures in the mutant mouse tottering when mainly brainstem structures and thalamic nuclei are activated (Noebels and Sidman, 1979). An increased rate of glucose utilization is observed in the basal ganglia of the rolling mouse Nagoya compared to asymptomatic heterozygotes, even during control periods with no abnormal behavioral manifestation (Kato et al., 1982).

The tolerance of DBA/2 mice for anoxia is low (Hamburgh and Vicari, 1960). This could result from the inability to utilize glucose adequately, but hyperthyroidism (see below) and other factors may be involved.

4.2. ATPases

A defect in oxidative phosphorylation in DBA/2 mice at 30 d of age was proposed by Abood and Gerard (1955), on the basis of in vitro measurements showing a reduced phosphate/oxygen ratio that was most marked when glutamate was the substrate. They reported ATPase activity to be reduced by 20% in DBA/2 mice compared with C57 mice.

Subsequently it was shown that Ca^{2+}-stimulated ATPase activity (and Mg^{2+} ATPase) is reduced in brain membrane preparations from DBA/2 mice compared with C57 mice (Rosenblatt et al., 1976; 1977), whereas Na/K ATPase activity does not differ between the strains. In cultured glial cells from DBA/2 mice, ecto-ATPase is deficient (Trams and Lauter, 1978).

Studies of total, Mg^{2+}, Na^{+}/K^{+}, and Ca^{2+} ATPase activity in brainstem homogenates from DBA/2J mice and in recombinant inbred strains derived from them (Palayoor and Seyfried, 1984a,b) show that total and Mg^{2+} ATPase are reduced in DBA/2J mice compared with C57 mice, but that there is no genetic association of this reduced activity with AGS susceptibility in the derived strains. Ca^{2+} ATPase is reduced in DBA/2J mice, however, and this reduction shows a clear correlation with audiogenic seizure susceptibility in the recombinant inbred strains (Palayoor and Seyfried, 1984b). Thus the gene *Caa*, which regulates Ca^{2+} ATPase activity, appears to importantly influence AGS susceptibility.

Other membrane abnormalities believed to be related to seizure susceptibility in DBA/2 mice include increased myelin levels (compared to resistant strains) (Seyfried et al., 1978; 1984b; Seyfried, 1979). The activity of a ganglioside metabolizing enzyme, β-D-galactosidase (EC 3.2.1.23) is markedly reduced in the inferior colliculus of DBA/2 mice compared to three seizure-resistant control strains (Wheeler et al., 1982).

4.3. Thyroid Hormone

Serum thyroxine (T_4) levels peak earlier (14 d) and are higher (from 7 to 18 d) in DBA/2 mice than in C57 BL/6J mice (Seyfried et al., 1979). Antithyroid treatments initiated 2–3 d prior to birth suppress AGS susceptibility in DBA/2 mice, and this effect can be reversed by postnatal thyroxine replacement (Seyfried et al., 1979; 1981). Excess thyroxine (between 5 and 8 d of age) leads to the appearance of AGS in 30% of C57 BL/6J (Seyfried et al., 1979). Study of the association between serum thyroxine concentration (at 14 d) and AGS susceptibility (at 21 d) in seven DBA × C57 recombinant inbred strains indicates that inherited differences in neonatal thyroxine levels are unlikely to be responsible for the seizure susceptibility (Seyfried et al., 1984a).

Measurement of radioiodide uptake into thyroid and brain indicates that anion transport into thyroid is increased in DBA/2 mice at the time of maximal seizure susceptibility (Engstrom et al., 1984).

The mechanism by which enhanced thyroxine during early development facilitates seizures is likely to involve enhanced synaptogenesis and increased sensitivity to sensory inputs (Macaione et al., 1984).

4.4. Pyridoxine

Pyridoxal phosphate is required as a coenzyme for numerous transaminases and decarboxylases involved in the metabolism of amino acids and other neurotransmitters. Dietary deficiency of pyridoxine was shown (by Coleman and Schlesinger, 1965) to enhance audiogenic seizure susceptibility in dilute mouse strains. This observation was extended to C57 BL/6J mice by Schlesinger and Lieff (1975), who also reported that supplemental pyridoxine-protected hybrid mice against AGS. They found no difference in the endogenous levels of pyridoxine in the brains of DBA/2J and C57 BL/6J mice.

In contrast Norris et al. (1985) have reported that feeding DBA/2J mice on a pyridoxine-deficient diet, with the addition of a metal chelator, *diminishes* AGS susceptibility, and that the ip injection of pyridoxal-5'-phosphate can facilitate the appearance of AGS responses. These data appear consistent with the report that the pyridoxal-5-phosphate concentration is 25% higher in the DBA mouse brain than in the control (CBA and Parkes) mouse brain (Chung and Cox, 1983).

4.5. Metals: Zinc, Magnesium, Copper

The content of zinc is increased in the whole brain of DBA/2 mice compared with CBA/a and Parkes mice by 2–4% (Chung and Johnson, 1983a). This difference is accentuated in regions with high Zn^{2+} content, such as the hippocampus and colliculi. The copper content of the brain in DBA/2 mice is also higher than that of CBA mice (5% higher) or Parkes mice (20% higher) (Chung and Johnson, 1983a).

Focal injection into the brain of 400 ng zinc (combined with ip injection of pyridoxal phosphate) makes CBA/Ca mice susceptible to audiogenic seizures (Chung and Johnson, 1983b).

Zinc ions inhibit glutamate decarboxylase activity and hence reduce the rate of GABA synthesis (Wu and Roberts, 1974). However there may also be effects of Zn^{2+} and Ca^{2+} on excitatory amino acid metabolism and receptors.

5. Biochemical Studies: Neurotransmitters

Measurements of neurotransmitter concentrations and of the activities of enzymes concerned in their synthesis or further metabolism, and of receptor binding sites, have principally concerned inhibitory and excitatory amino acid neurotransmitters and the monoamines, dopamine, noradrenaline, and serotonin (5-HT). Numerous differences between DBA mice and nonseizure strains have been reported. These are summarized below. Although they may contribute to the syndrome, none to date has been shown to be genetically linked to AGS.

5.1. Endogenous Levels of Excitatory and Inhibitory Neurotransmitter Amino Acid

Regional or whole brain levels of excitatory (glutamate and aspartate) and inhibitory (GABA, glycine, and taurine) amino acids have been compared in seizure-susceptible and -resistant strains of mice under resting conditions (Table 2). There is no difference in whole brain GABA levels between susceptible and resistant adult Swiss albino Rb mice (Simler et al., 1973; Maitre et al., 1974), whereas the whole brain GABA level in ep mice is elevated compared to resistant dd and gpc strains. When DBA/2 mice are compared to age-matched seizure-resistant controls before, during, and after the onset of maximal seizure susceptibility, no difference in whole brain GABA levels is observed at any age (Sykes and Horton, 1982), nor is there any difference in the levels of the inhibitory and excitatory amino acids in the cerebellum or regions concerned with the auditory pathway (auditory cortex, cochlear nuclei, inferior colliculi) (Davies and Owen, 1984). Using DBA/2 mice at a susceptible age (22 d), Toth and coworkers (1983) find unaltered whole brain levels of GABA, taurine, and aspartate, but increased levels of glycine (30%) and glutamate (20%) compared to the resistant BALB/c strain. Regional (cortical, hippocampal, striatal, cerebellar) GABA levels in adult DBA/2 mice are equal to the corresponding levels in the seizure-resistant Swiss CD-1 strain (Chapman et al., 1985), despite an earlier report of an increased (30%) whole brain GABA level in adult DBA/2 mice (Griffiths and Littleton, 1977). The cerebellar glycine level is elevated (40%) in adult DBA/2 mice (Chapman et al., 1985). Muramoto and coworkers (1981) have compiled differences in cerebellar amino acid levels between ataxic mutant mice (weaver, staggerer, nervous, reeler, rolling mouse

TABLE 2

Comparison of Resting Values of Brain Transmitter Amino Acid Levels in
Seizure-Susceptible and -Resistant Strains of Mice

Mouse strain			Amino acid Δ in audiogenic, %					
Audiogenic	Control	Brain region	GABA	TAU	GLY	ASP	GLU	Refs.
Ep		Whole	↑40[b]				↓30[c]	Naruse et al., 1960
SWISS Rb	Resistant Swiss Rb	Whole	→[a]					Simler et al., 1973
DBA/2	TO	Whole	→					Sykes and Horton, 1982
DBA/2	TO, C57BL	Whole	↑30			→		Griffiths and Littleton, 1977
DBA/2	BALB/C	Whole	→	→	↑30	→	↑20	Toth et al., 1983
DBA/2	BALB/C	Cerebellum, auditory path	→ →	→ →	→ →	→ →	→ →	Davies and Owen, 1984
DBA/2	SWISS CD-1	Cortex,	→	↑12	↓24	↓19	→	Chapman et al., 1985
		hippocampus,	→	→	→	↓31	↓25	
		striatum,	→	→	→	→	↓38	
		cerebellum	→	↑17	↑39	→	↑32	

[a]→, No significant change from control values.
[b]↑, Elevation compared to control values.
[c]↓, Reduction compared to control values.

Nagoya) and the asymptomatic heterozygote controls and report elevated glycine levels in all the five ataxic strains, whereas there is no uniform change in any of the other cerebellar amino acids. Audiogenic priming and sensitization are reported to decrease GABA levels. Audiogenic priming of the normally resistant C57 BL/6 strain produces a transient small decrease in whole brain GABA level (Sze, 1970). Sensitization of the CBA strain by the administration of Zn (intracerebral) and pyridoxal phosphate (ip) leads to decreased GABA, as well as increased glutamate and aspartate levels in the inferior colliculus (Chung and Johnson, 1983b; Chung et al., 1984).

During the clonic-tonic phase of sound-induced convulsions, the levels of GABA, aspartate, and glutamate decrease markedly in forebrains of A2G mice (Pasquini et al., 1968). A decrease in regional GABA levels is also observed following single (Simler et al., 1973) or repeated (Ciesielski et al., 1981) audiogenic seizures in susceptible Swiss Rb mice. During stimulus-induced convulsions in ep mice there are no significant changes in whole brain levels of GABA and glutamate (Naruse et al., 1960). Convulsions induced by ethanol withdrawal are more severe in DBA/2 than in C57 and TO strains of mice and are associated with increased brain levels of aspartate, and decreased GABA levels during the convulsive phase (Griffiths and Littleton, 1977).

5.2. Inhibitory Amino Acids

A wide range of compounds that enhance GABAergic function by different mechanisms offer potent protection against sound-induced seizures (Meldrum, 1979, 1985; Worms and Lloyd, 1981; Chapman et al., 1984a). Abnormalities of the GABA transmitter system of susceptible strains have therefore been sought as an explanation for the seizure-susceptibility, but no firm biochemical causal relationship has yet been established.

One seizure susceptible strain of mice, A2G, has a reduced rate of ^{14}C incorporation into GABA [30 min after (^{14}C)-glucose administration] (Al-Ani et al., 1970). However, no difference has been found in the activity of the GABA synthesizing enzyme, glutamic acid decarboxylase (EC 4.1.1.15), between DBA/2 mice and several seizure-resistant strains of mice (Tunnicliff et al., 1973; Sykes and Horton, 1982).

Differences in membrane GABA transport and binding systems have been reported between some seizure-susceptible and -resistant mouse strains. Whole forebrain slices of DBA/2 mice (at a seizure-susceptible age) have a decreased rate of K^+-stimulated GABA release when compared to slices from age-matched C57 mice (Hertz et al., 1974). Regional (cortex, pons-medulla, cerebellum) synaptosomal (^3H)-GABA uptake has been compared in DBA/2 and C57 B1/6 mice before, during and after the age of maximal seizure susceptibility in DBA/2 mice (Spyrou et al., 1984). Although there are regional and age-dependent variations in GABA uptake in both strains, there is no difference in the V_{max} of the GABA-uptake system between the two strains at any age or in any of the regions. The only strain difference observed at the susceptible age is for GABA uptake in cerebellar synaptosomes from DBA/2 mice. At a slightly older age (40–43 d) the synaptosomes isolated from pons-medulla of DBA/2 mice show a greater affinity for GABA than does the corresponding synaptosomal preparation from C57 B1/6 mice.

GABA and benzodiazepine binding also exhibit regional and age-dependent variation in both seizure-susceptible and -resistant strains (Ticku, 1979; Robertson, 1980; Horton et al., 1982; 1984). At all ages the number of (^3H)-GABA binding sites (both high- and low-affinity) in whole brain is lower, whereas the GABA affinity is higher in membrane preparations from DBA/2 mice compared to age-matched seizure-resistant controls (Ticku, 1979; Horton et al., 1982; 1984). Although there are several regional differences between strains, the only difference in regional GABA binding that correlates with the age of maximal seizure-susceptibility is a marked reduction in the GABA binding in the pons-medulla of DBA/2 when compared to corresponding preparations from seizure-resistant C57 mice (Horton et al., 1984). In contrast, there is an increased number of GABA binding sites [(^3H)-muscimol] in the brains, (especially in the brain stem) of the quaking mutants compared to the asymptomatic littermates (Maurin et al., 1980).

Both a decreased (Horton et al., 1982) and an increased (Robertson, 1980) number of benzodiazepine binding sites have been reported for whole brain of DBA/2 mice (at a seizure-susceptible age, compared to age-matched seizure-resistant controls, C57 BL/6J). In autoradiographic studies (Olsen et al., 1985a), a decrease in benzodiazepine receptor binding (relative to C57 mice or the seizure-resistant cross-bred strain D2 B6-Ias) has been observed in several midbrain/brainstem sites (substantia nigra, midbrain periaqueductal gray, caudal pons central gray, laterodorsal tegmental nucleus, and inferior colliculus). This decrease in binding results from a lower density of receptor sites. In seizure-susceptible gerbils a comparable regional deficit in benzodiazepine binding associated with a parallel deficit in bicuculline binding (to low-affinity GABA receptors) has been observed (Olsen et al., 1985b). The number of benzodiazepine binding sites in brains of quaking mice is larger than in asymptomatic littermates (Maurin et al., 1980).

Although elevated brain glycine concentrations protect against sound-induced seizures in DBA/2 mice (Toth et al., 1983), there have been no studies of possible abnormalities in the glycine transmitter system associated with seizure susceptibility.

Abnormalities in taurine transport and synaptosomal content have been linked to seizure-susceptibility in rats (Bonhaus and Huxtable, 1983; Bonhaus et al., 1984), but no corresponding information is available for mice.

5.3. Excitatory Amino Acids

There is a growing body of evidence for the central role played by excitatory amino acids in the initiation and spread of seizure activity (Meldrum and Chapman, 1983), and excitatory amino acid antagonists show potent anticonvulsant activity against sound induced seizures (*see* section 6.2). There is no information available as yet, however, concerning binding, transport, or turnover (or relevant enzyme activities) for aspartate, glutamate, or the sulphur-containing excitatory amino acids in seizure-susceptible mice.

5.4. Monoamines: Norepinephrine and Serotonin (5-HT)

Biochemical studies have led to the suggestion that a deficiency in monoaminergic transmission could contribute to seizure susceptibility in DBA/2 mice, but there is uncertainty about the data. Schlesinger et al. (1965) reported that total brain norepinephrine is 44% lower in DBA/2 mice than in C57 BL/6J mice at 21 d of age; similarly Kellogg (1976) found that the norepinephrine concentration in forebrain and in hindbrain was less (at 21 and 28 d of age) in DBA/2J compared with C57 BL/6 mice (with a greater than 50% reduction in hindbrain). Lints et al. (1980), however, studying whole brain levels of norepinephrine could find no significant differences between DBA/2J and C57 BL/6J mice (at 16, 20, or 28 d of age). Horton et al. (1984) measured binding of dihydroalprenolol, clonidine, and prazosin to a whole brain membrane preparations of DBA/2 and C57 mice. There are fewer prazosin (α_1 noradrenergic) binding sites in the DBA/2 mice at all ages.

The situation with regard to 5-HT content is similar. Thus, a reduced concentration of 5-HT in the brain of DBA/2 mice (relative to age-matched C57 BL/6J mice) was reported by Schlesinger et al. (1965) and Kellogg (1971). This difference was not found by Lints et al. (1980), however. McGeer et al. (1969) were also unable to find any differences in catecholamine or 5-HT content between DBA/2J and C57 mice.

A quantitative electron microscope study of synaptic boutons in the neocortex of infant mice (6 d of age) showed that there was an abnormality of monoaminergic terminals in the temporal cortex of DBA/2 mice; compared with Swiss S mice the proportion of such synapses was greatly reduced (Kristt et al., 1980).

5.5. Acetylcholine

Numerous biochemical observations suggest that cholinergic activity is enhanced in seizure-prone mice (relative to normal mice). In EP (El) mice the level of acetylcholine in the brain is 50% higher than in control (gpc) mice (Kurokawa et al., 1963). This is matched by a relatively greater cerebral choline acetyltransferase activity and reduced cholinesterase activity (Kurokawa et al., 1966). Acetylcholinesterase activity is 14% higher in the cortex of A2G mice (an AGS-prone strain) compared with C57/BL mice (Al-Ani et al., 1970). Cholineacetyltransferase activity is 14% higher in Frings mice than in CF mice (Shenoy et al., 1981).

Comparing DBA/2 mice with C57 mice, Ebel et al. (1973) found that acetylcholinesterase activity was significantly increased in the temporal cortex (3.98 ± 0.62 vs 2.26 ± 0.49) mmol/g protein/h and choline acetyltransferase was increased in both frontal and temporal cortex.

The properties of muscarinic acetylcholine receptors have been studied using (^{3}H)-quinuclidinyl benzilate as ligand in DBA/2J and C57 BL/6J mice (Aronstam et al., 1979). More QNB binding is found in the DBA hippocampus than in C57 hippocampus.

6. Pharmacological Studies
Relating to Neurotransmitters

6.1. GABA-Related Drugs

Enhanced GABA-mediated inhibition is a major mechanism for preventing or arresting clinical and experimental seizures, and there is strong electrophysiological, pharmacological, and biochemical evidence that many of the anti-epileptic drugs in common clinical use (e.g., benzodiazepine, barbiturates, and valproate) partially exert their anticonvulsant action through interaction with the GABA-system (Meldrum and Braestrup, 1984). Most clinically established antiepileptic drugs effectively block sound-induced seizures in DBA/2 mice (*see* Sutter et al., 1958b; Collins and Horlington, 1969; Worms and Lloyd, 1981; Chapman et al., 1984a, for summaries), as well as in other epilepsy-prone strains such as Swiss Rb, A2G, and Frings (*see* Lehmann, 1970).

The benzodiazepines are the most potent class of anticonvulsant compounds against sound-induced seizures in DBA/2 mice, with ED_{50} values against clonic convulsions ranging from 0.002 (lorazepam) to 4 (desmethylclobazam) mg/kg (*see* Chapman et al., 1984a). Acute administration of anticonvulsant doses of benzodiazepines has no effect on brain GABA levels in rodents (Chapman, 1984), but does inhibit GABA turnover (Bernasconi et al., 1982; Chapman, 1984). Benzodiazepines facilitate GABA-mediated inhibition by acting on the chloride channel (MacDonald and Barker, 1979). There is a good correlation between the anticonvulsant potency of the different benzodiazepines against audiogenic seizures in DBA/2 mice and the affinity of the same compounds for the GABA–benzodiazepine–receptor complex [inhibition of (^{3}H)-flunitrazepam binding] (Jensen et al., 1983). Certain β-carboline-derivatives (which bind to the benzodiazepine receptor) have anticonvulsant activity against audiogenic seizures in DBA/2 mice with ED_{50} values of around 0.01–0.5 mg/kg ip (ZK 91296, ZK 93426) (Meldrum and Braestrup, 1984; Jensen et al., 1984; Meldrum, 1984). Conversely, low doses of *proconvulsant* β-carboline derivatives, such as DMCM or β-CCM, will potentiate sound-induced seizures in DBA/2 mice using subthreshold sound stimulus (Jensen et al., 1983).

Protection against audiogenic seizures in mice following valproate is usually associated with an increased level of brain GABA (Simler et al., 1973; Schechter et al., 1978). Inhibition by valproate of the GABA-metabolizing enzymes, GABA-transaminase (EC 2.6.1.19) and succinic semialdehyde dehydrogenase (EC 1.2.1.16), can be demonstrated in rats. Subsequent studies have shown that valproate administration also affects other neurotransmitters, such as glycine and aspartate (*see* Chapman et al., 1982). Various short, branched-chain fatty acid analogs of valproic acid protect against sound-induced seizures in DBA/2 (Chapman et al., 1983; 1984b) and Swiss Rb (Maitre et al., 1974) mice following their systemic or intracerebroventricular administration. The anticonvulsant protection is linked, in a dose-dependent manner, to an increase in GABA level and a decrease in the aspartate levels in the brains of DBA/2 mice (Chapman et al., 1984b).

Many compounds that appear to act selectively on the GABA transmitter system (GABA agonists, GABA-transaminase inhibitors, GABA uptake inhibitors) have potent anticonvulsant action against sound-induced seizures in mice (Table 3), although systemic administration of massive doses of GABA itself (50 mmol/kg) fails to protect against audiogenic seizures in mice (Lehmann, 1970; Toth et al., 1983). Specific $GABA_A$-agonists such as muscimol and THIP are anticonvulsants in DBA/2 mice, although they produce myoclonus and paroxysmal EEG abnormalities in another animal model for reflex epilepsy: photically-induced seizures in genetically susceptible baboons. All of the GABA transaminase inhibitors listed in Table 3 elevate brain GABA levels and block sound-induced seizures. Only the (+)-isomer of γ-vinyl-GABA inhibits GABAtransaminase and exhibits significant anticonvulsant activity (Meldrum and Murugaiah, 1983).

The free acid forms of the GABA-uptake inhibitors, such as nipecotic acid, penetrate the blood–brain barrier poorly so that they are more effective anticonvulsants when injected intracerebroventricularly. Esterification of these acids increases their transport across the blood–brain barrier and their efficacy as anticonvulsants following ip injection. GABA transport carriers in neurons and glial cells have been evaluated by studying GABA transport in synaptosomes or brain slices (neuronal) and in cultures astrocytes (glial). Nipecotic acid is a potent inhibitor of both neuronal and glial GABA-uptake systems, whereas *cis*-4-OH-nipecotic acid and THPO [4,5,6,7-tetrahydroisooxazolo(4,5-c)pyridine-3-ol] are relatively specific inhibitors of the glial-uptake system, and DBA (2,4-diaminobutyric acid) is a relatively specific inhibitor for GABA uptake into neurons. Compounds inhibiting glial GABA uptake have anticonvulsant properties (Meldrum et al., 1982; Krogsgaard-Larsen et al., 1981; Croucher et al., 1983).

6.2. Glycine-Related Compounds

Valproate inhibits glycine metabolism and causes elevated levels of glycine in plasma and urine (*see* Chapman et al., 1982). It is not known, however, whether the anticonvulsant action of valproate is mediated in part through the glycine inhibitory transmitter system. Systemic administration of high levels of glycine (50 mol/kg,

TABLE 3
Protection Against Clonic Phase of Sound-Induced Seizures
in DBA/2 Mice by GABA-Related Compounds[a]

Anticonvulsant	ED_{50}, mg/kg, ip	Refs.
GABA agonists		
Muscimol	0.4	Worms and Lloyd, 1981; Meldrum et al., 1980
THIP	3–7	Worms and Lloyd, 1981; Meldrum et al., 1980
Baclofen	5–8	Worms and Lloyd, 1981; Meldrum et al., 1980
Progabide	50	Worms and Lloyd, 1981
GABA-T inhibitors		
AOAA	15–20	Worms and Lloyd, 1981; Palfreyman et al., 1981; Schlesinger et al., 1968
EOS	75–150 μg[b]	Anlezark et al., 1976a; Horton et al., 1977
Gabaculine	17–48	Worms and Lloyd, 1981; Palfreyman ct al., 1981, Schechter et al., 1979
Isogabaculine	16	Palfreyman et al., 1981; Schechter et al., 1979
EPP	21	Worms and Lloyd, 1981
γ-Acetylenic GABA	32–41	Worms and Lloyd, 1981; Palfreyman et al., 1981; Schechter et al., 1977
γ-Vinyl-GABA	600–1000	Palfreyman et al., 1981; Schechter et al., 1977; Meldrum and Murugaiah, 1983
Cycloserine	25[c]	Chung et al., 1984
GABA-uptake inhibitors		
Nipecotic acid	155	Horton et al., 1979
Nipecotic Et ester	47	Horton et al., 1979
Nipecotic pivaloyloxy Me ester	259	Meldrum et al., 1982
cis-4-OH-Nipecotic Me ester	191	Meldrum et al., 1982
THPO	114	Meldrum et al., 1982
2,4-Diamino-butyric acid	480	Horton et al., 1979

[a]Abbreviations: THIP, 4,5,6,7-tetrahydroisoxazolo[5,4-c]pyridin-301; AOAA, amino-oxyacetic acid; EOS, ethanolamine-O-sulfate; EPP, 5-ethyl-5-phenyl-2-pyrrolidone.

[b]Protective intracerebroventricular dose.

[c]Fully protective dose.

orally) protects against sound-induced seizures in DBA/2 mice (Toth et al., 1983).

6.3. Excitatory Amino Acid Antagonists

Receptors sensitive to excitatory amino acids can be divided into three groups: those preferentially activated by *N*-methyl-D-aspartate (NMDA), kainic acid, or quisqualic acid (Watkins and Evans, 1981). Potent antagonists, such as 2-amino-7-phosphonoheptanoic acid, have been synthesized to selectively block excitation resulting from NMDA (Evans et al., 1982), whereas relatively selective kainate/quisqualate antagonists are only now emerging (Jones et al., 1984b).

A number of the excitatory amino acid antagonists effectively suppress sound-induced seizures in DBA/2 mice (Table 4), as well as other types of experimental seizures (Meldrum and Chapman, 1983). Following intracerebroventricular administration, some of the excitatory amino acid antagonists are several orders of magnitude more potent than other anticonvulsants in suppressing clonic seizures in DBA/2 mice (Chapman et al., 1984a). However, because of the relatively poor uptake into the brain of the currently available antagonists, their potency is reduced following systemic administration when ED_{50} values more closely resembling those of other anticonvulsants are observed.

There appears to be a correlation between the anticonvulsant activity of these compounds and their relative potency as antagonists of the NMDA receptor (Meldrum and Chapman, 1983; Croucher et al., 1984b). As more selective antagonists for the kainate and quisqualate receptors are synthesized and tested for anticonvulsant activity, this generalization may have to be amended.

6.4. Monoamines

Drugs that deplete the brain of monoamines by inhibiting their synthesis or storage tend to facilitate audiogenic seizures. Thus, reserpine or tetrabenazine (depleting stores of monoamines) enhance seizures in DBA/2 mice (Lehmann, 1967). This effect can be diminished or reversed by the administration of L-DOPA (precursor for both dopamine and norepinephrine) or of 5-hydroxytryptophan (Boggan and Seiden, 1971; Boggan et al., 1971). α-Methyl-*p*-tyrosine (which inhibits tyrosine hydroxylase activity and thus blocks the synthesis of catecholamines) also facilitates audiogenic seizures (Schlesinger et al., 1970).

TABLE 4
Protection Against Clonic Phase of Sound-Induced Seizures in DBA/2 Mice
by Excitatory Amino Acid Antagonists

Antagonist	Ed$_{50}$, clonic		Refs.
	μmol, icv	mmol/kg, ip	
β-D-Aspartylaminomethylphosphonate (ASP-AMP)	0.0006	0.18	Jones et al., 1984a
γ-D-Glutamylaminomethylphosphonate (GLU-AMP)	0.0018	0.28	Jones et al., 1984a
2-Amino-7-phosphonoheptanoic acid (2-APH)	0.0018	0.04	Croucher et al., 1982
2-Amino-5-phosphonopentanoic acid (2-APV)	0.022	0.40	Jones et al., 1984a; Croucher et al., 1982
cis-2,3-Piperidinedicarboxylic acid (cis-2,3,-PIP)	0.017	0.52	Croucher et al., 1984a
γ-D-Glutamyl glycine (GDGG)	0.046	—	Croucher et al., 1982
γ-D-Glutamylaminomethylsulfonic acid (GAMS)	0.074	2.74	Croucher et al., 1984b
β-Kainic acid (β-KA)	0.09	2.0	Collins et al., 1984
β-Kainyl glycine (β-KA-GLY)	0.11	—	Collins et al., 1984
α-Kainyl Glycine (α-KA-GLY)	0.28	—	Collins et al., 1984
α-Kainylaminomethylphosphonate (α-KA-PHOS)	0.31	—	Collins et al., 1984
β-Kainylaminomethylphosphonate (β-KA-PHOS)	>1.5	—	Collins et al., 1984

TABLE 5
Monoaminergic Drugs Protecting Against Audiogenic Seizures
in DBA/2 Mice

Compound	Agonist	Dose, mg/kg	Refs.
L-DOPA	DA, NA	200	Boggan and Seiden, 1971; Jobe et al., 1983
Apomorphine	DA	1–10	Anlezark and Meldrum, 1975
N-Propylnorapo-morphine	DA	0.1	Anlezark et al., 1978
2,10,11-trihydroxy-*N*-propylnorapo-morphine	DA	2	Anlezark et al., 1981
Ergocornine	DA	2	Anlezark et al., 1976[b]
Bromocriptine	DA	10	Anlezark et al., 1976[b]
LSD 25	DA, 5-HT	9	Anlezark et al., 1976[b]
(+)-Amphetamine	DA, NA	15	Lehmann, 1967
Clonidine	NA a$_2$	0.2–0.4	Horton et al., 1980
Oxymetazoline	NA a$_2$	2.5–10.0	Horton et al., 1980
UK 14,304	NA a$_2$	0.6	Horton et al., 1980
L-5 Hydroxytryptophan	5-HT	330	Alexander and Kopeloff, 1976

Consistent with these effects of monoamine depletion are the observations in Table 5 showing that various agents that are either direct monoaminergic agonists or enhance the synaptic action of monoamines diminish audiogenic seizure responses. L-DOPA diminishes the occurrence of tonic extensor convulsions in DBA/2J mice, while increasing brain concentrations of both dopamine and norepinephrine. Inhibition of the conversion of L-DOPA to norepinephrine does not prevent the anticonvulsant effect, indicating that dopamine is critically involved (Dailey and Jobe, 1984). Direct dopaminergic agents, such as apomorphine and *n*-propyl-norapomorphine, can completely suppress the seizure response for 15–30 min following ip administration. This effect is prevented by prior administration of the dopamine antagonist, haloperidol. Some ergot alkaloids and their derivatives have a similar, but more sustained, action (but these are less specific in terms of site of action). The effect of serotoninergic agonists is less pronounced; LSD 25 is active only at high doses, and quipazine, 50 mg/kg, is only partially effective (Anlezark et al., 1978).

Among drugs acting on noradrenergic transmission, the most specific protective effect is produced by drugs that are agonists at the α_2 receptor. Yohimbine or piperoxan prevents this protective effect (in accordance with the assumption that it involves an agonist action at the α_2 receptor).

High doses of propranolol (16–32 mg/kg) produce partial protection against audiogenic seizures, but as this phenomenon is shown by both + and − stereoisomers of propranolol, it cannot be attributed to actions at either the β-adrenergic or serotoninergic receptors (Anlezark et al., 1979).

In contrast, in quaking mice, seizures induced by tactile stimulation are suppressed by yohimbine, and this effect is antagonised by clonidine or prazosin (Chermat et al., 1979; 1981).

7. Summary

Epilepsy occurs in many genetically determined syndromes in mice. The best-studied syndrome is that of sound-induced seizures in DBA/2 mice. The age-dependent seizure susceptibility is determined by three or more genes, one of which is associated with reduced activity of brain membrane Ca^{2+} ATPase. This defect might produce abnormalities in synaptic function that contribute to the seizure response. Many biochemical and pharmacological studies indicate that neurotransmitter function is altered in DBA/2 mice. Reported abnormalities concern amino acid neurotransmitters (GABA and glycine), the benzodiazepine receptor linked to the GABA-A recognition site, the monoamines (norepinephrine and serotonin), and acetylcholine. The significance of these abnormalities and their primary or secondary role in altered seizure susceptibility remains to be determined.

References

Abood, L. G. and Gerard, R. W.: A Phosphorylation Defect in the Brains of Mice Susceptible to Audiogenic Seizure, In: *Biochemistry of the Developing Nervous System* (H. Waelsh, ed.) Academic, New York, 1955.

Al-Ani, A. T., Tunnicliff, G., Rick, J. T., and Kerkut, A. G.: GABA production, acetylcholinesterase activity and biogenic amine levels in brains for mouse strains differing in spontaneous activity and reactivity. *Life Sci.* **9**: 21–27, 1970.

Alexander, G. J. and Gray, R.: Induction of convulsive seizures in sound sensitive albino mice: Response to various signal frequencies. *Proc. Soc. Exp. Biol. Med.* **140:** 1284–1288, 1972.

Alexander, G. J. and Kopeloff, L. M.: Audiogenic seizures in mice: Influence of agents affecting brain serotonin. *Res. Comm. Chem. Pathol. Pharmacol.* **14:** 437–447, 1976.

Anlezark, G. M. and Meldrum, B. S.: Effects of apomorphine, ergocornine and piribedil on audiogenic seizures in DBA/2 mice. *Br. J. Pharmacol.* **53:** 419–426, 1975.

Anlezark, G., Horton, R. W., Meldrum, B. S., and Sawaya, M. C. B.: Anticonvulsant action of ethanolamine-O-sulphate and di-*n*-propylacetate and the metabolism of γ-aminobutyric acid (GABA) in mice with audiogenic seizures. *Biochem. Pharmacol.* **25:** 413–417, 1976a.

Anlezark, G., Pycock, C., and Meldrum, B.: Ergot alkaloids as dopamine agonists: Comparison in two rodent models. *Eur. J. Pharmacol.* **37:** 295–302, 1976b.

Anlezark, G. M., Horton, R. W., and Meldrum, B. S.: Dopamine agonists and audiogenic seizures: The relationship between protection against seizures and changes in monoamine metabolism. *Biochem. Pharmacol.* **27:** 2821–2828, 1978.

Anlezark, G., Horton, R., and Meldrum, B.: The anticonvulsant action of the (−)− and (+)− enantiomers of propranolol. *J. Pharm. Pharmacol.* **31:** 482–483, 1979.

Anzelark, G., Marrosu, F., and Meldrum, B.: Dopamine Agonists in Reflex Epilepsy, In: *Neurotransmitters, Seizures and Epilepsy* (P. L. Morselli, K. G. Lloyd, W. Loscher, B. Meldrum, and E. H. Reynolds, eds.) Raven, New York, 1981.

Aronstam, R. A., Kellogg, C., and Abood, G. L.: Development of muscarinic cholinergic receptors in inbred strains of mice: Identification of receptor heretogeneity and relation to audiogenic seizure susceptibility. *Brain Res.* **162:** 231–241, 1979.

Bernasconi, R., Maitre, L., Martin, P., and Raschdorf, F.: The use of inhibitors of GABA-transaminase for the determination of GABA turnover in mouse brain regions: An evaluation of aminooxyacetic acid and gabaculine. *J. Neurochem.* **38:** 57–66, 1982.

Bevan, W.: Sound-precipitated convulsions: 1947 to 1954. *Psycholog. Bull.* **52:** 473–504, 1955.

Boggan, W. O. and Seiden, L. S.: Dopa reversal of reserpine enhancement of audiogenic seizure susceptibility in mice. *Physiol. Behav.* **6:** 215–217, 1971.

Boggan, W. O., Freedman, D. X., Lovell, R. A., and Schlesinger, K.: Studies in audiogenic seizure susceptibility. *Psychopharmacology* **20:** 48–56, 1971.

Bonhaus, D. W. and Huxtable, R. J.: The transport, biosynthesis and biochemical actions of taurine in genetic epilepsy. *Neurochem. Internat.* **5:** 413–419, 1983.

Bonhaus, D. W., Lippincott, S. E., and Huxtable, R. J.: Subcellular distribution of neuroactive amino acids in brains of genetically epileptic rats. *Epilepsia* **25:** 564–568, 1984.

Chapman, A. G.: Effect of Anticonvulsant Drugs on Brain Amino Acid Metabolism and GABA Turnover, In: *Neurotransmitters, Seizures, and Epilepsy, II,* (R. G. Fariello, P. L. Morselli, K. G. Lloyd, L. F. Quesney, and J. Engel, Jr., eds.) Raven, New York, 1984.

Chapman, A. G.: Cerebral Energy Metabolism and Seizures, In: *Recent Advances in Epilepsy* (T. A. Pedley and B. S. Meldrum, eds) vol. 2, Churchill Livingstone, Edinburgh, 1985.

Chapman, A. G., Keane, P. E., Meldrum, B. S., Simiand, J., and Vernieres, J. C.: Mechanism of anticonvulsant action of valproate. *Prog. Neurobiol.* **19:** 315–359, 1982.

Chapman, A. G., Meldrum, B. S., and Mendes, R.: Acute anticonvulsant activity of structural analogues of valproic acid and changes in brain GABA and aspartate content. *Life Sci.* **32:** 2023–2031, 1983.

Chapman, A. G., Croucher, M. J., and Meldrum, B. S.: Evaluation of anticonvulsant drugs in DBA/2 mice with sound-induced seizures. *Arzneim.-Forsch.* **34:** 1261–1264, 1984a.

Chapman, A. G., Croucher, M. J., and Meldrum, B. S.: Anticonvulsant activity of intracerebroventricularly administered valproate and valproate analogues. A dose-dependent correlation with changes in brain aspartate and GABA levels in DBA/2 mice. *Biochem. Pharmacol.* **33:** 1459–1463, 1984b.

Chapman, A. G., Cheetham, S. C., Hart, G. P., Meldrum, B. S., and Westerberg, E.: The effect of two convulsant β-carboline derivatives, DMCM and β-CCM, on regional neurotransmitter amino acid levels and on in vitro D-[^{3}H]-aspartate release in rodents. *J. Neurochem.* **45:** 370–381, 1985.

Chauvel, P., Louvel, J., Kurcewicz, I. and Debono, M.: Epileptic Seizures of the Quaking Mouse: Electroclinical Correlations, In: *Neurological Mutations Affecting Myelination* (N. Baumann, ed.) Elsevier, New York, 1980.

Chermat, R., Lachapelle, F., Baumann, N., and Simon, P.: Anticonvulsant effect of yohimbine in quaking mice: Antagonism by clonidine and prazosine. *Life Sci.* **25:** 1471–1476, 1979.

Chermat, R., Doare, L.,, Lachapelle, F., and Simon, P.: Effects of drugs affecting the noradrenergic system on convulsions in the quaking mouse. *Naunyn-Schmideberg's Arch Pharmacol.* **318:** 94–99, 1981.

Chung, S. H. and Cox, R. A.: Determination of pyridoxal phosphate levels in the brains of audiogenic and normal mice. *Neurochem. Res.* **10:** 1245–1249, 1983.

Chung, S.-H. and Johnson, M.: Divalent transition-metal ions (Cu^{2+} and Zn^{2+}) in the brains of epileptogenic and normal mice. *Brain Res.* **280:** 323–334, 1983a.

Chung, S.-H. and Johnson, M. S.: Experimentally induced susceptibility to audiogenic seizures. *Exp. Neurol.* **82:** 89–107, 1983b.

Chung, S.-H., Johnson, M. S., and Gronenborn, A. M.: L-Cycloserine: A potent anticonvulsant. *Epilepsia* **25:** 353–362, 1984.

Ciesielski, L., Simler, S., and Mandel, P.: Effect of repeated convulsive seizures on brain GABA levels. *Neurochem. Res.* **6:** 267–273, 1981.

Coleman, D. L. and Schlesinger, K.: Effect of pyridoxine deficiency on audiogenic seizure susceptibility in inbred mice. *Proc. Soc. Exp. Biol. Med.* **119:** 264–266, 1965.

Collins, R. L.: A new genetic locus mapped from behavioral variation in mice: Audiogenic seizure prone (*asp*). *Behav. Genet.* **1:** 99–109, 1970.

Collins, R. L.: Audiogenic Seizures, In: *Experimental Models of Epilepsy — A Manual for the Laboratory Worker* (D. P. Purpura, J. K. Penry, D. Tower, D. M. Woodbury, R. Walter, eds.) Raven, New York, 1972.

Collins, A. J. and Horlington, M.: A sequential screening test based on the running component of audiogenic seizures in mice, including reference compound PD_{50} values. *Br. J. Pharmacol.* **37:** 140–150, 1969.

Collins, J. F., Dixon, A. J., Badman, G., De Sarro, G., Chapman, A. G., Hart, G. P., and Meldrum, B. S.: Kainic acid derivatives with anticonvulsant activity. *Neurosci. Lett.* **51:** 371–376, 1984.

Croucher, M. J., Collins, J. F., and Meldrum, B. S.: Anticonvulsant action of excitatory amino acid antagonists. *Science* **216:** 899–901, 1982.

Croucher, M. J., Meldrum, B. S., and Krogsgaard-Larsen, P.: Anticonvulsant activity of GABA uptake inhibitors and their prodrugs following central or systemic administration. *Eur. J. Pharmacol.* **89:** 217–228, 1983.

Croucher, M. J., Meldrum, B. S., and Collins, J. F.: Anticonvulsant and proconvulsant properties of a series of structural isomers of piperidine dicarboxylic acid. *Neuropharmacology* **23:** 467–472, 1984a.

Croucher, M. J., Meldrum, B. S., Jones, A. W., and Watkins, J. C.: γ-D-Glutamylaminomethylsulphonic acid (GAMS) a ''kainic acid receptor'' antagonist, prevents sound-induced seizures in DBA/2 mice. *Brain Res.* **322:** 111–114, 1984b.

Dailey, J. W. and Jobe, P. C.: Effect of increments in the concentration of dopamine in the central nervous system on audiogenic seizures in DBA/2J mice. *Neuropharmacology* **23:** 1019–1024, 1984.

Davies, W. E. and Owen, C. D.: Amino acid levels in mice resistant and susceptible to audiogenic seizures (AGS). *Br. J. Pharmacol.* **82:** 254, 1984.

Ebel, A., Hermetet, J. C., and Mandel, P.: Comparative study of acetylcholinesterase and choline acetyltransferase enzyme activity in brain of DBA and C_{57} mice. *Nature, New Biology* **242:** 56–57, 1973.

Engstrom, F. L., Chow, S.-Y., Kemp, J. W., and Woodbury, D. M.: Radioiodine uptake in brain, CSF, thyroid, and salivary glands of audiogenic seizure mice. *Epilepsia* **25:** 518–525, 1984.

Evans, R. H., Francis, A. A., Jones, A. W., Smith, D. A. S., and Watkins, J. C.: The effects of a series of ω-phosphonic-α-carboxylic amino acids on electrically evoked and amino acid induced responses in isolated spinal cord preparations. *Br. J. Pharmacol.* **75:** 65–75, 1982.

Frings, H., Frings, M., and Hamilton, M.: Experiments with albino mice from stocks selected from predictable susceptibilities to audiogenic seizures. *Behavior* **9:** 44–52, 1956.

Fuller, J. L. and Collins, R. L.: Genetics of audiogenic seizues in mice: A parable for psychiatrists. *Semin. Psychiat.* **2:** 75–88, 1970.

Griffiths, P. J. and Littleton, J. M.: Concentrations of free amino acids in brains of mice of different strains during the physical syndrome of withdrawal from alcohol. *Br. J. Exp. Path.* **58:** 391–399, 1977.

Hall, C. S.: Genetic differences in fatal audiogenic seizures between two inbred strains of house mice. *J. Hered.* **38:** 2–6, 1947.

Hamburgh, M. and Vicari, F · A study of some physiological mechanisms underlying susceptibility to audiogenic seizures in mice. *J. Neuropath. Exp. Neurol.* **19:** 461–472, 1960.

Hare, J. E. and Hare, A. S.: Epileptiform mice, a new neurological mutant. *J. Heredity* **70:** 417–420, 1979.

Heller, A. H., Dichter, M. A., and Sidman, R. L.: Anticonvulsant sensitivity of absence seizures in the tottering mutant mouse. *Epilepsia* **25:** 25–34, 1983.

Hertz, L., Schousboe, A., Formby, B., and Lennox-Buchthal, M.: Some age-dependent biochemical changes in mice susceptible to seizures. *Epilepsia* **15:** 619–631, 1974.

Horton, R. W., Anlezark, G. M., Sawaya, M. C. B., and Meldrum, B. S.: Monoamine and GABA metabolism and the anticonvulsant action of di-*n*-propylacetate and ethanolamine-O-sulfate. *Eur. J. Pharmacol.* **41:** 387–397, 1977.

Horton, R., Anlezark, G., and Meldrum, B.: Noradrenergic influences on sound-induced seizures. *J. Pharmacol. Exp. Ther.* **214:** 437–442, 1980.

Horton, R. W., Collins, J. F., Anlezark, G. M., and Meldrum, B. S.: Convulsant and anticonvulsant actions in DBA/2 mice of compounds blocking the reuptake of GABA. *Eur. J. Pharmacol.* **59:** 79–83, 1979.

Horton, R. W., Prestwich, S. A., and Meldrum, B. S.: γ-Aminobuyric acid and benzodiazepine binding sites in audiogenic seizure-susceptible mice. *J. Neurochem.* **39:** 864–870, 1982.

Horton, R. W., Prestwich, S. A., and Jazrawi, S. P.: Neurotransmitter Receptor Binding in Genetically Seizure Susceptible Mice, In: *Neurotransmitters, Seizures and Epilepsy II* (R. G. Fariello, P. L. V. Morselli, K. G. Lloyd, L. F. Quesney, J. Engel, Jr., eds.) Raven, New York, 1984.

Ingram, D. K. and Corfman, T. P.: An overview of neurobiological comparisons in mouse strains. *Neurosci. Biobehav. Rev.* **4:** 421–435, 1980.

Jensen, L. H., Petersen, E. N., and Braestrup, C.: Audiogenic seizures in DBA/2 mice discriminate sensitively between low efficacy benzodiazepine receptor agonists and inverse agonists. *Life Sci.* **33:** 393–399, 1983.

Jensen, L. H., Petersen, E. N., Braestrup, C., Honore, T., Kehr, W., Stephens, D. N., Schneider, H., Seidelmann, D., and Schmiechen, R.: Evaluation of the β-carboline ZK 93426 as a benzodiazepine receptor antagonist. *Psychpharmacology* **83:** 249–256, 1984.

Jobe, P. C.: Pharmacology of Audiogenic Seizures, In: *Pharmacology of Hearing. Experimental and Clinical Bases* (R. D. Brown and E. A. Daigneault, eds.) Wiley, New York, 1981.

Jobe, P. C.: Neurotransmitters and epilepsy: An overview. *Fed. Proc.* **43:** 2503–2504, 1984.

Jobe, P. C. and Laird, H. E.: Neurotransmitter abnormalities as determinants of seizure susceptibility and intensity in the genetic models of epilepsy. *Biochem. Pharmacol.* **30:** 3137–3144, 1981.

Jobe, P. C., Woods, T. W., and Dailey, J. W.: Central nervous system dopamine, but not epinephrine suppresses audiogenic seizures in DBA/2J mice. *Fed. Proc.* **42:** 363, 1983.

Jones, A. W., Croucher, M. J., Meldrum, B. S., and Watkins, J. C.: Suppression of audiogenic seizures in DBA/2 mice by two new dipeptide NMDA receptor antagonists. *Neurosci. Lett.* **45:** 157–161, 1984a.

Jones, A. W., Smith, D. A. S., and Watkins, J. C.: Structure–activity relations of dipeptide antagonists of excitatory amino acids. *Neuroscience* **13:** 573–581, 1984b.

Kato, M., Hosokawa, S., Tobimatsu, S., and Kuroiwa, Y.: Increased local cerebral glucose utilization in the basal ganglia of the rolling mouse nagoya. *J. Cereb. Blood Flow Metab.* **2:** 385–393, 1982.

Kellogg, C.: Serotonin metabolism in the brains of mice sensitive or resistant to audiogenic seizures. *J. Neurobiol.* **2:** 209–219, 1971.

Kellogg, C.: Audiogenic seizures: Relation to age and mechanisms of monoamine neurotransmission. *Brain Res.* **106:** 87–103, 1976.

Kristt, D. A., Shirley, M. S., and Kasper, E.: Monoaminergic synapses in infant mouse neocortex: Comparison of cortical fields in seizure-prone and resistant mice. *Neuroscience* **5:** 883–891, 1980.

Krogsgaard-Larsen, P., Labouta, I. M., Meldrum, B., Croucher, M., and Schousboe, A.: GABA Uptake Inhibitors as Experimental Tools and Potential Drugs in Epilepsy Research, In: *Neurotransmitters, Seizures and Epilepsy* (P. L. Morselli, K. G. Lloyd, W. Loscher, B. S. Meldrum, and E. H. Reynolds, eds.) Raven, New York, 1981.

Kurokawa, M., Machiyama, Y., and Kato, M.: Distribution of acetylcholine in the brain during various states of activity. *J. Neurochem.* **10:** 341–348, 1963.

Kurokawa, M., Naruse, H., and Kato, M.: Metabolic studies on ep mouse, a special strain with convulsive disposition. *Prog. Brain. Res.* **21A,** 112–130, 1966.

Laird, H. E., Dailey, J. W., and Jobe, P. C.: Neurotransmitter abnormalities in genetically epileptic rodents. *Fed. Proc.* **43:** 2505–2509, 1984.

Lehmann, A.: Audiogenic seizures data in mice supporting new theories of biogenic amines mechanisms in the central nervous system. *Life Sci.* **6:** 1423–1431, 1967.

Lehmann, A. G.: Psychopharmacology of the Response to Noise, With Special Reference to Audiogenic Seizures in Mice, In: *Physiological Effects of Noise* (B. L. Welch and A. S. Welch, eds.) Plenum, New York, 1970.

Lints, C. E., Willott, J. F., Sze, P. Y., and Nenja, L. H.: Inverse relationship between whole brain monoamine levels and audiogenic seizure susceptibility in mice: Failure to replicate. *Pharmacol. Biochem. Behav.* **12:** 385–388, 1980.

Macaione, S., Giorgio, R. M., Nicotina, P. A., and Lentile, R.: Retina maturation following administration of thyroxine in developing rats: Effects on polyamine metabolism and glutamate decarboxylase. *J. Pharmacol. Neurochem.* **43:** 303–315, 1984.

MacDonald, R. L. and Barker, J. L.: Enhancement of GABA-mediated postsynaptic inhibition in cultured mammalian spinal cord neurons: A common mode of anticonvulsant action. *Brain Res.* **167:** 323–336, 1979.

Maitre, M., Ciesielski, L., and Mandel, P.: Effect of 2-methyl-2-ethyl caproic acid and 2-2-dimethyl valeric acid on audiogenic seizures and brain gamma aminobutyric acid. *Biochem. Pharmacol.* **23:** 2363–2368, 1974.

Maurin, Y., Le Saux, F., and Baumann, N.: Alteration of Neurotransmitter Receptor Binding in the Brain of Quaking Mice, In: *Neurological Mutations Affecting Myelination.* INSERM Symposium No. 14 (N. Baumann, ed.) Elsevier, Amsterdam, 1980.

Maxson, S. C.: Febrile convulsions in inbred strains of mice susceptible and resistant to audiogenic seizures. *Epilepsia* **21:** 637–645, 1980.

Maxson, S. C., Fine, M. D., Ginsburg, B. S., and Koniecki, D. L.: A mutant for spontaneous seizures in C57 BL/10Bg mice. *Epilepsia* **24:** 15–24, 1983.

McGeer, E. G., Ikeda, H., Asakura, T., and Wada, J. A.: Lack of abnormality in brain aromatic amines in rats and mice susceptible to audiogenic seizure. *J. Neurochem.* **16:** 945–950, 1969.

Meldrum, B. S.: Convulsant Drugs, Anticonvulsants and GABA-Mediated Neuronal Inhibition, In: *GABA-Neurotransmitters* Alfred Benzon Symposium 12 (P. Krogsgaard-Larsen, J. Scheel-Kruger, and H. Kofod, eds) Munksgaard, Copenhagen, 1979.

Meldrum, B. S.: Proconvulsant and anticonvulsant actions in photosensitive baboons of β-carboline derivatives. *Neuropharmacology* **23:** 845–846, 1984.

Meldrum, B. S.: GABA and Other Amino Acids, In: *Handbook of Experimental Pharmacology* (H.-H. Grey and D. Janz, eds.) vol. 74, Springer Verlag, Berlin, 1985.

Meldrum, B. S., and Chapman, A. G.: Excitatory Amino Acids and Anticonvulsant Drug Action, In: *Glutamine, Glutamate, and GABA in the Central Nervous System* (L. Hertz, E. Kvamme, E. G. McGeer, and A. Schousboe, eds.) Alan R. Liss, New York, 1983.

Meldrum, B. S. and Murugaiah, K.: Anticonvulsant action in mice with sound-induced seizures of the optical isomers of γ-vinyl-GABA. *Eur. J. Pharmacol.* **89:** 149–152, 1983.

Meldrum, B. S. and Braestrup, C.: GABA and the Anticonvulsant Action of Benzodiazepines and Related Drugs, In: *Actions and Interactions of GABA and Benzodiazepines* (N. G. Bowery, ed.) Raven, New York, 1984.

Meldrum, B., Pedley, T., Horton, R., Anlezark, G., and Franks, A.: Epileptogenic and anticonvulsant effects of GABA agonists and GABA uptake inhibitors. *Brain Res. Bull.* **5,** (suppl. 2): 685–690, 1980.

Meldrum, B. S., Croucher, M. J., and Krogsgaard-Larsen, P.: GABA-Uptake Inhibitors as Anticonvulsant Agents, In: *Problems in GABA Research* (Y. Okada and E. Roberts, eds.) Excerpta Medica, Amsterdam, 1982.

Muramoto, O., Kanazawa, I., and Ando, K.: Neurotransmitter abnormality in rolling mouse nagoya, an ataxic mutant mouse. *Brain Res.* **215:** 295–304, 1981.

Naruse, H., Kato, M., Kurokawa, M., Haba, R., and Yabe, T.: Metabolic defects in a convulsive strain of mouse. *J. Neurochem.* **5:** 359–369, 1960.

Noebels, J. L.: Analysis of inherited epilepsy using single locus mutations in mice. *Fed. Proc.* **38:** 2405–2410, 1979.

Noebels, J. L. and Sidman, R. L.: Inherited epilepsy: Spike-wave and focal motor seizures in the mutant mouse tottering. *Science* **204:** 1334–1336, 1979.

Nonaka, R., Nakamoto, Y., and Suzuki, J.: Distribution of 2-deoxy-D-[1-^{14}C]-glucose (2DG) in normal and epileptic mice brains. *Jap. J. Neuropsychopharm.* **2:** 419–423, 1980.

Norris, D. K., Murphy, R. A., and Chung, S. H.: Alteration of amino acid metabolism in epileptogenic mice by elevation of brain pyridoxal phosphate. *J. Neurochem.* **44:** 1403–1410, 1985.

Olsen, R. W., Jones, I., Seyfried, T. N., McCabe, R. T., and Wamsley, J. K.: Benzodiazepine receptor binding deficit in audiogenic seizure-susceptible mice. *Fed. Proc.* (abst.) **44(4):** 1106, 1985a.

Olsen, R. W., Wamsley, J. K., McCabe, R. T., Lee, R. J., and Lomax, P.: Benzodiazepine/γ-aminobutyric acid receptor deficit in the midbrain of the seizure susceptible gerbil. *Proc. Natl. Acad. Sci. USA* **82:** 6701–6705, 1985b.

Palayoor, S. T. and Seyfried, T. N.: Genetic study of cationic ATPase activities and audiogenic seizure susceptibility in recombinant inbred and congenic strains of mice. *J. Neurochem.* **42:** 529–533, 1984a.

Palayoor, S. T. and Seyfried, T. N.: Genetic association between Ca^{2+}-ATPase activity and audiogenic seizures in mice. *J. Neurochem.* **42:** 1771–1774, 1984b.

Palfreyman, M. G., Schechter, P. J., Buckett, W. R., Tell, G. P., and Koch-Weser, J.: The pharmacology of GABA-transaminae inhibitors. *Biochem. Pharmacol.* **30:** 817–824, 1981.

Pasquini, J. M., Salomone, J. R., and Gomez, C. J.: Amino acid changes in the mouse brain during audiogenic seizures and recovery. *Exp. Neurol.* **21:** 245–256, 1968.

Robertson, H. A.: Audiogenic seizures: Increased benzodiazepine receptor binding in a susceptible stain of mice. *Eur. J. Pharmacol.* **66:** 249–252, 1980.

Rosenblatt, D. E., Lauter, C. J., and Trams, E. G.: Deficiency of a Ca^{2+} ATPase in brains of seizure prone mice. *J. Neurochem.* **27:** 1299–1304, 1976.

Rosenblatt, D. E., Lauter, C. J., and Trams, E. G.: ATPases in animal models of epilepsy. *J. Mol. Med.* **2:** 137–144, 1977.

Sax, D. A., Hirano, A., and Shofer, R. J.: Staggerer, a neurological murine mutant: An electron microscopic study of the cerebellar cortex in the adult. *Neurology* **18:** 1093–1100, 1968.

Schechter, P. J., Tranier, Y., Jung, M. J., and Bohlen, P.: Audiogenic seizure protection by elevated brain concentration in mice: Effects of γ-acetylenic GABA and γ-vinyl GABA, two irreversible GABA-T inhibitors. *Eur. J. Pharmacol.* **45:** 319–328, 1977.

Schechter, P. J., Tranier, Y., and Grove, J.: Effect of *n*-dipropylacetate on amino acid concentrations in mouse brain: Correlations with anticonvulsant activity. *J. Neurochem.* **31:** 1325–1327, 1978.

Schechter, P. J., Tranier, Y., and Grove, J.: Gabaculine and isogabaculine: in vivo biochemistry and pharmacology in mice. *Life Sci.* **24:** 1173–1182, 1979.

Schlesinger, K. and Uphouse, L. L.: Pyridoxine dependency and central nervous system excitability. *Adv. Biochem. Psychopharmacol.* **4:** 105–140, 1972.

Schlesinger, K. and Lieff, B.: Levels of pyridoxine and susceptibility to electroconvulsive and audiogenic seizures. *Psychopharmacology* **42:** 27–32, 1975.

Schlesinger, K. and Sharpless, S. K.: Audiogenic Seizures and Acoustic Priming, In: *Psychopharmacogenetics* Plenum, New York, 1975.

Schlesinger, K., Boggan, W., and Freedman, D. X.: Genetics of audiogenic seizures: I. Relation to brain serotonin and norepinephrine in mice. *Life Sci.* **4:** 2345–2351, 1965.

Schlesinger, K., Boggan, W., and Freedman, D. X.: Genetics of audiogenic seizures: II. Effects of pharmacological manipulation of brain serotonin, norepinephrine and gamma-aminobutyric acid. *Life Sci.* **7:** 437–447, 1968.

Schlesinger, K., Boggan, W. O., and Freedman, D. X.: Genetics of audiogenic seizures: III. Time response relationships between drug administration and seizure susceptibility. *Life Sci.* **9:** 721–729, 1970.

Schreiber, R. A.: Developmental changes in brain glucose, glycogen, phosphocreatine, and ATP levels in DBA/2J and C57BL/6J mice, and audiogenic seizures. *J. Neurochem.* **37:** 655–661, 1981.

Schreiber, R. A. and Ungar, A. L.: Glucose protects DBA/2J mice from audiogenic seizures: Correlation with brain glycogen levels. *Psychpharmacology* **84:** 128–131, 1984.

Seyfried, T. N.: Audiogenic seizures in mice. *Fed. Proc.* **38:** 2399–2404, 1979.

Seyfried, T. N.: Convulsive Disorders, In: *The Mouse in Biomedical Research* (H. L. Foster, J. D. Small, and J. C. Fox, eds.) Academic, New York, 1982.

Seyfried, T. N., Glaser, G. H., and Yu, R. K.: Cerebral, cerebellar, and brain stem gangliosides in mice susceptible to audiogenic seizures. *J. Neurochem.* **31:** 21–27, 1978.

Seyfried, T. N., Glaser, G. H., and Yu, R. K.: Thyroid hormone influence on the susceptibility of mice to audiogenic seizures. *Science* **205:** 598–600, 1979.

Seyfried, T. N., Glaser, G. H., and Yu, R. K.: Thyroid hormone can restore the audiogenic seizure susceptibility of hypothyroid DBA/2J mice. *Exp. Neurol.* **71,** 220–225, 1981.

Seyfried, T. N., Glaser, G. H., and Yu, R. K.: Genetic analysis of serum thyroxine content and audiogenic seizures in recombinant inbred and congenic strains of mice. *Exp. Neurol.* **83:** 423–428, 1984a.

Seyfried, T. N., Bernard, D., Mayeda, F., Macala, L., and Yu, R. K.: Genetic analysis of cerebellar lipids in mice susceptible to audiogenic seizures. *Exp. Neurol.* **84:** 590–595, 1984b.

Seyfried, T. N., Glaser, G. H., Yu, R. K., and Palayoor, S. T.: Inherited Convulsive Disorders in Mice, In: *Basic Mechanisms of the Epilepsies* 2nd ed., (A. V. Delgado-Escueta, A. A. Ward, and D. M. Woodbury, eds.) Raven, New York.

Shenoy, A. K., Bakhit, C., Swinyard, E. A., and Gibb, J. W.: Choline acetyltransferase and glutamic acid decarboxylase activities in audiogenic seizure susceptible (Frings) mouse brain. *Fed. Proc.* **40:** 271, 1981.

Simler, S., Ciesielski, L., Maitre, M., Randrianarisoa, H., and Mandel, P.: Effect of sodium *n*-dipropylacetate on audiogenic seizures and brain γ-aminobutyric acid levels. *Biochem. Pharmacol.* **22:** 1701–1708, 1973.

Spyrou, N. A., Prestwich, S. A., and Horton, R. W.: Synaptosomal [³H]-GABA uptake and [³H]-nipecotic acid binding in audiogenic seizure susceptibility (DBA/2) and resistant (C57B1/6) mice. *Eur. J. Pharmacol.* **100:** 207–210, 1984.

Suter, C., Klingman, W. O., Lacy, O. W., Boggs, D., Marks, R., and Coplinger, C. B.: Sound-induced seizures in animals. *Neurology* **8** (suppl. 1): 117–120, 1958a.

Suter, C., Klingman, W. O., Boggs, D., Lacy, O. W., Marks, R., and Coplinger, C. B.: Sound-induced seizures in animals: The efficiency of certain anticonvulsants in controlling sound-induced seizures in DBA/2 mice. *Neurology* **8,** (suppl. 1): 121–124, 1958b.

Suzuki, J. and Nakamoto, Y.: Seizure patterns and electroencephalograms of El mouse. *Electroencephalog. Clin. Neurophysio.* **43:** 299–311, 1977.

Suzuki, J., Nakamoto, Y., and Shinkawa, Y.: Local cerebral glucose utilization in epileptic seizures of the mutant El mouse. *Brain Res.* **226,** 359–363, 1983.

Sykes, C. C. and Horton, R. W.: Cerebral glutamic acid dehydrogenase activity and γ-aminobutyric acid concentrations in mice susceptible or resistant to audiogenic seizures. *J. Neurochem.* **39:** 1489–1491, 1982.

Sze, P. Y.: Neurochemical Factors in Auditory Stimulation and Development of Susceptibility to Audiogenic Seizures, In: *Physiological Effects of Noise* (B. L. Welch and A. S. Welch, eds.) Plenum, New York, 1970.

Ticku, M. K.: Differences in γ-aminobutyric acid receptor sensitivity in inbred strains of mice. *J. Neurochem.* **33:** 1135–1138, 1979.

Toth, E., Lajtha, A., Sarhan, S., and Seiler, N.: Anticonvulsant effects of some inhibitory neurotransmitter amino acids. *Neurochem. Res.* **8:** 291–302, 1983.

Trams, E. G. and Lauter, C. J.: Ecto-ATPase deficiency in glia of seizure-prone mice. *Nature* **271:** 270–271, 1978.

Tunnicliff, G., Wimer, C. C., and Wimer, R. E.: Relationships between neurotransmitter metabolism and behavior in seven inbred strains of mice. *Brain Res.* **61:** 428–434, 1973.

Vicari, E. M.: Fatal convulsive seizures in the DBA mouse strain. *J. Physiol.* **32:** 79–97, 1951.

Watkins, J. C. and Evans, R. H.: Excitatory amino acid transmitters. *Ann. Rev. Pharmacol. Toxicol.* **21:** 165–204, 1981.

Wheeler, D. F., Contreras, N. E. I. R., and Bachelard, H. S.: DNA content and enzymic activities in the auditory regions of seizure-susceptible and nonsusceptible strains of mice. *Neurochem. Res.* **7:** 1075–1088, 1982.

Worms, P. and Lloyd, K. G.: Functional Alterations of GABA Synapses in Relation to Seizures, In: *Neurotransmitters, Seizures and Epilepsy* (P. L. Morselli, K. G. Lloyd, W. Loscher, B. Meldrum, and E. H. Reynolds, eds.) Raven, New York, 1981.

Wu, J. Y. and Roberts, E.: Properties of brain glutamate decarboxylase: Inhibition studies. *J. Neurochem.* **23:** 759–767, 1974.

The Spontaneously Epileptic Mongolian Gerbil

Peter Lomax, Randall J. Lee,
and Richard W. Olsen

1. History of the UCLA Colony

The Mongolian gerbil (*Meriones unguiculatus*) was first described to the western scientific community by Milne-Edwards (1867). The species was introduced to the United States in 1954 by Schwentker (*see* Schwentker, 1972) and these animals formed the basis of the colony developed by Tumblebrook Farms, West Brookfield, Massachusetts, currently the major breeders and suppliers. Gerbils were first bred at UCLA by Rich in 1967 (Rich, 1968). Abnormal behaviors were soon noted in some of the animals and were recognized as epileptiform in nature by Loskota, who decided to study the gerbil as a model of the epilepsies; this research subsequently constituted the basis of his dissertation for the degree of doctor of philosophy at UCLA (Loskota, 1974).

The UCLA colony was derived from seizure-sensitive (SS) and seizure-resistant (SR) parental types selected from randomly bred animals. Selective breeding by a "closed colony" technique of mating progeny of the original three pairs of SS animals for about 18 generations yielded the WJL/UC SS strain. The STR/UC SR strain was bred from four SR pairs in the same way (Loskota et al., 1974a).

2. Description of the Behavioral Seizures

When a gerbil of the SS strain is removed from its home cage and placed in a novel environment, abnormal motor behavior occurs within 20–30 s, which may progress to a myoclonic-tonic seizure. The seizures vary in severity, although when fully developed, they remain fairly constant for that animal.

The severity of seizures is rated on a seven-point scale according to the degree of motor involvement:

Grade 0: No seizures occurring within 5 min of free exploratory running in the test area

Grade 1: Rapid twitching of the vibrissae and flattening of the pinnae against the head

Grade 2: Normal coordinated locomotion is interrupted by static midstride postures with twitching of whiskers and ears

Grade 3: Vibrissae and pinnae twitch, motion is arrested, head and body are lowered to the floor, and myoclonic body jerks occur

Grade 4: The seizure progresses as in grade 3, then the animal stands with opisthotonus with the limbs splayed. The forelimbs jerk and the animal jumps vertically. This is followed by a tonic phase with a contracted posture and extension of the limbs. Straub tail and slow jerking of the head are seen. A quiescent recovery period follows

Grade 5: A fully developed clonic-tonic seizure with body rollover. This is followed by a period of postictal depression of varying duration

Grade 6: A series of seizures progressing to death (''status epilepticus'')

3. Ontogenesis of the Seizure Diathesis

The onset of behavioral seizures in young animals tested from d 30 (the age of weaning) was between 44 and 60 d. The severity of the seizures increases with age and reaches a stable score between 120 and 180 d. Thereafter, the severity remains fairly constant throughout the animal's life. There have not yet been any reports of studies of the brain during this period of seizure onset. By selecting animals for mating that had appropriate seizure scores, it has been possible to create a colony containing gerbils with the full range of scores. Thus, for example, one could select subjects with grade

4–5 seizures to screen an anticonvulsant, or grade 1–2 for a potential convulsant. This allows the generation of significant screening data on groups with as few as six animals, although generally we have used groups of about 10 (Loskota and Lomax, 1974; Ten Ham et al., 1975).

4. Electroencephalographic Correlates of Seizures

The *sine qua non* of epileptic phenomena is spontaneous, recurrent, self-sustaining discharges of the CNS. EEG recordings were made from freely moving gerbils of the SS strain during seizures of varying severity (Loskota and Lomax, 1975). The localization of seizure activity in various brain areas and the generalization of this activity could be correlated with the motor manifestations. Paroxysmal bursting was also recorded in the parietal cortex of SS animals in which no concomitant motor activity was observed. No such abnormal EEG activity has been seen in animals that do not exhibit motor seizures.

5. Regional Brain Distribution of 5-HT and GABA

In preliminary studies, 5-HT and GABA levels were compared in SR and SS animals (Loskota, 1974). No significant differences were found in whole brain 5-HT levels between the two strains. At 6 mo of age, brain 5-HT concentrations (506 ng/g) were lower than those in 12-mo-old animals (668 ng/g) ($p < 0.01$).

Lower GABA levels ($p < 0.02$) were found in the parietal lobe of SS gerbils; in the temporal lobe and thalamus the levels were lower, but failed to reach significance.

6. Brain Uptake of Dopamine and Norepinephrine

The initial rates of uptake of (^{3}H)-DA into synaptosomes of SR and SS gerbil striatum, thalamus, hypothalamus, cerebral cortex, and whole brain, and the uptake of (^{3}H)-NE into whole brain of these strains, were investigated (Loskota, 1974). In none of these experiments were any differences in K_m and V_{max} noted between the two strains.

7. Brain Dopamine Activity and Seizures

Cox and Lomax (1976) studied the effects of modifying brain biogenic amine activity on spontaneous seizures in the gerbil. Groups of animals were treated with compounds that deplete NE, NE plus DA, and 5-HT. Depleting NE reduced the seizure severity, whereas lowering 5-HT was without effect. Brain NE was depleted by diethyldithiocarbamate (DDC, which inhibits DA beta-hydroxylase), which will increase DA levels. When DA was increased by DDC and the functional depletion of NE was prevented with DL-theodihydroxyphenylserine (which is converted directly to NE), the seizure severity was again decreased. DA receptor agonists, apomorphine and bromocriptine, also reduced the mean seizure scores. The time course of these effects correlated with that for the stereotypic behavior, seen in the rat, which is mediated specifically at DA receptors. The presynaptic dopamine receptors appear to mediate the anticonvulsant effect based on studies involving specific DA receptor agonists (Lee and Lomax, 1983a). These data indicate that a search for differences in regional levels, or turnover rates of these amines, particularly DA, between SS and SR gerbils, might indicate the genetic deficiency underlying the seizure phenomenon.

8. Endogenous Opioids and Seizures

Over the past few years there has developed a growing interest in the possible role of neuroendocrine peptides and peripheral hormones in CNS function. These peptides exhibit high affinity for their binding sites (dissociation constants of the order of $10^{-8}-10^{-10}$ M). Consequently, the peptides would need to be present only in very low concentrations to trigger a cellular response, and they would tend to have a longer duration of action than the amine neurotransmitters. These observations have given rise to the concept of "neuromodulation"; there is as yet no consensus, however, on what defines modulation, and evidence for nonsynaptic release of peptides in the mammalian CNS is lacking.

The general concept of modulation has been applied to epileptic phenomena, particularly for explaining the long periods between seizures in the epileptic brain. Gerbils of the SS strain exhibit a relative refractory period of 48–168 h after each seizure when tested repeatedly. The animals show a similar decrease in incidence and severity of seizures if subjected to mild physical restraint prior to testing. On the other hand, exposure to an ambient temperature of 11 °C for 180 min increased the severity of subsequent seizures (Bajorek, 1981). Of the several neuropeptides, beta-endorphin and the enkephalins have been postulated to play a role in the pathogenesis of the epilepsies and to mediate postictal depression in kindled rats (Frenk et al., 1979). These could be released in response to stress and so modify the seizure state.

Intracerebroventricular (icv) injection of beta-endorphin (0.1–3 μg) into SS gerbils reduced the incidence and severity of both the motor manifestations and the preceding high voltage focal spiking and the accompanying seizure activity in the cortical EEG. This "anticonvulsant" effect was prevented by prior administration of naloxone (1 mg/kg, ip) (Bajorek and Lomax, 1982).

The effects of prototypic agonists for mu (morphine), kappa (ketocyclazocine), and sigma (*N*-allylnormetazocine) opioid receptors were tested similarly. Each opioid decreased the incidence and severity of the seizures. The anticonvulsant effects of morphine and ketocyclazocine were blocked by naloxone, but those of *N*-allylnormetazocine were unaffected. Each of the agonists caused a characteristic spectrum of abnormal behaviors. The study indicates that the opioid anticonvulsant action is not specific to one type of opioid receptor, although receptor-specific behaviors were observed (Lee et al., 1984).

beta-Endorphin, dynorphin, met-enkephalin, and morphine were injected, through chronically implanted guides, into a lateral ventricle of SS gerbils. All animals injected with beta-endorphin demonstrated tonic extension of their extremities with trunk opisthotonus following drug administration. After a period of 3–118 s, the animals resumed normal exploratory behavior. During the periods of abnormal behavior, the EEG showed a tendency to synchrony at 5–6 Hz. Dynorphin occasionally produced high-amplitude slow wave spiking accompanied by immobility lasting 10–240 s. met-Enkephalin produced recurrent spikes of 200–500 μV amplitude with a frequency of 0.15–0.33 Hz. The morphine-treated gerbils showed recurrent spikes similar to those induced by met-enkephalin, with the addition of high-frequency paroxysmal spiking that lasted 5–30 s. No characteristic motor seizures were observed during the spiking episodes. Naloxone blocked the EEG synchrony and abnormal behaviors induced by beta-endorphin, but had no effect on the responses to the other opioids. These results suggest that "opioid seizures" are not the same as the natural epileptiform attacks in the SS gerbils in respect to cortical EEG and motor behavior (Lee et al., 1983).

In order to ascertain the role of opioids in the pathogenesis of seizures, the effects of morphine injected icv were compared in SS and SR gerbils. In both strains the characteristic EEG pattern of recurrent intermittent spikes of 200–500 μV amplitude and 0.15–0.33 Hz frequency, interrupted by high-frequency paroxysmal spiking, was observed. No motor seizures were seen. Thus, there appears to be a distinct difference between the opiate "seizures" (which can be induced in many species, including SR gerbils) and the epileptiform attacks in the SS strain.

Naltrexone was used to investigate the possibility that endogenously released opioid peptides may modulate seizures during the relative refractory period after an initial seizure. Naltrexone pretreatment affected neither the incidence nor the severity of the seizures during the first testing session. When the animals were tested for seizures a second time, 1.75 h later, the naltrexone pretreated animals evinced a significantly greater incidence and severity than did controls (Lee et al., 1983).

Further studies were undertaken to determine whether a spontaneous seizure in SS gerbils would modify the threshold for seizures induced chemically by pentylenetetrazole (PTZ) or by electroshock (MES) (Lee and Lomax, 1984). For those animals in which a spontaneous seizure had been induced prior to administration of PTZ, the dose–response curve was shifted to the right of that for nonseized animals. The ED_{50} for the control, nonseized group was 32 mg/kg (95% confidence limits: 30.7–34.1), whereas for the postseizure group it was 60 mg/kg (95% confidence limits: 57.8–62.2). If the animals were treated with naltrexone prior to the induction of a natural seizure, the postictal inhibitory effect on PTZ-induced convulsions was attenuated. No effect of a spontaneous seizure on MES seizures could be detected; however, it is difficult to obtain graded responses to electroshock in SS gerbils. In the rat, MES-induced seizures have been reported to be modified by opioids (Tortella et al., 1984).

Taken together, these data support the hypothesis that endogenous opioids released during an epileptic attack, or in response to nonspecific stress, can modulate subsequent seizure propensity and play a physiological role in preventing recurrent seizures (i.e., status epilepticus). As a corollary to this suggestion, it is possible that an abnormality in brain peptide release, or in their membrane binding sites, could underlie the epileptic diathesis in the SS gerbil.

9. ACTH and Seizures

Nonspecific stress has long been implicated in the pathogenesis of epileptic phenomena. Such varied factors as lack of sleep, emotional disturbances, over exertion, and intellectual strain have been seen to precipitate attacks in epileptic patients. These considerations led to speculation that the pituitary-adrenal axis may play a role in the initiation of seizures, or in the modulation of their expression, in the SS gerbils.

ACTH was administered directly into a lateral ventricle of SS gerbils prepared with a chronically implanted cannula. With a dose of 5 μg, there was reduction in seizure incidence (by 28%), but this was only marginally significant. The mean seizure score and seizure duration were, however, decreased. This dose of ACTH also caused periods of abnormal behavior characterized by increased grooming and episodes of cataleptic like immobility.

The large dose of ACTH required to produce these effects, coupled with the unusual behavior, suggests that these represent nonspecific pharmacological, rather than functional, effects (Bajorek, 1981).

10. Adrenocortical Hormones and Seizures

The gerbil adrenal cortex is approximately three times the mass of that in the rat—presumably a reflection of the adaptation of this species to an arid environment. There are two major glucocorticoids, cortisol and 19-hydroxy-11-deoxycortisol (Oliver and Peron, 1964). Injection of cortisol (1 and 2 mg/kg 1 h prior to testing) led to a dose-dependent decrease in seizure severity. In control experiments the diluent (water/ethanol) alone caused some reduction in the seizure scores. However, the anticonvulsant effect was also manifest with the water-soluble succinate salt. The anticonvulsant effect of cortisol could be demonstrated at 1 h after injection, but at 24 h no significant reduction of severity was seen. Cortisone (2 mg/kg) did not affect the seizure expression (Bajorek, 1981).

The anticonvulsant action of cortisol was unexpected, since previous reports had identified it as a proconvulsant in the rat (Woodbury, 1952). The duration of the effect of cortisol (up to 1 h after injection) suggests that it may be active during the refrac-

tory period (Lomax et al., 1984). Radioimmunassay of plasma corticoids in SS gerbils demonstrated an increase from 0.2 to 0.4 μg/mL
60 s after handling stress; the majority of the animals had a seizure
during this time period (Bajorek, 1981). It seems likely that the
adrenocortical hormones may be of importance in regulating the
postictal refractory period, possibly by the activation of processes
such as transcriptional synthetic mechanisms or other genomic
systems. These have been postulated to be induced by the action
of adrenal hormones on neuronal systems (McEwen, 1977).

11. Thyrotropin-Releasing Hormone and Seizures

Thyrotropin-releasing hormone (TRH) has been shown to have
physiological activity in the CNS, in addition to being the releasing factor for thyroid-stimulating hormone (TSH) from the adenohypophysis. TRH has been shown to exert a variety of behavioral
effects after central or peripheral administration in several species.
Excitation, increased motor activity, changes in body temperature,
anorexia, tremor, and EEG arousal have been described (*see* review
by Nemeroff et al., 1979).

Central injection of TRH (10 μg, icv) significantly increased the
mean seizure scores in SS gerbils, and the duration of the seizures
was increased by a factor of 10. Spiking activity in the cortical EEG
correlated with the motor seizures. Some interictal spiking that was
not associated with any identifiable behaviors was also seen. Abnormal behavior, such as stereotyic movements, head or body jerks,
tremors and fixed staring, were observed after TRH (10 μg, icv);
at 50 μg, locomotor activity was markedly reduced and abnormal
spikes and spindles appeared on the cortical EEG. Increased foot
stomping (which is a normal arousal behavior in the gerbil) was
prominently increased after injection of TRH (Bajorek, 1981; Bajorek et al., 1983).

Thyroid hormone levels will be increased after administration
of TRH and these can lower seizure thresholds. However, this effect
is only appreciable after chronic administration of the hormones
(T_3 or T_4). It is likely, therefore, that the changes in seizure
parameters seen after central administration of TRH are caused by
a direct action of the peptide on the CNS and are the consequence
of a nonspecific general arousal.

12. Arginine Vasopressin and Seizures

Arginine vasopressin (AVP) is involved in the maintenance of body fluid osmolality, blood volume, and blood pressure. There is now increasing evidence that the peptide may also have other functions. Recent reports (Kasting, 1982; Kasting et al., 1981a) have suggested that AVP may be an endogenous antipyretic and a mediator of febrile convulsions. Immunohistochemical techniques revealing extrahypothalamic AVP-containing projections, terminating in neural structures other than the neurohypophysis, have led to the suggestion that the peptide may have a neurotransmitter or neuromodulator function in the CNS.

In SS gerbils, injection of AVP (1 or 5 μg, icv, and 0.01–1.0 mg/kg, sc) caused dose-related falls in core temperature. Stereotypic scratching and body shakes were observed after central, but not systemic, injections. By both routes of administration, AVP caused significant decreases in seizure incidence and severity, and the animals appeared to be sedated (Lee and Lomax, 1983b).

The sedated, flaccid appearance following sc injection of AVP may be related to general depression and reduced muscle tone. The anticonvulsant effect could be secondary to this nonspecific depression, even at doses that did not produce overt muscular weakness. Alternatively, AVP may indeed have anticonvulsant properties. This view is supported by the observation that vasotocin, an AVP analog, protects rats from PTZ-induced convulsions (Kasting et al., 1981b). These results in the SS gerbil do not lend any support to the hypothesis that AVP is involved in the pathogenesis of infantile febrile convulsions.

13. Role of the GABA Receptor/Ionophore Complex in Seizures

The development of radioligand binding techniques for brain homogenates has allowed receptor sites for neurotransmitters and drugs to be assayed. From such studies specific receptor binding sites for GABA have been defined. These receptors, which are characterized on the basis of sensitivity to the agonist muscimol and the antagonist bicuculline, are coupled to chloride channels and modulated by barbiturates and by benzodiazepines (*see* Olsen and Snowman, 1982). Specific benzodiazepine receptor sites have also been identified by radioligand binding. The convulsant picrotoxin also demonstrates specific binding to brain membranes, related to its action as an antagonist of GABA-activated chloride channels, but at a site distinct from the GABA recognition site. Picrotoxin binding sites are competitively inhibited by barbiturates and are therefore designated as picrotoxin/barbiturate receptors. The three classes of drug-binding sites appear to be closely coupled in a single protein complex, as envisioned in Fig. 1 (Olsen et al., 1984a).

The binding sites for picrotoxin-like compounds are probably involved in their convulsant actions by acting as antagonists of GABA postsynaptic inhibitory chloride channels. These sites also appear to mediate the convulsant action of pentylenetetrazole and bemegride. Similarly, the interaction of depressant drugs, such as the anxiolytic pyrazolopyridines and hypnotic and sedative barbiturates, with these convulsant receptor sites appears to reflect their ability to potentiate GABA function. The reaction of the clinically effective anticonvulsants with GABA receptor binding is less certain; because of their low potency, and because they have other actions on the neural membrane, it is difficult to assign recep-

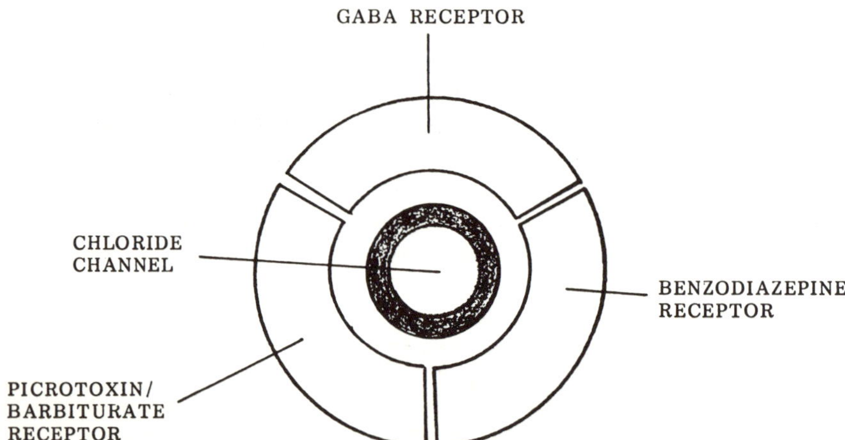

Fig. 1. The GABA–receptor–ionophore complex is characterized as comprising three receptor sites: the GABA receptor; the benzodiazepine receptor; the picrotoxin/barbiturate receptor. The chloride ion channel is associated with, or is part of, the receptor complex. Chloride ion permeability may be potentiated (by CNS depressant drugs) or reduced (by CNS excitatory drugs). Endogenous or exogenous compounds may modulate the receptor complex (for further details, *see* text and Olsen, 1981).

tor sites with confidence. The case for the benzodiazepines is less ambiguous and the GABA-linked high-affinity receptor sites appear to mediate their anticonvulsant action (for review, *see* Olsen et al., 1984a).

If, indeed, drugs that augment GABA-mediated inhibition are active as anticonvulsants, it is a reasonable supposition that a functional GABA deficit might be implicated in some forms of epilepsy. Seizures can result from blockade of the biosynthesis of GABA, of synaptic release of GABA, or of the postsynaptic response to GABA at the level of the receptor sites, the associated chloride channel, or at the modulatory sites for benzodiazepines or barbiturate/ picrotoxin (Fig. 1) Olsen, 1981).

As discussed above (section 5) GABA levels were lower in the parietal cortex and thalamus of SS gerbils compared to animals of the SR strain. We therefore examined the levels of benzodiazepine/ GABA receptors in the brain of SS gerbils. Brains were divided into four regions for GABA receptor assay; brainstem, cerebellum, cortex, and the "rest of the brain." Animals (six SS and six SR) were compared on a blind basis for (^{3}H)-GABA binding and barbiturate enhancement. No significant differences were found for cortex, cerebellum, or "rest of the brain"; there was a marginally significant deficit in the pons-medulla in the SS animals. In another series of animals (^{3}H)-muscimol binding was used to assay GABA receptors in SS gerbils within 5 min after the onset of a seizure. No differences were found in comparison to SR animals in any of the brain areas. Benzodiazepine receptors were assayed in these same animals and lower binding was demonstrated in the SS thalamus/midbrain (26%, $p < 0.01$), whereas a nonsignificant decrease was observed in the hippocampus. Further dissection of the brain revealed that this deficit was mainly localized to the caudal midbrain/colliculus. However, a Scatchard plot of (^{3}H)-flunitrazepam binding at seven ligand concentrations failed to show any significant differences between SS and SR in either K_d or B_{max} in total midbrain. In another group of SS animals, sacrificed 5 min after a seizure, decreased binding was found in the pons-medulla and the striatum. Thus, there was a trend toward lower benzodiazepine and GABA receptor binding in the brainstem of SS gerbils.

The relatively large volume of tissue necessary for the receptor assays renders it difficult to unmask small differences in receptor binding in discrete brain regions, because of dilution by normal tissue in grossly dissected samples. In order to achieve finer localization of the distribution and density of benzodiazepine binding sites, we turned to the technique of autoradiography of brain slices (Olsen et al., 1984b), paying particular attention to the mesencephalon (Olsen et al., 1983c). Cryosotat sections of the midbrain were labeled with (^{3}H)-flunitrazepam under conditions known to result in specific labeling of benzodiazepine receptors. The grain density in the autoradiograms was scanned by a microscope and a microphotometer interfaced with a computer. These studies confirmed the differences in benzodiazepine receptor binding between SR and SS gerbils; they showed a 20% deficit in grain density in the substantia nigra and a 12% deficit in the periaqueductal gray matter in SS animals. There were no significant differences in the retrosplenial cortex, hippocampus, or superior colliculus. The grain density was 19% greater in the interpenduncular nuclei of the SS strain. Comparing SS animals before and after a seizure, there was a significant increase in binding in several areas (substantia nigra, interpenduncular nucleus, retrosplenial cortex) following the seizure; however, post-seizure gerbils still had significantly less binding than SR in the substantia nigra and the periaqueductal gray. This increase in binding following a seizure could explain the lack of a difference in the thalamus/midbrain homogenate binding when SR and SS (postseizure) animals were compared. It may also be relevant to other reports of similar increases in benzodiazepine binding following seizures in the gerbil (Asano and Mizutani, 1980).

Grain density was measured in 250 μm^2 sections of the substantia nigra pars reticulata, for seven concentrations of the radioactive ligand, and converted to moles of specific benzodiazepine receptor binding by comparison to a standard curve. Scatchard analysis of the data showed a significant decrease in the number of binding sites for SS compared to SR, with no change in binding affinities.

These studies have demonstrated that the number of benzodiazepine receptors (and also, presumably, of low-affinity GABA receptors) is decreased in mesencephalic regions of the SS gerbil. The deficit in the substantia nigra and periaqueductal gray could be caused by a decreased number of GABA-receptive neurons or by reduced receptor proteins on each neuron. In either case, a deficit in GABA-mediated inhibition would result and this could contribute to the seizure diathesis in these animals. This view is consistent with the efficacy of anticonvulsants, which augment GABA activity (Loskota and Lomax, 1975).

The role of the midbrain periaqueductal gray in seizure susceptibility is unknown, but it might be speculated that it is involved in the transmission of the afferent stimuli that produce the pathological response; the substantia nigra could be involved in the circuitry leading to the motor expression of the seizures in the SS gerbil. This structure is involved in seizures induced by electroshock, convulsant compounds, and kindling (*see* Gale, 1983). Anticonvulsants that augment GABA are effective when injected directly into the substantia nigra; the anticonvulsant effect of GABA-augmenting drugs administered systemically can be correlated with an increase in nigral nerve ending pools of GABA; lesions of the substantia nigra reduce motor and limbic seizures. Thus, a deficit of GABA/benzodiazepine receptor activity in this part of the CNS could contribute to seizure susceptibility.

These observations of lower receptor binding in the gerbil genetic model of epilepsy are consistent with the GABA hypothesis. Increasing evidence for changes in GABA-mediated inhibitory synaptic transmission in some types of epilepsy may be an important clue to understanding the basic mechanisms of the human disorder.

14. Conclusions

The variety of models and experimental preparations for the study of the epilepsies exceeds the variety of expression of the epilepsies themselves. Many of the models may be classified as "invasive" preparations in that the seizure phenomenon is induced by some manipulation of an otherwise normal brain. Models, such as the spontaneously epileptic gerbil, whose causality appears to derive primarily from genetic and developmental abnormalities would appear to offer greater scope for studying the etiology, ontogenesis, and pathogenesis of human idiopathic epilepsies, as distinct from secondary epilepsies. It should be borne in mind, however, that each of the human epilepsies may well have a specific model most appropriate for its study. Other genetic models are described in other sections of this book.

Compared to some genetic models, the gerbil offers some specific advantages: it is a small, relatively inexpensive animal that breeds rapidly and is suitable for genetic and neurochemical studies; on the other hand, it has a skull that is sufficiently large to allow for stereotaxic surgery for the chronic implantation of electrodes for depth recording, or cannulae for intracranial administration of drugs to precise locations with the aid of a stereotaxic atlas; such an atlas of the gerbil brain has been prepared (Loskota et al., 1974b).

From the studies reviewed above, the endogenous opioid system and the GABA–receptor complex would seem to offer profitable areas for further research into the underlying abnormalities of the epileptic state in the SS strain. Studies are currently in progress to examine the developmental time course of receptor differences in respect to the ontogenesis of the seizure state, in the well-defined time frame of 44–180 d (*see* section 3), and their relationship to the severity of the seizures.

References

Asano, T. and Mizutani, A.: Brain benzodiazepine receptors and their rapid changes after seizures in the Mongolian gerbil. *Jap. J. Pharmacol.* **30:** 783–788, 1980.

Bajorek, J. G.: Neuroendocrine modulation of seizures in the Mongolian gerbil (doctoral dissertation). Los Angeles, University of California, 1981.

Bajorek, J. G., Lee, R. J., Lomax, P., and Delgado-Escueta, A. V.: Neuropeptides as possible endogenous modulators of seizure behavior in a genetic model of epilepsy. *Abst. Int. Sym. Epilepsy,* American Epilepsy Society, 1983.

Bajorek, J. G. and Lomax, P.: Modulation of spontaneous seizures in the Mongolian gerbil: Effects of β-endorphin. *Peptides* **3:** 83–86, 1982.

Cox, B. and Lomax, P.: Brain amines and spontaneous epileptic seizures in the Mongolian gerbil. *Pharmacol. Biochem. Behav.* **4:** 263–267, 1976.

Frenk, H., Engel, J., Ackermann, R. F., Shavit, Y., and Liebeskind, J. C.: Endogenous opioids may mediate post-ictal behavioral depression in amygdaloid-kindled rats. *Brain Res.* **167:** 435–440, 1979.

Gale, K.: The Role of the Substantia Nigra in the Anticonvulsant Actions of GABAergic Drugs, In: *Neurotransmitters, Seizures and Epilepsy,* vol. 2, (R. G. Fariello, J. Engel, P. L. Morselli, and L. F. Quesney, eds.) Raven, New York, 1983.

Kasting, N. W.: Vasopressin: A homeostatic factor in the febrile process. *Neurosci. Biobehav. Rev.* **6:** 215–222, 1982.

Kasting, N. W., Veale, W. L., Cooper, K. E., and Lederis, K.: Vasopressin may mediate febrile convulsions. *Brain Res.* **213:** 327–333, 1981a.

Kasting, N. W., Veale, W. L., and Cooper, K. E.: Vasotocin protects rats against convulsions induced by pentylenetetrazole. *Experientia* **37:** 1001–1002, 1981b.

Lee, R. J., Bajorek, J. G., and Lomax, P.: Similar anticonvulsant, but unique behavioral effects of opioid agonists in the seizure sensitie Mongolian gerbil. *Neuropharmacology* **23:** 517–524, 1984.

Lee, R. J., Bajorek, J. G., and Lomax, P.: Opioid peptides and seizures in the spontaneously epileptic Mongolian gerbil. *Life Sci.* **33:** 567–570, 1983.

Lee, R. J. and Lomax, P.: Dopaminergic receptors in the modulation of seizures in the Mongolian gerbil. *Proc. West. Pharmacol. Soc.* **26:** 273–276, 1983a.

Lee, R. J. and Lomax, P.: Thermoregulatory, behavioral and seizure modulatory effects of AVP in the gerbil. *Peptides* **4:** 801–805, 1983b.

Lee, R. J. and Lomax, P.: The effect of spontaneous seizures on pentylenetetrazole and maximal electroshock induced seizures in the Mongolian gerbil. *Eur. J. Pharmacol.,* **106:** 91–96, 1984.

Lomax, P., Bajorek, J. G., and Lee, R. J.: Neuroendocrine modulation of seizures in the Mongolian gerbil (abst.). *9th Int. Cong. Pharmacol.* 1984.

Loskota, W. J.: The Mongolian gerbil (*Meriones unguiculatus*) for the study of the epilepsies and anticonvulsants (doctoral dissertation). Los Angeles, University of California, 1974.

Loskota, W. J. and Lomax, P.: The Mongolian gerbil as an animal model for the study of the epilepsies: Anticonvulsant screening. *Proc. West. Pharmacol. Soc.* **17:** 40–45, 1974.

Loskota, W. J. and Lomax, P.: The Mongolian gerbil (*Meriones unguiculatus*) as a model for the study of the epilepsies: EEG records of the seizures. *Electroencephal. Clin. Neurophysiol.* **38:** 597–604, 1975.

Loskota, W. J., Lomax, P., and Rich, S. T.: The gerbil as a model for the study of the epilepsies: Seizure patterns and ontogenesis. *Epilepsia* **15:** 109–119, 1974a.

Loskota, W. J., Lomax, P., and Verity, M. A.: *A Stereotaxic Atlas of the Mongolian Gerbil Brain* (Meriones unguiculatus). Ann Arbor Science, Ann Arbor, Michigan, 1974b.

McEwen, B. S.: Adrenal steriod feedback on neuroendocrine tissues. *Ann. NY Acad. Sci.* **297:** 568–579, 1977.

Milne-Edwards, A.: Observations sur quelques mammiferes du nord de la Chine. *Ann. Sci. Nat.* (Zool.) **7:** 375–377, 1867.

Nemeroff, C. B., Loosen, P. T., Bissette, G., Manberg, P. J., Wilson, I. C., Lipton, M. A., and Prange, A. J.: Pharmaco-behavioral effects of hypothalamic peptides in animals and man: Focus on thyrotropin-releasing hormone and neurotensin. *Psychoneuroendocrinology* **3:** 279–310, 1979.

Oliver, J. T. and Peron, F. G.: 19-Hydroxy-11-deoxycortisol, a major steriod secreted by the adrenal gland of the Mongolian gerbil. *Steroids* **4:** 351–363, 1964.

Olsen, R. W.: GABA–benzodiazepine–barbiturate receptor interactions. *J. Neurochem.* **37:** 1–13, 1981.

Olsen, R. W. and Snowman, A. M.: Chloride-dependent enhancement by barbiturates of GABA receptor binding. *J. Neurosci.* **2:** 1812–1823, 1982.

Olsen, R. W., Wamsley, J. K., Lee, R. J., and Lomax, P.: The Benzo-diazepine/barbiturate/GABA Receptor–Chloride Ionophore Complex in a Genetic Model for Generalized Epilepsy, In: *Basic Mechanisms of the Epilepsies* (A. V. Delgado-Escueta, A. A. Ward, and D. M. Woodbury, eds.) Raven, New York, 1984a.

Olsen, R. W., Snowhill, E. W., and Wamsley, J. K.: Autoradiographic localization of low affinity GABA receptors with [^{3}H]-bicuculline methochloride. *Eur. J. Pharmacol.* **99:** 247–248, 1984b.

Olsen, R. W., Wamsley, J. K., Lee, R. J., and Lomax, P.: Alterations in Benzodiazepine/GABA Receptor–Chloride Ion Channel Complex in the Seizure-Sensitive Mongolian Gerbil, In: *Neurotransmitters, Seizures and Epilepsy* vol. 2 (R. G. Fariello, J. Engel, P. L. Morselli, and L. F. Quesney, eds.) Raven, New York, 1983.

Rich, S. T.: The Mongolian gerbil (*Meriones unguiculatus*) in research. *Lab. An. Care* **18:** 235–243, 1968.

Schwentker, V.: *The Gerbil, an Annotated Bibliography* Tumblebrook Farms, West Brookfield, 1972.

Ten Ham, M., Loskota, W. J., and Lomax, P.: Acute and chronic effects of Δ^9-tetrahydrocannabinol on seizures in the gerbil. *Eur. J. Pharmacol.* **31:** 148–152, 1975.

Tortella, F. C., Long, J. B., Robles, L., and Holaday, J. W.: Physiological tolerance to endogenous opioid activation: Inhibition of seizure protection in morphine tolerant rats (abstract). *Fed. Proc.* **43:** 936, 1984.

Woodbury, D. M.: Effect of adrenocortical steriods and adrenocorticotrophic hormone on electroshock seizure threshold. *J. Pharmacol. Exp. Ther.* **105:** 27–36, 1952.

The Genetically Epilepsy-Prone Rat

Hugh E. Laird II and Phillip C. Jobe

1. Introduction

One of the key elements in developing a treatment for any disorder is understanding its cause. As simple and logical as that statement sounds, it nevertheless represents the major obstacle to providing effective therapy in most disorders. Indeed, such is the case for the neurological disorder called epilepsy. In part, this can be explained by the fact that epilepsy is not a specific disease, but a family of neurological disorders that is defined by the clinical symptoms displayed by the patient. The plethora of symptoms observed in epilepsy is the result of the seizure process affecting different brain regions. The fact that different areas of the central nervous system can be and are affected suggests a multifactorial cause for this family of seizure disorders.

If there are multiple factors that can cause epilepsy, then the selection of an appropriate model system in which one can study the pathophysiology of the epileptic state is of paramount importance. As the reader can tell from the chapters in this book, the number of model systems used to study seizure disorders is legion. In fact, the search for the "ideal" model for epilepsy continues unabated. This search has followed two basic approaches. In the first, an "epileptic-like" state is artificially produced in the animal and the pathophysiological and pharmacological consequences are evaluated. In the second approach, animal strains with a genetic predisposition to seizures are developed and the molecular bases for the seizure-prone state are studied. Both of these approaches have yielded and will continue to yield valuable information regarding epilepsy. Nevertheless, it is important that the investigator carefully define the experimental questions to be asked so that appropriate model systems may be selected. For example, questions on the molecular bases for seizure susceptibility are more likely to be answered using genetic models of epilepsy. On the other hand, studies seeking information on the mechanisms of seizure initiation and/or propagation may be addressed in either artificial or genetic epilepsy models.

We have chosen for our research a genetic model of epilepsy. The genetically epilepsy-prone rat (GEPR) possesses an inborn hypersensitivity to a variety of seizure-inducing stimuli. By carefully breeding these animals to achieve genetic homogeneity, we can use this model to search for central nervous system abnormalities associated with their seizure disorder. Such a model permits a methodical differentiation of those molecular defects responsible for the epileptic state from those that are epiphenomenon.

2. Development of the GEPR

Originally the GEPRs were called audiogenic seizure-susceptible (AGS) rats since the animals were selected for breeding purposes based on their susceptibility to sound-induced seizures (Jobe et al., 1973a,b; Consroe et al., 1979). From the original colony of Sprague-Dawley-derived AGS rats developed by A. L. Picchioni and L. Chin at the University of Arizona in 1958, the authors have selectively bred two lines of GEPRs (GEPR/UAZ) that differ in their seizure characteristics as follows. The *moderate seizure* line displays two wild running phases separated by a short period of quiescence. The last running episode is terminated with a generalized clonic convulsion (Fig. 1, pattern A). These animals have an audiogenic response score (ARS) (*see* Jobe et al. 1973a, for details of this scoring system) of 2 and are called *GEPR-2s*. The *severe seizure* line has only one running phase, which terminates in a tonic extensor convulsion (Fig. 1, pattern D). Animals from this line have an ARS of 9 and are labeled *GEPR-9s*.

One of the authors has independently developed a Sprague-Dawley-derived *moderate seizure* line of GEPRs, which displays one wild running phase terminating in a generalized clonic convulsion (Fig. 1, pattern A). These rats are awarded an ARS of 3 and labeled *GEPR-3s*. Each of these lines/colonies are being bred by a brother/sister mating paradigm to achieve genetic homogeneity. Breeding pairs for each of the colonies are selected for their consistency of

Fig. 1. Convulsive patterns of genetically epilepsy-prone rats. (A) Convulsive endpoint consists of generalized clonus involving forelimbs, hindlimbs, pinnae, and/or vibrissae. (B) Convulsive endpoint consists of tonic flexion of neck, trunk, and forelimbs with clonus of hindlimbs. (C) Convulsive endpoint is similar to (B), except hindlimbs are in partial tonic extension (i.e., tonic extension of thighs and legs with clonus of feet). (D) Convulsive endpoint is similar to (C) except hindlimbs are in complete tonic extension (i.e., animal in maximal convulsion).

seizure response and latency to onset of initial seizure activity (running) and convulsion. Latency values are measured as the elapsed time from the onset of the sound stimulus to the display of the behavior, running or convulsion (Laird, 1974).

Even though the original breeding stock for the AGS rat colonies was selected for its sensitivity to audiogenic seizure, work to date suggests that these animals are not only susceptible to sound-induced seizures, but, in addition, are more susceptible to a variety of other seizure-inducing stimuli (Laird and Huxtable, 1978; Jobe and Laird, 1981; Laird et al., 1984). For example, GEPRs exhibit convulsions in response to hyperthermia (Jobe et al., 1982) and have lower thresholds for convulsions caused by electroshock, pentylenetetrazol, and bicuculline (*see* Table 1) (Laird and Huxtable, 1978). GEPR-2s and GEPR-9s have lower thresholds for minimal, maximal, and intracerebral electroshock (Duplisse et al., 1973; Laird and Huxtable, 1978). The intracerebral electroshock study showed GEPR-9 subjects had markedly lower electroshock thresholds in the ventral cochlear nuclei, inferior colliculi, medial geniculate

TABLE 1

Comparison of Electroshock and Chemoshock Thresholds
in GEPR and Non-GEPR[a]

| | Threshold values | |
Treatment	GEPR	Non-GEPR
Min EST[b]	16 mA[d]	24 mA
Max EST	26 mA[d]	66 mA
IET (IC)	46 μA[d]	186 μA
IET (VCN)	45 μA[d]	1000 μA
IET (MG)	93 μA[d]	463 μA
IET (RF)	95 μA[d]	189 μA
PTZ (IV)	11 mg/kg[d]	20 mg/kg
STRY (IV)	39 μg/kg	42 μg/kg
BIC (IV)	17 μg/kg	17 μg/kg
BIC (IC)[c]	200 ng[d]	500 ng

[a]From Laird and Huxtable, 1978. Abbreviations: min EST, minimal electroshock threshold; max EST, maximal electroshock threshold; IET, intracerebral electroshock threshold; IC, inferior colliculi; VCN, ventral cochlear nuclei; MG, medial geniculate; RF, reticular formation; PTZ, pentylenetetrazol; STRY, strychnine; BIC, bicuculline; IV, intravenous injection.

[b]From Duplisse et al., 1973.

[c]From Duplisse, 1976.

[d]$p < 0.05$ when compared to non-GEPR.

nuclei, and reticular formation in comparison to controls (*see* Table 1) (Laird and Huxtable, 1978). Furthermore, these workers confirmed the observation of Duplisse (1976) that electrical stimulation of auditory nuclei in the GEPRs produced a seizure behaviorally identical to the seizure induced by sound. The lower pentylenetetrazol seizure thresholds were observed in the GEPR-2s and GEPR-9s lines, but not the GEPR-3 line (Laird and Huxtable, 1978, Reigel et al., 1984). Interestingly, the lower threshold to bicuculline was detected when injected directly into the inferior colliculi (Duplisse, 1976), but not when given systemically (Table 1) (Laird and Huxtable, 1978; Reigel et al., 1984).

Other evidence suggests that the GEPRs are abnormally sensitive (95% of the animals respond with seizures) to pure tone stimuli (usual sound stimuli is a mixed frequency—*see* Jobe et al., 1973a) of 12 KHz (100 dB, 100 ms duration, 2/s) with lower or higher frequencies of pure tone much less effective in eliciting seizures (e.g., <20% response) (Gehlbach and Faingold, 1984). In addition, the GEPRs have been reported to be more sensitive to at least three other types of seizure stimuli. First, GEPR-9s are highly susceptible to seizures induced by intracerebroventricular (icv) injections of morphine (Stewart et al., 1984; Reigel et al., 1985). Not only are the GEPR-9s more sensitive to ICV morphine seizures, but the convulsions elicited are far more severe (progressing to tonic extensor convulsions) than those observed in nonepileptic controls. Second, the GEPR is more sensitive to barbiturate withdrawal convulsions (Bourn and Garrett, 1983). These workers showed that the nonaudiogenic-seizure-responsive progeny of GEPRs became susceptible to sound-induced seizures during barbiturate withdrawal, whereas similarly treated progeny of nonepileptic controls did not. Finally, Savage et al. (1985) have shown that the early stages of kindling development are accelerated in both the GEPR-3s and GEPR-9s, whereas the later stages of kindling are accelerated only in the GEPR-9.

In addition to an increased sensitivity to various seizure-inducing stimuli, the GEPRs have other characteristics that are different from other "normal" Sprague-Dawley-derived rats. For example, Ebadi et al. (1985) have found an increase in the pyridoxal phosphate content of the cerebellum along with a decrement in the cortex of the GEPR-3s. These findings are associated with differences in pyridoxal kinase activity in the basal ganglia, cerebellum, hippocampus, and cortex, with the pyridoxal kinase activity lower in the first two structures and higher in the remainder. Since either an excess or a deficit in brain pyridoxal phosphate can cause seizures, these observations in the GEPRs are provocative. However, much work remains to be done to determine what role, if any, a defect in the pyridoxal phosphate system may play in regulating the seizure-prone state in the GEPRs.

Glucose utilization in the brains of the GEPRs has been compared to that in brains from the nonepileptic controls (McEachron et al., 1984; Waterhouse et al., 1985). These workers used the 2-deoxyglucose method of Sokoloff et al. (1977), which correlates the rate of neuronal glucose metabolism with neuronal activity. Using this technique, the GEPRs were shown to have an increased glucose utilization in the cerebellar cortex, deep cerebellar nuclei, vestibular nuclei, brain stem, superior olive, inferior colliculus, medial geniculate, and hippocampus. In addition, there was a greater right/left asymmetry in brain glucose uptake in the GEPRs. These data suggest a tonic neuronal hyperactivity within selected brain areas of the GEPRs. Such neuronal excitability could certainly contribute to the enhanced seizure susceptibility observed in the epileptic rat.

The response of the GEPRs to anticonvulsant drugs is another area that has been extensively investigated. In general, all clinically useful anticonvulsant agents are effective in preventing sound-induced seizures in the GEPRs (Consroe et al., 1979; 1980; Dailey and Jobe, 1985). The anticonvulsant action of phenobarbital in the GEPRs, as in the human, was not dependent on the sedative action of the drug (Consroe et al., 1980). These same workers showed that anticonvulsants useful in treating major motor seizures (grand mal) had high protective indices (defined as the neurotoxic dose for 50% of the animals divided by the anticonvulsant dose for 50% of the animals), whereas drugs more effective in absence and myoclonic seizures had the lowest protective indices in the GEPRs. Dailey and Jobe (1985) have observed three response patterns in the GEPR-3s and GEPR-9s to the established anticonvulsants. First, carbamazepine and phenytoin were more effective in blocking audiogenic seizures in the GEPR-9s than in the GEPR-3s. Second, ethosuximide and phenobarbital were equally effective anticonvulsants in GEPR-3s and GEPR-9s. Finally, valproic acid was more effective in blocking seizures in GEPR-3s than in GEPR-9s. These investigators suggested that the three anticonvulsant response patterns indicated different mechanisms of anticonvulsant drug action. Furthermore, the three response patterns observed in the GEPRs show that this model can be used to separate anticonvulsant agents into three clinically useful classes. The first class contains agents useful in treating major motor seizures, whereas the second group contains agents useful in treating absence seizures. The third class of agents have broad-spectrum anticonvulsant activity.

In addition to showing that the GEPRs exhibit a differential response to clinically useful anticonvulsant drugs, Dailey and Jobe (1985) have shown that these animals are sensitive to the anticonvulsant action of antidepressants. The anticonvulsant and proconvulsant profiles of three antidepressant drugs—imipramine, amitriptyline, and desipramine—were evaluated in the GEPR-3s and GEPR-9s. The anticonvulsant response pattern for the tricyclic antidepressants in the GEPRs was similar to that observed for carbamazepine and phenytoin, which means that tricyclic antidepressants are more effective in blocking sound-induced seizures in the GEPR-9s than in the GEPR-3s. These data suggest that the tricyclic antidepressants carbamazepine and phenytoin may exert their anticonvulsant action by increasing the length of time norepinephrine remains in the synaptic cleft (*see* section 3.1.1.1). Suffice it to say that this observation suggests a common mechanism of anticonvulsant action for carbamazepine, phenytoin, and the tricyclic antidepressants.

The proconvulsant action of the tricyclics was characterized in GEPR-3s, GEPR-9s, and nonepileptic controls. Each type of rat responded to toxic doses of the tricyclic antidepressants with clonic convulsions that differed behaviorally from those seen during sound-induced seizures (Dailey and Jobe, 1985). The proconvulsant dose in 50% of the animals was similar in the three types of animals tested. These data show that the GEPRs is certainly no more sensitive to the proconvulsant action of these agents than are nonepileptic controls.

The last topic of research to be discussed in this section is the general area of brain lesion studies. These studies fall into two types. One used the technique to elucidate which brain regions of the GEPRs were involved in mediating sound-induced seizures (Duplisse, 1976), whereas the second sought an understanding of the brain regions responsible for the various components of the behavioral convulsion displayed by the GEPRs during the sound-induced seizure episode (Browning et al., 1985). Both of these approaches provided valuable information on the brain structures of importance in modulating audiogenic seizures in the GEPRs. Duplisse (1976) observed that bilateral electrolytic lesions of either the ventral cochlear nuclei or the inferior colliculi prevented sound-induced seizures in GEPR-2s. In contrast, lesions of the medial geniculate or superior olivary complex had slight to no effect on the audiogenic seizure activity in these animals. Duplisse suggested that the inferior colliculi and ventral cochlear nuclei were important structures in the genesis of sound-induced seizures in the GEPRs.

Recently, Browning et al. (1985) showed that bilateral mechanical lesions of the nucleus reticularis pontis oralis (a midbrain nucleus) and pontine reticular formation produced a marked reduction in sound-induced seizure severity in GEPR-9s. These lesions abolished all tonic components of the convulsion (i.e., blocked hindlimb extension, forelimb extension, and, in some cases, body clonus) in the GEPR-9s. In addition, identical lesions in the GEPR-3s abolished the clonic convulsions usually observed in these animals (Browning et al., 1985). The one structure consistently destroyed in these experiments was the nucleus reticularis pontis oralis. Therefore, this structure seems to be important in regulating the type of convulsive activity displayed by the GEPRs in response to sound-induced seizures. Lesions of this structure do not affect other parameters of the audiogenic seizure behavior, such as the latency to onset running. Another interesting observation by these workers was that unilateral lesions of the nucleus reticularis pontis oralis did not inhibit the severity of the convulsive process in the GEPRs. These observations open new areas for neurochemical studies to discern which transmitter abnormalities may be associated with this midbrain/pontine nuclei.

3. Neurotransmitter Systems in the GEPR

In the following sections we will present a discussion of the evidence available on the role of various central transmitter systems in regulating the seizure process in the GEPRs. This discussion will focus on the monoaminergic and amino acid neurotransmitter systems since these areas have received the greatest attention over the years. Our intent is to present the data obtained in this model in as much detail as possible so that other investigators may critically analyze the function of the respective transmitter systems in regulating the epileptic state in the GEPRs. Finally, we will suggest directions for future research on transmitter systems in this model of epilepsy.

3.1. Monoaminergic Transmitter Systems

There is an extensive body of evidence showing that central nervous system monoaminergic transmitters regulate seizure activity in genetically epileptic rodents (Laird et al., 1984; Jobe and Laird, 1981). In the GEPRs the evidence supporting a regulatory role for norepinephrine is quite substantial. In addition, other studies have provided evidence of a similar role for serotonin. On the other hand, no such body of evidence has been obtained for dopamine or acetylcholine. To some degree this latter statement reveals the volume of work that remains to be done to elucidate the role of dopamine and acetylcholine in regulating the seizure process in the GEPRs.

3.1.1. Catecholamines

The noradrenergic and dopaminergic systems in the central nervous system have been investigated using pharmacological and pathophysiological approaches. In the pharmacological approach, treatments are designed to either increase or decrease the availability of norepinephrine and/or dopamine in the synaptic cleft, while monitoring the changes in the epileptic state of the animal. In contrast, the pathophysiological approach seeks to understand what neurochemical abnormalities, if any, are associated with the seizure-prone state in the GEPRs. Both of the experimental approaches have made significant contributions to our knowledge about the roles the noradrenergic and dopaminergic transmitter systems play in the regulation of the seizure state in the GEPRs.

3.1.1.1. PHARMACOLOGICAL STUDIES. The pharmacological studies have used treatments that produced decreases in the tissue concentrations (synaptic levels) of the catecholamines and/or treatments that caused an increase in tissue (synaptic) levels of the catecholamine(s). The results of these studies are summarized in Tables 2 and 3. In general, pharmacological manipulations that decrease central nervous system norepinephrine synaptic activity increased seizure severity. For example, seizure severity was increased when the neuronal content of norepinephrine was reduced by interference with vesicular storage with reserpine or Ro 4-1284 (Picchioni et al., 1963; Jobe, 1970; Jobe et al., 1973a), by destruction of noradrenergic nerve terminal with a neurotoxin like 6-hydroxydopamine (Bourn et al., 1972; 1977; 1978), by electrolytic lesions of the locus ceruleus (Bourn, 1974), by displacement with a false transmitter (Jobe, 1970; Jobe et al., 1973a), or by inhibition

TABLE 2
Relationship of Decrements in Brain NE and DA to the
Severity of Sound-Induced Seizures in the GEPR[a]

Treatment	Catecholamine content		Seizure severity
	NE	DA	
Reserpine (ICV)	– – –	– –	+ + +
Reserpine (IP)	– – –	– – –	+ + +[b]
Ro 4-1284	– – –	– – –	+ + +
Ro 4-1284 + aMPT	– – –	– –	+ + + (Prolonged)
Ro 4-1284 + DEDTC	– – –	+/–	+ + + (Prolonged)
Ro 4-1284 + Disulfiram	– – –	+/–	+ + + (Prolonged)
6-OH-DA	– – –	– – –	+ + +[c]
6-OH-DA + DMI	– – –	+/–	+ + +[c]
aMPT	– – –	– – –	+/–
aMMT	– – –	– –	+ +
aMPT + AMMT	– – –	– –	+ +
Cold stress	+/–	+/–	+/–
aMPT + cold stress	– – –	– – –	+ +

[a]From Jobe et al., 1973a. Abbreviations: IP, intraperitoneal; ICV, intracerebroventricular; aMPT, alpha-methyl-*p*-tyrosine; aMMT, alpha-methyl-*m*-tyrosine; 6OH-DA, 6-hydroxydopamine; DMI, desmethylimipramine; DEDTC, diethyldithiocarbamate; –, slight decrease; – –, moderate decrease; – – –, marked decrease; +/–, no change; +, slight increase, + +, moderate increase; + + +, marked increase.

[b]From Picchioni et al., 1963.

[c]From Bourn et al., 1972; 1978.

of norepinephrine synthesis combined with a stress-induced depletion of neurotransmitter from storage vesicles (Jobe et al., 1973a). All of these treatments seem to have one factor in common for the enhancement in seizure severity, namely a reduction to a critical level of central norepinephrine.

The question of whether decrements in neuronal content of dopamine augments the increase in seizure severity cannot be addressed fully by the data given in Table 2. Nevertheless, it is apparent that seizure severity is not influenced to any great degree by moderate to large deficits in neuronal dopamine content. For example, when a large deficit in dopamine was produced by 6-hydroxydopamine without any change in neuronal norepine-

TABLE 3
Action of L-DOPA and Dopamine on Seizure Severity in the GEPR[a]

| Treatment | Catecholamine levels | | Seizure severity |
	NE	DA	
Iproniazid + L-DOPA	+ + +	+ + +	– – –
Iproniazid + DEDTC + L-DOPA	+/–	+ + +	+/–
Iproniazid + Dopamine	+ + +	+ + +	– – –
Iproniazid + Dopamine + DEDTC	+/–	+ + +	+/–

[a]From Ko et al., 1982. Abbreviation: DEDTC, Diethyldithiocarbamate.

phrine, an increase in seizure severity did not occur (Bourn et al., 1978). Additional evidence distinguishing between noradrenergic and dopaminergic influences on seizure severity in the GEPR, has been obtained with time-course studies using inhibitors of catecholamine synthesis and a catecholamine-depleting agent, Ro 4-1284 (a reserpine-like agent). Table 2 shows that persistent inhibition of norepinephrine synthesis coupled with depletion of neuronal norepinephrine prolongs the enhancement of seizure severity produced by depletion of vesicular norepinephrine content by Ro 4-1284 (Jobe, 1970; Jobe et al., 1973a). Further, this increase in seizure severity was protracted by inhibition of norepinephrine synthesis at either the tyrosine hydroxylase (EC 1.14.16.2) or dopamine-beta-hydroxylase (EC 1.14.17.1.) step. Tyrosine hydroxylase is the rate-limiting enzyme in the synthesis of both norepinephrine and dopamine, whereas the enzyme dopamine-beta-hydroxylase catalyzes the conversion of dopamine to norepinephrine. In contrast, Ro 4-1284-treatment coupled with inhibition of dopamine-beta-hydroxylase permitted neuronal dopamine to return to control levels while the increase in seizure severity persisted. Thus, the critical factor in prolonging the seizure severity associated with the above treatment is the decrement in neuronal norepinephrine levels and not the decrements in neuronal dopamine content.

Another approach that has been used to study the role of the catecholaminergic systems in the GEPR is to pharmacologically increase the synaptic content and/or activity of norepinephrine and/or dopamine while observing what changes occur in the epileptic state. Examples of such agents would include drugs that block the reuptake of norepinephrine or dopamine into the nerve terminal and precursors of norepinephrine, dopamine, and receptor agonists. Drugs from each of these categories have been tested against seizures in the GEPRs. In each case, seizure intensity has been reduced (Jobe, 1970; Jobe et al., 1973a; 1984b). Furthermore, this anticonvulsant effect seems to be mediated through the noradrenergic system rather than the dopaminergic system. This conclusion is based on the following observations. First, seizure severity in the GEPRs is strongly inhibited by drugs that are more selective in blocking reuptake of norepinephrine (Jobe, 1970; Jobe et al., 1973a; 1984b). Second, L-DOPA or dopamine is capable of reducing seizure activity in the GEPRs only when converted to norepinephrine (Table 3) (Ko et al., 1982). Finally, alpha-1, beta-1, and beta-2 adrenergic receptor agonists were effective in blocking epileptic activity in the GEPRs (Ko et al., 1983). In contrast, clonidine, an alpha-2 receptor agonist, had a biphasic action in the epileptic rats (Ko et al., 1983), increasing seizure severity at low doses, and seizure blocking activity at high doses.

The studies detailed above have investigated the role of the noradrenergic and dopaminergic transmitter systems in regulating seizure intensity in the GEPRs. To date only one pharmacological study has provided an insight into the function of the catecholaminergic transmitter systems in regulating seizure susceptibility in this model. Jobe et al. (1981) addressed the question of whether a defect in noradrenergic transmission resulted in a seizure-prone state by experimentally reducing brain norepinephrine and testing animals for the development of seizure susceptibility. In this study, Ro 4-1284, an agent that reduces levels of stored tissue monoamines, was given to two groups of rats: (1) normal offspring of nonepileptic control parents, and (2) nonseizure-susceptible offspring of GEPR parents. After treatment with Ro 4-1284, each group was tested for susceptibility to sound-induced seizures (under control conditions, both groups were negative to this test). The drug treatment caused a marked increase in the incidence (65%) of audiogenic seizures in the progeny of the GEPR parents, whereas only a small number (17%) of the progeny of nonepileptic control parents became susceptible to seizures. Since Ro 4-1284 treatment caused a significant reduction in brain norepinephrine, dopamine, and serotonin content, it seems that deficits in central monoamines in rats, coupled with some other genetic seizure-prone factor(s), leads to the seizure-susceptible state. Other studies must be done to answer the question of which of the monoamines (norepinephrine, dopamine, or serotonin) are more important determinants in modulating the seizure-susceptible state in the GEPRs. In addition, investigations designed to elucidate the other genetic "factor(s)" remain to be conducted.

3.1.1.2. PATHOPHYSIOLOGICAL STUDIES. The pharmacological studies described above indicate that experimentally induced decrements or increments in noradrenergic activity will produce reciprocal changes in seizure severity and, under some conditions, in seizure susceptibility in the GEPRs. These studies do not, however, address the issue of the innate state of the catecholaminergic transmitter systems in the GEPRs. One of the early studies seeking an answer to this question found that the GEPR-2s had uniformly lower endogenous levels of norepinephrine throughout the CNS (Table 4) when compared to nonepileptic rats (Laird, 1974). Subsequently, a variety of other studies have confirmed that the central noradrenergic system is abnormal in the GEPRs. For example, norepinephrine decrements have been found in the telencephalon, hypothalamus-thalamus, midbrain, pons-medulla, and spinal cord of all lines of GEPRs (Jobe et al., 1983a,b). Interestingly, the GEPR-3 line has an increment of norepinephrine in the cerebellum.

TABLE 4
Comparison of Brain Norepinephrine Content
in GEPR-2s and Control Subjects[a]

Brain area	GEPR-2, ng/g	Control, ng/g
CH	270 ± 10[b]	430 ± 30
MB	300 ± 10[b]	560 ± 20
HT	1740 ± 120[b]	2930 ± 160
PM	420 ± 40[b]	710 ± 40
CB	180 ± 10	210 ± 20

[a]Abbreviations: CH, cerebral hemispheres; MB, midbrain; HT, hypothalamus; PM, pons/medulla; CB, cerebellum. Each value is the mean $\pm$ SEM of 10 animals.
[b]$p < 0.05$ when compared to control.

As pointed out above, pharmacological data indicate that noradrenergic abnormalities may be responsible for determining seizure susceptibility in the GEPRs. Because of this possibility, the brains of GEPRs were examined to determine whether innate abnormalities in this catecholaminergic system could account for both types of seizure regulation. Data from our current projects suggest that the abnormal state of seizure susceptibility of the GEPRs could be caused by noradrenergic deficits in some, but not all, parts of the brain (Dailey et al., 1985a; Jobe et al., 1982a,b; 1985; Laird et al., 1983; and unpublished data). Additional data from these same investigations indicate that noradrenergic decrements in even fewer brain areas are eligible as candidates for regulation of seizure intensity in the GEPRs.

In our evolving studies, four types of noradrenergic states have been identified in brains of seizure naive GEPRs. In the first type, neurochemical indices of noradrenergic activity are of equal magnitudes in nonepileptic subjects, GEPR-3s, and GEPR-9s. Thus, noradrenergic deficits do not appear to exist in the GEPRs in these particular areas. In the second type, noradrenergic deficits of equal magnitude exist in both kinds of GEPRs. In the third type, the deficits are graded; that is, the highest degree of noradrenergic activity is present in nonepileptic subjects, an intermediate level of activity is characteristic of GEPR-3s, and the least amount of activity exists in GEPR-9s. In the fourth type, a noradrenergic increment is evident in the GEPR-3s, whereas a decrement is present in GEPR-9s.

The first type of noradrenergic condition is characteristic of the striatum (Jobe et al., 1984b). In this structure, norepinephrine concentration is not abnormal in either the GEPR-3 or GEPR-9. This lack of defect is consistent with the hypothesis that noradrenergic terminals in the striatum are not responsible for seizure predisposition in the GEPR. Neither enhanced seizure susceptibility nor intensity appear to be dependent on noradrenergic status.

The second type of noradrenergic state is found in the hippocampus (Jobe et al., 1984b), hypothalamus (Jobe et al., 1984c), and inferior colliculus (Jobe et al., 1985). Norepinephrine concentrations in these structures of the GEPR are abnormally low with the deficits being of equal magnitude in both the GEPR-3 and GEPR-9. Behaviorally, both the GEPR-3 and GEPR-9 are susceptible to seizures triggered by stimuli that fail to initiate seizures in nonepileptic subjects. The common behavioral trait and the common magnitude of concentration deficits in the two types of GEPRs suggest that noradrenergic deficiencies in the hippocampus, hypothalamus, and inferior colliculus may be determinants of susceptibility.

In contrast, these three areas of noradrenergic defect do not appear to regulate seizure severity in the GEPRs. If noradrenergic deficits in the hippocampus, hypothalamus, and inferior colliculus were directly responsible for allowing severe seizures in the GEPR-9s and for preventing severe seizures in the GEPR-3s, the deficits in the severe seizure subjects would be greater than in the moderate seizure animals.

The third type of noradrenergic condition characterizes the cortex and associated structures (Jobe et al., 1984b), thalamus (Jobe et al., 1985), superior colliculus and associated structures, and the pons/medulla (unpublished observations). In these areas, norepinephrine concentrations are progressively less from nonepileptic subjects, to GEPR-3s, to GEPR-9s. These graded deficits suggest that noradrenergic terminals in these areas may be important in the regulation of seizure severity, as well as seizure susceptibility. The deficits in these areas could determine susceptibility because both GEPR-3s and GEPR-9s share the neurochemical and seizure-susceptibility defects. These noradrenergic deficits could determine seizure severity because they are greater in GEPR-9s than in GEPR-3s. Therefore, in the cortex, thalamus, superior colliculus, and pons/medulla the moderate deficits of the GEPR-3s may allow for the occurrence of seizure, but prevent the appearance of the severe toxic extensor convulsions. In contrast, severe noradrenergic deficits in these brain areas in the GEPR-9s may not only allow for the occurence of environmentally induced seizures, but facilitate the development of maximal degrees of seizure severity.

The fourth type of noradrenergic state is present in the cerebellum. Norepinephrine concentrations in this region of the brain are abnormally high in the GEPR-3s, but not in the GEPR-9s (Jobe et al., 1985). Indeed, cerebellar norepinephrine concentrations in the GEPR-9s may be abnormally low. This pattern of abnormality suggests a role for cerebellar noradrenergic terminals in the regulation of seizure severity but not susceptibility. Perhaps the noradrenergic increment in GEPR-3s prevents the moderate seizure animals from experiencing the more severe tonic extensor convulsions characteristic of the GEPR-9s. This particular noradrenergic increment clearly does not have the capacity to negate susceptibility to seizures that occur in response to an acoustical stimulus.

These arguments represent an interim interpretation of data relative to the seizure-regulating roles of noradrenergic terminals in specific brain areas of the GEPRs. Alternative hypotheses may also explain the abnormalities. The areas that we have set forth as possible sites of noradrenergic regulation of susceptibility and/or intensity require further analyses for validation of their roles in seizure mechanisms.

Abnormalities in norepinephrine concentrations in GEPR-3s and GEPR-9s do not appear to exist because of previous seizure episodes sustained by the animals (Dailey and Jobe, 1983). The animals used in these experiments had not been exposed to a seizure-provoking stimulus. Consequently, we believe that they were naive to seizures. This idea is subject to at least one caveat. Some animals in the GEPR-9 colony sustain seemingly unprovoked or spontaneous convulsions. Such seizures are rare. Among a population of about 400 animals, spontaneous seizures are observed by trained technicians and animal-care personnel about once or twice per mo. Spontaneous convulsions have never been observed in the GEPR-3 colony.

The possibility that a neurochemical defect may exist prior to a seizure episode in an epileptic animal and therefore be part of the underlying seizure-prone state of the animal is an important issue. Precautions must be taken to distinguish between these types of abnormalities and those that might occur as a result of seizure activity. Otherwise a neurochemical defect that results from seizure episodes might be mistaken as a candidate for one that causes either seizure predisposition or seizure initiation. Safeguards against such misinterpretations are certainly important relative to the noradrenergic systems of the brain. Dailey and colleagues (1982) have observed that tyrosine hydroxylase activity is significantly altered by audiogenic seizures in GEPRs. Deficits in activity of this enzyme preexist seizures, whereas following seizure episodes, the activity is abnormally high.

Other studies have shown that the norepinephrine turnover rate (an estimate of norepinephrine synthesis rate under a steady-state condition) is lower in all major brain areas except the cerebellum of GEPR-9s, whereas norepinephrine turnover rate is lower in telecephalon, hypothalamus/thalamus, and midbrain, but not the pons-medulla of GEPR-3s (Laird, 1983; Jobe et al., 1984a). These data show that not only do the GEPRs have lower endogenous levels of brain norepinephrine, but the biosynthesis of this neurotransmitter is also abnormally low. The significance of these observations to the epileptic state of these animals remains to be determined, i.e., what are the correlations between the magnitude of the defect in norepinephrine synthesis and seizure severity?

Dopamine concentrations in the GEPR-3s were normal, whereas slight deficits were detectable in the midbrain, pons-medulla, and corpus striatum of the GEPR-9s (Jobe et al., 1982b; Laird, 1983). However, numerous pharmacological observations argue against a major functional role for dopamine in seizure regulating in the GEPRs.

As pointed out above, midbrain tyrosine hydroxylase activity has been shown to be markedly elevated, whereas monoamine oxidase activity was normal in GEPR-3s with previous seizure experience (Dailey et al., 1982). Subsequently, these workers investigated tyrosine hydroxylase activity in the brains of GEPR-3s without a previous history of seizure. Under seizure-free conditions, midbrain tyrosine hydroxylase activity was significantly lower in the GEPR-3s in comparison to nonepileptic subjects. A greater decrement in midbrain enzyme activity was observed in seizure-naive GEPR-9s. Thus, there seemed to be a correlation between the degree of reduced tyrosine hydroxylase activity in the midbrain of the seizure-free GEPR-3s and GEPR-9s and the lower brain norepinephrine content. The importance of these differences in midbrain tyrosine hydroxylase activity is difficult to judge since the major noradrenergic cell bodies in the brain lie in the pons-medulla and not the midbrain. In order to understand what relevance these differences in enzyme activity have to the pathogenesis of the epileptic state in GEPRs, additional studies must be directed toward evaluating tyrosine hydroxylase activity in the locus ceruleus.

Workers from our laboratories have evaluated the alpha and beta adrenergic receptor systems in the brains of the GEPRs (Ko, 1981; Waters, 1983; Nicoletti et al., 1986). These experiments have revealed no differences in the number of, or affinity for, the beta-adrenergic binding sites in the brains of the GEPRs (3 or 9) and the normal subjects. With respect to the alpha-1 and alpha-2 adrenergic receptors, the picture is a little less consistent. For example, Ko (1981) and Booker et al. (1986) have shown no difference in alpha-1 receptor populations between GEPRs (3 and 9) and control using WB-4101 as the binding ligand. By comparison, Nicoletti et al. (1986), using prazosin as the binding ligand, found a slight but significant reduction in the number of alpha-1 recognition sites (31% fewer) with no change in affinity in the frontal cortex of GEPR-9s and control rats. Similar differences in alpha-1 adrenergic binding sites were not found in the hippocampus of the GEPR-9s (Table 5). Further, these workers found no abnormality in alpha-2 adrenergic binding sites (3-H clonidine was used as the ligand for these studies) in either the frontal cortex or hippocampus of GEPR-9 subjects.

In addition to determining the number of alpha-1 binding sites present in selected areas of the GEPRs brain, Nicoletti et al. (1986) evaluated the stimulation of inositol phospholipid hydrolysis caused by norepinephrine. In this procedure the accumulation of (^{3}H)-

TABLE 5
Specific (^{3}H)-Prazosin and (^{3}H)-Clonidine Binding to
Synaptic Membranes of Frontal Cortex and Hippocampus
From GEPR-9 or Control Subjects[a]

	Prazosin binding, fmol/mg protein		Clonidine binding, fmol/mg protein	
	FC	HP	FC	HP
Controls	95 ± 3 (10)	46 ± 5 (5)	107 ± 12 (5)	103 ± 8 (5)
GEPR-9	66 ± 2[b] (10)	44 ± 2 (5)	127 ± 18 (5)	108 ± 7 (5)

[a]Abbreviations: FC, frontal cortex; HP, hippocampus. Each value represents the mean ± SEM of 5–10 separate determinations (number in parenthesis), each one relative to one rat.

[b]$p < 0.01$, when compared to control values. The specific α-1- and α-2-adrenergic receptor bindings were assayed at concentrations of 4 and 10 nM of (^{3}H)-prazosin and (^{3}H)-clonidine, respectively.

inositol-1-phosphate (I-1-P) is measured in lithium-treated brain slices prelabeled with (^{3}H)-inositol according to the method of Berridge et al. (1982), as modified by Roth et al. (1984).

No differences were found in the basal formation of I-1-P in the frontal cortex, hippocampus, corpus striatum, and inferior colliculus obtained from GEPR-9s and normal subjects (Table 6). However, norepinephrine-stimulated I-1-P production was markedly reduced in the frontal cortex of the GEPR-9s. This defect in norepinephrine-stimulated I-1-P formation seems to be restricted to frontal cortex, since it was not observed in the hippocampus, corpus striatum, or inferior colliculus of the GEPR-9s (Table 6).

There is a growing body of evidence that the alpha-1 adrenergic receptor for norepinephrine is coupled to the metabolism of membrane inositol phospholipids (Michell, 1975; Brown et al., 1984). As a consequence, stimulation of the alpha-1 receptor mediates an activation of phospholipase C (EC 3.1.4.3.) which leads phosphotidylinositol-1-4-phosphate and phosphotidylinositol-1-4,5-diphosphate to be hydrolyzed to yield inositol-1-phosphate and diacylglycerol. Inositol-1-phosphate and diacylglycerol are thought to function as second messengers producing a series of events including mobilization of intracellular calcium, arachidonic acid release, and protein kinase C activation (Michell, 1975; Nishizuka 1984).

Thus, the decrement of norepinephrine-stimulated I-1-P production observed in the frontal cortex of the GEPR-9s suggests that there is a defect in the activation of the alpha-1 adrenergic receptors. Furthermore, this defect is not the result of an abnormality in the intrinsic phosphotidylinositol metabolism since no differences in either basal or carbamylcholine-stimulated I-1-P formation have been found in the frontal cortex of these animals (Nicoletti et al., 1986). The coexistence of fewer alpha-1 adrenergic binding sites and decreased formation of norepinephrine-induced I-1-P suggests that these events are causally related. However, the possibility remains that there may be a defect in coupling of the alpha-1 receptor to phospholipase C, which functions as a transducer. Indeed, such a defect could account for the difference in magnitude of decrement seen in norepinephrine-induced I-1-P formation and the modest reduction in density of alpha-1 receptor sites in the frontal cortex (compare Tables 5 and 6).

TABLE 6

(^{3}H)-Inositol-1-Phosphate Formation Produced by Norepinephrine and Carbamylcholine in Slices From Frontal Cortex and Hippocampus of Controls and GEPR-9s[a]

Receptor agonist	Frontal cortex, cpm/mg protein		Hippocampus, cpm/mg protein	
	Control	GEPR-9	Control	GEPR-9
None	396 ± 21 (10)	467 ± 21 (10)	497 ± 16 (6)	484 ± 17 (6)
NE, $10^{-6}M$	1056 ± 32 (6)	529 ± 26[b] (6)	1039 ± 122 (3)	1455 ± 187 (3)
$10^{-5}M$	3849 ± 167 (6)	2115 ± 91[b] (6)	4703 ± 223 (4)	4956 ± 150 (4)
$10^{-4}M$	4715 ± 402 (7)	2305 ± 122[b] (8)	6003 ± 323 (5)	5643 ± 215 (6)
Carbachol, $10^{-4}M$	924 ± 65 (6)	813 ± 59 (6)	1731 ± 47 (6)	1843 ± 61 (6)

[a]From Nicoletti et al., 1986. Values represent the mean ± SEM of 3–10 separate determinations (number in parentheses).

[b]$p < 0.01$ when compared to control rats.

These observations show that, at least in the frontal cortex of the GEPR-9s, the postsynaptic response to alpha-1 receptor stimulation is obtunded. Since alpha-1 agonists are anticonvulsant in the

GEPRs, activation of cortical alpha-1 receptors must exert an inhibitory action on neuronal hyperactivity. We can speculate that in the GEPR-9s, the alpha-1 inhibitory system is defective and this defect facilitates the seizure-prone state of these animals. In contrast, Ko (1981) and Booker et al. (1986) did not observe a difference in alpha-1 receptor density in the GEPR-3s. These workers used WB-4101 as the ligand for characterizing the alpha-1 receptor. It may be that WB-4101 and prazosin bind with different affinities to the alpha-1 receptors or that the GEPR-3s do not have a lower density of alpha-1 receptors like the GEPR-9s. Nevertheless, additional experiments are needed to compare prazosin binding and norepinephrine-induced I-1-P formation in selected brain regions of all types of GEPRs. Furthermore, the studies described above in the GEPR-9s were conducted in animals that had been previously convulsed (last convulsion was at least 4 wk prior to prazosin or I-1-P experiments). It may be that some of the effects observed may be the result of the previous seizure episodes. Obviously, this question must be answered. Much work remains to be done to elucidate what other postsynaptic defects may be present in the GEPRs and to characterize the functional significance of such defects to the seizure state.

3.1.2. Serotonin

The role of the central serotonergic transmitter system in regulating the seizure state in the GEPRs has been studied in a manner similar to that used to study the noradrenergic transmitter system. The major difference is that less work has been done on elucidating the function of the central serotonergic system in modulating the epileptic state of the GEPRs.

3.1.2.1. PHARMACOLOGICAL STUDIES. Various procedures have been used to alter central serotonin transmission, and the subsequent effects of these treatments on seizure activity in the GEPRs have been evaluated. In general, when storage or synthesis of serotonin was inhibited by pharmacological methods, a marked increase in the intensity of the sound-induced seizure occurred (Jobe et al., 1973b; Jobe, 1981; Laird et al., 1974). For example, treatment with *p*-chlorophenylalanine (PCPA), an inhibitor of tryptophan hydroxylase (the rate-limiting enzyme in serotonin synthesis), or Ro 4-1284 produced an increase in the severity of the audiogenic convulsions and reduced whole and regional brain levels of serotonin (Jobe et al., Laird, 1974). Ro 4-1284 produced a greater augmentation of seizure intensity than did PCPA, but this may be explained to some degree by the observation that Ro 4-1284 leads to a greater depletion of norepinephrine. Thus, the difference in capacities of Ro 4-1284 and PCPA to enhance the severity of convulsion in the GEPRs is a function of the differing abilities of these agents to deplete central stores of serotonin, norepinephrine, and/or dopamine (Jobe et al., 1973b).

PCPA time-course studies showed that this agent caused an increase in audiogenic seizure severity, as indicated by a decrease in latency for convulsion and an increase in seizure severity, both of which persisted for approximately 1 wk (Laird, 1974). Neurochemical evaluations revealed that PCPA caused a persistent preferential depletion of serotonin in cerebral hemispheres, midbrain, pons/medulla, and hypothalamus. These studies showed a good correlation between serotonin depletion in the four regions of the brain and an increase in seizure severity. However, the two regions of the brain that exhibited the most persistent correlation were the cerebral hemispheres and the midbrain. In contrast, norepinephrine was modestly but significantly reduced in the cerebral hemispheres and hypothalamus only on d 3 of the 9-d study. Thus, the seizure-enhancing effect of PCPA was correlated best with depletion of serotonin in the cerebral hemispheres and midbrain and was independent of changes in central norepinephrine content.

Other studies have provided evidence that a decrease in serotonin levels in the brain is functionally associated with the enhancement of sound-induced seizures. These experiments studied the interaction between iproniazid, a monoamine oxidase inhibitor, and Ro 4-1284, and found that iproniazid administered before Ro 4-1284 inhibited the increase in seizure intensity observed when Ro 4-1284 was given alone (Jobe et al., 1973b). The iproniazid treatment reduced the Ro 4-1284-induced depletion of central serotonin, but did not inhibit the decrements in central norepinephrine and dopamine. Further, pretreatment with PCPA completely blocked the effect of iproniazid on Ro 4-1284-induced increase in seizure intensity and decrease in central serotonin levels. Therefore, the enhancement of seizure intensity caused by Ro 4-1284 seems to depend, at least in part, on lower brain levels of serotonin. However, the decrements in brain norepinephrine levels probably contribute to the Ro 4-1284 action to some degree.

In another study, GEPR-2s were treated with *p*-chloromethamphetamine (PCMA), a chlorinated amphetamine derivative, which causes selective depletion of neuronal serotonin levels with only a slight effect on catecholamine levels (i.e., a modest and transient elevation) (Pletscher et al., 1970; Miller et al., 1970). These studies showed that PCMA had a biphasic effect on sound-induced seizures (Laird et al., 1974). During its early phase of activity (1 and 24 h), it exerted an anticonvulsant effect, as indicated by an increase in latency for convulsion and a decrease in seizure severity. During its later phase of activity (48 and 72 h), it caused a marked enhancement of audiogenic seizure, as indicated by a decrease in latency for convulsion and a marked increase in seizure severity. The neurochemical data from the PCMA studies showed that serotonin was markedly depleted in the cerebral hemispheres and midbrain and moderately lower in the pons/medulla. In contrast, norepinephrine concentration was not reduced in any region of the brain at either time period. In fact, norepinephrine levels tended to be elevated in various regions of the brain. Although there is a good correlation between depletion of serotonin in the cerebral hemispheres and midbrain with increased seizure severity during the 48- and 72-h time periods, there seems to be a lack of correlation during the 24-h time period. This apparent lack of correlation may be explained by the anticonvulsant effects of methamphetamine (Lehman, 1970; Pfeifer and Galambos, 1967; Rudzik and Johnson, 1970). Thus, the anticonvulsant effect seen during the first 24 h after treatment with PCMA can be explained by the direct action of the drug. The proconvulsant effect caused by serotonin depletion becomes dominant only after the direct effect of PCMA is gone. By comparison, at the 48-h time period, there was a good correlation between serotonin depletion and enhancement of audiogenic seizure activity. Furthermore, because the cerebral hemispheres and midbrain showed the greatest decline in serotonin levels, it was suggested that these structures were important sites for the inhibitory function of serotonin in the GEPR-2s (Laird et al., 1974). Interestingly, the seizure enhancement observed 48 h after PCMA occurred despite modest but statistically significant increments of norepinephrine levels in the midbrain and hypothalamus.

3.1.2.2. PATHOPHYSIOLOGICAL STUDIES. GEPR-2s, with a history of convulsions, have abnormally low endogenous levels of serotonin in the cerebral hemispheres and midbrain when compared to nonepileptic subjects (Laird, 1974). In the seizure-experienced GEPR-3, serotonin concentrations are abnormally low in the telencephalon, hypothalamus/thalamus, midbrain, and pons/medulla (Jobe et al., 1982a).

In seizure-naive subjects, the pattern of serotonergic abnormalities that emerges from comparisons between control subjects,

GEPR-3s and GEPR-9s are similar to those of the noradrenergic systems. In one area, the cerebellum, serotonin concentrations may be normal in both GEPR-3s and GEPR-9s. In most assays, serotonin levels in the cerebella of GEPRs have not been significantly different from control concentrations (Jobe et al., 1983; and unpublished data). Normal serotonin levels in the cerebellar terminal fields suggest that this area of the brain is a site where serotonergic transmission is not responsible for the seizure-prone state of the GEPRs. Validation of this possibility through the use of other indices of serotonergic activity is essential.

In the midbrain, data to date suggest that the naive GEPR-3s and GEPR-9s are actually similar to seizure-experienced GEPR-3s in that a serotonergic deficit of the midbrain characterizes all three types of animals (Jobe et al., 1983; Dailey et al., 1985b).

In the pons/medulla, serotonin concentrations are abnormally low in the GEPR-3s, but probably not in the GEPR-9s. Such a serotonin concentration deficit in the pons/medulla of the GEPR-3s, but not the GEPR-9s, suggests that this neurochemical aberration does not cause the appearance of seizure susceptibility nor regulate seizure severity.

Serotonin deficits in the telencephalon and the hypothalamus/thalamus of the seizure-naive GEPR-3s appear equal to those of the corresponding brain areas in the seizure-naive GEPR-9s (Jobe et al., 1983; Dailey et al., 1985b; unpublished data). Behaviorally, both GEPR-3s and GEPR-9s are susceptible to seizures precipitated by stimuli that will not trigger seizures in nonsusceptible rats. Consequently, serotonergic deficits of equal magnitude in the brain structures of the GEPR-3s and GEPR-9s would be possible determinants of susceptibility. In contrast, deficits of equal magnitude would not be responsible for causing moderate seizures in GEPR-3s and severe seizure in GEPR-9s.

3.1.3. Acetylcholine

Central cholinergic function in regulating seizure activity in the GEPR has received little attention. This situation arises more from the lack of pharmacological tools and level of experimental difficulty in studying the central cholinergic system than from the lack of interest or importance of this transmitter system in this model.

Jobe (1981) showed that young epileptic rats killed in liquid nitrogen had lower whole brain levels of acetylcholine (ACh) durng audiogenic convulsions. More recently, Laird et al. (1986) determined innate levels of choline and acetylcholine in several brain regions of GEPR-9s and compared them with control rats. No differences in brain choline content were observed, but there was an increase in acetylcholine levels in the thalamus and striatum of GEPR-9s. In addition, a decrease in acetylcholinesterase activity was found in the midbrain of the GEPR-9s. Interestingly, the cholinergic cell bodies in the midbrain project to the thalamus and striatum. However, these data must be interpreted with care since the GEPR-9s used in these experiments had a previous history of multiple seizure episodes that may have had an enduring effect on the central cholinergic system. This latter statement has received some support from a recent experiment in which acetylcholine levels were determined in thalamic tissue from control rats and GEPR-9s with no previous history of seizure episodes. No differences in thalamic acetylcholine content were found in control and nonconvulsed GEPR-9s (Laird et al., 1986). Additional studies are needed to evaluate the relationship between previous seizure episodes and the functional status of the central cholinergic system.

Other studies have evaluated the postsynaptic component of the central cholinergic system. Speth et al. (1979) determined specific quinuclidinyl benzilate (QNB) binding in four regions (inferior colliculi, hippocampus, striatum, and cortex) of the brains from GEPR-9s and controls and found no differences. Recently, Nicoletti et al. (1986) determined the functional integrity of the central muscarinic receptors by comparing carbamylcholine-stimulated I-1-P formation in the frontal cortex, hippocampus, and striatum obtained from epileptic and normal rats. Again, carbamylcholine-stimulated I-1-P formation was not different between the two types of rats. It is apparent that much work remains to be done on the role of the central cholinergic transmitter system in modulating seizure activity in the GEPRs.

3.2. Amino Acid Transmitter Systems

The role of central amino acid transmitter systems in epilepsy remains to be defined. Nevertheless, there is a growing body of evidence that the excitatory amino acid transmitter, glutamic acid, and inhibitory amino acid transmitters, like gamma-aminobutyric acid (GABA) and glycine, have an important function in regulating central neuronal activity (for reviews, *see* Wood, 1975; Tower, 1976; Emson, 1978). Defects within either the excitatory or inhibitory amino acid transmitter systems could lead to a hyperexcitable state that could increase the susceptibility to seizure development. The validity of the preceding statement is shown by the various reports from the literature in which treatments that interfere with synaptic transmission in the inhibitory amino acid transmitter systems cause seizures (Wood, 1975), whereas similar blockade of excitatory amino acid transmission can produce an anticonvulsant action (Croucher et al., 1982).

3.2.1. Excitatory Amino Acid Transmitters

To date, few studies have explored the role of the excitatory amino acids in the epileptic state of the GEPR. Of the few investigations that have taken place, two have provided suggestive evidence that the glutamic acid transmitter system in the GEPR is abnormal. For example, Mills et al. (1985) have shown an increase in the number of hippocampal-specific (^{3}H)-glutamic acid binding sites in the GEPRs. These workers studied specific (^{3}H)-glutamic acid binding in hippocampal membrane preparations from GEPR-3s, GEPR-9s, and controls, and showed a graded increase in the number of specific glutamic acid binding sites from 12 pmol/mg protein in controls to 15 pmol/mg protein in the GEPR-3s, and 20 pmol/mg protein in the GEPR-9s. Scatchard analysis of the binding data revealed that the graded increase in the number of specific glutamic acid binding sites in the GEPR-3s and GEPR-9s was not associated with a change in the apparent binding affinity constant (Mills et al., 1985).

A complimentary observation has been made by Lehmann and Huxtable (1985). These workers have shown an increased potassium-stimulated release of labeled glutamic acid from previously loaded stores in the hippocampus of GEPR-9s. In this study, the investigators preloaded hippocampal tissues with (^{3}H)-glutamic acid using a dialysis fiber implanted *in situ* and then regulated and monitored the composition of the extracellular fluid environment around the fiber. By this technique, they found a marked increase in the amount of hippocampal (^{3}H)-glutamic acid released in response to potassium stimulation in the GEPR-9s.

Other studies have compared the innate concentrations of free glutamic acid and aspartate in the cortex and inferior colliculi of GEPR-9s and controls (Huxtable and Laird, 1978) and found no differences. Further experiments sought to determine whether there were unique correlations between free amino acid concentrations in the cortex and inferior colliculi of GEPR-9s and controls (Huxtable et al., 1982). Again, no abnormalities were found in free amino acids in the epileptic rat brain. Finally, Bonhaus and Huxtable (1984) studied the transport of glutamic acid into brain synaptosomes prepared from whole rat brain obtained from GEPR-9s and control rats. No differences between epileptic and normal brain were found in the rate or amount of glutamic acid transported into the synaptosomes.

Based on these few studies, we see emerging suggestive evidence that the GEPRs have an abnormality in the glutamic acid transmitter system in the hippocampus. Much additional work must be done to extend these observations to other brain regions. In addition, it is necessary to determine whether these glutamic acid abnormalities are caused by previous seizure episodes or represent an innate defect.

3.2.2. Inhibitory Amino Acid Transmitters

In contrast to the excitatory amino acid transmitters, more work has been done to characterize the role of the inhibitory amino acid transmitter systems in modulating the epileptic state associated with the GEPRs. Nevertheless, there are significant gaps in our information concerning the function of GABA, glycine, and taurine in the control of seizure activity in the GEPR. At this time there is a growing interest among epilepsy researchers in using the GEPR as a model system to answer questions about the function of these various inhibitory amino acid transmitters in the regulation of epileptogenesis.

3.2.2.1. PHARMACOLOGICAL STUDIES. One of the earliest studies seeking information on the function of GABA in the GEPRs was that of Duplisse (1976). In this study, GABA was injected in graded doses bilaterally into the inferior colliculi of GEPR-2s, and the animals were tested for susceptibility to sound- and bicuculline-induced seizures. The GABA injections produced a dose-related decrease in susceptibility to seizure episodes from both sound and bicuculline. In these experiments, bicuculline-induced seizures were produced by injecting the convulsant chemical bilaterally into the inferior colliculi 5 min after GABA administration. GABA injections alone produced no overt behavioral changes in the GEPR or control rats. Interestingly, bicuculline injections directly into the inferior colliculi produced seizures behaviorally identical to those caused by exposure of the GEPRs to the sound stimulus (Duplisse, 1976). In addition to his experiments with GABA, Duplisse (1976) performed dose–response experiments with bicuculline injected bilaterally into the inferior colliculi to compare seizure susceptibility and intensity between GEPR-2s and controls. Not only were the

GEPR-2s more susceptible to bicuculline-induced seizures (Table 1), but their seizures were more severe (i.e., the GEPR-2s displayed a higher percentage of tonic extensor convulsions) than the controls.

These data provide suggestive evidence that GABA is functionally important in regulating seizure genesis and propagation, at least in the inferior colliculi. First, the injection of GABA into the inferior colliculi of GEPR-2s shows that the pharmacological application of this agent will attenuate sound-induced seizures, apparently by acting on endogenous GABA receptors. Second, blockade of GABA-A receptor sites in the inferior colliculi by bicuculline elicits seizure behavior in GEPRs and controls identical to that observed in the GEPR during exposure to the sound stimulus. Finally, the greater sensitivity of the GEPR-2s to seizures induced by bicuculline injected into inferior colliculi suggests an inherent pre- or postsynaptic GABA abnormality at this site. Recently, other workers (Roberts et al., 1984) have shown an abnormally large number of GABA cell bodies in the inferior colliculi (*see* section 3.2.2.2), which provides further support for the idea that the GABAergic system in the inferior colliculi is functionally aberrant in the GEPRs. Further support for a functional defect in the GABA system of the inferior colliculi has been provided by Gehlbach and Faingold (1984). Larger iontophoretic doses of GABA and flurazepam were needed to depress neuronal firing in the inferior colliculi of GEPRs than in controls.

Laird and Huxtable (1976) reported that intracerebroventricular injections of GABA, taurine, and glycine will, in a dose-dependent fashion, inhibit sound-induced seizures in GEPR-9s. Of these amino acids only taurine produced a significant increase in latency for convulsion over the entire dose range (4.8, 9.6, and 19.2 μmol/animal). By comparison, GABA produced an elevation in this parameter only at the highest dose, 19.2 μmol.

In addition, these workers showed that taurine inhibited, in a dose-dependent fashion, the severity of sound-induced seizures by 31, 66, and 80%, respectively. In contrast, the 9.6 and 19.2 μmol doses of GABA reduced seizure severity by 37 and 46%, respectively. Behaviorally, all doses of taurine caused dose-related reductions in locomotor activity and increases in ataxia. On the other hand, GABA caused these effects only at the highest dose and to a lesser degree. Therefore, the anticonvulsant action of these amino acids by the intracerebroventricular route may result from nonspecific neuronal depression. However, since the lowest anticonvulsant doses of GABA and taurine produced little, if any, neurological deficits, it seems that the anticonvulsant effect and the nonspecific neuronal depressant effect may be separate actions (Laird and Huxtable, 1978). Support for this idea was provided by the effect on intracerebral electroshock thresholds of bilateral injections of taurine into the inferior colliculi of the GEPR-9s (Laird and Huxtable, 1978). These experiments showed that the injection of 200 nmol of taurine into the inferior colliculi raised the electroshock threshold in this structure by over 60% in the GEPR-9s, but had no effect on this parameter in controls. Furthermore, the taurine action was slow in developing (onset of effect was 12 min), but long-lasting (thresholds were significantly elevated for 18 d after the injection). In spite of the effect of taurine on intra-cerebral electroshock threshold in the GEPR-9s, it did not modify the severity of the audiogenic convulsion (i.e., tonic extensor convulsions in the GEPR-9s were not affected by taurine). Similarly, taurine injections did not affect the severity of the electroshock convulsion in nonepileptic controls. These data support a selective effect for taurine at the site of injection to retard initiation of seizure activity in the GEPRs. Since this action of taurine was observed only in the GEPRs, it is tempting to suggest that this agent temporarily corrects some neurochemical defect at this site (Laird and Huxtable, 1978).

There is a growing body of evidence supporting a neuromodulatory role for taurine in the brain (Bonhaus et al., 1983). Nevertheless, the exact mechanism by which taurine produces its anticonvulsant action is poorly understood, but may be related to calcium binding to membranes (Lazarewicz et al., 1985).

The pharmacological data provides suggestive evidence that there may be inborn defects in the GABAergic transmitter system and possibly in the function of the neuromodulator taurine in the GEPR. By comparison, the evidence of a role for glycine is sparse. For example, the anticonvulsant action of glycine when injected intracerebroventricularly in this model was not impressive (Laird and Huxtable, 1976). In addition, the GEPRs were not more sensitive to strychnine-induced seizures (Table 1) than nonepileptic controls (Laird and Huxtable, 1978). Since strychnine is considered to elicit its convulsant action by blockade of glycine receptors in the spinal cord and possibly the lower brain stem (DeFeudis, 1978), it seems unlikely from the evidence available that the GEPRs have a marked defect in the glycinergic transmitter system.

3.2.2.2. PATHOPHYSIOLOGICAL STUDIES. Huxtable and Laird (1978) found no differences in free amino acid concentrations nor in ratios between glutamate, glutamine, taurine, GABA, or glycine in the inferior colliculi and cortex of GEPR-9s and nonepileptic controls. Although the absolute tissue concentrations and ratios of the amino acids are normal, more recent studies have found abnormalities in other components of these transmitter systems in the GEPR that may have marked functional consequences. For example, Bonhaus et al. (1982) and Bonhaus and Huxtable (1984) showed a decrease in high-affinity taurine transport in the brain and platelets of GEPR-9s. Bonhaus and Huxtable (1984) suggest that the defect in high-affinity taurine transport in the brain contributes to the seizure-susceptible state in the GEPR. These workers did not find a similar defect in GABA or glutamate transport.

GABA binding experiments have been done in controls, GEPR-3s, and GEPR-9s, and showed graded increase in the number of low- and high-affinity binding sites for the GABA-A receptor ligand, muscimol, with no change in the apparent binding affinity (Booker et al., 1983). These experiments were conducted on previously convulsed GEPRs. Thus, other GABA binding experiments need to be done to determine whether these abnormalities characterize nonconvulsed subjects. If these abnormalities in GABA receptors exist in the seizure-naive state, they can logically be postulated as neurochemical factors responsible for seizure predisposition in the GEPR.

Another closely related set of experiments has characterized the benzodiazepine binding sites in the GEPRs (Mimaki et al., 1984; Tacke and Braestrup, 1984). The two groups report different results. Mimaki et al. (1984) observed an abnormally high number of high-affinity benzodiazepine binding sites in whole brain membrane preparations with no change in the apparent affinity constant in nonconvulsed GEPR-9s. In fact, there was no difference between the number of benzodiazepine binding sites in nonconvulsed and convulsed GEPR-9s. In contrast, Tacke and Braestrup found no difference in specific benzodiazepine binding in membrane preparations from cortex, hippocampus, brainstem, and cerebellum between GEPR and controls. In addition, they found no differences in specific binding for the convulsant benzodiazepine receptor ligand, methyl-6,7-dimethoxy-4-ethyl carboline-3-carboxylate, or the chloride channel blocker, t-butylbicyclophosphorothionate. However, Tacke and Braestrup (1984) found a decrement in muscimol stimulated (^{3}H)-diazepam binding in the GEPR. Recently, Booker et al. (1986) have replicated the observation of Mimaki et al. (1984) and found that the number of specific benzodiazepine binding sites in the cortex is elevated in the GEPRs. Nevertheless, the importance of the increments in specific benzodiazepine binding to the epileptic state in the GEPRs remains to be defined.

Roberts et al. (1984) made an intriguing observation that the inferior colliculi of the GEPR-9s had 200–300% more glutamic acid decarboxylase-positive neurons (GABA neurons) than controls. The increase in glutamic acid decarboxylase-positive neurons was associated with small-to-medium-size neurons. Under-standing the function of these GABA neurons is important since the electrophysiological data discussed above (sections 3.2.2.1 and 3.2.2.2) suggest a hyporesponsive system with normal GABA content at the level of the inferior colliculi. It is apparent that we have many questions to answer before the role of the various inhibitory amino acid transmitter systems in the GEPR can be understood.

4. Conclusions

The GEPRs are a valuable model of epilepsy. They provide the experimenter with a system in which questions about seizure susceptibility, initiation, propagation, and severity may be studied. Furthermore, the genetic homogeneity found in the various lines of GEPRs give the researcher the additional advantage of control over that parameter in the experimental paradigm. Through the use of this model, valuable information has been obtained on the role of various neurotransmitter systems in the regulation of the epileptic state.

Some general observations seem in order with respect to the work that has been completed in the GEPR model. First, experimentation with the transmitter systems discussed above is insufficient to provide a definitive understanding of their function in the epileptic state. Second, the scope and depth of scientific inquiry in the model needs to be increased. Third, it is important that the new information obtained in the GEPR be compared and evaluated against previous data obtained in the GEPR and other epilepsy models. Fourth, care must be exercised to separate neurochemical changes caused by an inborn predisposition to seizures from those resulting from previous seizure episodes before any causal relationships are postulated. Fifth, the experimenter should carefully determine the scientific question(s) being asked so the most appropriate type(s) of experimental model(s) of epilepsy can be selected. Finally, a need exists to compare the GEPR with other models of epilepsy within the same laboratory.

Extensive data on the GEPRs support a major role for the central noradrenergic transmitter system in modulating seizure intensity and, at least to some degree, seizure susceptibility. The central noradrenergic system has defects presynaptically in tissue content, turnover rate, and biosynthetic enzyme (tyrosine hydroxylase) activity and postsynaptically in number and function of alpha-1 adrenergic receptor sites. However, several gaps in our knowledge of this system remain. For example, information is needed on whether the lower tissue content of norepinephrine reflects less storage or fewer numbers of noradrenergic neurons. Postsynaptically, the beta-adrenergic receptor-mediated effects on cyclic-AMP production need to be fully characterized, as do the alpha-1 adrenergic receptor-mediated effects on phosphotidylinositol formation. Furthermore, the function of the noradrenergic "second messenger" systems (cyclic-AMP and phosphotidylinositol) needs to be characterized in all lines of the GEPRs.

The function of the dopaminergic transmitter system in regulating the seizure-dependent state in the GEPRs still awaits elucidation. Available data indicate that this system does not regulate severity of sound-induced seizure. However, the interaction of the dopaminergic transmitter system with other transmitter systems in regulating the seizure process in the GEPRs has not been systematically explored. In addition, the postsynaptic component of the central dopaminergic system has not been characterized to date. Thus, we await further evidence on the central dopaminergic transmitter system to understand its function in the epileptic state of the GEPRs.

Enough evidence has been obtained in the GEPRs to suggest a role for the central serotonergic transmitter system in seizure modulation in this model. However, the level of exploration of this transmitter system in the GEPRs is still rudimentary. The seizure-regulating function of the central serotonergic system seems to be more subtle and/or the etiologically relevant defect more restricted than for the noradrenergic system. It is likely that the neural tracts or terminal areas in the cortex and midbrain of this system functionally interact with similar projections and terminals of the noradrenergic system. This interaction augments the activity of this latter system. Thus, decreases in serotonergic function would exacerbate similar deficits in noradrenergic function and enhance seizure activity. Such a functional relationship is supported by the evidence available on this system in the GEPRs. Obviously, substantial additional evidence to support or refute such a role for serotonin is required.

At this point, it is difficult to assess the role of the central cholinergic system in the epileptic state of the GEPRs. Our understanding of this system in these animals is so limited that any statement suggesting a potential function must be viewed as conjecture. From a scientific point of view, however, characterization of central cholinergic function in epilepsy is an important endeavor.

One of the new and exciting areas of research in the GEPRs is understanding the function of the central amino acid neurotransmitter systems in modulating seizure episodes in these animals. The body of evidence is modest, but impressive, in suggesting its importance in the pathogenesis of the epileptic condition in the GEPRs. New evidence suggests that defects in the glutaminergic and GABAergic neurotransmitter systems and the taurine modulatory system play a role in the seizure state associated with the GEPRs. However, additional work remains to be done before the importance of these observations can be fully assessed. Work to define the biosynthesis, storage, and postsynaptic activity of these amino acid transmitters in the GEPRs is important for the future.

The amino acid transmitter systems and the noradrenergic transmitter system represent the more fruitful areas for future research, since treatments that modify these systems elicit major changes in the seizure process in the GEPRs. Nevertheless, work on the dopaminergic, serotonergic, and cholinergic transmitter systems should continue to provide an understanding of how these systems interact and/or function with other transmitters systems in modifying the seizure-prone state of the GEPRs. Finally, an important area for future research is the peptide neurotransmitters, since these agents may play a critical role in the control of seizure activity in the GEPRs as seen by the abnormal sensitivity of these epileptic animals to the proconvulsant effect of intracerebroventricular morphine. This fertile area of research is worthy of intense investigation in the GEPRs as well as other models of epilepsy.

In our opinion, the knowledge to be gained from investigating the function of neurotransmitters systems in epilepsy is critical to understanding the epileptic state. Moreover, such studies should provide better insights into the function of transmitter systems in regulating normal neuronal activity and information processing within the central nervous system.

References

Berridge, M. J., Downes, C. P., and Hanley, M. R.: Lithium amplifies agonist dependent phosphatidylinositol responses in brain and salivary glands. *Biochem. J.* **206**: 587–595, 1982.

Bonhaus, D. W. and Huxtable, R. J.: Seizure-susceptibility and decreased taurine transport in the genetically epileptic rat. *Neurochem. Int.* **6**(3): 365–368, 1984.

Bonhaus, D., Laird, H. E., and Huxtable, R. J.: Decreased taurine transport in epileptic rats. *Proc. West. Pharmacol. Soc.* **25**: 25–28, 1982.

Bonhaus, D. W., Laird, H. E., Mimaki, T., Yamamura, H. I., and Huxtable, R. J.: Possible Bases for the Anti-Convulsant Action of Taurine, In: *Sulfur Amino Acids Biochemical and Clinical Aspects* (K. Kuriyama, R. J. Huxtable, and H. I. Wata, eds.) Liss, New York, 1983.

Booker, J. G., Jobe, P. C., Dailey, J. W., and Lane, J. D.: Cortical GABA-receptors in the genetically epilepsy prone rat. *Fed. Proc.* **42**: 365 (abst.), 1983.

Booker, J. G., Dailey, J. W., Jobe, P. C., and Lane, J. D.: Cerebral cortical GABA and benzodiazepine binding sites in genetically seizure prone rats. *Life Sci.* 1986 (accepted for publication).

Bourn, W. M.: Modulation of audiogenic seizures by cortical norepinephrine in the rat (doctoral dissertation). Tuscon, University of Arizona, 1974.

Bourn, W. M., Chin, L., and Picchioni, A. L.: Enhancement of audiogenic seizure by 6-hydroxydopamine. *J. Pharm. Pharmacol.* **24**: 913–914, 1972.

Bourn, W. M., Chin, L., and Picchioni, A. L.: Effect of neonatal 6-hydroxydopamine treatment on audiogenic seizures. *Life Sci.* **21**: 701–705, 1977.

Bourn, W. M., Chin, L., and Picchioni, A. L.: The role of dopamine in sound-induced convulsions. *J. Pharm. Pharmacol.* **30**: 800–801, 1978.

Bourn, W. M. and Garrett, R. L.: Increased susceptibility of audiogenic rats to barbital withdrawl convulsions. *Pharmacol. Biochem. Behav.* **19**: 839–841, 1983.

Brown, E., Kendall, D. A., and Nahorski, S. R.: Inositol phospholipid hydrolysis in rat cerebral cortical slices: 1. Receptor characterization. *J. Neurochem.* **42**: 1379–1387, 1984.

Browning, R. A., Nelson, D. K., Mogharreban, N., Jobe, P. C., and Laird II, H. E.: Effect of midbrain and pontine tegmental lesions on audiogenic seizures in genetically epilepsy-prone rats. *Epilepsia* **26**(2): 175–183, 1985.

Consroe, P., Kudray, K., and Schmitz, R.: Acute and chronic antiepileptic drug effects in audiogenic seizure-susceptible rats. *Exp. Neurol.* **70**: 626–637, 1980.

Consroe, P. C., Picchioni, A. L., and Chin, L.: Audiogenic seizure susceptible rats. *Fed. Proc.* **38**: 2411–2416, 1979.

Croucher, M. J., Collins, J. F., and Meldrum, B. S.: Anticonvulsant action of excitatory amino acid antagonists. *Science* **21**: 899–901, 1982.

Dailey, J. W. and Jobe, P. C.: Noradrenergic abnormalities in the genetically epilepsy-prone rat: Do they cause or result from seizures? *Soc. Neurosci.* **9**(1): 400, 1983.

Dailey, J. W. and Jobe, P. C.: Anticonvulsant drugs and the genetically epilepsy-prone rat. *Fed. Proc.* **44**: 2640–2644, 1985.

Dailey, J. W., Battarbee, H. D., and Jobe, P. C.: Enzyme activities in the central nervous system of the epilepsy-prone rat. *Brain Res.* **231**: 225–230, 1982.

Dailey, J. W., Reigel, C. E., Jr., Ko, K. H., Acurio, M. T., Penny, J. E., and Jobe, P. C.: Abnormalities in brain norepinephrine levels in genetically epilepsy-prone rats. *Soc. Neurosci.* (Absts.), **11**: 1316, 1985a.

Dailey, J. W., Reigel, C. E., Jr., Woods, T. W., Penny, J. E., and Jobe, P. C.: Types of serotonergic abnormalities characteristic of genetically epilepsy-prone rats (GEPRs). *Fed. Proc.* **44**: 1107, 1985b.

DeFeudis, F. V.: Central glycine-receptors. *Gen. Pharmacol.* **9**: 139–144, 1978.

Duplisse, B. R.: Mechanism of susceptibility of rats to audiogenic seizure (doctoral dissertation). Tucson, University of Arizona, 1976.

Duplisse, B. R., Picchioni, A. L., Chin, L., and Consroe, P.: Electroshock seizure thresholds: Differences between audiogenic and non-audiogenic rats. *J. Int. Res. Commun.* **1**: 9, 1973.

Ebadi, M., Jobe, P. C., and Laird II, H. E.: Variation of pyridoxal phosphate and pyridoxal kinase in brains of genetically epilepsy-prone rats. *Epilepsia* **26**: 353–359, 1985.

Emson, P. C.: Biochemical and Metabolic Changes in Epilepsy, In: *Taurine and Neurological Disorders* (A. Barbeau and R. J. Huxtable, eds.) Raven, New York, 1978.

Gehlback, G. and Faingold, C. L.: Audiogenic seizures and effects of GABA and benzodiazepines on inferior colliculus neurons in the genetically epilepsy prone rat. *Neurosci. Abst.* **10**: 569, 1984.

Huxtable, R. J. and Laird II, H. E.: Are amino acid patterns necessarily abnormal in epileptic brains? Studies on the genetically seizure-susceptible rat. *Neurosci. Lett.* **10**: 341–345, 1978.

Huxtable, R. J., Laird, H., Bonhaus, D., and Thies, A. C.: Correlations between amino acid concentrations in brains of seizure-susceptible and seizure-resistant rats. *Neurochem. Int.* **4**(1): 73–78, 1982.

Jobe, P. C.: Relationship of brain metabolism to audiogenic seizure in the rat (doctoral dissertation). Tucson, University of Arizona, 1970.

Jobe, P. C.: Pharmacology of Audiogenic Seizures, In: *The Pharmacology of Hearing: Experimental and Clinical Bases* (R. D. Brown and E. A. Daigneault, eds.) Wiley, New York, 1981.

Jobe, P. C., Brown, R. D., and Dailey, J. W.: Effect of Ro 4-1284 on audiogenic seizure susceptibility and intensity in epilepsy-prone rats. *Life Sci.* **28**: 2031–2038, 1981.

Jobe, P. C. and Laird II, H. E.: Neurotransmitter abnormalities as determinants of seizure susceptibility and intensity in the genetic models of epilepsy. *Biochem. Pharmacol.* **30**: 3137–3144, 1981.

Jobe, P. C., Laird II, H. E., and Dailey, J. W.: Innate abnormalities in CNS serotonin concentrations in genetically epilepsy-prone rats (GEPRs). *Absts. Soc. Neurosci.* **9**: 399, 1983.

Jobe, P. C., Laird II, H. E., Ko, K. H., Ray, T., and Dailey, J. W.: Abnormalities in monoamine levels in the central nervous system of the genetically epilepsy-prone rat. *Epilepsia* **23**: 359–366, 1982a.

Jobe, P. C., Laird II, H. E., Woods, T. W., and Dailey, J. W.: Noradrenergic abnormalities in two independently developed strains of genetically epilepsy-prone rats. *Neurosci. Abst.* **8**: 1982b.

Jobe, P. C., Picchioni, A. L., and Chin, L.: Role of brain norepinephrine on audiogenic seizure in the rat. *J. Pharmacol. Exp. Ther.* **184**: 1–10, 1973a.

Jobe, P. C., Picchioni, A. L., and Chin, L.: Role of brain 5-hydroxytryptamine in audiogenic seizure in the rat. *Life Sci.* **13**: 1–13, 1973b.

Jobe, P. C., Ko, K. H., and Dailey, J. W.: Abnormalities in norepinephrine turnover rate in the central nervous system of the genetically epilepsy-prone rat. *Brain Res.* **290**: 357–360, 1984a.

Jobe, P. C., Woods, T. W., and Dailey, J. W.: Catecholamine concentrations in the hippocampus, striatum, and the remaining telencephalon of the genetically epilepsy-prone rat (GEPR). *Neuro. Abs.* **10**: 409, 1984b.

Jobe, P. C., Woods, T. W., Reigel, C. E., and Dailey, J. W.: Biogenic amine concentrations in the hypothalamus of the genetically epilepsy-prone rat (GEPR). *Pharmacologist* **26**(3): 220, 1984c.

Jobe, P. C., Reigel, C. E., Jr., Woods, T. W., Penny, J. E., and Dailey, J. W.: Do noradrenergic deficits in the genetically epilepsy-prone rat (GEPR) correlate with seizure susceptibility or severity? *Fed. Proc.* **44**: 1107, 1985.

Ko, K.: Monoamine transmission in the regulation of seizures in genetically epilepsy-prone rat (doctoral dissertation). Shreveport, Louisiana State University, 1981.

Ko, K., Dailey, J. W., and Jobe, P. C.: Effects of increments in norepinephrine concentrations on seizure intensity in the genetically epilepsy-prone rats. *J. Pharmacol. Exp. Ther.* **222**: 662–669, 1982.

Ko, K., Dailey, J. W., and Jobe, P. C.: Seizure intensity changes caused by adrenergic agonists in genetically epilepsy-prone rats (GEPR). *Fed. Proc.* **42**: 364, 1983.

Laird II, H. E.: 5-Hydroxytryptamine as an inhibitory modulator of audiogenic seizures (doctoral dissertation). Tucson, University of Arizona, 1974.

Laird II, H. E.: Abnormal concentrations and turnover rates of norepinephrine and dopamine in brains of genetically seizure-susceptible rats. *Epilepsia* **24**: 107, 1983.

Laird, H. E., Chin, L., and Picchioni, A. L.: Enhancement of audiogenic seizure intensity by *p*-chloromethamphetamine. *Proc. West. Pharmacol. Soc.* **17**: 46–50, 1974.

Laird, H. E., Dailey, J. W., and Jobe, P. C.: Innate abnormalities in norepinephrine levels in genetically epilepsy-prone rats (GEPRs). *Fed. Proc.* **42**: 363, 1983.

Laird II, H. E., Dailey, J. W., and Jobe, P. C.: Neurotransmitter abnormalities in genetically epileptic rodents. *Fed. Proc.* **43**: 2505–2509, 1984.

Laird II, H. E., Hadjicostantinou, M., and Neff, N. H.: Abnormalities in the central cholinergic transmitter system of the genetically epilepsy prone rat. *Life Sci.* **39**(9): 783–787, 1986.

Laird II, H. E. and Huxtable, R. J.: Effect of Taurine on Audiogenic Seizure Response in Rats, In: *Taurine* (R. Huxtable and A. Barbeau, eds.) Raven, New York, 1976.

Laird, H. E. and Huxtable, R.: Taurine and Audiogenic Epilepsy, In: *Taurine and Neurological Disorders* (A. Barbeau and R. J. Huxtable, eds.) Raven, New York, 1978.

Lazarewicz, J. W., Noremberg, K., Lehmann, A., and Hamburger, A.: Effects of taurine on calcium binding and accumulation in rabbit hippocampal and cortical synaptosomes. *Neurochem. Int.* **7**(8): 421–427, 1985.

Lehman, A.: Psychopharmacology of the Response to Noise, With Special Reference to Audiogenic Seizure in Mice, In: *Physiological Effects of Noise* (B. L. Welch and A. S. Welsch, eds.) Plenum, New York, 1970.

Lehman, A. and Huxtable, R. J.: (Personal communications), 1985.

McEachron, D. L., Jobe, P. C., Smith, W. K., Schlusselberg, D., Woodward, D. J., and Waterhouse, B. D.: 2-DG uptake patterns in the CNS of genetically epilepsy-prone rats: A computer facilitated analysis. *Neurosci. Abst.* **10**: 409, 1984.

Michell, R. H.: Inositol phospholipid and cell surface receptor function. *Biochem. Biophys. Acta* **415**: 81–147, 1975.

Miller, F. P., Cox, R. H., Snodgrass, W. R., and Maickel, R. P.: Comparative effects of *p*-chlorophenylalanine, *p*-chloroamphetamine and *p*-chloro-*N*-methylamphetamine on rat brain, norepinephrine, serotonin and 5-hydroxyindole-3-acetic acid. *Biochem. Pharmacol.* **19**: 435–442, 1970.

Mills, S. A., Reigel, C. E., Jobe, P. C. and Savage, D. D.: Increase in the number of hippocampal (^{3}H)-glutamate binding sites in rats inbred for a genetic predisposition to acoustic stimulus induced seizures. *Neurosci. Abst.* **11**: 1315, 1985.

Mimaki, T., Yabuuchi, H., Laird, H., and Yamamura, H. I.: Effects of seizures and antiepileptic drugs on benzodiazepine receptors in rat brain. *Ped. Pharmacol.* **4**: 205–211, 1984.

Nicoletti, F., Barbaccia, M. L., Iadarola, M., Pozzi, O., and Laird II., H. E.: Abnormality of alpha-1 adrenergic receptors in the frontal cortex of epileptic rats. *J. Neurochem.* **46**: 270–273, 1986.

Nishizuka, Y.: Turnover of inositol phospholipids and signal transduction. *Science* **225**: 1365–1370, 1984.

Pfeifer, A. K. and Galambos, E.: The effect of *d1-p*-chloroamphetamine on the susceptibility to seizures and on the monoamine level in brain and heart of mice and rats. *J. Pharm. Pharmacol.* **19**: 400–402, 1967.

Picchioni, A., Chin, L., Choisser, D., and Breitner, C.: 5-Hydroxytryptamine (5-HT) and norepinephrine (NE): Relationship to audiogenic seizure. *Pharmacologist* **5**: 238, 1963.

Pletscher, A., Da Prada, M., and Burkard, W. P.: The Effect of Substituted Phenylethylamines on the Metabolism of Biogenic Monoamines, In: *International Symposium on Amphetamines and Related Compounds*, (E. Costa and S. Garattini, eds.) Raven, New York, 1970.

Reigel, C. E., Jobe, P. C., Woods, T. W., and Dailey, J. W.: A GABAergic convulsant profile in the genetically epilepsy-prone rat. *Neurosci. Abst.* **10**: 408, 1984.

Reigel, C. E., Jr., Stewart, J. J., Rinaudo, J. W., Dailey, J. W., and Jobe, P. C.: Extraordinary small doses of intracerebroventricular morphine produces tonic extensor convulsions in the genetically epilepsy-prone rat. *Fed. Proc.* **44**: 1106, 1985.

Roberts, R. C., Ribak, C. E., Peterson, G. M., and Oertel, W. H.: Increased numbers of GABAergic neurons in the inferior colliculus of the genetically epilepsy-prone rat. *Neurosci. Abst.* **10**: 410, 1984.

Roth, B. L., Nakaki, T., Chuang, D.-M., and Costa, E.: Aortic recognition sites for serotonin (5HT) are coupled to phospholipase C and modulate phosphatidylinositol turnover. *Neuropharmacol.* **23**: 1223–1225, 1984.

Rudzik, A. D. and Johnson, G. A.: Effect of Amphetamine and Amphetamine Analogs on Convulsive Thresholds, In: *International Symposium on Amphetamines and Related Compounds* (E. Costa and S. Garrattini, eds.) Raven, New York, 1970.

Savage, D. D., Reigel, C. E., and Jobe, P. C.: The development of angular bundle kindled seizures is accelerated in rats with a genetic predisposition to audiogenic seizures. *Neurosci. Abst.* **10**: 1319, 1985.

Sokoloff, L., Reivich, M., Kennedy, C., Desrosiers, M. H., Patlak, C. S., Pettigrew, K. D., Sakurada, O., and Shinohara, M.: The (14-C)-deoxyglucose method for the measurement of local cerebral glucose utilization: Theory, procedure, and normal values in the conscious and anesthetized albino rat. *J. Neurochem.* **28**: 897–916, 1977.

Speth, R., Laird II, H. E., and Yamamura, H. I.: Unpublished observations, 1979.

Stewart, J. J., Rinaudo, J. D., Reigel, C. E., Jobe, P. C., and Dailey, J. W.: Intracerebroventricular morphine produces convulsions in genetically epilepsy-prone rats. *Neurosci. Abst.* **10**: 408, 1984.

Tacke, U. and Braestrup, C.: A study on benzodiazepine receptor binding in audiogenic seizure-susceptible rats. *Acta Pharmacol. Toxicol.* **55**: 252–259, 1984.

Tower, D. A.: GABA and Seizures: Clinical Correlates in Man, In: *GABA in Nervous System Function* (E. Roberts, T. N. Chase, and D. B. Tower, eds.) Raven, New York, 1976.

Waterhouse, B. D., McEachron, D. L., and Woodward, D. J.: Personal communications, 1985.

Waters, S.: Serotonin receptor binding: Characterization of serotonin-1 receptor subpopulations and determination of the effect of neurotoxic chemicals on ligand binding interaction. (Master's thesis). Tucson, University of Arizona, 1983.

Wood, J. D.: The role of gamma-aminobutyric acid in the mechanism of seizures. *Prog. Neurobiol.* **5**: 77–95, 1975.

The Epileptic Chickens

D. D. Johnson and J. M. Tuchek

1. Epileptic Chicken as a Seizure Model

1.1. Genetics and Seizure Patterns in Epileptic Chickens

The discovery by R. D. Crawford of an autosomal recessive mutation in domestic fowl producing a strain of chickens with a high seizure susceptibility added a new and potentially valuable model for the study of epilepsy (Crawford, 1970). As a graduate student, Dr. Crawford had been studying behavioral genetics in the Fayoumi breed. Following graduation, he moved to the Agriculture Canada Research Station in Kentville, Nova Scotia. In order to continue his studies, eggs from the Fayoumi breed were obtained from the University of Massachusetts. When these eggs were hatched, two chicks were culled since they were deemed to have congenital loco, a genetic condition widespread in domestic fowl. It is now believed that these chicks were probably epileptic. A technician assigned to care for the newly hatched chicks reported that Fayoumi chicks did indeed have a strange behavior, particularly when subjected to abnormal stimuli. Within a few days after hatching, two males and three females were identified as having violent convulsions, and these provided the base stock for the epileptic population. Early studies by Crawford at the University of Saskatchewan investigating the mode of inheritance included crosses with other breeds in an attempt to locate the mutation on the chromosome map. Tests have been conducted to determine linkage relationships to seven autosomal marker genes. These markers are included in four of five well known autosomal linkage groups. The epi mutant was found to segregate independently of all the markers. Because initial populations were very small, the mutation was lost from the pure Fayoumi breed. The stock presently used was derived

95

from crosses used to measure gene linkage relationships and hence incorporates a very broad genetic base and is called ''synthetic'' epileptic because of its crossbreed origin (Crawford, 1982).

In all early studies designed to determine the mode of inheritance, seizures were induced by frightening the animals by banging on the sides of the cages. Using this methodology, the trait of high seizure susceptibility was reported to be an autosomal recessive mutation with incomplete penetrance, since there was a 9% deficiency of epileptic chicks from reciprocal mating of epileptics with heterozygotes (Crawford, 1970).

In 1974 Crichlow and Crawford reported that seizures could be evoked in homozygotes (epileptics) by intermittent photic stimulation, the optimum frequency being 14 flashes/s (Crichlow and Crawford, 1974). Heterozygotes (carriers) did not respond. Using this methodology to reexamine the mode of inheritance, it was concluded that the epi mutant has complete penetrance (Crawford, 1982).

In epileptics subjected to intermittent photic stimulation the seizure could be divided into three phases. In phase 1, usually beginning within 30 s after initiation of the photic stimulus, the head and neck are slowly rotated and arched back and upward. Phase 2 is characterized by extension of the wings with some loss of balance. Phase 3 consists of violent flapping of the wings with clonic movements of the legs resulting in thrashing and tumbling motions. The motor seizure does not contain a tonic component. Once evoked, phase 3 may continue for several minutes and, in some birds, may reoccur in status-like form without further stimulation.

In studies in which anticonvulsant activity was being measured, two types of data were obtained. First, the effect of the drug on the incidence of seizures (% seizures) was recorded. Second, since some drugs may reduce the severity of seizures without reducing the incidence, the phase to which each seizure progressed was recorded using numerical values 1–3. The total seizure score for each group of chickens was then obtained. This number was divided by the number of chickens per group to yield the ''seizure index,'' which became a measure of the overall seizure severity in control or treated groups. The latter data can be analyzed for statistical significance using nonparametric statistical methods.

Seizures can also be evoked in epileptic chicks by hyperthermia (Lee, 1973; Johnson et al., 1983). In 2–5-d-old chicks, two seizure types can be evoked by using microwave diathermy to elevate body temperature (Johnson et al., 1983). The initial seizure pattern mimics that described above when elicited by intermittent photic stimulation. The second seizure occurring after a further elevation of body temperature is characterized by rapid clonus followed by rigid tonic extension.

1.2. Electrophysiological Studies

Epileptic fowl have an abnormal interictal EEG characterized by slow wave, high voltage spike and wave activity. Upon exposure to intermittent photic stimulation, this activity disappears and is replaced by bilateral spiking at the same frequency as the stimulus. The spiking increases in amplitude and signals the onset of the overt motor seizure. Following phase 3 of the seizure, EEG activity is reduced. A variable period of post-ictal depression ensues, during which the bird is unresponsive to IPS, and EEG is relatively silent (Crichlow and Crawford, 1974).

1.3. Pharmacological Studies

The value of any model of human pathology is dependent on how closely the syndrome in the model parallels that in humans. Estimates of the plasma concentrations at which all major anticonvulsant drugs are effective in controlling human epilepsies are now known. The development of the epileptic chicken as a valid pharmacological model of human epilepsy was therefore based on the premise that, to provide a suitable model, the major anticonvulsants must not only reduce the incidence of seizures in response to intermittent photic stimulation, but must do so at plasma (and presumably brain) concentrations approximating those required in the appropriate type of human epilepsy. It is also widely known that certain types of epilepsies are treatable with specific classes of drugs, whereas other classes of drugs are ineffective. Thus, in screening the available anticonvulsant agents it should also be possible to determine whether intermittent photic stimulation-evoked seizures represent a pharmacological model of any particular type of human epilepsy.

The efficacy of a number of anticonvulsants in abolishing or reducing the incidence of seizures in response to photic stimulation is shown in Table 1. Phenobarbital (Johnson et al., 1977), phenytoin (Davis et al., 1978), primidone (Johnson et al., 1978), dipropylacetic acid (valproate), and benzodiazepines, and trimethadione were all found to have anticonvulsant activity. Ethosuximide,

Table 1
Epileptic Chicken as a Pharmacological Model of Human Epilepsy

Drug	Range of effective plasma concentrations, μg/mL[a]	
	Epileptic chicken	Humans
Phenobarbital	4–14	10–30
Phenytoin	8–15[b]	10–20
Primidone	22 (Metabolized to pheno-barbital PEMA inactive)	
Ethosuximide	Inactive	
Trimethadione	210–442[b,c]	700
Dipropylacetic acid	105–245	
Benzodiazepines		
Clonazepam	ED_{50} = 0.025 mg/kg	
Diazepam	ED_{50} = 0.5 mg/kg	
Chlordiazepoxide	ED_{50} = 1.5 mg/kg	

[a]The lower value of the mean plasma concentration at which a significant reduction in seizure incidence was first observed.

[b]Complete protection not obtained because of appearance of toxicities.

[c]Expressed as TMO and DMO.

an agent of choice in therapy of absence seizures in humans, was without effect (Davis et al., 1978). The range of plasma concentrations within which a significant reduction in seizure incidence occurred in epileptic chicken vs that range generally regarded as effective in human epilepsies is also shown in Table 1. Inspection of Table 1 indicates that for phenobarbital, phenytoin, primidone, and valproate the required plasma concentrations for seizure control are similar to those required in human epilepsies. As in humans, primidone was metabolized to phenobarbital in the epileptic chickens, but primidone itself was shown to have anticonvulsant activity. Phenylethylmalonamide (PEMA), a second metabolite of primidone, was found to be inactive. Similarly trimethadione was metabolized to dimethadione and both compounds were active. Inspection of Table 1 suggests that in terms of their response to known anticonvulsant drugs, epileptic chickens provide a pharmacological model for those human epilepsies historically referred to as grand mal epilepsies. Carbamazepine also prevents seizures in response to photic stimulation, but the effective plasma concentrations have not been determined. The finding that ethosuximide was inactive indicates that if this model was to be used for screening chemicals for anticonvulsant activity, a second model would have to be included to determine potential activity against absence seizures. When seizures were evoked by hyperthermia, phenobarbital was effective in delaying seizures, but valproate and phenytoin were ineffective (Johnson et al., 1983).

2. Brain Monoamines and Seizure Susceptibility in Epileptic Chickens

2.1. Interictal Norepinephrine, Dopamine, and 5-Hydroxytryptamine Concentrations

The report of a lowered electroconvulsive threshold in mice when monoamines were depleted by reserpine (Chen et al., 1954) stimulated interest in the role of central monoamines in seizure mechanisms. The majority of the literature on this subject prior to 1972 indicated that normal concentrations of norepinephrine, dopamine, and 5-hydroxytryptamine suppressed seizure activity in experimental models of epilepsy. Drugs interferring with the metabolism or reuptake of monoamines tended to inhibit seizures, whereas drugs interferring with storage, synthesis, or release of monoamines increased seizure susceptibility (Maynert, 1969; Lovell, 1971). 6-Hydroxydopamine-induced neurotoxic deletions of central catecholaminergic neurons enhanced the seizure severity in audiogenic seizure-susceptible rats (Bourn et al., 1972), potentiated metrazol-induced seizures, and facilitated the production of seizures by kindling (Corcoran et al., 1974; Arnold et al., 1973; McIntyre et al., 1979). Deficiencies of brain amines were reported in audiogenic seizure-susceptible mice (Schlesinger and Uphouse, 1972). More recent data on the role of these amines in various other models of epilepsy are given in other chapters of this book. These data provided the impetus for initiating studies to determine the possible role of abnormalities in norepinephrine, dopamine, and 5-hydroxytryptamine concentrations or functions in the production of the high seizure susceptibility in epileptic chicken. The concentrations of norepinephrine, dopamine, and 5-hydroxytryptamine were therefore determined in various areas of the brains of epileptic fowl sacrificed during interictal periods and compared to concentrations existing in their nonepileptic heterozygote hatchmates. Brain norepinephrine, dopamine, and 5-hydroxytryptamine were deter-

Table 2

Comparison of Brain Monoamine Concentrations (μg/g) in Adult Epileptic and Nonepileptic Chicken

Brain region	5-Hydroxytryptamine		Dopamine		Norepinephrine	
	Epileptic	Nonepileptic	Epileptic	Nonepileptic	Epileptic	Nonepileptic
Cerebral hemisphere (n = 20–22)	0.94 ± 0.02[a]	1.02 ± 0.02	0.50 ± 0.01[a]	0.56 ± 0.02	0.34 ± 0.01[a]	0.29 ± 0.02
Optic lobes (n = 19–24)	0.94 ± 0.02	0.72 ± 0.01	—	—	0.39 ± 0.02	0.42 ± 0.02
Midbrain-pons-medulla (n = 10–12)	1.11 ± 0.02	1.04 ± 0.04	0.19 ± 0.01	0.17 ± 0.01	0.89 ± 0.06	0.94 ± 0.02
Cerebellum (n = 10–12)	0.26 ± 0.01	0.22 ± 0.01	—	—	0.31 ± 0.02	0.32 ± 0.01

[a]Significantly different from nonepileptic $p < 0.05$. (Values are mean ± SE).

mined spectrophotofluorometrically using the o-phthaladehyde method for 5-hydroxytryptamine (Maickel et al., 1968) and the iodine oxidation procedure for norepinephrine and dopamine (Jacobowitz et al., 1967). When chickens were pretreated with tryptophan (see below), the butanol supernatant used for extracting 5-hydroxytryptamine was washed with borate buffer to remove 5-hydroxytryptophan, which would interfere with the 5-hydroxytryptamine determination (Brodie et al., 1966). Data from these experiments are shown in Table 2. During interictal periods the concentrations of 5-hydroxytryptamine and dopamine were significantly lower in cerebral hemispheres from epileptic chickens, whereas the concentration of norepinephrine was significantly higher (Johnson et al., 1981). No significant differences were noted in other brain areas. The concentrations of dopamine in the optic lobes and cerebellum were below the limits of detection by the assay procedure.

In order to determine the possible role of low 5-hydroxytryptamine concentrations in producing the high seizure susceptibility, brain 5-hydroxytryptamine concentrations were elevated by administering the precursor amino acid L-tryptophan, the monoamine oxidase inhibitor phenelzine, or a combination of L-tryptophan plus phenelzine. The effects of these treatments on brain 5-hydroxytryptamine concentrations, seizure susceptibility and seizure severity (index) are shown in Table 3. Tryptophan alone

Table 3

Effect of Elevating Brain 5-HT Concentrations on Seizure Susceptibility in Response to Photic Stimulation

Treatment	% Increase in 5-hydroxytryptamine concentration			Seizures, %	Seizure index
	Cerebral hemispheres	Optic lobes	Midbrain-pons-medulla		
Control				100	3
L-Tryptophan, 150 mg/kg	26.6[a]	—	6.9	100	3
Phenelzine, 100 mg/kg	51.5[a]	20	60.9[a]	75	2.4
L-Tryptophan 150 mg/kg, plus phenelzine 100 mg/kg	73[a]	53[a]	99[a]	100	3

[a]$p < 0.05$ with respect to the control group. Note: $n = 5$ chickens per group.

produced a significant increase in 5-hydroxytryptamine concentrations only in the cerebral hemispheres, but did not alter the seizure susceptibility or severity. Phenelzine increased 5-hydroxytryptamine in both the hemispheres and the midbrain pons–medulla, but had no significant effect on seizure incidence or severity. The combination of phenelzine plus tryptophan produced marked increases in the 5-hydroxytryptamine concentrations in the hemispheres optic lobes and midbrain pons–medulla, but with no concomitant effect on seizure susceptibility.

Two hours following L-DOPA administration (250 mg/kg ip) to epileptic chickens, dopamine concentrations were significantly increased in the cerebral hemispheres and midbrain pons–medulla (Table 4). These increases in dopamine concentrations did not reduce seizure susceptibility or severity.

Table 4

Effect of L-DOPA (250 mg/kg, ip) on DA Concentrations
(μg/g; Mean $\pm$ SEM), Seizure Susceptibility and Seizure Severity (Index)
in Epileptic Chickens in Response to Photic Stimulation

	Dopamine concentration			
	Cerebral hemispheres	Midbrain-pons-medulla	Seizures, %	Seizure index
Control	0.45 ± 0.03	0.19 ± 0.03	100	3
L-DOPA 250 mg/kg	2.08 ± 0.42^a	1.11 ± 0.34^a	100	3

[a]$p < 0.05$, $n = 5$ per group.

The role of norepinephrine was investigated by administering the receptor-blocking agents phenoxybenzamine (50 mg/kg), phentolamine (50, 75, and 100 mg/kg), or propranolol (5 mg/kg) to groups of 10 chickens. The response of the treated chickens to photic stimulation was determined and compared to appropriate controls 1.5, 3, 5, and 7 h following drug administration. Neither the α-adrenergic receptor-blocking agents phenoxybenzamine or phentolamine, or the β-receptor blocking agent propranolol reduced seizure susceptibility or severity of seizures (Johnson et al., 1981). These pharmacological data indicate a lack of involvement of noradrenergic mechanisms in the seizure process in epileptic chickens.

These data suggest that even though the 5-hydroxytryptamine and dopamine concentrations are low in the epileptic chicken, these amines are not involved in the etiology of the high seizure susceptibility, since elevating the amine levels did not affect the seizure responses. It was concluded that factors other than the primary genetic defect may account for the observed abnormalities in monoamine levels. Since epileptic chickens have grossly abnormal interictal EEG activity (Crichlow and Crawford, 1974), it is possible that the continuous abnormal neural activity produces secondary changes in neurotransmitter levels. Since these studies were conducted on adult chickens, it is also possible that spontaneous seizures and/or seizures evoked to determine the phenotype may produce persistent biochemical changes.

2.2. Monoamine Turnover Rates in Epileptic and Nonepileptic Chickens

Since in vivo estimation of the turnover rates of biogenic amines may provide more information about neuronal activity than measurements of amine concentrations, the turnover rates of 5-hydroxytryptamine, dopamine, and norepinephrine were also determined. If it is assumed that a steady state exists wherein the synthesis and transport of the biogenic amine into a metabolic compartment is equal to its efflux and metabolism, then it may be assumed that the amine turnover rate is equal to its synthesis rate. Thus, if a tracer dose of tritiated precursor amino acid is given, the label will initially be incorporated into the corresponding biogenic amine in a linear fashion (Neff et al., 1971).

Following the administration of (^{3}H)-L-tryptophan as the precursor to 5-hydroxytryptamine, no significant differences were detected in the 5-hydroxytryptamine turnover rate between epileptic and nonepileptic chickens. Data obtained in these experiments confirmed the earlier observation that 5-hydroxytryptamine concentrations are significantly lower in epileptic chickens (Table 5).

Table 6 shows that after injection of (^{3}H)-tyrosine, norepinephrine turnover rate was found to be significantly lower ($p < 0.005$) in epileptic chickens than in their nonepileptic heterozygote hatchmates. No differences were found in the dopamine turnover rates.

The above observation of a low norepinephrine turnover rate in epileptic chickens when compared to their nonepileptic heterozygote hatchmates, together with the finding of elevated norepinephrine levels in epileptics, suggests that there may be a lower rate of neuronal firing in noradrenergic pathways. It is noteworthy that norepinephrine levels are abnormally low and turnover rates are also reduced in the genetically epilepsy-prone rat (Jobe and Laird,

Table 5
Tryptophan, 5-Hydroxytryptamine Concentrations
and 5-Hydroxytryptamine Turnover Rates
in Epileptic and Nonepileptic Chickens

	Tryptophan, $\mu g/g$	5-Hydroxy-tryptamine, $\mu g/g$	5-Hydroxytryptamine turnover rate estimate (SA 5-hydroxytryptamine/SA tryptophan)
Epileptic (n = 8)	2.77 ± 0.55	0.945 ± 0.067[a]	1.106 ± 0.173
Nonepileptic (carrier) (n = 8)	2.47 ± 0.40	1.310 ± 0.068	1.38 ± 0.167

[a]$p < 0.01$, SA = specific activity; each value is the mean of the number of values shown in brackets ± SEM.

1981). In addition, an inherited deficiency in noradrenergic transmission resulting in an increased seizure susceptibility appears to be consistent with pharmacological data that suggest that reducing norepinephrine concentrations (e.g., with reserpine) increases seizure susceptibility. Jobe and coworkers (1981) postulated that a CNS monoaminergic deficit causes the appearance of the epilepsy-prone state in rats that carry some other genetic seizure trait. Decreased monoaminergic transmission in itself was thought to be insufficient to cause a high seizure susceptibility. The role of norepinephrine in producing the elevated seizure susceptibility in audiogenic seizure-susceptible mice is controversial (Jobe and Laird, 1981). However, in the epileptic baboon intracerebroventricular (icv) norepinephrine suppresses photically-induced seizures (Altshuler et al., 1976). Taken together, the data revealing lower norepinephrine turnover rates in the epileptic chicken and reported deficits in noradrenergic transmission in many other genetic models of epilepsy are consistent with the hypothesis that decreased transmission at synapses where this amine is the neurotransmitter results in increased seizure susceptibility. The data on epileptic chickens is not sufficient to suggest that the decreased norepinephrine turnover rate is a primary genetic defect or a consequence of the epileptic state. Studies have not been conducted to determine if selective depletion of norepinephrine would further increase seizure susceptibility in epileptic chickens.

Table 5

Norepinephrine and Dopamine Turnover Rates in Epileptic and Carrier Chickens

	Tyrosine, $\mu g/g$	NE, $\mu g/g$	DA, $\mu g/g$	NE turnover, SA NE/ SA Tyr	DA turnover, SA DA/ SA Tyr
Epi (n = 10)	15.55 ± 0.85	0.732 ± 10.047[a]	0.723 ± 0.039	1.773 ± 0.117[a]	0.9682 ± 0.105
Non-epi (n = 10)	12.33 ± 0.51	0.577 ± 0.52[a]	0.834 ± 0.032	2.546 ± 0.234	0.862 ± 0.066

[a]$p < 0.005$; NE, norepinephrine; DA, dopamine.

3. Brain Amino Acids and Seizure Susceptibility in Epileptic Chickens

3.1. Brain Amino Acid Concentrations in Epileptic Chickens

Although rather simplistic, one plausible hypothesis that may explain the underlying cause of epilepsy is that an imbalance between excitatory and inhibitory mechanisms exists in the brain in this neurological condition (Avoli, 1983). Among the neurotransmitters, the dicarboxylic acids, glutamic acid and aspartic acid, are the most abundant excitatory compounds in many brain areas. These amino acids have been shown to produce excitatory effects when applied microiontophoretically to various brain regions (Watkins and Evans, 1981; Wyler and Ward, 1980). Furthermore, excitatory compounds that act at dicarboxylic amino acid receptors are convulsant when injected focally into the brain and some produce seizures when given systemically in high doses (Johnston, 1973). Excitatory amino acid antagonists such as phosphonic derivatives of aliphatic amino acids have been shown to be potent anticonvulsants in audiogenic seizure-susceptible mice. The anticonvulsant potencies of these derivatives parallel their relative potencies as antagonists to the excitation caused by iontophoretic application of N-methyl-d-aspartate (Croucher et al., 1982). The role of γ-aminobutyric acid (GABA) as an inhibitory neurotransmitter is generally accepted (Krnjevic, 1974). The involvement of GABA in convulsive disorders including human epilepsies has been suggested by several authors (Wood, 1975; Tower, 1976; Emson, 1978). Convulsions can be elicited experimentally by decreasing brain GABA levels through the use of agents such as isonicotinic acid or derivatives of diaminopropionic acid, which inhibit glutamic acid decarboxylase (GAD), an enzyme responsible for GABA synthesis (Wood, 1975; Kazi and Rao, 1983). Conversely, drugs that elevate brain GABA concentrations by inhibiting enzymes in the GABA degradative pathway have been shown to be potent anticonvulsant agents. Examples of these drugs include the GABA transaminase inhibitors, γ-vinyl GABA and γ-acetylenic GABA (Schechter et al., 1977; Loscher, 1981; Meldrum, 1981).

Since various studies have suggested that amino acid metabolism is altered in either experimentally induced or naturally occurring convulsive disorders (*see* van Gelder et al., 1980), the amounts of aspartic acid, glutamic acid, glutamine, taurine, and GABA in the brains of epileptic chickens were examined using high-performance liquid chromatographic (HPLC) techniques. Taurine was included in the study since it is known to produce inhibitory effects when applied iontophoretically, although its role as an inhibitory neurotransmitter is unknown (*see* Pasantes-Morales et al., 1982). Carrier and epileptic chickens were decapitated and their brains were quickly removed. The cerebral hemispheres, optic lobes, and cerebella were dissected out and frozen in liquid nitrogen. The amino acids, aspartic acid, glutamic acid, glutamine, taurine, and GABA were extracted with 75% ethanol and quantified using fluorescent HPLC techniques as described by Jones et al., 1981.

When comparisons of the amounts of amino acids in the various brain regions between adult carrier and epileptic chickens were made, significant differences were seen only in GABA concentrations in the cerebral hemispheres. The GABA content was higher in the epileptic cerebral hemispheres (3.65 μmol/g tissue) than in the corresponding region from carrier (heterozygote) chickens (2.57 μmol/g) (Tuchek et al., in preparation). These results are similar to those reported by Kurokawa and coworkers for a convulsive strain of mouse (Kurokawa et al., 1966). No differences were seen in GABA content in the optic lobes or cerebellum, and no differences were seen in any of the other amino acids studied in the three brain regions. Interestingly, when the brains of 1-d-old or 6-wk-old chicks were examined for amino acid content, no differences in GABA levels in any of the brain regions, including the cerebral hemispheres, were found when comparisons between carriers and epileptics were made. The brain content of GABA in the 1-d or 6-wk-old carrier and epileptic chick (approximately 2.40 mol/g) was similar to that found in adult carrier chickens.

3.2. Pharmacology of Drugs Affecting GABA Metabolism in Epileptic Chickens

Although the GABA content in epileptic cerebral hemispheres was found to be higher than in cerebral hemispheres from carrier chickens, it is apparently not sufficiently elevated to control the seizure process in these animals. Drugs capable of increasing brain GABA levels are good anticonvulsants in the epileptic chicken. One example of such a drug is γ-acetylenic GABA (amino-4-hex-5-ynoic acid), which increases brain GABA levels by inhibiting GABA transaminase, an enzyme responsible for the degradation of this inhibitory neurotransmitter. When this drug was administered iv (10 mg/kg) into 6-wk-old epileptic chickens, a significant anticonvulsant effect was apparent within 4 h after its administration, as indicated in Table 7. In these experiments, intermittent photic stimulation (IPS) at 14 pulses/s was used to induce seizures. Table 7 also indicates that the anticonvulsant effect of 10 mg/kg γ-acetylenic GABA appeared to be slightly less at 6 h in comparison to 4 h, suggesting that the effect of the drug was dissipating. Table 8 shows the effect of γ-acetylenic GABA (10 mg/kg) on amino acid levels in the cerebral hemispheres, optic lobes, and cerebella taken from epileptic chickens at 2, 4, and 6 h after the drug was given intravenously. Although the dose of γ-acetylenic GABA used in these experiments appeared to significantly elevate GABA levels at 4 h in all the three brain regions studied, it had no apparent effect on

Table 7

Effect of γ-Acetylenic GABA (10 mg/kg) on IPS-induced Seizures in 6-wk-old Epileptic Chickens

	Hours after injection, 10 mg/kg			
	Control, %	2, %	4, %	6, %
Seizure incidence	100	100	40	40
Seizure index (severity)	3	2	0.4^a	1.2^a

$^a n = 5$; $p < 0.01$; Seizure index as defined in section 1.1.

the levels of the other amino acids. At 6 h postinjection, GABA levels decreased toward control concentrations (particularly in the cerebral hemispheres and cerebellum) and correlated with a diminished anticonvulsant effect at this hour in comparison to that at 4 h (Table 7). Thus, epileptic chickens can be protected from intermittent photic stimulation-induced seizures if brain GABA levels are sufficiently elevated. In view of these results, it is interesting to speculate on the extent that GABA concentrations must be elevated and in which particular brain regions must it be elevated in order to prevent the seizure-inducing process. As indicated in the previous sections, adult epileptic chickens have 42% more GABA in the cerebral hemispheres than carrier chickens do, and still have seizures. In comparison, γ-acetylenic GABA treatment (Table 8) in 6-wk-old epileptics increased GABA levels in the cerebral hemispheres by only 49% at 4 h, and this elevation afforded a significant protection against seizures (normally, GABA levels

Table 8
Effect of γ-Acetylenic GABA on Amino Acid Levels
in Brains of Epileptic Chickens

Brain area	Time after treatment, h	Amino acid, μmol/g tissue				
		Asp	Glu	Gln	Tau	GABA
Cerebral hemispheres	Control	3.38 (0.09)	12.92 (0.33)	9.26 (0.24)	3.78 (0.43)	2.96 (0.09)
	2	3.15 (0.16)	12.75 (0.32)	8.72 (0.55)	3.94 (0.19)	4.64[b] (0.61)
	4	2.97 (0.12)	12.58 (0.27)	8.70 (0.32)	3.86 (0.16)	4.42[b] (0.45)
	6	3.05 (0.15)	12.33 (0.26)	8.18 (1.27)	3.70 (0.23)	3.13 (0.76)
Optic lobes	Control	4.47 (0.13)	9.45 (0.36)	9.04 (0.64)	2.11 (0.21)	4.27 (0.22)
	2	3.83 (0.11)	8.62 (0.15)	8.00 (0.60)	2.10 (0.14)	6.38[b] (0.73)
	4	3.75 (0.30)	8.82 (0.16)	8.00 (0.18)	2.10 (0.16)	6.38[b] (0.84)
	6	4.15 (0.48)	9.00 (0.19)	8.15 (0.99)	2.03 (0.12)	6.33[b] (0.68)
Cerebellum	Control	3.52 (0.82)	11.23 (2.02)	10.33 (2.17)	3.94 (0.17)	1.77 (0.14)
	2	3.07 (0.08)	9.70 (0.14)	8.68 (1.05)	4.08 (0.04)	2.36[a] (0.42)
	4	3.09 (0.05)	10.27 (0.31)	9.36 (0.23)	3.92 (0.57)	2.37[a] (0.45)[a]
	6	2.95 (0.13)	10.02 (0.30)	8.60 (0.93)	3.87 (0.14)	1.81 (0.16)

[a]$p < 0.05$.

[b]$p < 0.01$. Each value represents mean (standard deviation in parentheses) of three animals. Controls represent saline-injected epileptic chickens. Asp, Aspartate; Glu, Glutamate; Gln, Glutamine; Tau, Taurine.

in epileptics and carriers are the same at this age). This discrepancy may be partly explained by the observation that at 4 h, treatment with the GABA transaminase inhibitor elevated GABA not only in the cerebral hemispheres, but in the optic lobes (74%) and the cerebellum (33%) as well. This would indicate that it is important to raise GABA levels in other regions of the brain as well as in the cerebral hemispheres in order to protect against the development of seizures in this hereditary model of epilepsy. Perhaps more importantly, the anticonvulsant action of increasing GABA levels is dependent upon the cellular compartment in which GABA levels are increased. Presumably, increasing GABA concentrations in the ''nerve terminal'' GABA pool would have greater anticonvulsant-like actions than increasing GABA concentrations in the ''metabolic'' GABA pool (*see* Iadarola and Gale, 1979). It has been shown by several authors that the administration of γ-acetylenic GABA to animals in vivo can increase the concentration of GABA found in synaptosomes (nerve terminal GABA pool) prepared from these animals (Meldrum, 1981; Loscher, 1981). Increases in the nerve terminal GABA pool may have occurred in our experiments, although GABA concentrations in synaptosomes were not determined in chickens treated with γ-acetylenic GABA. However, when GABA concentrations were determined in synaptosomal preparations obtained from untreated adult epileptic and carrier chickens, the data indicated that the GABA concentrations were the same in both preparations. On the basis of these results, it can be concluded that the elevated GABA concentrations seen in the cerebral hemispheres of adult epileptic chickens are present in the metabolic rather than the nerve terminal GABA pool and therefore make little contribution to controlling the seizure process.

In summary, our studies on brain amino acid concentrations in epileptic chickens have indicated that levels of aspartate, glutamate, and taurine were similar to those in carrier chickens, but that the GABA content was higher in the cerebral hemispheres of adult epileptic chickens. This increase in GABA appeared to be in the nonsynaptosomal or metabolic pool and could be indicative of a compensatory response in the adult epileptic chicken brain. The increase in GABA content may be the result of an enhanced rate of synthesis and/or a decreased rate of release and degradation of GABA. However, this compensation is obviously not complete since adult epileptic chickens have seizures readily. Based on the observation that the total amount of GABA was elevated only in the cerebral hemispheres of the epileptic chicken, it may be speculated that this is the region of the brain that is altered, resulting in the epileptic condition. Perhaps related to this speculation is the observation that the adult epileptic cerebral hemispheres are significantly larger (approximately 40%) than cerebral hemispheres from carrier chickens (Johnson and Davis, 1983). This increase in cerebral hemisphere size may be due to a reactive gliosis, which commonly occurs in many convulsive disorders (Willmore and Rubin, 1981; Pollen and Trachtenberg, 1970).

GABA levels in 1-d- or 6-wk-old chickens were the same as those seen in the brains of carrier hatchmates. Epileptic chickens treated with γ-acetylenic GABA (10 mg/kg) had significantly elevated amounts of GABA in all brain regions studied at 4 h after injection, and this correlated with the marked protection against intermittent photic stimulation-induced seizures. Investigations on possible GABA receptor alterations in epileptic chickens are currently being undertaken in our laboratory. Studies on the benzodiazepine receptor have indicated no abnormalities in this receptor system in epileptic chickens (Fisher et al., 1985).

Acknowledgment

This research was supported by a grant from the Medical Research Council of Canada, grant number MA 8604.

References

Altshuler, H. L., Killam, E. K. and Killam, K. F.: Biogenic amines and the photomyoclonic syndrome in the baboon, *Papio papio. J. Pharmacol. Exp. Ther.* **196:** 156–166, 1976.

Arnold, P. S., Facine, R. V., and Wise, R. A.: Effects of atropine, reserpine, 6-hydroxydopamine and handling on seizure development in the rat. *Exp. Neurol.* **40:** 457–470, 1973.

Avoli, M.: Is Epilepsy a Disorder of Inhibition or Excitation? In: *Epilepsy: An Update on Research and Therapy* (G. Nistico, R. Di Perri, and H. Meinardi, eds.) Alan Liss, New York, 1983.

Bourn, W. M., Chin, L., and Picchioni, A. L.: Enhancement of audiogenic seizures by 6-hydroxydopamine. *J. Pharm. Pharmacol.* **24:** 913–914, 1972.

Brodie, B. B., Comer, M. S., Costa, E., and Diabac, A.: The role of brain serotonin in the mechanism of the central action of reserpine. *Exp. Ther.* **152:** 340–349, 1966.

Chen, G., Ensor, C. R., and Bohner, B. A.: A facilitative action of reserpine on the central nervous system. *Proc. Soc. Exp. Biol. Med.* **86:** 507–510, 1954.

Corcoran, M. E., Fibiger, H. C., McCaughran, J. A., and Wada, J. A.: Potentiation of amygdaloid kindling and metrazol-induced seizures by 6-hydroxydopamine in rats. *Exp. Neurol.* **45:** 118–133, 1974.

Crawford, R. D.: Epileptiform seizures in domestic fowl. *J. Hered.* **61:** 185–188, 1970.

Crawford, R. D.: Genetics and Behavior of the epi Mutant Chicken, In: *The Brain and Behavior of the Fowl* (Takanori Ookawa, ed.) Japan Scientific Societies Press, Tokyo, 1982.

Crichlow, E. C. and Crawford, R. D.: Epileptiform seizures in domestic fowl. II. Intermittent light stimulation and the electroencephalogram. *Can. J. Physiol. Pharmacol.* **52:** 424–429, 1974.

Croucher, M. J., Collins, J. F., and Meldrum, B. S.: Anticonvulsant action of excitatory amino acid antagonists. *Science* **216:** 899–901, 1982.

Davis, H. L., Johnson, D. D., and Crawford, R. D.: Epileptiform seizures in domestic fowl. VII. Plasma phenytoin concentrations and anticonvulsant activity. *Can. J. Physiol. Pharmacol.* **56:** 310–315, 1978.

Emson, P. C.: Biochemical and Metabolic Changes in Epilepsy, In: *Taurine and Neurological Disorders* (A. Barbeau and R. Huxtable, eds.) Raven, New York, 1978.

Fisher, R. E., Davis, L. G., Tuchek, J. M., Johnson, D. D., and Crawford, R. D.: Benzodiazepine receptors and seizure susceptibility in epileptic fowl. *Can. J. Physiol. Pharmacol.* **63:** 85–88, 1985.

Iadarola, M. J. and Gale, K.: Dissociation between drug-induced increases in nerve terminal and non-nerve terminal pools of GABA in vivo. *Eur. J. Pharmacol.* **59:** 125–129, 1979.

Jacobowitz, D., Cooper, T., and Barner, N. B.: Histochemical and chemical studies of the localization of adrenergic and cholinergic nerves in normal and denervated cat hearts. *Circ. Res.* **20:** 289–298, 1967.

Jobe, P. C. and Laird, H. E.: Neurotransmitter abnormalities as determinants of seizure susceptibility and intensity in the genetic models of epilepsy. *Biochem. Pharmacol.* **30:** 3137–3144, 1981.

Jobe, P. C., Brown, R. D., and Dailey, J. W.: Effect of Ro 4-1284 on audiogenic seizure susceptibility and intensity in epilepsy-prone rats. *Life Sci.* **28:** 2031–2038, 1981.

Johnson, D. D.: Epileptiform seizures in domestic fowl. VIII. Anticonvulsant activity of primidone and its metabolites phenobarbital and phenylethlmalonamide. *Can. J. Physiol. Pharmacol.* **56:** 630–633, 1978.

Johnson, D. D., Crawford, K. D. A., and Crawford, R. D.: Febrile seizures in epileptic chicks: The effects of phenobarbital phenytoin and valproate. *Can. J. Neurological Sciences* **10:** 96–99, 1983.

Johnson, D. D., Davis, H. L., Bailey, D. G., and Crawford, R. D.: Epileptiform seizures in domestic fowl. VI. Plasma phenobarbital concentrations and anticonvulsant activity. *Can. J. Physiol. Pharmacol.* **55:** 848–854, 1977.

Johnson, D. D., Jaju, A. T., Ness, L., Richardson, J. S., and Crawford, R. D.: Brain norepinephrine, dopamine and 5-hydroxytryptamine concentration abnormalities and their role in the high seizure susceptibility of epileptic chickens. *Can. J. Physiol. Pharmacol.* **50:** 144–179, 1981.

Johnson, D. D. and Davis, H. L.: Drug Responses and Brain Biochemistry of the epi mutant chicken, In: *The Brain and Behavior of the Fowl* (T. Ookawa, ed.) Japan Scientific Societies Press, Tokyo, 1983.

Johnston, G. A. R.: Convulsions induced in 10-day-old rats by intraperitonial injection of monosodium glutamate and related excitant amino acids. *Biochem. Pharmacol.* **22:** 137–140, 1973.

Jones, B. N., Paabo, S., and Stein, S.: Amino acid analysis and enzymatic sequence determination of peptides by an improved o-phthaladehyde precolumn labelling procedure. *J. Liq. Chrom.* **4:** 565–586, 1981.

Kazi, S. J. and Rao, S. L. N.: β-N-γ-Glutamyl diaminopropionic acid, a convulsant and an inhibitor of brain glutamate decarboxylase. *J. neurochem.* **41:** 244–245, 1983.

Krnjevic, K.: Chemical nature of synaptic transmission in vertebrates. *Physiol. Rev.* **54:** 518–540, 1974.

Kurokawa, M., Naruse, H., and Kato, M.: Metabolic studies on epi mouse, a special strain with convulsive predisposition. *Prog. Brain Res.* **21A:** 112–130, 1966.

Lee, K. E.: Studies of behavioral and physiological bases of genetically controlled epileptiform seizures in domestic fowl (doctoral dissertation). University of Saskatchewan, Saskatoon, Saskatchewan, 1973.

Loscher, W.: Effect of inhibitors of GABA aminotransferase on the metabolism of GABA in brain tissue and synaptosomal fractions. *J. Neurochem.* **36:** 1521–1527, 1981.

Lovell, R. A.: Some Neurochemical Aspects of Convulsions, In: *Handbook of Neurochemistry 6* (A. Lajtha, ed.) Plenum, New York, 1971.

Maickel, R. P., Cox, R. H., Saillant, J., and Millar, F. P.: A method for the determination of serotonin and norepinephrine in discrete areas of rat brain. *Int. J. Neuropharmacol.* **7:** 275–281, 1968.

Maynert, E. W.: The role of biochemical and neurochemical factors in the laboratory evaluation of antiepileptic drugs. *Epilepsia* **10:** 145–162, 1969.

McIntyre, D. C., Sarri, M., and Pappes, B. A.: Potentiation of amygdaloid kindling in adult or infant rats by injections of 6-hydroxydopamine. *Exp. Neurol.* **63:** 527–544, 1979.

Meldrum, B. S.: New GABA-Related Anticonvulsant Drugs: Animal Tests and Clinical Efficacy, In: *Advances in Epileptology: XIIth Epilepsy International Symposium* (M. Dam, L. Gram, and J. K. Penry, eds.) Raven, New York, 1981.

Neff, N. H., Spand, P. F., Cropdetti, A., Wang, C. T., and Costa, E.: A simple procedure for calculating the synthesis rate of norepinephrine, dopamine and serotonin in rat brain. *J. Pharmacol. Exp. Ther.* **176:** 701–709, 1971.

Pasantes-Morales, H., Arzate, N. E., and Cruz, C.: The role of taurine in nervous tissue: Its effects on ionic fluxes. *Adv. Exp. Med. Bio.* **139:** 273–292, 1982.

Pollen, D. A. and Trachtenberg, M. C.: Neuroglia: Gliosis and focal epilepsy. *Science* **167:** 1252–1263, 1970.

Schechter, P. J., Tranier, Y., Jung, M. J., and Bohlen, P.: Audiogenic seizure protection by elevated brain GABA concentration in mice: Effects of γ-acetylenic GABA and vinyl GABA, two irreversible GABA-T inhibitors. *Eur. J. Pharmacol.* **45:** 319–328, 1977.

Schlesinger, K. and Uphouse, L. L.: Pyridoxine Dependency and Central Nervous System Excitability, In: *Advances in Biochemical Psychopharmacology* vol. 4 (E. Costa and P. Greengard, eds.) Raven, New York, 1972.

Tower, D. B.: GABA and Seizures: Clinical Correlates in Man, In: *GABA in Nervous System Function* (E. Roberts, T. N. Chase,, and D. B. Tower, eds.) Raven, New York, 1976.

Tuchek, J. M., Turnbull, J. L., Johnston, D. D., and Crawford, R. D.: Brain amino acid concentrations in epileptic fowl. *J. Neurochem.* (in preparation).

van Gelder, N. M., Janjua, N. A., Metrakos, K., MacGibbon, B., and Metrakos, J. D.: Plasma amino acids in 3 sec spike-wave epilepsy. *Neurochemm. Res.* **5:** 659–671, 1980.

Watkins, J. C. and Evans, R. H.: Excitatory amino acid transmitters. *Ann. Rev. Pharmacol. Toxicol.* **21:** 165–204, 1981.

Willmore, L. J. and Rubin, J. J.: Antiperioxidant pretreatment and iron-induced epileptiform discharges in the rat: EEG and histopathologic studies. *Neurology* **31:** 63–69, 1981.

Wood, J. D.: The role of γ-aminobutyric acid in the mechanism of seizures. *Prog. Neurobiol.* **5:** 77–95, 1975.

Wyler, A. R. and Ward, A. A.: A Window to Brain Mechanisms, In: *Epilepsy* (J. S. Lockard and A. A. Ward, eds.) Raven, New York, 1980.

Biochemical and Pharmacologic Studies of Neurotransmitters in the Kindling Model

James O. McNamara, Douglas W. Bonhaus,
Barbara J. Crain, Randy L. Gellman,
and Cheolsu Shin

1. Introduction

The kindling phenomenon was discovered in the late 1960s by Graham Goddard and colleagues (Goddard, 1967; Goddard et al., 1969). Although the seizure-inducing potential of focal electrical stimulation of brain had been previously reported by other investigators (Delgado and Sevillano, 1961), Goddard and colleagues recognized the potential importance of this phenomenon, demonstrated its permanence, and initiated a systematic characterization. The possibility that study of kindling may provide insight into mechanisms of both neuronal plasticity and human epilepsy has led to extensive investigation of this model. We have recently published a comprehensive and critical review of all aspects of the kindling model (McNamara et al., 1985); the reader is referred to this document for additional detail of subjects only briefly covered in the present manuscript.

The primary focus of this chapter is to provide a critical review of pharmacologic and biochemical studies of putative neurotransmitters in the kindling model. A brief overview of the kindling phenomenon and approaches to elucidating basic mechanisms of kindling have also been included. A recent publication from this laboratory (McNamara, 1984) outlined a strategy for integrating studies of neurotransmitters with data obtained from these other approaches and emphasized how these investigations should be mutually reinforcing.

2. The Kindling Phenomenon

The term kindling refers to a phenomenon in which repeated administration of an initially subconvulsive electrical stimulus results in progressive intensification of seizure activity culminating in a generalized seizure. In rats stimulated in the amygdala, the initial stimulus often elicits focal electrical seizure activity [after-discharge recorded on electroencephalogram (EEG)] without overt behavioral seizure activity. Subsequent stimulations induce the development of overt seizures, generally evolving through the following classes as described by Racine (1972b): (1) facial clonus; (2) head nodding; (3) forelimb clonus; (4) rearing, and (5) rearing and falling. The behavior observed in classes 1 and 2 is similar to that found in human complex partial or limbic seizures; the behavior in later classes is consistent with complex partial seizures spreading to secondarily generalized clonic motor seizures. Hereafter, the term limbic seizure will be used to refer to behavior in which the rat is immobilized with or without associated facial clonus or head nodding; these behaviors are defined as limbic seizures when correlated with afterdischarge recorded from the bipolar electrode in the amygdala. The term motor seizure will be used to refer to the clonic or tonic activity of the extremities during a class 4 or 5 seizure.

Once the enhanced sensitivity has developed, as evidenced by a class 5 seizure, the animal is said to be ''kindled.'' This effect is long lasting, since animals left unstimulated for as long as 12 mo will respond to one of the first two electrical stimuli with a class 5 seizure (Goddard et al., 1969). Thus the term ''kindling'' implies that the change is permanent.

Kindling can be induced by electrical stimulation of many, but not all, sites in the brain. The amygdala is the structure most commonly used because relatively few stimulations are required to induce kindling; other limbic structures often used include the entorhinal cortex, hippocampus, and pyriform cortex. Kindled seizures have not been elicited with electrodes placed in superior colliculus, reticular formation, or cerebellum.

Kindling can be established in frogs (Morrell and Tsura, 1976), reptiles (Rial and Gonzales, 1978), mice (Leech and McIntyre, 1976), rats (Goddard et al., 1969), rabbits (Tanaka, 1972), dogs (Wauquier et al., 1979), cats (Wada and Sato, 1974), rhesus monkeys (Goddard et al., 1969), and baboons (Wada and Osawa, 1976). Although some differences exist, the behavioral patterns of amygdala-kindled seizures are remarkably similar in many of these species. Differences exist in the sensitivity to kindling among strains within a species (Racine et al., 1973), as well as among species. Kindling is readily established in a seizure-sensitive primate, the *Papio papio* baboon.

The electrical stimulus must induce local afterdischarge in order for kindling to develop. However, subthreshold stimulations cause a lowering of the afterdischarge threshold (Racine, 1972a). If the current intensity is set slightly below the afterdischarge threshold, repeated stimulation will lower the threshold and can lead to afterdischarge production and subsequent kindling.

A key criterion for an animal model of epilepsy is that seizures occur spontaneously, since epilepsy by definition implies the presence of spontaneous seizures (as opposed to seizures elicited by an electrical stimulus). Most kindling experiments are terminated after elicitation of a single or several class 5 motor seizures. Spontaneous generalized motor seizures are infrequently observed in rats stimulated to this extent. However, continued periodic stimulation of amygdala, hippocampus, or entorhinal cortex of the rat led to spontaneous seizures (Pinel and Rovner, 1978). In contrast to the 15–30 stimulations required to elicit a single class 5 seizure, a mean of 348 stimulations resulted in spontaneous motor seizures in all subjects. Among subjects exhibiting at least three spontaneous class 5 seizures, spontaneous seizures persisted for as long as 7 mo following termination of the stimulation, suggesting that the epilepsy was permanent.

Study of clinically useful anticonvulsants against kindled seizures could provide an opportunity to validate the model, if the therapeutic indices in the model matched the human condition. Albright and Burnham (1980) have analyzed the dose–response relationship of standard anticonvulsants in suppressing fully kindled seizures in rats. The effectiveness of anticonvulsants against fully kindled seizures does correspond in a general way to the effectiveness of these drugs in human epilepsy. More definitive conclusions cannot be made for two reasons: (1) the absence of detailed information on toxicity and drug concentrations in the kindling model preclude calculation of therapeutic indices; and (2) detailed information on the relative efficacy of standard anticonvulsants against limbic and motor seizures in humans is not available.

3. Approaches to Understanding Basic Mechanisms of Kindling

3.1. Kindling Network

Identification of the network of brain structures containing the alterations responsible for kindling would facilitate the search for the underlying mechanisms. Knowledge of the structures subserving initiation and propagation of kindled seizures might in turn provide a clue to the structures responsible for kindling. These two questions have been addressed by a number of experimental approaches: (1) analysis of propagation of metabolic (2-deoxyglucose) and electrical (EEG) changes during kindled seizures; (2) study of transfer, a phenomenon in which establishment of kindling by electrical stimulation in one brain region results in fewer stimulations required to establish kindling in a second region (Goddard et al., 1969); (3) study of effects of lesions on kindling development or on fully kindled seizures; (4) study of kindling antagonism, a phenomenon in which induction of kindling by periodic stimulation of two distinct limbic sites — rather than a single site — results in a consistent elicitation of class 4 or 5 seizures from only one of the sites (Duchowny and Burchfiel, 1981); and (5) study of neurotransmitters with both pharmacologic and biochemical approaches — these results will be considered in section 3.4.

These various studies have provided a framework for elucidating the structures subserving generation of kindled seizures and maintenance of kindling itself. Analyses of the functional anatomy of kindled seizures with 2-deoxyglucose and EEG monitoring have provided a catalog of limbic, basal ganglia, and brainstem sites that are candidate structures for seizure generation (Engel et al., 1978; McNamara et al., 1985). Kindled seizure threshold has been found to be elevated following lesions of substantia nigra (McNamara et al., 1984) and the hippocampus (Yoshida, 1984) or its connections (Gellman and McNamara, 1984), a result which suggests that these structures may contribute to the generation or propagation of kindled seizures.

Studies of transfer, kindling antagonism, and the effects of lesions on kindling development have provided limited insight into the network responsible for kindling. The transfer experiments indicate that alterations associated with — and perhaps causally related to — kindling are not restricted to the stimulated focus, but also involve some synaptically remote targets. Some of these targets may reside in the brainstem (McIntyre, 1975). The inhibitory effects of hippocampal lesions (Dasheiff and McNamara, 1982; Savage et al., 1984b) on amygdala kindling development suggest that hemispheric sites remote from the stimulating electrode might be key members of the kindling network. Taken together the findings from all these experimental approaches raise the possibility that the responsible alterations reside in multiple structures within a network. Solution of this complex problem will require experiments using correlative lesion, electrophysiologic, and pharmacologic techniques. Data obtained with anatomic and biochemical techniques can guide and reinforce these approaches.

3.2. Morphologic Studies

Kindling represents a permanent and dramatic alteration of neuronal function that is likely to be accompanied by morphologic alterations at cellular or subcellular sites. That such changes do not involve overt pathologic differences between kindled and control animals such as neuronal loss, gliosis, or edema at the stimulating electrode tip has been repeatedly demonstrated (Brotchi et al., 1978; Goddard, 1972; Goddard and Douglas, 1975). Among subcellular sites at which key changes might transpire, synapses constitute a leading candidate. At least three classes of alterations could occur: formation of new synapses, loss of synapses, and alteration of existing synapses. Modification of existing synapses seems the most likely, but searches for alterations of this sort in kindled animals have been largely unsuccessful to date. A single preliminary report described ultrastructural observations in the primary and secondary cortical sites in a transcallosal kindling model (Racine and Zaide, 1978). In a blinded examination of cortical layers one and two, Racine and Zaide found bilateral 40% increases in axonal terminal size without changes in terminal number. Qualitatively, they also reported synaptic terminals "wrapping around" dendritic spines and possible increases in spine size in the kindled animals. Similar experiments in more simplified and better-defined anatomic regions (i.e., entorhinal cortex and hippocampal formation) may well disclose the altered synaptic morphology suggested in these preliminary experiments by Racine and Zaide. Interpretations of the functional significance for kindling of any such changes must be made cautiously, since evidence of modified neuronal transmission secondary to repetitive stimulation may imply only that a given synapse is active during a kindled seizure and not that the kindling phenomenon has been localized to a given set of synapses.

3.3. Cellular Electrophysiologic Studies

Long-term potentiation (LTP) is a phenomenon involving a prolonged increase in synaptic efficacy following repetitive electrical stimulation. The mechanism underlying this phenomenon remains unclear, despite extensive investigations by numerous laboratories (Douglas et al., 1982; Levy and Stewart, 1983; McNaughton, 1982). Goddard and Douglas (1975) proposed that LTP of excitatory synaptic communication is the cellular basis of kindling. Evidence supporting this hypothesis includes: (1) both kindling and LTP involve long-lasting enhanced responsiveness to a fixed electrical stimulus; (2) trains of stimulations effective in producing LTP are effective in producing kindling; and (3) kindling can be induced by repetitive stimulations of pathways that have shown potentiation.

Several laboratories have begun to test this hypothesis by measuring synaptic efficacy of monosynaptic connections in kindled animals (Maru et al., 1982; Racine et al., 1983; Giacchino et al., 1984; King et al., 1985). Two of these studies (Maru et al., 1982; Racine et al., 1983) have described development of LTP at least transiently during the development of kindling. However, experiments to date have failed to demonstrate the presence of permanent LTP *uniformly* paralleling the permanence of the kindled condition. Although disproving the hypothesis that LTP is the physiologic basis of kindling is a difficult task, the preliminary evidence suggests that mechanisms in addition to or other than LTP may be responsible. The most tenable hypothesis at present perhaps is that LTP may somehow contribute to the development of kindling rather than subserve its permanence.

3.4. Studies of Neurotransmitters

Considerable work has been focused on the biochemistry and pharmacology of central nervous system neurotransmitters in the kindling model. This section provides a critical perspective of this work.

3.4.1. Acetylcholine

Acetylcholine is a putative neurotransmitter of mammalian brain with mainly excitatory effects. These excitatory actions have been the basis of a series of pharmacologic and biochemical experiments investigating the role of cholinergic neurotransmission in kindling.

Two lines of research suggest a role for cholinergic neurotransmission in seizures. First, local injection of carbamycholine, a cholinergic agonist, into the amygdala at 48-h intervals resulted in progressively more intense seizures in a pattern similar to those of electrically kindled seizures (Vosu and Wise, 1975; Wasterlain and Jonec, 1983). Injections into the caudate and hippocampus were also effective. This kindling-like effect exhibited pharmacologic properties consistent with muscarinic receptor activation (Wasterlain and Jonec, 1983). Second, in anesthetized rats kindling-like stimulations (biphasic square wave pulses, 0.1 ms duration, 100 Hz) produced a long-lasting neuronal supersensitivity to the excitatory actions of acetylcholine. Burchfiel et al. (1979) found that a single stimulation of the fornix produced supersensitivity of hippocampal CA1 pyramidal cells to iontophoretically applied acetylcholine. The supersensitivity persisted for at least 4 h, but was preceded by a 40–60-min period of subnormal sensitivity immediately following the afterdischarge. These findings may be of significance to kindling since the ability to evoke a subsequent afterdischarge was found to be positively correlated to the sensitivity of the CA1 pyramidal cells to acetylcholine. Thus, one requisite for evoking afterdischarge in this pathway may be the sensitivity of CA1 neurons to this excitatory neurotransmitter. These findings are consistent with the hypothesis that a progressive increase in neuronal sensitivity to acetylcholine underlies the development of kindling. However, the permanence of this neuronal supersensitivity has not been determined nor has a temporal relationship between development of kindling and development of neuronal supersensitivity been established.

These studies suggest that the cholinergic system may be involved in the propagation of seizures. However, cholinergic neurotransmission does not appear to be essential to the development of kindling or the expression of kindled seizures. Systemic administration of the muscarinic antagonist atropine was reported to produce only a small increase in the number of stimulations required to kindle Sprague-Dawley rats (Arnold et al., 1973; Albright et al., 1979); in two other strains of rats, atropine had no effect on the rate of kindling (Corcoran et al., 1976; Blackwood et al., 1982). Systemic treatment with the cholinesterase inhibitor parathion did not modify the rate of kindling (Joy et al., 1981). Atropine did not suppress seizures in previously kindled animals (Arnold et al., 1973).

These pharmacologic studies suggest that muscarinic cholinergic neurotransmission is not essential for the development of
kindling or the expression of kindled seizures. However, biochemical studies of cholinergic neurotransmission do show alterations after kindling.

Activity of the synthetic enzyme, choline acetyltransferase, was
found to be decreased bilaterally in frontal cortex, amygdala, and
hippocampus of amygdala-kindled Royal Victoria hospital hooded
rats 2 wk after the final kindling seizures (Noda et al., 1982). Similarly Dasheiff and McNamara (1980) found a decrease in activity
of this enzyme in the dentate gyrus ipsilateral to the stimulated
amygdala of kindled Sprague-Dawley rats 1 d after the final kindling
stimulation. A decrease in the activity of this enzyme suggests that
kindled seizures produce a compensatory decrease in the rate of
synthesis of acetylcholine.

In other biochemical studies, the muscarinic cholinergic receptor
number was found to be decreased in brains of kindled rats (Byrne
et al., 1980; Dasheiff and McNamara, 1980; Dashieff et al., 1982;
Noda et al., 1982). However, this decrease in muscarinic receptor
number was not associated with the permanent kindled seizure-
susceptibility as the alterations disappeared soon after the final
kindling stimulation (Dasheiff et al., 1982; Blackwood et al., 1982;
Noda et al., 1982). The finding that seizures elicited by mechanisms
other than kindling similarly decreased muscarinic receptor number
(Dasheiff et al., 1982; Lerer et al., 1983) suggests that the decline
in receptor number is a phenomenon of seizures in general and
not kindling *per se*. The decrease in muscarinic receptors in the hippocampal formation has been localized to the somata and dendritic
trees of the dentate granule cells (Savage et al., 1983) and selectively involves a subpopulation of receptors with a low affinity for
the cholinergic agonist carbachol (Savage and McNamara, 1982).
The seizure-induced decline in receptor number does not depend
on extrinsic cholinergic afferents, since the decline was detectable
in animals kindled after destruction of the septal cholinergic afferents to the hippocampal formation (Dasheiff and McNamara,
1980). Instead, the seizure-induced decline in receptor number appears to be dependent on entorhinal afferents, since knife cuts of
the afferents arising from entorhinal cortex prevented the decline
(Savage et al., 1984b). This represents a form of heterologous receptor regulation, since the entorhinal afferent is not cholinergic. In
view of the excitatory actions of acetylcholine, the seizure-induced
decline in muscarinic receptor number, like the change in choline

acetyltransferase activity, may be a compensatory mechanism designed to stabilize granule cell excitability.

In summary, repeated activation of muscarinic receptors can produce a kindling-like response, and seizure propagation through the hippocampus may be affected by the sensitivity of hippocampal neurons to acetylcholine. However, blockade of these receptors has no major effect on the rate of kindling or on kindled seizures. Moreover, the biochemical changes found in the cholinergic system of kindled rats are likely to be compensatory responses to seizures in general. Thus although these studies did not elucidate the underlying basis of kindling, the results of these studies do provide insight into how the cholinergic system may function during seizures.

3.4.2. Monoamines

The catecholamine and indoleamine systems are composed of a few small brainstem nuclei containing neurons whose axons ramify widely. These axons project to large areas of the brain, including much of the limbic system, and electrophysiological evidence suggests that they are inhibitory in most target areas (Foote et al., 1983; Wang and Aghaganian, 1977; Segal and Bloom, 1976; Moore and Bloom, 1978; Moore and Bloom, 1979). In addition to directly inhibiting neurons, the biogenic amines may modulate responses to other neurotransmitters (Woodward et al., 1979; Foote et al., 1983; Waterhouse et al., 1982). With these characteristics, a small group of monoaminergic neurons could synchronously alter neuronal excitability in widespread areas and possibly regulate susceptibility to seizures. In fact the evidence presented in this section implicates both norepinephrine and serotonin in the control of kindled seizure susceptibility.

3.4.2.1. NOREPINEPHRINE. Depletion of forebrain norepinephrine (NE) dramatically facilitates kindling development. Injections of the neurotoxin 6-hydroxydopamine (6-OHDA) into the ascending noradrenergic bundle, but not into the dopaminergic fibers, accelerated the kindling process (Corcoran and Mason, 1980). Lesioned animals required less than half as many stimulations to elicit the first class 5 seizure, compared to controls. These lesions lowered forebrain levels of NE to less than 10% of control values, but did not change the dopamine (DA) content. Similarly, intracerebroventricular (icv) administration of 6-OHDA, which depleted forebrain NE to less than 20% of control levels, also reduced the number of stimulations needed to evoke the first class 5 seizure by 50% (Arnold et al., 1973; McIntyre et al., 1979; McIntyre and Edson, 1981; Corcoran et al., 1974). Electrolytic or mechanical lesions of the ascending NE bundle reduced NE levels to less than 25% of control and facilitated kindling development (Ehlers et al., 1980; Araki et al., 1983a).

A number of observations on the lesion-induced facilitation of kindling are consistent across lesioning methods. First, the degree to which kindling development is enhanced is positively correlated with the extent of NE depletion in the forebrain. At least 50% of the NE must be depleted to detect any facilitation, and subsequently the greater the depletion, the greater the facilitation (Ehlers et al., 1980; McIntyre and Edson, 1981). Second, NE depletion facilitated kindling regardless of the site of seizure initiation. Kindling is augmented to a similar extent in animals stimulated in the anterior neocortex, amygdala, or hippocampus (McIntyre and Edson, 1982; Altman and Corcoran, 1983; Araki et al., 1983a). Third, seizures elicited in the NE-depleted animals are identical to controls in duration and threshold, and the kindling effect is permanent (Mohr and Corcoran, 1981). Finally, the facilitation is most apparent in the transition from the short afterdischarges not associated with motor seizures (class 0–1) to long afterdischarges eliciting generalized behavioral seizures (classes 2–5).

Although it is clear that NE depletion of the forebrain facilitates kindling, the anatomic specificity of this effect is not known. Depletion of NE at the stimulation site (amygdala) by local injection of 6-OHDA significantly enhanced kindling; 30% few stimulations were required to kindle 6-OHDA-injected animals than vehicle-injected controls (McIntyre, 1980). The lesions reduced amygdala NE content to 30–40% of control, but did not alter local content of DA or serotonin (5-HT). Monoamine levels in the remaining portion of the brain were unchanged from control. Thus it appears that depletion of NE in the amygdala facilitates amygdala-kindling development. However, it is important to determine whether depletion of NE at the site of stimulation or whether depletion of NE in the amygdala regardless of the site of stimulation is responsible for the facilitation.

Consistent with lesion experiments, pharmacological depletion of the catecholamines and 5-HT by reserpine or the shorter-acting drug, tetrabenazine, facilitated kindling development (Arnold et al., 1973; Wilkison and Halpern, 1979). A significant reversal of the kindling effect occurred when monoaminergic function returned. Thus an animal that responded to a stimulation with a class 5 seizure while under the influence of reserpine only responded with a class 2 seizure when the effects of the drug had dissipated. Additional stimulations were required to evoke subsequent class 5 seizures. Depletion of catecholamines by treatment with the tyrosine hydroxylase inhibitor alpha-methyl-*p*-tyrosine (AMPT) during kindling produced conflicting results; it either inhibited afterdischarge development (Wilkison and Halpern, 1979), had no effect (Araki et al., 1983b), or facilitated seizure development (Callaghan and Schwark, 1979). The reason for this discrepancy is not clear. However, the results with AMPT do not diminish the significance of the overwhelming evidence that indicates that depletion of NE facilitates kindling development.

Pharmacological blockade of postsynaptic adrenergic receptors during kindling development has been used to explore the molecular mechanisms underlying the facilitation seen in NE-depleted animals. In contrast to the clear evidence obtained in the NE-depletion experiments, the data from the few studies utilizing adrenergic receptor antagonists to impair adrenergic neurotransmission demonstrate less dramatic effects on kindling development. Thus daily pretreatment with either the alpha-adrenergic antagonist, phenoxybenzamine, or the beta-adrenergic antagonist, propranolol, did not consistently alter the rate of kindling (Callaghan and Schwark, 1979; Peterson et al., 1981).

A number of explanations can be invoked to account for the inconsistency observed between the effects of pharmacological blockade of adrenergic receptors and the effects of NE depletion on kindling development. First, it is possible that coadministration of alpha- and beta-adrenergic antagonists is necessary to mimic the effect of the depletions. Second, it is possible that the adrenergic antagonist drugs were not tested at appropriate dosages or were not selective for the relevent sites of action. A third possibility is that the depletions cause some other effect not directly related to adrenergic neurotransmission mediated through alpha and beta receptors.

The ability of enhanced NE neurotransmission to influence kindling development has also been studied. Pargyline, a monoamine oxidase inhibitor, suppressed the increase in AD duration following multiple kindling stimulations (Wilkison and Halpern, 1979). Pretreatment with desmethylimipramine (DMI), a drug that blocks NE uptake into terminals and releases stored NE, also slowed kindling development (McIntyre et al., 1982a). However, both of these drugs exert similar influences over NE and 5-HT; thus the results can not be unambiguously attributed to enhanced adrenergic function. Xylazine, an alpha-adrenergic agonist with some selectivity for alpha$_2$ receptor sites, either slightly facilitated (0.3 mg/kg) or slightly suppressed (3.0 mg/kg) kindling development (Joy et al., 1983). Clonidine, an alpha agonist with similar specificity, had no effect on kindling development (Callaghan and Schwark, 1979). The interpretation of these results is complicated by the multiple influences alpha$_2$ agonists have on adrenergic neurotransmission. The beta-adrenergic drugs have not been studied to date.

In short, depletion of forebrain NE dramatically facilitates kindling development. The results of direct pharmacological manipulations of the adrenergic receptors, although not contradictory, do not clarify the mechanism by which the NE depletion acts. Further experimentation utilizing selective agonists and antagonists in appropriate dosages are required to elucidate the influence of the adrenergic system on kindling development.

Depletion of forebrain NE has no effect on seizure expression in a fully kindled animal. Injections of 6-OHDA into the ascending NE bundles of previously kindled rats did not alter the duration or the behavioral rank of subsequent seizures (Westerberg et al., 1984). Pharmacological depletion of NE by reserpine likewise failed to have a profound influence on kindled seizures. Seizures elicited in reserpine-treated animals were frequently more severe, including episodes of tonic extension of the limbs. However, there was no change in seizure duration or threshold (Arnold et al., 1973; Corcoran et al., 1974). Blockade of adrenergic receptors with the alpha-adrenergic antagonist, phenoxybenzamine, or the beta-adrenergic antagonist, propranolol, also failed to alter seizure expression (Ashton et al., 1980).

In contrast, the few studies on the effects of direct adrenergic agonists suggest that enhancing adrenergic neurotransmission may inhibit kindled seizures. Clonidine and xylazine, both alpha-agonists, decreased seizure duration by approximately 50% in response to a test stimulation (Ashton et al., 1980; Joy et al., 1983). However, lower doses of xylazine (0.3 mg/kg) had a proconvulsant effect in that it lowered the seizure threshold. Beta adrenergic agonists have not been studied.

Acute treatment with tricyclic antidepressants, such as amitriptyline, imipramine, or nortriptyline decreased seizure duration in response to a threshold stimulation (Babington and Wedeking, 1973; Knobloch et al., 1982). Since these drugs block the reuptake of both NE and 5-HT, and presumably increase the content of transmitter in the synaptic cleft, this study suggests that NE may inhibit seizure expression in a kindled animal. However, the lack of transmitter specificity precludes conclusive interpretation.

In summary, the information available supports the hypothesis that adrenergic neurotransmission is not essential for kindled seizure expression. Decreasing NE by lesions or by receptor blockade does not alter seizure expression. It is unclear whether augmenting endogenous adrenergic activity can inhibit kindled seizures. More direct investigation is needed to test this possibility.

The observation that depletion of NE facilitates kindling development, whereas fully kindled seizures are insensitive to the same treatment, suggests that the kindling process is a result of altered adrenergic neurotransmission. Accordingly, many biochemical investigations attempting to identify a molecular alteration in the adrenergic system that could differentiate a kindled brain from a control have been undertaken. No consistent changes in biochemical indices of adrenergic presynaptic function have been found in kindled animals.

Norepinephrine content has been measured in brain regions of kindled animals by several investigators. At the earliest time point examined, 7 d following the last seizure, Callaghan and Schwark (1979) reported significant decreases in NE content in the hippocampus, limbic cortex, midbrain, and frontal cortex, but no changes in the hypothalamus, brainstem, or striatum. Several investigators found unchanged levels in the amygdala, striatum, limbic cortex, and hippocampus 3–4 wk following the last kindled seizure (Engel and Sharpless, 1977; Blackwood, 1981; Burnham et al., 1981). The only long-lasting alteration documented was a 40% increase in hypothalamic NE (Burnham et al., 1981). Correlating the early declines measured after kindled seizures with alterations in NE levels seen following seizures evoked by chemoconvulsants or electroshock could help determine the contribution of seizure activity alone in producing the early declines.

Basal and potassium-evoked release of NE from nine brain regions, including the hypothalamus, was not significantly different between control and kindled animals 3 wk following the last seizure (Kant et al., 1980). Four weeks following the last seizure, tyrosine hydroxylase activity, the rate-limiting step in NE biosynthesis, did show a significant decline (33%) in the stimulated amygdala, but not in the hippocampus, hypothalamus, thalamus, striatum, or cortex (Farjo and Blackwood, 1978). The decline in tyrosine hydroxylase activity would be expected to parallel a decline in NE turnover.

However, a subsequent study by these authors performed at the same time point, 4 wk after the last seizure, found no alteration in the turnover or content of NE in the amygdala or hippocampus (Blackwood, 1981). Similarly, forebrain NE turnover was also unchanged in kindled rats 7 d after the last seizure (Wilkison and Halpern, 1980). In light of the contradictory data, the functional significance of the reduced tyrosine hydroxylase activity in the stimulated amygdala and the increased content of NE reported for the hypothalamus 4 wk after the last kindled seizure is unclear.

Long-lasting alterations in adrenergic receptor binding sites have been reported. Similar decreases in beta-adrenergic receptor antagonist binding following kindling have been described by two independent laboratories (McNamara, 1978a; McIntyre and Roberts, 1983). A 20% decrease of specific binding of (^{3}H)-dihydroalprenolol (DHA) was found in the stimulated amygdala 3 d following the last kindled seizure (McNamara, 1978a). Alterations in binding could not be detected in the surrounding structures or in the amygdalas of electrode-implanted unstimulated controls. The decline was not detectable at the earliest time point examined, 15 h following the last seizure, was decreased by 20% at 3 d and then returned to baseline levels by 7 d (Byrne et al., 1980; McNamara, 1978b; McNamara et al., 1980). Binding of DHA was also found to be decreased in the amygdala and frontal cortex of kindled animals 23 d after the last seizure when measured by McIntyre and Roberts (1983). No other time points were examined; thus reversal was not demonstrated. This long-lasting reduction in beta-adrenergic receptor binding sites could be partially reversed by a single kindled seizure 1 d prior to sacrifice, a result that suggests that seizure activity can alter beta-adrenergic receptors in specific areas of the brain. Consistent with this hypothesis is the observation that repeated administration of electroconvulsive shock decreases the beta-adrenergic receptor density in the cortex and hippocampus, but not in striatum, cerebellum, or hypothalamus (Kellar et al., 1981). The variable time course of receptor binding declines between the two studies may be caused by the different interseizure intervals used to induce kindling. The meaning of the receptor alteration remains unknown. However, its transient nature suggests that it alone can not account for the permanence of kindling. The necessity of experiencing multiple seizures over a relatively long period of time in order to induce the change suggests that seizure activity alone also can not be responsible.

In summary, there have been no consistent, long-lasting changes in the presynaptic indices of adrenergic function described to date. Postsynaptically, the number of beta-adrenergic receptors in the amygdala and frontal cortex do appear to decrease transiently in response to kindled seizures.

3.4.2.2. DOPAMINE. Pharmacological manipulations of the dopaminergic system do not consistently alter the rate of kindling development. Impairment of DA neurotransmission by selectively depleting DA with direct injection of 6-OHDA into the ascending mesocortical dopaminergic projection of animals pretreated with DMI did not alter the rate of kindling (Corcoran and Mason, 1980). Although this dopamine depletion was incomplete, equivalent decreases in NE did facilitate kindling development (Ehlers et al., 1980; McIntyre and Edson, 1981). In contrast, daily pretreatment with low doses of the DA receptor antagonists, haloperidol or pimozide (<0.5 mg/kg), facilitated kindling development (Gee et al., 1983; Sato et al., 1979). Paradoxically, neither higher doses (0.5–10.0 mg/kg) of haloperidol (Le Gal La Salle, 1980; Gee et al., 1983; Babington and Wedeking, 1973) nor daily administration of a dopaminergic agonist, apomorphine or bromocriptine (Callaghan and Schwark, 1979; Gee et al., 1983; Stock et al., 1983), changed the number of stimulations required to elicit the first class 5 seizure. The absence of a dose-dependent effect of acute administration of DA receptor antagonists and the lack of effect of neurotoxin-mediated destruction of DA neurons suggests that reduction of DA neurotransmission has no major effect on kindling development.

Little is known about how drugs acting upon the dopamine system affect seizure expression in kindled animals. The few studies done to date have quantified AD duration in response to a supra-threshold stimulation of unknown relation to the threshold, a relatively insensitive method of detecting drug effects. Apomorphine in low doses (<1 mg/kg) did not change the motor response to a test stimulation (Corcoran et al., 1974; Ashton et al., 1980) and haloperidol (1–10 mg/kg) had no effect or very slightly decreased the seizure response (Babington and Wedeking, 1973; Ashton et al., 1980; Stock et al., 1983). Even though maximally sensitive test methods were not employed, the negative results tentatively suggest that DA is not important in the expression of a kindled seizure.

A number of studies have sought kindling-induced alterations in biochemical indices of DA function. Regional measurements of DA content, release, and turnover produced partially conflicting information. At 1 (Stock et al., 1983) and 7 d (Callaghan and Schwark, 1979) following the last seizure, no changes in DA content were observed in the hippocampus, striatum, hypothalamus, amygdala, or limbic cortex. A significant decline in DA turnover rate was reported for the entire forebrain 7 d following the last seizure (Wilkison and Halpern, 1980). At 3–4 wk following the last seizure, a decline in DA content was reported for the hypothalamus (Burnham et al., 1981), and either no change (Blackwood, 1981) or a decline (Engel and Sharpless, 1977) observed in the stimulated amygdala. However, at the same time point, 1 mo postkindling, there was no difference reported between control and kindled animals in DA turnover rate in the amygdala (Blackwood, 1981). Additionally, basal DA release from the hypothalamus and amygdala were not altered 1 mo after the last kindled seizure nor was potassium-stimulated release affected (Kant et al., 1980). The measurements of DA content in hippocampus, striatum, and limbic cortex were unchanged from control at 1 mo postkindling (Burnham et al., 1981). Thus, no compelling evidence of permanent changes in the presynaptic indices of DA neurotransmission was documented.

Postsynaptic DA function has been assessed with receptor binding and DA-stimulated adenylate cyclase measurements. Dopamine receptor binding, as measured with (^{3}H)-haloperidol in membranes prepared from striatum of kindled rats sacrificed 30 min after the last seizure, was not different from implanted, handled controls (Ashton et al., 1980). Using (^{3}H)-spiroperidol, Gee et al. (1979) found significant but complex alterations in receptor binding in membranes prepared from the amygdala and frontal cortex, but not from the striatum of kindled animals 1 d following the last seizure. Gee et al. (1980) also studied the effects of kindling on regional DA-sensitive adenylate cyclase (AC) activity. At 3–4 h following the last seizure, basal and DA-stimulated AC activity in membrane preparations from amygdala and frontal cortex, but not striatum, were decreased by 25–40% compared to operated controls. The percent increases caused by DA stimulation did not change from control values. The regionally specific declines in DA receptor binding paralleled the declines in DA-sensitive AC activity. In neither the binding nor adenylate cyclase studies were later time points examined, nor were controls for recent seizure experience included. Thus it is not possible to determine whether the alterations measured were specific to kindling or whether they were a response to multiple seizures.

In summary, the data on the whole do not appear to support the hypothesis that dopaminergic neurons are important in the development or permanence of kindling or in the expression of a kindled seizure.

3.4.2.3. SEROTONIN. In the experiments reported to date, the rate of kindling development is not altered by manipulations of serotonergic neuronal function. Discrete electrolytic lesions of the midbrain raphe nuclei had no effect on the number of stimulations required to produce either hippocampal or amygdala kindling (Araki et al., 1983a). These lesions lowered 5-HT content to 25–40% of control values in forebrain regions, but did not appreciably alter NE levels. A second study used radiofrequency lesions of the raphe nuclei and decreased by 35% the number of amygdaloid stimulations required to kindle (Racine and Coscina, 1979). However, these radiofrequency lesions encompassed a very large area of the midbrain and the pons. Thus, the absence of a lesion-control group with comparable tissue destruction involving surrounding structures but sparing the raphe makes it impossible to sort out raphe-specific effects from effects caused by involvement of adjacent structures and from nonspecific lesion effects.

Pharmacological reduction of serotonergic neuronal function by destruction of 5-HT neurons with icv injections of the neurotoxin 5,7-dihydroxytryptamine (Araki et al., 1983b) or depletion of 5-HT with the synthesis inhibitor *p*-chlorophenylalanine (PCPA) prior to the initiation of kindling had no effect on the number of stimulations required to elicit the first class 5 seizure (Wilkison and Halpern, 1979; Racine and Coscina, 1979; Araki et al., 1983b). Repeated administration of the PCPA daily 24 h prior to each stimulation produced conflicting results; both an increase (Stach et al., 1981) and a decrease (Munkenbeck and Schwark, 1982) in the rate of kindling acquisition have been reported. Overall, reduction of 5-HT neurotransmission has no consistent effect on kindling development.

In two experiments designed to augment serotonergic neuronal function, no consistent effects on kindling development have been demonstrated. Daily injections of the 5-HT precursor 5-hydroxytryptophan (5-HTP) facilitated kindling in rats (Munkenbeck and Schwark, 1982), but not in rabbits (Stach et al., 1981) and daily pretreatment with the direct 5-HT agonist quipazine had no effect on seizure acquisition in rats (Bowyer, 1982).

In conclusion, impairment of serotonergic neurotransmission by selective depletion of 5-HT or destruction of the raphe nuclei does not affect kindling development. Similarly, the available evidence suggests that enhancement of 5-HT neurotransmission also fails to influence kindling development.

Pharmacological manipulations of the serotonergic system have produced conflicting results in experiments on fully kindled animals. At least two factors contribute to these inconsistencies. First, the drugs available to study 5-HT are not as selective as those available for other neurotransmitter systems. Second, there appears to be a species difference in that rabbits are more sensitive to the 5-HT drugs than are rats. Augmenting 5-HT neurotransmission with high dose 5-HTP (25 mg/kg in rabbits, 100 mg/kg in rats) slightly increased AD duration in both species (Stach et al., 1981; Ashton et al., 1980). Treatment with a variety of direct-acting receptor agonists prior to a test stimulation has given conflicting results. Femoxetine enhanced seizure expression in rabbits as evidenced by an increased AD duration (Stach et al., 1981), but quipazine had no effect in rabbits. Fluoxetine, a 5-HT uptake inhibitor, had an anticonvulsant effect in that it increased AD threshold in rats (Siegal and Murphy, 1979). Depletion of 5-HT with PCPA decreased the behavioral rank of kindled seizures in rabbits, but had no effect in rats (Stach et al., 1981; Siegel and Murphy, 1979). Blockade of 5-HT receptors with the antagonists cyproheptadine in rabbits (Stach et al., 1981) and metergoline or mianserin in rats (Ashton et al., 1980) decreased motor seizure duration by 20–30%.

In contrast to the ambiguous results of pharmacologic augmentation of 5-HT neurotransmission, electrical stimulation in the area of the raphe nucleus, but not in adjacent sites, produced a profound anticonvulsant effect, quantified both as an increase in seizure threshold in cats (Siegal and Murphy, 1979) and as a decrease in AD duration after a threshold stimulation in rats (Kovacs and Zoll, 1974). The method of electrical stimulation employed, low-frequency pulses for 1 h prior to testing, has been previously demonstrated to increase 5-HT release in the forebrain (Aghajanian et al., 1967; Shields and Eccleston, 1972). These studies suggest that the 5-HT-rich median and dorsal raphe nuclei may be capable of exerting inhibitory influences over the structures responsible for kindled seizures. If this is true, then the ambiguities of the pharmacological studies may indicate that the specific drugs tested were unable to effectively regulate endogenous 5-HT neurotransmission.

Limited work has been done to elucidate biochemical alterations in the 5-HT system in kindled animals. Decreased 5-HT and 5-HIAA content in the midbrain, but not hypothalamus, amygdala, or brainstem, 7 d after the last seizure has been reported (Munkenbeck and Schwark, 1982). Since no later time points were examined, it is unknown if the change is permanent. The observation that 30 min following the last kindled seizure (^{3}H)-spiperone binding was no different from control in membranes prepared from hippocampus, frontal cortex, or olfactory area (Ashton et al., 1980) is likewise potentially reflecting the status of the 5-HT system immediately postictally, rather than examining any long-term sequelae of the kindling process.

The available information suggests that serotonergic neurotransmission is not essential for the development, expression, or permanence of kindling. Of significant note is the observation that augmentation of the dorsal and median raphe nuclei activity by electrical stimulation can impair seizure expression. It is possible that enhancement of serotonergic function may have anticonvulsant affects. Further work is clearly indicated to test this possibility.

3.4.3. Opioid Peptides

Endogenous opioid peptides are present in neurons in many nuclei of the limbic system and basal ganglia including nucleus accumbens, caudate, putamen, substantia nigra, amygdala, hippocampal formation, pyriform cortex, and entorhinal cortex (Watson et al., 1982). Since many of these same regions have been implicated in kindling, the question naturally arises as to whether the development or the expression of kindled seizures can be modified by opiate administration. In fact, under various experimental conditions, including kindling, opiates and the opioid peptides may have convulsant, proconvulsant, or anticonvulsant effects. Key variables include the route of drug administration, the particular seizure model employed, the receptor selectivity of the ligand chosen, and the drug dose and have been reviewed elsewhere (McNamara et al., 1985).

3.4.3.1. EFFECTS OF OPIATES ON THE DEVELOPMENT OF KINDLING. In most studies, systemically administered opiate agonists and antagonists have shown little effect on the development of kindling. In particular, the development of amygdaloid afterdischarges and the number of stimuli required to produce kindling were unaffected by analgesic doses of morphine [10 mg/kg (Post et al., 1979)]. However, this dose was lower than the doses that have generally had either proconvulsant (Foote and Gale, 1983) or anticonvulsant effects in other seizure models (Cowan et al., 1979; Foote and Gale, 1983; Urca and Frenk, 1980; 1982). Additional studies using higher doses (50 mg/kg, ip) would be necessary to rule out the possibility that kindling development can be regulated by morphine administration. However, at such doses morphine could be acting at sites other than mu opiate binding sites, so the mechanism of its effect would need to be examined carefully.

The development of amygdaloid kindling has also been studied in rats receiving either the mixed agonist–antagonist cyclazocine (mu antagonist, kappa and sigma agonist) or the kappa agonist ethylketocyclazocine [EKC; actually a kappa, mu, and delta agonist (Chang et al., 1981)] prior to each of the first ten kindling stimuli only. Although the cyclazocine-treated group showed significant retardation in the development of afterdischarges and in the severity of behavioral seizures, two effects that might have resulted from the drug's elevation of seizure threshold, there was no difference in either drug group in the total number of stimuli required to kindle the animals (Albertson et al., 1984). It is possible that continued treatment prior to each kindling stimulus would retard kindling development.

In contrast to the limited studies of the effects of opioid agonists on kindling development, a number of groups have shown that naloxone does not alter the development of afterdischarges recorded from the amygdala during kindling (Corcoran and Wada, 1979; Hardy et al., 1980; Post et al., 1979). Although some experiments have shown a decrease in the number of stimuli required to produce kindling, this effect is neither dose-dependent nor consistently reproducible (Corcoran and Wada, 1979; Hardy et al., 1980; Post et al., 1979).

In summary, there seems to be little evidence that the development of kindling can be systematically manipulated by systemic administration of either opiate agonists or antagonists, although the number of studies utilizing agonists is quite limited. It is also important to note that systemic administration of drugs with multiple, possibly mutually antagonistic, sites of action is not necessarily a sensitive tool for detecting the presence of localized changes in opioid-mediated neuronal communication.

3.4.3.2. EFFECTS OF OPIATES ON KINDLED SEIZURES. The effects of systemically administered morphine on kindled seizures are dose-dependent, with little effect demonstrated at low doses (Caldecott-Hazard et al., 1982; Frenk et al., 1979; Hardy et al., 1980; Le Gal La Salle et al., 1977; Stone et al., 1982). At higher doses (20–50 mg/kg, ip), animals displayed fewer class 4 or 5 seizures (Albertson et al., 1984; Caldecott-Hazard et al., 1982; Le Gal La Salle et al., 1977). However, this behavioral effect was not accompanied by a consistent alteration in the electrical component, since both prolongation (Caldecott-Hazard et al., 1982; Le Gal La Salle et al., 1977) and reduction (Albertson et al., 1984) of the afterdischarge durations were reported. These effects were stereospecific (Le Gal La Salle et al., 1977), reversed by naloxone, and not associated with a simple elevation in the afterdischarge threshold (Albertson et al., 1984). The cataleptic state, but not the alterations in afterdischarge duration, were also produced by LY146104, an enkephalin with a high affinity for mu binding sites (Caldecott-Hazard et al., 1982).

The effects of delta- and epsilon-specific agonists have not been examined. However, there is a single study on the effects of kappa and sigma opiate agonists and mixed agonist–antagonists (Albertson et al., 1984). The kappa agonists EKC and ketocyclazocine reduced behavioral severity and variably reduced afterdischarge duration. Since these results were similar to those of morphine in the same study, and since these ligands bind to mu receptors (Chang et al., 1981), it is possible that these drugs were acting at least partially through mu binding sites. However, the effects of the kappa ligands were only naloxone-reversible when supra-threshold (not threshold) stimuli were used. The sigma agonist SKF 10,047 decreased the afterdischarge duration, and its effects were reversed by naloxone. The mixed agonists–antagonists cyclazocine and cyclorphan elevated seizure threshold and decreased the behavioral seizure without affecting the afterdischarge duration. Except for cyclorphan's effect on seizure threshold, these effects were not naloxone-reversible, but were in fact potentiated by naloxone (Albertson et al., 1984).

Finally it should be pointed out that the effects of naloxone alone are not completely defined. In general, naloxone has no effect on the ictal events after a single stimulus (Albertson et al., 1984; Frenk et al., 1979; Kelsey and Belluzzi, 1982). However, in a different strain of rats, naloxone decreased both the severity of the behavioral seizure and the afterdischarge duration after a single stimulus (Stone et al., 1982). These results in fully kindled animals are in contrast to those obtained during the development of kindling, where, in some experiments, naloxone reduced the number of amygdala stimulations required to produce class 5 seizures (Corcoran et al., 1979; Hardy et al., 1980).

In summary, there is some evidence that systemically administered morphine can modify the behavioral expression of fully kindled seizures by decreasing the clonic component without a corresponding effect on the afterdischarge duration. Morphine's site of action in producing this effect is unknown. The effects of delta and epsilon agonists are unknown, and studies of kappa and sigma agonists have just begun.

3.4.3.3. EFFECTS OF OPIATES ON POSTICTAL EVENTS. Perhaps the most striking effects of systemically administered morphine on kindling are the prolongation of the postictal behavioral and EEG depression and the increase in interictal and postictal spike frequency. These occurred at analgesic doses that had no effect on the rate of kindling development or on the expression of kindled seizures (Albertson et al., 1984; Frenk et al., 1979; Hardy et al., 1980; Le Gal La Salle et al., 1977; Stone et al., 1982). The increased spike frequency correlated with the catalepsy produced by the opiate and was naloxone-reversible (Albertson et al., 1984; Le Gal La Salle et al., 1977; Stone et al., 1982). The kappa ligands EKC and ketocyclazocine, which also bind to mu receptors, had effects similar to those of morphine (Albertson et al., 1984).

These data raise the possibility that endogenous opioids may be released during kindling seizures and be responsible for the various postictal events (Frenk et al., 1979). If this hypothesis is true, one would predict that opiate antagonists would prevent postictal depression, and in some (Frenk et al., 1979; Stone et al., 1982) but not all (Albertson et al., 1984; Caldecott-Hazard et al., 1983; Hardy et al., 1980) studies, naloxone alone has indeed significantly shortened the postictal depression. The duration of the postictal depression was also decreased in morphine-tolerant animals and after repeated elicitation of full behavioral convulsions (Yitzhaky et al., 1982). Naloxone also shortened the postictal inhibition of subsequent kindled seizures when stimuli were given at 10-, but not 60-min, intervals (Kelsey and Belluzzi, 1982).

In summary, morphine, a mu agonist with little or no effect on the rate of kindling or on the expression of fully kindled seizures, exacerbates the spontaneous interictal spiking and the postictal behavioral depression seen after kindled seizures, and does so in a naloxone-sensitive fashion. Furthermore, the behavioral arrest, twitching, and wet-dog shakes seen after icv morphine (Frenk et al., 1978; Henriksen et al., 1978) are similar to the postictal behavior after a kindled seizure. Thus, it seems quite possible that endogenous opioids may mediate the postictal behavioral depression and interictal spiking found in kindled animals. The site or sites of this postulated enhanced opioid neuronal communication remain to be determined.

3.4.3.4. BIOCHEMISTRY OF OPIOIDS. Numerous brain regions that are likely to be important in the development and maintenance of kindling contain one or more endogenous opioid peptides. These include the nucleus accumbens, caudate, putamen, substantia nigra, amygdala, hippocampal formation, pyriform cortex, and entorhinal cortex (Watson et al., 1982). Radioimmunoassays of opioid peptide levels have therefore been performed on samples from most of these areas after amygdala-kindling.

Levels of both met- and leu-enkephalin were elevated 40% in both cerebral hemispheres 28 h after the last kindled seizure (Vindrola et al., 1981a). The rise in leu-enkephalin levels during the kindling process began first in the stimulated hemisphere and then spread bilaterally, whereas met-enkephalin levels were unaltered until the animals had experienced at least five class 5 kindled seizures (Vindrola et al., 1981b). Similarly, levels of met^5-enkephalin-arg^6-gly^7-leu^8, an enkephalin peptide that is contained in the preproenkephalin A precursor molecule and found in neurons containing met- and leu-enkephalin, were elevated 70 and 100% in hippocampus and substantia nigra, respectively, at 24 h after the last kindled seizure (Iadarola et al., 1986).

Using specimens from amygdala-kindled rabbits, Przewlocki et al. (1983b) found three- to four fold elevations in hippocampal formation of immunoreactive dynorphin A 1-13 and smaller increases in alpha-neoendorphin and leu-enkephalin, but not beta-endorphin, 24 h after the last kindled seizure. Hippocampal dynorphin A levels remained elevated 1 mo later. The change was specific for hippocampus, with no changes in dynorphin immunoreactivity in striatum, hypothalamus, pituitary, thalamus, medulla/pons, cortex, or cerebellum. Whether the elevations in leu-enkephalin and alpha-neoendorphin levels demonstrated at 24 h showed similar persistence and spatial selectivity was not determined.

In contrast to the result in rabbits, hippocampal dynorphin A1-8 levels were actually decreased in rats 24 h after the last kindled seizure, with no change in substantia nigra (Iadarola et al., 1986). Whether this change is permanent, and whether it reflects differences between dynorphin A1-8 and dynorphin A1-13, other methodological differences, or species differences remains to be determined.

The kindled-seizure-related increases in many, but not all, opioid peptides could be associated with the kindling phenomenon itself. However, the possibility that they are responses to repeated seizures must be seriously considered since hippocampal dynorphin immunoreactivity is increased 24 h after intra-amygdala-kainic acid-induced seizures (Lason et al., 1983) and since met-enkephalin levels are increased in numerous brain regions in many seizure models for up to 2 wk (Green et al., 1978; Hong et al., 1979; 1980; Lason et al., 1983).

There is only one study of opiate receptors in kindling. In that experiment, binding to mu and delta opiate receptors was examined autoradiographically in the rat hippocampal formation 24 h, 7 d, and 28 d after the last amygdala-kindled seizure (Crain et al., 1985). Specific binding of the mu ligand (^{125}I)-FK 33824 was transiently decreased by approximately 20–25% in many lamina, particularly in CA1, 24 h after the last kindled seizure, but returned to control values by 7 d. Decreases in specific binding to delta receptors by the delta agonist (^{125}I)-DADL in the presence of the unlabeled mu agonist PL-032 were found only in CA1 and the dorsal blade of the dentate gyrus at 24 h. At 7 d, however, binding was decreased throughout the hippocampal formation. By 28 d, it returned to control values. Since the decreased binding to mu and delta receptors is transient, it is more likely to be a response to the seizures rather than part of the mechanism underlying the permanent kindled state. However, this hypothesis would need to be tested by examining binding after seizures produced by other methods.

In order to interpret the functional significance of the various changes in levels of opioid peptides and their binding sites in a given brain region, the electrophysiological effect of the peptide and its rates of turnover and release must also be known. For example, iontophoresis of enkephalins into hippocampal slices increases pyramidal cell firing (Dingledine, 1981; Zieglgansberger et al., 1979) and hippocampal leu-enkephalin levels are elevated after amygdala kindling (Przewlocki et al., 1983b). If this elevation reflects continued increased synthesis and release, leu-enkephalin may be playing a convulsant role in the hippocampus of kindled animals. Since leu-enkephalin binds to both mu and delta receptors, the transient decreases in mu and delta binding (Crain et al., 1985) may reflect down-regulation of the receptor in response to an increased efflux of enkephalin during kindled seizures. If, however, the elevation in leu-enkephalin levels represents primarily decreased release, leu-enkephalin might be part of an anticonvulsant mechanism designed to decrease pyramidal cell firing.

Determining the functional significance of the dramatic permanent increase in rabbit hippocampal dynorphin content (Przewlocki et al., 1983b) is important because, to date, this is one of the few long-lasting biochemical alterations documented in kindling. However, the task is hampered by the lack of sufficient data. The first issue is how increased dynorphin levels reflect changes in synthesis and release or both. The second is whether there is an associated receptor change. Dynorphin is found principally in the mossy fibers (McGinty et al., 1982), but mu and delta opiate binding sites are relatively scarce in stratum lucidum (Crain et al., 1985) the principal site of mossy fiber terminals (Meibach and Maayani, 1980), and kappa binding sites are preferentially located in stratum moleculare of the guinea pig dentate gyrus (Foote and Maurer, 1982). The third problem is whether dynorphin increases or decreases pyramidal cell firing (Henriksen et al., 1982; Walker et al., 1982) or acts to raise seizure thresholds (Przewlocka et al., 1983a). Finally, there is the question of whether all of dynorphin's effects, particularly its inhibitory effects, are mediated through opiate receptors (Henriksen et al., 1982; Walker et al., 1982).

In summary, biochemical studies of endogenous opioid levels in kindled seizures show persistent elevations in dynorphin immunoreactivity in hippocampal formation. Alpha-neo-endorphin levels, but not beta-endorphin levels, are also elevated for at least 24 h after the last kindled seizure. Studies of mu and delta opiate receptors show transient decreases in binding that return to control levels 7 d and 28 d respectively, after the last kindled seizure. How these ligand and receptor changes are related to alterations in opioid neuuronal communication will require further experiments. Pharmacologic experiments suggest that the altered neuronal communication is most likely associated with postictal events, rather than the kindled seizures themselves.

3.4.4. Somatostatin

Somatostatin is a peptide contained in discrete neuronal populations in mammalian brain. This peptide exhibits excitatory actions upon iontophoretic application to CNS neurons (Havlicek and Friesen, 1979).

One pharmacologic study suggests that somatostatin may contribute to the expression of kindled seizures (Higuchi et al., 1983). Systemic administration of cysteamine (beta-mercaptoethylamine), a compound that depletes CNS somatostatin (Sagar et al., 1982), suppressed kindled seizures at 24 and 48 (but not 4) h after administration. This anticonvulsant action diminished over the ensuing 8 d. The effect of cysteamine may be related to the depletion of brain somatostatin, since the time course of seizure suppression coincides with the apparent time course of the depletion. Intracerebroventricular administration of anti-sera raised against somatostatin also suppressed kindled seizures, but to a much lesser extent. Although preliminary, these findings suggest that somatostatin may be involved in seizure expression in this model.

Biochemical studies demonstrated increased somatostatin immunoreactivity (ranging from 20 to 100%) in animals sacrificed 2 mo following establishment of amygdaloid kindling (Kato et al., 1983). These elevations were present in the amygdala, sensorimotor cortex, and pyriform cortex bilaterally, as well as in entorhinal cortex ipsilateral to the stimulated amygdala. If this increased immunoreactivity indeed reflects increased numbers of somatostatin molecules, the time course raises the possibility that somatostatin is somehow related to the maintenance of kindling. To examine the role of somatostatin in kindling it will be necessary to know the consequences of pharmacologic manipulations of somatostatin on kindling (e.g., the use of other drugs, demonstration of dose dependency, and so on) and to determine whether the altered steady-state content reflects increased or decreased turnover and release of somatostatin.

3.4.5. Excitatory Amino Acids

Acidic amino acids are distributed throughout the brain and have predominately excitatory actions. Glutamate and aspartate are two dicarboxylic amino acids that have been implicated in certain seizure disorders. These findings form the basis of several investigations into the role of excitatory amino acids in kindling.

Initial pharmacologic studies demonstrated that icv administration of glutamate or aspartate amino acid antagonists suppressed kindled seizures. Of four amino acid analogs tested, 2-amino-5-phosphono-valeric acid and 2-amino-7-phosphonoheptanoic acid were the most effective in suppressing seizure duration and intensity. However, since information regarding the amount of current used relative to threshold, the reversibility of the effect, and the effects of vehicle alone were not provided in this report, these results must be considered preliminary (Peterson et al., 1983).

Kindling of the angular bundle is associated with a 40% increase in the binding of (^{3}H)-glutamate to a quisqualate-sensitive site in hippocampal synaptic membranes (Savage et al., 1982; 1984c). This increase was transient since it was found in animals sacrificed at 24 h, but not 28 d, after completion of kindling. No changes were found in (^{3}H)-glutamate binding to an ibotenate-sensitive site. The transient nature of the increase in glutamate binding indicates that this alteration does not subserve the maintenance of kindled seizure susceptibility (Savage et al., 1984c). Whether it might contribute to the development of kindling or represent a consequence of stimulation-induced seizures is presently unknown.

Kainic acid is an excitatory dicarboxylic amino acid capable of producing limbic seizures upon administration to animals. Displacement studies have disclosed the presence of kainic acid binding sites that possess characteristics compatible with those of receptors that mediate the neuroexcitatory and convulsive actions of this compound. In the hippocampal formation, these binding sites are most abundant by far in stratum lucidum of area CA3, the lamina defined by termination of the dentate granule cell axon terminals. Savage et al. (1984a) used quantitative autoradiographic methods to demonstrate a 50% reduction in the specific binding of (^{3}H)-kainic acid in stratum lucidum of CA3 1 d following completion of kindling from either amygdala or angular bundle. In view of the potent neuroexcitatory action of kainic acid on CA3 hippocampal pyramidal cells, a decrease in the number or affinity of kainic acid receptors might be expected to reduce hippocampal excitability and thus inhibit the development of subsequent seizures.

Localization of kainic acid receptors in the CA3 area (i.e., on mossy fiber terminals or pyramidal cell dendrites) and identification of the nature and function of the endogenous compound that interacts with kainic acid receptors will shed light on the functional significance of this receptor alteration. This information, together with knowledge about the time course and permanence of the altered binding and the effects of seizures other than those induced by kindling, will be necessary to understand the role of kainic acid receptors in kindling.

3.4.6. GABA

γ-Aminobutyric acid (GABA) is probably the principal inhibitory neurotransmitter in mammalian brain. Benzodiazepines are potent anticonvulsants that exert their effets, at least in part, through an interaction with benzodiazepine receptors (BZR). Extensive evidence suggests that BZRs are coupled to GABA receptors (GABA-R) in a macromolecular complex that also includes a chloride ion channel; benzodiazepines enhance GABA's ability to open the chloride ion channel. The BZR–GABA-R complex also contains the barbiturate receptor site. Barbiturates also enhance GABA's opening of the chloride ion channel. Many studies have focused their attention on the pharmacology and biochemistry of these receptor systems in kindling, under the premise that kindling may induce, or be the result of, a change in inhibitory synaptic function.

3.4.6.1. PHARMACOLOGY. The function of the BZR–GABA-R complex can be pharmacologically manipulated in a variety of ways. Enhancement of GABA neurotransmission can be achieved indirectly by administration of benzodiazepines and barbiturates. In addition, GABA-mediated neurotransmission can be directly enhanced in two ways: (1) administering GABA receptor agonists, and (2) increasing available GABA by inhibiting its degradative enzyme, GABA transaminase.

Benzodiazepines and pentobarbital slowed the rate of kindling development (Wise and Chinerman, 1974; Racine et al., 1975; Kalichman et al., 1981). The delay caused by the benzodiazepines could be reversed by a specific benzodiazepine receptor antagonist CGS 8216 (Robertson et al., 1984), a result that suggests that the anticonvulsant actions are exerted at benzodiazepine receptors. Use of a single high dose (1500 mg/kg) of γ-vinyl GABA (GVG), a GABA transaminase inhibitor, retarded kindling development assessed during the period when GVG increases nerve terminal GABA (Shin et al., 1984). Use of lower doses with daily dosage intervals and a different kindling paradigm did not affect the development of kindling (Kalichman et al., 1981).

Benzodiazepines suppressed fully kindled seizures (Babington and Wedeking, 1973; Wise and Chinerman, 1974; Racine et al., 1975; McIntyre et al., 1982b). This suppression was reversed by a benzodiazepine receptor antagonist, Ro 15-1788 (Albertson et al., 1982; Le Gal La Salle and Feldblum, 1983a). Pentobarbital also blocked fully kindled seizures, potentially through an action at the barbiturate receptor site coupled to GABA receptors (Kalichman et al., 1982; Bowyer et al., 1983).

Systemic treatment with progabide, a GABA receptor agonist, suppressed fully kindled seizures at high doses (Joy et al., 1984), but not at lower doses (Kalichman et al., 1982). Systemic treatment with the GABA transaminase inhibitors GVG, γ-acetylenic GABA (Myslobodsky et al., 1979; Myslobodsky and Valenstein, 1980; Kalichman et al., 1982; Bowyer et al., 1983), and aminooxyacetic acid (Le Gal La Salle, 1980) likewise suppressed the clonic motor component of fully kindled seizures. However, disagreement exists as to whether systemic GVG suppresses focal afterdischarges and limbic seizures in this model. In our hands, with careful attention to the relation of test current strength to generalized seizure threshold, and with adequate drug doses at appropriate intervals, limbic as well as motor kindled seizures were suppressed by systemic GVG (Shin et al., 1984).

The anticonvulsant efficacy of GABA-enhancing agents has also been assessed with focal intracerebral applications. Microinjection of either muscimol or GVG into the substantia nigra suppressed both limbic and clonic motor seizures triggered from the amygdala, olfactory structures, or entorhinal cortex (Le Gal La Salle et al., 1983b; McNamara et al., 1984; 1983). Microinjection of GVG into the amygdala suppressed kindled seizures triggered by hippocampal stimulation (Le Gal La Salle and Feldblum, 1983b), although, in this study, animals served as their own controls. Concurrent saline injection controls were not used. Microinjection of gabaculine, a GABA transaminase inhibitor, into substantia inominata ipsilateral to the stimulated amygdala was reported to suppress amygdala-kindled seizures (Okamoto and Wada, 1984). This study, however, failed to demonstrate reversibility and spatial selectivity adequately. Together these data further support the idea that pharmacologic enhancement of GABA-mediated neurotransmission suppresses both limbic and motor seizures.

Inhibiting GABA-mediated neurotransmission by the antagonists bicuculline or 3-mercaptopurine or picrotoxin did not affect the fully kindled seizure response, but the testing paradigm was not designed to detect proconvulsant effects (Bowyer et al., 1983; Kalichman et al., 1982).

Taken together, the data on benzodiazepines, barbiturates, and drugs acting directly on GABA receptors or indirectly through inhibiting GABA inactivation support the idea that enhanced GABA transmission will suppress both kindled seizures and kindling development. This, in turn, supports the hypothesis that GABA-mediated neurotransmission may play an important role in the mechanisms underlying kindling.

3.4.6.2. BIOCHEMISTRY. Based on the pharmacologic data implicating the GABA neuronal system in the kindling model, many studies have searched for biochemical changes that can underlie the development and/or permanence of kindling.

Kindled seizures caused transient increases (40%) of BZR binding in the hippocampal formation of kindled rats (Shin et al., 1983; Valdes et al., 1982). These increases were probably caused by seizures themselves and did not represent a molecular alteration specific to kindling, since electroshock seizures also induced these changes and since the increases were no longer present at 7 and 28 d after the last kindled seizure. These increases were localized to a distinct neuronal population, i.e., dentate granule cells, on the basis of data from radiohistochemistry and lesion studies (Valdes et al., 1982).

Tuff et al. (1983a) reported increased BZR binding in hippocampus and amygdala and decreased binding in caudate 2 wk after the last kindled seizure. However, Niznik et al. (1983) found no change in BZR binding in amygdala, hippocampus, or striatum 2 wk after the last kindled seizure; reductions of BZR binding were found in the ipsilateral frontal cortex and hypothalamus at this point and also 60 d after the last kindled seizure (Burnham et al., 1983). These conflicting results may be caused in part by differences in the number of motor seizures before sacrifice, in areas that are sampled with different dissection methods, in membrane washing methods, and in timing of sacrifice. Furthermore, the effects of seizures themselves were not examined by these authors.

Radiohistochemical studies of GABA-R disclosed a 20% increase in (^{3}H)-muscimol binding in the dentate molecular layer of hippocampus of rats sacrificed 24 h after the last kindled seizure (Shin et al., 1983). The distribution of the increase in silver grain counts in the hippocampal formation coincided with the somata and dendrites of the granule cells and thus was likely localized to these cells. However, the increase was transient, paralleling the time course of BZR binding alterations, and therefore was probably secondary to seizures. Similarly, Tuff et al. (1983b) found no alteration in (^{3}H)-GABA or (^{3}H)-muscimol binding in any region using well-washed membranes 2 wk after the last kindled seizure. This is consistent with the idea that the increase at 24 h is transient.

Since GABA increases the affinity of BZR for benzodiazepines (Karobath and Sperk, 1979), the interaction between GABA and BZR binding has also been examined in kindled animals. Enhancement by GABA of BZR binding in dentate membranes from kindled rats was attenuated in animals sacrificed 24 h after the last kindled seizure (Fanelli and McNamara, 1983). This reduced responsiveness of BZR binding to GABA stimulation seems paradoxical since (^{3}H)-muscimol binding to the GABA-R in dentate was increased at the same time. However, direct (^{3}H)-muscimol binding measures the high-affinity GABA-R, whereas the GABA regulation of BZR binding reflects the occupancy of the low-affinity GABA-R by muscimol or GABA. Therefore, one possible explanation for the paradox is that the increase in (^{3}H)-muscimol binding may represent the desensitized GABA-R, which in turn cannot effect BZR enhancement. The desensitization may be caused by increased amounts of GABA in the synaptic cleft released as a consequence of seizures. Alternatively, the BZR may be in a conformational state reflecting maximal activation after the seizures and not capable of further regulation by added GABA.

Studies of (^{3}H)-GABA release from hippocampal slices disclosed increased release in slices from entorhinal cortex kindled rats (Liebowitz et al., 1978). Whether this was a calcium-dependent release was not tested. The extent to which this represented synaptic release of GABA is unknown. Direct measurement of GABA concentration in frontal cortex, cerebellum, or brainstem 1 wk after the last kindled seizure revealed no significant changes (Fabisiak and Schwark, 1982).

In summary, the bulk of the data support the idea that the BZR–GABA-R changes represent a consequence of kindled seizures rather than the cause of kindling. The direction of the receptor alterations, namely an increase, together with the increased release of GABA, suggest the presence of increased inhibition. The localization of the receptor changes to the granule cells predicts that these cells in particular will exhibit enhanced inhibition. Interestingly, King et al. (1984) found increased inhibition—which is probably GABA mediated—of dentate granule cells in hippocampal slices prepared from kindled rats sacrificed 24 h after the last seizure. Likewise, Túff et al. (1983a) found increased inhibition of granule cells in vivo—presumably recurrent and GABA-mediated—following kindling. There is a strong likelihood that the molecular alterations, although not the molecular basis of kindling, probably represent one aspect of a carefully orchestrated response of these cells to seizures, namely an enhancement of inhibition. Thus, it can be speculated that this increased inhibition of dentate granule cells is designed to reduce excitatory synaptic input to CA3 pyramidal cells and thereby reduce the likelihood of further seizures.

4. Perspective

Studies of neurotransmitters have not yet provided a coherent explanation of the kindling phenomenon. Nonetheless many of these experiments have generated valuable information, some of which is relevant to kindling and some of which is relevant to experimental approaches to epilepsy in general.

                                    McNamara et al.

Some of these studies have reinforced the importance of an experimental approach that seeks to localize to a discrete cellular population those molecular changes induced by seizure states. Radiohistochemical localization of kindled seizure-induced alterations of muscarinic cholinergic (Savage et al., 1983), GABA (Shin et al., 1983), and benzodiazepine (Valdes et al., 1982) receptor changes to dentate granule cells of hippocampal formation permitted addressing three key questions. First, are the dentate granule cells members of the network of brain structures subserving kindling? Second, which afferent(s) of the granule cells is (are) responsible for the seizure-induced changes? Third, do these molecular changes result in altered physiology of the affected cells?

To determine whether the dentate granule cells might be members of the kindling network, the effects of removal of dentate granule cells on kindling development were examined. The injection of colchicine, a quite selective neurotoxin for dentate granule cells, into the dentate gyrus retarded the rate of amygdala kindling development by approximately 40% (Dasheiff and McNamara, 1982). Injection of colchicine into neocortex did not modify the rate of kindling development. Microsection of the entorhinal cortex, a lesion that destroys excitatory entorhinal afferents to the dentate granule cells and hippocampal formation, also markedly retarded the rate of amygdala kindling (Savage et al., 1984b). Moreover, knifecuts in neocortex did not modify the rate of kindling development. Together, these studies, which employ different lesion methods, but with a similar end result of presumed reduction of neuronal transmission through hippocampal formation, raise the possibility that the dentate granule cells and hippocampal formation subserve a facilitatory role in amygdala kindling development. Importantly, the anatomic localization of molecular changes induced by kindled seizures provided the impetus for these investigations.

Localizing the seizure-induced reduction of muscarinic receptor number to the dentate granule cells permitted investigation of the granule cell afferent responsible for the molecular changes. The receptor change was not dependent on extrinsic cholinergic afferents of the granule cells, since removal of the major cholinergic afferent by septal lesion did not prevent the seizure-induced decline (Dashieff and McNamara, 1980). The seizure-induced decline appears to depend on the principal excitatory afferent of the granule cells, since microsection of entorhinal afferents of the granule cells prevents the decline (Savage et al., 1984b). The entorhinal afferent is not cholinergic and therefore this appears to represent a form of heterosynaptic regulation of receptors. This has led to the hypothesis that the mechanism responsible for the receptor decline involves repeated depolarization of the granule cells. These findings with muscarinic receptors may serve as a model for understanding the mechanism of some of the host of alterations of neurotransmitters described in many seizure states.

Localizing the benzodiazepine and GABA receptor increases to the dentate granule cells permitted determining whether these molecular changes result in altered physiology of the granule cells. King et al. (1984) compared the physiology of granule cells in hippocampal slices removed from kindled and control animals. These experiments demonstrated the presence of increased inhibition of the granule cells in the kindled animals. The increased inhibition was almost certainly synaptic and GABA-mediated. It seems reasonable to speculate that these electrophysiologic changes are caused, at least in part, by the increased benzodiazepine and GABA receptor binding of the granule cells and the increased GABA release found in hippocampal slices of kindled rats (Leibowitz et al., 1978). These findings demonstrate the feasibility of an experimental approach that seeks to determine the functional significance of seizure-related biochemical changes. Indeed it underscores the importance of devising biochemical approaches to epilepsy that include a specified framework for ultimately determining the functional significance of seizure-related molecular changes at a cellular level.

Biochemical studies of neurotransmitters in kindling to date appear to have taught us more about the brain's response to seizures than the molecular basis of kindling. One promising finding regarding the molecular basis of kindling is the increased number of quisqualate-sensitive glutamate binding sites in hippocampal formation membranes of kindled rats (Savage et al., 1984c). Identical biochemical findings have been reported in hippocampal slices undergoing long-term potentiation, and indeed these biochemical changes have been advanced as the molecular basis of long-term potentiation (Lynch et al., 1982). If correct, these findings in kindled rats may reflect a biochemical index of long-term potentiation that, in turn, might contribute to the development, but not permanence, of kindling. The time course of both long-term potentiation and glutamate receptor changes in kindled rats is more compatible with a role in the development than in the maintenance of kindling.

Results of pharmacologic studies suggest potentially fruitful approaches to biochemical studies of this model. Pharmacologic regulation of GABA and norepinephrine have had much more profound effects on kindling than regulation of other putative neurotransmitters, such as acetylcholine, dopamine, or serotonin, suggesting that biochemical analysis of these transmitters in kindling may be particularly rewarding. However, this may reflect the potency and selectivity of the pharmacologic tools available to some extent. Indeed development of new probes for study of excitatory amino acids and the host of recently discovered peptides should have a major impact on future directions of investigations.

References

Aghajanian, G. K., Rosecrans, J. A., and Sheard, M. H.: Serotonin: Release in forebrain by stimulation of midbrain raphe. *Science* **156:** 402–403, 1967.

Albertston, T. E., Bowyer, J. F., and Paule, M. G.: Modification of the anticonvulsant efficacy of diazepam by Ro 15-1788 in the kindled amygdaloid seizure model. *Life Sci.* **31:** 1597–1601, 1982.

Albertson, T. E., Joy, R. M., and Stark, L. G.: Modification of kindled amygdaloid seizures by opiate agonists and antagonists. *J. Pharmacol. Exp. Ther.* **228:** 620–627, 1984.

Albright, P. S. and Burnham, W. M.: Development of a new pharmacological seizure model: Effects of anticonvulsants on cortical- and amygdala-kindled seizures in the rat. *Epilepsia* **21:** 681–689, 1980.

Albright, P. S., Burnham, W. M., and Okazaki, M.: Effect of atropine sulfate on amygdaloid kindling in the rat. *Exp. Neurol.* **66:** 409–412, 1979.

Altman, I. M. and Corcoran, M. E.: Facilitation of neocortical kindling by depletion of forebrain noradrenaline. *Brain Res.* **270:** 174–177, 1983.

Araki, H., Aihara, H. Watanabe, S., Ohta, H., Yamamoto, T., and Ueki, S.: The role of noradrenergic and serotonergic systems in the hippocampal kindling effect. *Jpn. J. Pharmacol.* **33:** 57–64, 1983a.

Araki, H., Aihara, H., Watanabe, S., Yamamoto, T., and Ueki, S.: Effects of reserpine, and α-methyl-*p*-tyrosine, *p*-chlorophenylalanine and 5,7-dihydroxytryptamine on the hippocampal kindling effect in rats. *Jpn. J. Pharmacol.* **33:** 1177–1182, 1983b.

Arnold, P. S., Racine, R. J., and Wise, R. A.: Effects of atropine, reserpine, 6-hydroxydopamine, and handling on seizure development in the rat. *Exp. Neurol.* **40:** 457–470, 1973.

Ashton, D., Leysen, J. E., and Wauquier, A.: Neurotransmitters and receptor binding in amygdaloid kindled rats: Serotonergic and noradrenergic modulatory effects. *Life Sci.* **27:** 1547–1556, 1980.

Babington, R. G. and Wedeking, P. W.: The pharmacology of seizures induced by sensitization with low intensity brain stimulation. *Pharmacol. Biochem. Behav.* **1**: 461–467, 1973.

Blackwood, D.: The Role of Noradrenaline and Dopamine in Amygdaloid Kindling, In: *Neurotransmitters, Seizures and Epilepsy* (P. L. Morselli, K. G. Lloyd, W. Loscher, B. Meldrum, and E. H. Reynolds, eds.) Raven, New York, 1981.

Blackwood, D. H. R., Martin, M. J., and Howe, J. G.: A study of the role of the cholinergic system in amygdaloid kindling in rats. *Psychopharmacology* **76**: 66–69, 1982.

Bowyer, J. F.: Phencyclidine inhibition of the rate of development of amygdaloid kindled seizures. *Exp. Neurol.* **75**: 173–183, 1982.

Bowyer, J. F., Albertson, T. E., and Winters, W. D.: Cortical kindled seizures: Modification by excitant and depressant drugs. *Epilepsia* **24**: 356–367, 1983.

Brotchi, J., Tanaka, T., and Leviel, V.: Lack of activated astrocytes in the kindling phenomenon. *Exp. Neurol.* **58**: 119–125, 1978.

Burchfiel, J. L., Duchowny, M. S., and Duffy, F. H.: Neuronal supersensitivity to acetylcholine induced by kindling in the rat hippocampus. *Science* **204**: 1096–1098, 1979.

Burnham, W. M., King, G. A., and Lloyd, K. G.: Extra-focal catecholamine levels in "kindled" rat forebrains. *Prog. Neuropsychopharmacol.* **5**: 537–541, 1981.

Burnham, W. M., Niznik, H. G., Okazaki, M. M., and Kish, S. J.: Binding of [³H] flunitrazepam and [³H]Ro 5-4864 to crude homogenates of amygdala-kindled rat brain: Two months post-seizure. *Brain Res.* **279**: 359–362, 1983.

Byrne, M. C., Gottlieb, R., and McNamara, J. O.: Amygdala kindling induces muscarinic cholinergic receptor declines in a highly specific distribution within the limbic system. *Exp. Neurol.* **69**: 85–98, 1980.

Caldecott-Hazard, S., Shavit, Y., Ackermann, R. F., Engel, J., Jr., Frederickson, R. C. A., and Liebeskind, J. C.: Behavioral and electrographic effects of opioids on kindled seizures in rats. *Brain Res.* **251**: 327–333, 1982.

Caldecott-Hazard, S., Yamagata, N., Hedlund, J., Camacho, H., and Liebeskind, J. C.: Changes in simple and complex behaviors following kindled seizures in rats: Opioid and nonopioid mediation. *Epilepsia* **24**: 539–547, 1983.

Callaghan, D. A. and Schwark, W. S.: Involvement of catecholamines in kindled amygdaloid convulsions in the rat. *Neuropharmacology* **18**: 541–545, 1979.

Chang, K. J., Hazum, E., and Cuatrecasas, P.: Novel opiate binding sites selective for benzomorphan drugs. *Proc. Natl. Acad. Sci. USA* **78**: 4141–4145, 1981.

Corcoran, M. E. and Mason, S. T.: Role of forebrain catecholamines in amygdaloid kindling. *Brain Res.* **190:** 473–484, 1980.

Corcoran, M. E. and Wada, J. A.: Naloxone and the kindling of seizures. *Life Sci.* **24:** 791–795, 1979.

Corcoran, M. E., Fibiger, H. C., McCaughran, J. A., Jr., and Wada, J. A.: Potentiation of amygdaloid kindling and metrazol-induced seizures by 6-hydroxydopamine in rats. *Exp. Neurology* **45:** 118–133, 1974.

Corcoran, M. E., Wada, J. A., and Wake, A.: A failure of atropine to retard amygdaloid kindling. *Exp. Neurol.* **51:** 271–275, 1976.

Cowan, A. Geller, E. B., and Adler, M. W.: Classification of opioids on the basis of change in seizure threshold in rats. *Science* **206:** 465–467, 1979.

Crain, B. J., Chang, K. J., and McNamara, J. O.: A quantitative in vitro autoradiographic study of mu and delta opioid binding in the hippocampal formation of kindled rats. *Soc. Neurosci. Abs.* **11:** 1323, 1985.

Dasheiff, R. M. and McNamara, J. O.: Intradentate colchicine retards the development of amygdala kindling. *Ann. Neurol.* **11:** 347–352, 1982.

Dasheiff, R. M. and McNamara, J. O.: Evidence for an agonist independent down regulation of hippocampal muscarinic receptors in kindling. *Brain Res.* **195:** 345–353, 1980.

Dasheiff, R. M., Savage, D. D., and McNamara, J. O.: Seizures downregulate muscarinic cholinergic receptors in hippocampal formation. *Brain Res.* **235:** 327–334, 1982.

Delgado, J. M. R. and Sevillano, M.: Evolution of repeated hippocampal seizures in the cat. *Electroencephalogr. Clin. Neurophysiol.* **13:** 722–733, 1961.

Dingledine, R.: Possible mechanisms of enkephalin action on hippocampal CA1 neurons. *J. Neurosci.* **1:** 1022–1035, 1981.

Douglas, R. M., Goddard, G. V., and Riives, M.: Inhibitory modulation of long-term potentiation: Evidence for a postsynaptic locus of control. *Brain Res.* **240:** 259–272, 1982.

Duchowny, M. S. and Burchfiel, J. L.: Facilitation and antagonism of kindled seizure development in the limbic system of the rat. *Electroencephalogr. Clin. Neurophysiol.* **51:** 403–416, 1981.

Ehlers, C. L., Clifton, D. K., and Sawyer, C. H.: Facilitation of amygdaloid kindling in the rat by transecting ascending noradrenergic pathways. *Brain Res.* **189:** 274–278, 1980.

Engel, J., Jr. and Sharpless, N. S.: Long-lasting depletion of dopamine in the rat amygdala induced by kindling stimulation. *Brain Res.* **136:** 381–386, 1977.

Engel, J., Jr., Wolfson, L., and Brown, L.: Anatomical correlates of electrical and behavioral events related to amygdaloid kindling. *Ann. Neurol.* **3:** 538–544, 1978.

Fabisiak, J. P. and Schwark, W. S.: Cerebral free amino acids in the amygdaloid kindling model of epilepsy. *Neuropharmacology* **21**: 179–182, 1982.

Fanelli, R. J. and McNamara, J. O.: Kindled seizures result in decreased responsiveness of benzodiazepine receptors to γ-aminobutyric acid (GABA). *J. Pharmacol. Exp. Ther.* **226**: 147–150, 1983.

Farjo, J. B. and Blackwood, D. H. R.: Reduction in tyrosine hydroxylase activity in the rat amygdala induced by kindling stimulation. *Brain Res.* **153**: 423–426, 1978.

Foote, F. and Gale, K.: Morphine potentiates seizures induced by GABA antagonists and attenuates seizures induced by electroshock in the rat. *Eur. J. Pharmacol.* **95**: 259–264, 1983.

Foote, R. W. and Maurer, R.: Autoradiographic localization of opiate kappa-receptors in the guinea pig brain. *Eur. J. Pharmacol.* **85**: 99–103, 1982.

Foote, S. L., Bloom, F. E., and Ashton-Jones, G.: Nucleus locus ceruleus: new evidence of anatomical and physiological specificity. *Physiol. Rev.* **63**: 844–914, 1983.

Frenk, H., Engel, J., Jr., Ackermann, R. F., Shavit, Y., and Liebeskind, J. C.: Endogenous opioids may mediate post-ictal behavioral depression in amygdaloid kindled rats. *Brain Res.* **167**: 435–440, 1979.

Frenk, H., Urca, G., and Liebeskind, J. C.: Epileptic properties of leucine- and methionine-enkephalin: Comparison with morphine and reversibility by naloxone. *Brain Res.* **147**: 327–337, 1978.

Gee, K. W., Hollinger, M. A., Bowyer, J. F., and Killan, E. K.: Modification of dopaminergic receptor sensitivity in rat brain after amygdaloid kindling. *Exp. Neurol.* **66**: 771–777, 1979.

Gee, K. W., Killam, E. K., and Hollinger, M. A.: Effects of haloperidol-induced dopamine receptor supersensitivity on kindled seizure development. *J. Pharmacol. Exp. Ther.* **225**: 70–76, 1983.

Gee, K. W., Killam, E. K., Hollinger, M. A., and Giri, S. N.: Effect of amygdaloid kindling on dopamine-sensitive adenylate cyclase activity in rat brain. *Exp. Neurol.* **70**: 192–199, 1980.

Gellman, R. L. and McNamara, J. O.: Neuronal network subserving the kindling model of epilepsy: The Role of the fimbria. *Soc. Neurosci. Abstr.* **10**: 343, 1984.

Giacchino, J. L., Somjen, G. G., Frush, D. P., and McNamara, J. O.: Lateral entorhinal cortical kindling can be established without potentiation of the entorhinal-granule cell synapse. *Exp. Neurol.* **86**: 483–492, 1984.

Goddard, G. V.: Development of epileptic seizures through brain stimulation at low intensity. *Nature* **214**: 1020–1021, 1967.

Goddard, G. V.: Long Term Alteration Following Amygdaloid Stimulation, In: *The Neurobiology of the Amygdala* (B. Eleftherious, ed.) Plenum, New York, 1972.

Goddard, G. V. and Douglas, R. M.: Does the engram of kindling model the engram of normal long term memory? *Can. J. Neurol. Sci.* **2:** 385–394, 1975.

Goddard, G. V., McIntyre, D. C., and Leech, C. K.: A permanent change in brain function resulting from daily electrical stimulation. *Exp. Neurol.* **25:** 295–330, 1969.

Green, A. R., Peralta, E., Hong, J. S., Mao, C. C., Atterwill, C. K., and Costa, E.: Alterations in GABA metabolism and met-enkephalin content in rat brain following repeated electroconvulsive shocks. *J. Neurochem.* **31:** 607–611, 1978.

Hardy, C., Panksepp, J., Rossi, J., III, and Zolovick, A. J.: Naloxone facilitates amygdaloid kindling in rats. *Brain Res.* **194:** 293–297, 1980.

Havlicek, V. and Friesen, H. G.: Comparison of Behavioral Effects of Somatostatin and Beta-Endorphin in Animals, In: *Central Nervous System Effects of Hypothalamic Hormones and Other Peptides* (R. Collu, A. Barbeau, J. R. Ducharme, and J. G. Rochefort, eds.) Raven, New York, 1979.

Henriksen, S. J., Bloom, F. E., McCoy, F., Ling, N., and Guillemin, R.: Beta-endorphin induces nonconvulsive limbic seizures. *Proc. Natl. Acad. Sci. USA* **75:** 5221–5225, 1978.

Henrikson, S. J., Chouvet, G., and Bloom, F. E.: *In vivo* cellular responses to electrophoretically applied dynorphin in the rat hippocampus. *Life Sci.* **31:** 1785–1788, 1982.

Higuchi, T., Sikand, G. S., Kato, N., Wada, J. A., and Friesen, H. G.: Profound suppression of kindled seizures by cysteamine: Possible role of somatostatin to kindled seizures. *Brain Res.* **288:** 359–362, 1983.

Hong, J. S., Gillin, J. C., Yang, H.-Y. T., and Costa, E.: Repeated electroconvulsive shocks and the brain content of endorphins. *Brain Res.* **177:** 273–278, 1979.

Hong, J. S., Wood, P. L., Gillin, J. C., Yang, H.-Y. T., and Costa, E.: Changes in hippocampal met-enkephalin content after recurrent motor seizures. *Nature* **285:** 231–232, 1980.

Iadarola, M. J., Shin, C., McNamara, J. O., and Yang, H.-Y. T.: Changes of dynorphine, enkephalin and cholecystokinin content of hippocampus and substantia nigra after amygdala kindling. *Brain Res.* **365:** 185–191, 1986.

Joy, R. M., Stark, L. G., Gordon, L., Peterson, S. L., and Albertson, T. E.: Chronic cholinesterase inhibition does not modify amygdaloid kindling. *Exp. Neurol.* **73:** 588–594, 1981.

Joy, R. M., Albertson, T. E., and Stark, L. G.: An analysis of the actions of progabide, a specific GABA receptor agonist, on kindling and kindled seizures. *Exp. Neurol.* **83:** 144–159, 1984.

Joy, R. M., Stark, L. G., and Albertson, T. E.: Dose-dependent proconvulsant and anticonvulsant actions of the alpha$_2$ adrenergic agonist, xylazine, on kindled seizures in the rat. *Pharmacol. Biochem. Behav.* **19:** 345–350, 1983.

Kalichman, M. W., Burnham, W. M., and Livingston, K. E.: Pharmacological investigation of gamma-aminobutyric acid (GABA) and fully-developed generalized seizures in the amygdala-kindled rat. *Neuropharmacology* **21:** 127–131, 1982.

Kalichman, M. W., Livingston, K. E., and Burnham, W. M.: Pharmacological investigation of γ-aminobutyric acid (GABA) and the development of amygdala-kindled seizures in the rat. *Exp. Neurol.* **74:** 829–836, 1981.

Kant, G. J., Meyerhoff, J. L., and Corcoran, M. E.: Release of norepinephrine and dopamine from brain regions of amygdaloid-kindled rats. *Exp. Neurol.* **70:** 701–705, 1980.

Karobath, M. and Sperk, G.: Stimulation of benzodiazepine receptor binding by γ-aminobutyric acid. *Proc. Natl. Acad. Sci. USA* **76:** 1004–1006, 1979.

Kato, N., Higuchi, T., Friesen, H. G., and Wada, J. A.: Changes of immunoreactive somatostatin and beta-endorphin content in rat brain after amygdaloid kindling. *Life Sci.* **32:** 2415–2422, 1983.

Kellar, K. J., Cascio, C. S., Bergstrom, D. A., Butler, J. A., and Iadarola, P.: Electroconvulsive shock and reserpine: Effects on beta-adrenergic receptors in rat brain. *J. Neurochem.* **37:** 830–836, 1981.

Kelsey, J. E. and Belluzzi, J. D.: Endorphin mediation of post-ictal effects of kindled seizures in rats. *Brain Res.* **253:** 337–340, 1982.

King, G. L., Dingledine, R., Giacchino, J. L., and McNamara, J. O.: Abnormal neuronal excitability in hippocampal slices from kindled rats. *J. Neurophys.* **54:** 1295–1304, 1985.

Knobloch, L. C., Goldstein, J. M., and Malick, J. B.: Effects of acute and subacute antidepressant treatment on kindled seizures in rats. *Pharmacol. Biochem. Behav.* **17:** 461–465, 1982.

Kovacs, D. A. and Zoll, J. G.: Seizure inhibition by median raphe nucleus stimulation in rat. *Brain Res.* **70:** 165–169, 1982.

Lason, W., Przewlocka, B., and Przewlocki, R.: The effect of gamma-hydroxybutyrate and anticonvulsants on opioid peptide content in the rat brain. *Life Sci.* **33:** (supl I.) 599–602, 1983.

Leech, C. K. and McIntyre, D. C.: Kindling rates in inbred mice: An analog to learning? *Behav. Biol.* **16:** 439–452, 1976.

Le Gal La Salle, G.: Inhibition of kindling-induced generalized seizures by aminooxyacetic acid. *Can. J. Physiol. Pharmacol.* **58:** 7–11, 1980.

Le Gal La Salle, G. and Feldblum, S.: Reversal of the anticonvulsant effects of diazepam on amygdaloid-kindled seizures by a specific benzodiazepine antagonist: Ro 15–1788. *Eur. J. Pharm.* **86:** 91–93, 1983a.

Le Gal La Salle, G., Calvino, B., and Ben-Ari, Y.: Morphine enhances amygdaloid seizures and increases inter-ictal spike frequency in kindled rats. *Neurosci. Lett.* **6:** 255–260, 1977.

Le Gal La Salle, G., Kaijima, M., and Feldblum, S.: Abortive amygdaloid kindled seizures following microinjection of γ-vinyl-GABA in the vicinity of substantia nigra in rats. *Neurosci. Lett.* **36:** 69–74, 1983b.

Lerer, B., Stanley, M., Demetriou, S., and Gershon, S.: Effect of electroconvulsive shock on muscarinic cholinergic receptors in rat cerebral cortex and hippocampus. *J. Neurochem.* **41:** 1680–1683, 1983.

Levy, W. B. and Steward, O.: Temporal contiguity requirements for long-term associative potentiation/depression in the hippocampus. *Neuroscience* **8:** 791–797, 1983.

Liebowitz, N. R., Pedley, R. A., and Cutler, R. W. P.: Release of γ-aminobutyric acid from hippocampal slices of the rat following generalized seizures induced by daily electrical stimulation of entorhinal cortex. *Brain Res.* **138:** 369–373, 1978.

Lynch, G., Halpain, S., and Baudry, M.: Effects of high-frequency synaptic stimulation on glutamate receptor binding studied with a modified in vitro hippocampal slice preparation. *Brain Res.* **244:** 101–111, 1982.

Maru, E., Tatsuno, J., Okamoto, J., and Ashida, H.: Development and reduction of synaptic potentiation induced by perforant path kindling. *Exp. Neurol.* **78:** 409–424, 1982.

McGinty, J. F., Henriksen, S. J., Goldstein, A., Terenius, L., and Bloom, F. E.: Opioid peptide identity and localization in hippocampus. *Life Sci.* **31:** 1797–1800, 1982.

McIntyre, D. C.: Split-brain rat: Transfer and interference of kindled amygdala convulsions. *Can. J. Neurol. Sci.* **2:** 429–437, 1975.

McIntyre, D. C.: Amygdala kindling in rats: Facilitation after local amygdala norepinephrine depletion with 6-hydroxydopamine. *Exp. Neurol.* **69:** 395–407, 1980.

McIntyre, D. C. and Edson, N.: Facilitation of amygdala kindling after norepinephrine depletion with 6-hydroxydopamine in rats. *Exp. Neurol.* **74:** 748–757, 1981.

McIntyre, D. C. and Edson, N.: Effect of norepinephrine depletion on dorsal hippocampus kindling in rats. *Exp. Neurol.* **77:** 700–704, 1982.

McIntyre, D. C. and Roberts, D. C. S.: Long-term reduction in beta-adrenergic receptor binding after amygdala kindling in rats. *Exp. Neurol.* **82:** 17–24, 1983.

McIntyre, D. C., Edson, N., Chao, G., and Knowles, V.: Differential effect of acute vs chornic desmethylimipramine on the rate of amygdala kindling in rats. *Exp. Neurol.* **78:** 158–166, 1982a.

McIntyre, D. C., Pusztay, W., and Edson, N.: Effect of flurazepam on kindled amygdala convulsion in catecholamine-depleted rats. *Exp. Neurol.* **77:** 78–85, 1982b.

McIntyre, D. C. Saari, M., and Pappas, B. A.: Potentiation of amygdala kindling in adult or infant rats by injections of 6-hydroxydopamine. *Exp. Neurol.* **63:** 527–544, 1979.

McNamara, J. O.: Selective alterations of regional beta-adrenergic receptor binding in the kindling model of epilepsy. *Exp. Neurol.* **61:** 582–591, 1978a.

McNamara, J. O.: Muscarinic cholinergic receptors participate in the kindling model of epilepsy. *Brain Res.* **154:** 415–420, 1978b.

McNamara, J. O.: Role of neurotransmitters in seizure mechanisms in the kindling model of epilepsy. *Fed. Proc.* **54:** 2516–2520, 1984.

McNamara, J. O., Byrne, M., Dasheiff, R. M., Fitz, J. G.: The kindling model of epilepsy: A review. *Prog. Neurobiol.* **15:** 139–159, 1980.

McNamara, J. O., Galloway, M. T., Rigsbee, L. C., and Shin, C.: Evidence implicating substantia nigra in regulation of kindled seizure threshold. *J. Neurosci.* **4:** 2410–2417, 1984.

McNamara, J. O., Rigsbee, L. C., and Galloway, M. T.: Evidence that substantia nigra is crucial to neural network of kindled seizures. *Eur. J. Pharmaco.* **86:** 485–486, 1983.

McNamara, J. O., Bonhaus, D. W., Crain, B. J., Gellman, R. L., Giacchino, J. L., and Shin, C.: The kindling model of epilepsy: A critical review. *CRC Crit. Rev. Neurobiol.* **1:** 341–392, 1985.

McNaughton, B.: Long-term synaptic enhancement and short-term potentiation in rat fascia dentata act through different mechanisms. *J. Physiol.* **324:** 249–262, 1982.

Meibach, R. C. and Maayani, S.: Localization of naloxone-sensitive [^{3}H]-dihydromorphine binding sites within the hippocampus of the rat. *Eur. J. Pharmacol.* **68:** 175–179, 1980.

Mohr, E. and Corcoran, M. E.: Depletion of noradrenaline and amygdaloid kindling. *Exp. Neurol.* **72:** 507–511, 1981.

Moore, R. Y. and Bloom, F. E.: Central catecholamine neuron systems: Anatomy and physiology of the dopamine systems. *Ann. Rev. Neurosci.* **1:** 129–169, 1978.

Moore, R. Y. and Bloom, F. E.: Central catecholamine neuron systems: Anatomy and physiology of the norepinephrine and epinephrine systems. *Ann. Review Neursci.* **2:** 113–168, 1979.

Morrell, F. and Tsura, N.: Kindling in the frog: Development of spontaneous epileptiform activity. *Electroencephalogr. Clin. Neurophysiol.* **40:** 1–11, 1976.

Munkenbeck, K. E. and Schwark, W. S.: Serotonergic mechanisms in amygdaloid-kindled seizures in the rat. *Exp. Neurol.* **76:** 246–253, 1982.

Myslobodsky, M. S., Ackerman, R. F., and Engel, J.: Effects of γ-acetylenic GABA and γ-vinyl GABA on metrazol-activated, and kindled seizures. *Pharmacol. Biochem. Behav.* **11:** 265–271, 1979.

Myslobodsky, M. S. and Valenstein, E. S.: Amygdaloid kindling and the GABA system. *Epilepsia* **21:** 163–175, 1980.

Niznik, H. B., Kish, S. J., and Burnham, W. M.: Decreased benzodiazepine receptor binding in amygdala-kindled rat brains. *Life Sci.* **33:** 425–430, 1983.

Noda, Y., Wada, J. A., and McGeer, E. G.: Lasting influence of amygdaloid kindling on cholinergic neurotransmission. *Exp. Neurol.* **78:** 91–98, 1982.

Okamoto, M. and Wada, J. A.: Reversible suppression of amygdaloid kindled convulsion following unilateral gabaculine injection into the substantia innominata. *Brain Res.* **305:** 389–392, 1984.

Peterson, S. L., Albertson, T. E., Stark, L. G., Joy, R. M., and Gordon, L. S.: Cumulative after-discharge as the principal factor in the acquisition of kindled seizures. *Electroencephalogr. Clin. Neurophysiol.* **51:** 192–200, 1981.

Peterson, D. W., Collins, J. F., and Bradford, H. F.: The kindled amygdala model of epilepsy: Anticonvulsant action of amino acid antagonists. *Brain Res.* **275:** 169–172, 1983.

Pinel, J. P. J. and Rovner, L. I.: Electrode placement and kindling-induced experimental epilepsy. *Exp. Neurol.* **58:** 335–346, 1978.

Post, R. M., Davenport, S., Pert, A., and Squillace, K. M.: Lack of effect of an opiate agonist and antagonist on the development of amygdala kindling in the rat. *Commun. Psychopharmacol.* **3:** 185–190, 1979.

Przewlocka, B., Stala, L., Lason, W., and Przewlocki, R.: The effect of various opiate receptor agonists on the seizure threshold in the rat. Is dynorphin an endogenous anticonvulsant? *Life Sci.* **33:** (Supl. I) 595–598, 1983a.

Racine, R. J.: Modification of seizure activity by electrical stimulation: I. After-discharge threshold. *Electroencephalogr. Clin. Neurophysiol.* **32:** 269–279, 1972a.

Racine, R. J.: Modification of seizure activity by electrical stimulation. II. Motor seizure. *Electroenceph. Clin. Neurophysiol.* **32:** 281–294, 1972b.

Racine, R. J., Burnham, W. M., Gartner, J. G., and Levitan, D.: Rates of motor seizure development in rats subjected to electrical brain stimulation: Strain and interstimulation interval effects. *Electroenceph. Clin. Neurophysiol.* **35:** 553–556, 1973.

Racine, R. and Coscina, D. V.: Effects of midbrain raphe lesions or systemic *p*-chlorophenylalanine on the development of kindled seizures in rats. *Brain Res. Bull.* **4:** 1–7, 1979.

Racine, R. and Zaide, J.: A Further Investigation Into the Mechanisms Underlying the Kindling Phenomenon, In: *Limbic Mechanisms: The Continuing Evolution of the Limbic System Concept* (K. E. Livingston and O. Hornykiewicz, eds.) Plenum, New York, 1978.

Racine, R., Livingston, K., and Joaquin, A.: Effects of procaine hydrochloride, diazepam, and diphenylhydantoin on seizure development in cortical and subcortical structures in rats. *Electroencephalogr. Clin. Neurophysiol.* **38:** 355–365, 1975.

Racine, R. J., Milgram, N. W., and Hafner, S.: Long-term potentiation phenomena in the rat limbic forebrain. *Brain Res.* **260:** 217–231, 1983.

Rial, R. V. and Gonzales, J.: Kindling effect in the reptilian brain: Motor and electrographic manifestations. *Epilepsia* **19:** 581–587, 1978.

Robertson, H. A., Riives, M., Black, D. A. S., and Peterson, M. R.: A partial agonist at the anticonvulsant benzodiazepine receptor: Reversal of the anticonvulsant effects of Ro 15-1788 with CGS-8216. *Brain Res.* **291:** 388–390, 1984.

Sagar, S. M., Landry, D., Millard, W. J., Badge, T. M., Arnold, M. A., and Martin, J. B.: Depletion of somatostatin-like immunoreactivity in the rat central nervous system by cysteamine. *J. Neurosci.* **2:** 225–231, 1982.

Sato, M., Hikasa, N., and Otsuki, S.: Experimental epilepsy, psychosis and dopamine receptor sensitivity. *Biol. Psychiat.* **14:** 537–540, 1979.

Savage, D. D. and McNamara, J. O.: Kindled seizures selectively reduce a subpopulation of [³H] quinuclidinyl benzilate binding sites in rat dentate gyrus. *J. Pharm. and Exp. Ther.* **222:** 670–673, 1982.

Savage, D. D., Dasheiff, R. M., and McNamara, J. O.: Kindled seizure-induced reduction of muscarinic cholinergic receptors in rat hippocampal formation: Evidence for localization to dentate granule cells. *J. Comp. Neurol.* **221:** 106–112, 1983.

Savage, D. D., Nadler, J. V., and McNamara, J. O.: Reduced kainic acid binding in rat hippocampal formation after limbic kindling. *Brain Res.* **323:** 128–131, 1984a.

Savage, D. D., Rigsbee, L. C., and McNamara, J. O.: Knife cuts of entorhinal cortex: Effects on development of amygdaloid kindling and seizure-induced decrease of muscarinic cholinergic receptors. *J. Neurosci.* **5:** 408–413, 1984b.

Savage, D. D., Werling, L. L., Nadler, J. V., and McNamara, J. O.: Selective and reversible increase in the number of quisqualate-sensitive glutamate binding sites on hippocampal synaptic membranes after angular bundle kindling. *Brain Res.* **307:** 332–335, 1984c.

Savage, D. D., Werling, L. L., Nadler, J. V., and McNamara, J. O.: Selective increase in L-[³H] glutamate binding to a quisqualate-sensitive site on hippocampal synaptic membranes after angular bundle kindling. *Eur. J. Pharmacol.* **85:** 225–256, 1982.

Segal, M. and Bloom, F. E.: The action of norepinephrine in the rat hippocampus. II. Activation of the input pathway. *Brain Res.* **107:** 499–511, 1976.

Shields, P. J. and Eccleston, D.: Effects of electrical stimulation of rat midbrain on 5-hydroxytryptamine synthesis as determined by a sensitive radioisotope method. *J. Neurochem.* **19:** 265–272, 1972.

Shin, C., Legg, S., and McNamara, J. O.: Systemic γ-vinyl GABA retards kindling development and suppresses kindled seizures. *Neurosci. Abst.* **10:** 343, 1984.

Shin, C., Pedersen, H., and McNamara, J. O.: Parallel alterations in GABA and benzodiazepine receptors in fascia dentata in the kindling model of epilepsy: A radiohistochemical study. *Soc. Neurosci. Abstr.* **9:** 487, 1983.

Siegal, J. and Murphy, G. J.: Serotonergic inhibition of amygdala-kindled seizures in cats. *Brain Res.* **174:** 337–340, 1979.

Stach, R., Lazarova, M. B., and Kacz, D.: Serotonergic mechanism in seizures kindled from the rabbit amygdala. *Naunyn Schmiedebergs Arch. Pharmacol.* **316:** 56–58, 1981.

Stock, G., Kummer, P., Stumpf, H., Zenner, K., and Sturm, V.: Involvement of dopamine in amygdaloid kindling. *Exp. Neurol.* **80:** 439–450, 1983.

Stone, W. S., Eggleton, C. E., and Berman, R. F.: Opiate modification of amygdaloid-kindled seizures in rats. *Pharmacol. Biochem. Behav.* **16:** 751–756, 1982.

Tanaka, T.: Progressive changes of behavioral and electroencephalographic responses to daily amygdaloid stimulation in rabbits. *Fukuoka Acta. Med.* **63:** 152–163, 1972.

Tuff, L. P., Racine, R. J., and Adamec, R.: The effects of kindling on GABA-mediated inhibition in the dentate gyrus of the rat. I. Paired-pulse depression. *Brain Res.* **277:** 79–90, 1983a.

Tuff, L. P., Racine, R. J., and Mishra, R. K.: The effects of kindling on GABA-mediated inhibition in the dentate gyrus of the rat. II. Receptor binding. *Brain Res.* **277:** 91–98, 1983b.

Urca, G. and Frenk, H.: Pro- and anticonvulsant action of morphine in rats. *Pharmacol. Biochem. Behav.* **13:** 343–347, 1980.

Urca, G. and Frenk, H.: Systemic morphine blocks the seizures induced by intracerebroventricular (i.c.v.) injections of opiates and opioid peptides. *Brain Res.* **246:** 121–126, 1982.

Valdes, F., Dasheiff, R. M., Birmingham, F., Crutcher, K. A., and McNamara, J. O.: Benzodiazepine receptor increases following repeated seizures: Evidence for localization to dentate granule cells. *Proc. Natl. Acad. Sci. USA* **79:** 193–197, 1982.

Vindrola, O., Briones, R., Asai, M., and Fernandex-Guardiola, A.: Amygdaloid kindling enhances the enkephalin content in the rat brain. *Neurosc. Lett.* **21:** 39–43, 1981a.

Vindrola, O., Briones, R., Asai, M., and Fernandez-Guardiola, A.: Brain content of Leu^5- and Met^5-enkephalin changes independently during the development of kindling in the rat. *Neurosci. Lett.* **26:** 125–130, 1981b.

Vosu, H. and Wise, R. A.: Cholinergic seizure kindling in the rat: Comparison of caudate, amygdala, and hippocampus. *Behav. Biol.* **13:** 491–495, 1975.

Wada, J. A. and Osawa, T.: Spontaneous recurrent seizure state induced by daily electric amygdaloid stimulation in Senegalese baboons (*Papio Papio*). *Neurology* **26:** 273–286, 1976.

Wada, J. A. and Sato, M.: Generalized convulsive seizures induced by daily electrical stimulation of the amygdala in cats: Correlative electrographic and behavioral features. *Neurology* **24:** 565–574, 1974.

Walker, J. M., Moises, H. C., Coy, D. H., Baldrighi, G., and Akil, H.: Nonopiate effects of dynorphin and des-tyr-dynorphin. *Science* **218:** 1136–1138, 1982.

Wang, R. Y. and Aghaganian, G. K.: Inhibition of neurons in the amygdala by dorsal raphe stimulation: Mediation through a direct serotonergic pathway. *Brain Res.* **120:** 85–102, 1977.

Wasterlain, C. G. and Jonec, V.: Chemical kindling by muscarinic amygdaloid stimulation in the rat. *Brain Res.* **271:** 311–323, 1983.

Waterhouse, B. D., Moises, H. C., Yeh, H. H., and Woodward, D. J.: Norepinephrine enhancement of inhibitory synaptic mechanisms in cerebellum and cerebral cortex: Mediation by beta adrenergic receptors. *J. Pharmacol. Exp. Ther.* **221:** 495–506, 1982.

Watson, S. J., Khachaturian, H., Akil, H., Coy, D. H., and Goldstein, A.: Comparison of the distribution of dynorphin systems and enkephalin systems in brain. *Science* **218:** 1134–1136, 1982.

Wauquier, A., Ashton, D., and Melis, W.: Behavioral analysis of amygdaloid kindling in beagle dogs and the effects of clonazepam, diazepam, phenobarbital, diphenylhydantoin, and flunarizine on seizure manifestation. *Exp. Neurol.* **64:** 579–586, 1979.

Westerberg, V., Lewis, J., and Corcoran, M. E.: Depletion of noradrenaline fails to affect kindled seizures. *Exp. Neurol.* **84:** 237–246, 1984.

Wilkison, D. M. and Halpern, L. M.: Turnover kinetics of dopamine and norepinephrine in the forebrain after kindling in rats. *Neuropharmacology* **18:** 219–222, 1980.

Wilkison, D. M. and Halpern, L. M.: The role of biogenic amines in amygdalar kindling: Local amygdalar afterdischarge. *J. Pharmacol. Exp. Ther.* **211:** 151–158, 1979.

Wise, R. A. and Chinerman, J.: Effects of diazepam and phenobarbital on electrically-induced amygdaloid seizures and seizure development. *Exp. Neurol.* **45:** 355–363, 1974.

Woodward, D. J., Moises, H. C., Waterhouse, B. D., Hoffer, B. J., and Freedman, R.: Modulatory actions of norepinephrine in the central nervous system. *Federation Proc.* **38:** 2109–2116, 1979.

Yitzhaky, J., Frenk, H., and Urca, G.: Kindling-induced changes in morphine analgesia and catalepsy: Evidence for independent opioid systems. *Brain Res.* **237:** 193–201, 1982.

Yoshida, K.: Influences of bilateral hippocampal lesions upon kindled amygdaloid convulsive seizure in rats. *Physiol. Behav.* **32:** 123–126, 1984.

Zieglgansberger, W., French, E. D., Siggins, G. R., and Bloom, F. E.: Opioid peptides may excite hippocampal neurons by inhibiting adjacent inhibitory interneurons. *Science* **205:** 415–417, 1979.

In Vitro Models of Epilepsy

Roger D. Traub, Robert K. S. Wong,
and Richard Miles

1. Introduction

In this chapter, we shall review what is known about the cellular mechanisms of epileptic events in vitro. Our main concern is to understand the electrical events that neurons generate during a fit, and to understand what kinds of interactions (synaptic and non-synaptic) between neurons underlie the fit. We further seek to comprehend which intrinsic properties of neurons are required for epilepsy or that at least facilitate epileptogenesis. Our method consists of making experimental observations of seizure phenomena in the hippocampal slice, attempting to reproduce the observations with simulation models on a large computer, and testing the models with further experiments. We shall emphasize underlying mechanisms that are as independent as possible of physiological and pharmacological details. This provides a clear conceptual framework in which to interpret diverse experiments. Such an approach is necessary because epilepsy involves understanding the behavior of a *population* of neurons. Our work has lead to models that illuminate the underlying mechanisms of two types of epileptic events: interictal spikes and tonic seizures.

1.1. Definition of Epilepsy

Following the tradition of Jackson (1931), we define epileptic behavior in a population of neurons to occur when the membrane potentials of the constituent neurons are abnormally synchronized. Since some degree of synchrony underlies many neuronal phenomena (making it possible, for example, to observe rhythmic EEG events), this definition hinges on what is meant by ''abnormal." In general, no precise distinction between normal and abnormal degrees of synchrony may exist; however, during in vitro epileptic events, virtually all of the pyramidal cells behave in a synchronized, stereotyped manner. Our definition does not distinguish between events (a) in which neurons synchronously and *phasically* generate depolarizing bursts (e.g., interictal spikes), or (b) in which neurons collectively develop sustained depolarizations—to the point where transmembrane potentials virtually vanish—along with massive changes in extracellular ion concentrations (e.g., spreading depression). Both types of synchronized behavior occur in brain slices (Snow et al., 1983), but it is not yet known if spreading depression occurs during clinical seizure states in man. We consider in this chapter only *phasic* synchronized events. One must be cautious in extending our definition of epilepsy from in vitro to in vivo epilepsy. Thus, it is not known, at the cellular level, if synchrony of neuronal bursting occurs across large regions of cortex during events that are clearly epileptic, i.e., tonic-clonic seizures. Only *local* synchrony is known to occur in both in vitro and in vivo epileptic events (see below).

The technical advantages of using in vitro methods have been reviewed elsewhere (Pedley and Traub, 1985) as has the basic anatomical structure of the hippocampal slice (Andersen, 1975; Andersen et al., 1971). However, anatomical discussions usually omit one critical aspect of hippocampal circuitry that has been demonstrated by electrophysiological (but not anatomical) techniques: the existence, at least within CA3, of recurrent excitatory synaptic connections (MacVicar and Dudek, 1980; Miles and Wong, 1983, 1986a,b). Two additional concepts are of especial importance for epilepsy in the hippocampal slice. First, within particular regions of the slice, such as CA1 or CA3, neurons have relatively uniform anatomical and electrophysiological properties. For example, within the CA1 region, most of the cells are pyramidal cells, and CA1 pyramids have characteristic responses to somatic, dendritic, or antidromic stimulation (Schwartzkroin, 1975; 1977). A similar statement holds for the CA3 region (Wong and Prince, 1981). For studies on epilepsy, this uniformity of cellular responsiveness is especially striking since convulsant agents block a major component of GABA-mediated synaptic inhibition, so that the responses of inhibitory interneurons can be ignored. The second important concept concerning epilepsy in vitro is that microdissected pieces of the hippocampal slice, containing as few as about 1000 cells (Miles et al., 1984) can generate epileptiform events. This means that computer models of epilepsy can contain almost as many "neurons" as occur in the real biological system. It is now feasible to simulate models with several hundred "neurons."

2. Hippocampal Pyramidal Cells

2.1. Intrinsic Properties

CA3 pyramidal cells produce 1–2 ms action potentials, singly or in pairs. In the resting slice, spontaneous excitatory and inhibitory synaptic events occur (Brown et al., 1979; Miles and Wong, 1984). In addition, CA3 cells in the resting slice display so-called intrinsic bursts that consist of a series of two or more action potentials riding on a depolarizing envelope. The burst may contain one or more slow (about 10 ms) action potentials that are, at least in part, Ca^{2+}-mediated (Wong and Prince, 1981). These bursts are truly intrinsic in the sense that dissociated hippocampal neurons can produce them (Numann et al., 1982). Bursts can be generated both in or near the soma and in the apical dendrites (Wong et al., 1979). Intrinsic bursts are not an artifact of the slice, since they also occur in vivo (Kandel and Spencer, 1961). Spontaneous bursts occur asynchronously in different cells at rates of about 0.25–2.0 Hz (Wong and Prince, 1981; Hablitz and Johnston, 1981), the rate depending on the membrane potential; more depolarized membrane potentials lead to higher rates of spontaneous bursting. Of importance is the fact that a long-lasting (i.e., about 50 ms) burst can be evoked by intracellular stimuli as brief as 5 ms. Bursts are followed by a prolonged hyperpolarization lasting as long as 1 or 2 s (Wong and Prince, 1981; Alger and Nicoll, 1980; Hotson and Prince, 1980; Hablitz, 1981).

As in other neurons, electrogenesis in these cells is determined by the interaction between diverse synaptic inputs with the passive and active membrane properties. By "passive properties" we mean the distributed membrane leakage conductance and capacitance. By "active properties" we mean properties determined by specific ionic channels, each possessing a history-dependent conductance that is controlled by transmembrane potential, intra- or extracellular divalent cation concentration, or other factors (transmitters, modulators, and so on). Passive electrotonic properties of CA3 neurons have been studied by several groups of investigators (Turner and Schwartzkroin, 1983; Johnston, 1981; Brown et al., 1981). These cells have dendrites of electrotonic length about 0.5

(basilar dendrites) to about 1.0 (apical dendrites) space constants (Turner and Schwartzkroin, 1983). One of Rall's (1962) requirements for collapsing the dendritic tree into an electrically equivalent cylinder is approximately satisfied: at branch points, $d^{3/2} = d_1^{3/2} + d_2^{3/2}$, where d, d_1, and d_2 are the diameters of parent and two daughter branches respectively (Turner and Schwartzkroin, 1983). However, another of Rall's requirements is not satisfied, since different branches of a given dendrite may have widely different electrotonic lengths (Turner and Schwartzkroin, 1983). Analyses of electrotonic properties based purely on electrical measurements without consideration of the detailed anatomy must therefore be interpreted cautiously. Input resistance of CA3 pyramids ranges from 20 to about 40 MΩ in the hands of different investigators, with membrane time constants typically 20–40 ms. It is not known if membrane resistivity is the same over different regions of the membrane.

The active properties of CA3 neurons are complex. There is a classical TTX-sensitive fast Na$^+$ channel, a delayed rectifier K$^+$ channel, and a Ca^{2+}-mediated K$^+$ channel. There is a Ca^{2+} channel capable of producing slow action potentials in the presence of TTX (Wong and Prince, 1981) and a (possibly different) Ca^{2+} channel that contributes to a slow inward current present between bursts (Brown and Griffith, 1983; Johnston et al., 1980). CA1 neurons— and probably CA3 neurons—possess an A current (a K$^+$ current inactivated by depolarization and deinactivated by hyperpolarization) (Gustafsson et al., 1982), and an M-current (a slow K$^+$ current activated by depolarization and blocked by acetylcholine (Halliwell and Adams, 1982)). There is also a current, possibly carried by K$^+$ ions, induced by GABA and resistant to convulsant agents that contributes to the so-called delayed IPSP (Newberry and Nicoll, 1984). Data from dissociated cells suggest the existence of a slow inward Na$^+$ current (R. Numann and R. K. S. Wong, unpublished data).

Quantitative voltage-clamp data are being assembled on these diverse channels (Numann et al., 1982; Clark and Wong, 1983), but it is not yet possible to reproduce all of the intrinsic membrane behaviors of these neurons from first principles with quantitative accuracy. We have been able to reproduce the phenomenology of bursting with a computer model that is qualitatively correct and that makes specific testable predictions (Traub, 1982).

CA1 cells differ from CA3 cells in that their somas have less tendency to generate bursts; however, CA1 apical dendrites do have intrinsic burst-generating capacity (Wong and Traub, 1982). In response to steady somatic injected currents, CA1 cells usually generate trains of action potentials that adapt and then stabilize in frequency (Schwartzkroin, 1978; Gustaffson and Wigstrom, 1981; Madison and Nicoll, 1982).

2.2. Synaptic Organization in the Hippocampal Slice

The synaptic interconnections most important for epileptogenesis are the excitatory axons between CA3 cells (Traub and Wong, 1982). These axons have not been identified anatomically, but are known to occur by virtue of physiological experiments. These experiments involve simultaneous intracellular recording from a pair of cells, shown not to be electrotonically coupled, such that one or more action potentials in one cell reproducibly cause EPSPs or action potentials in the other cell. These connections are hard to find by randomly selecting pairs of cells for penetration: MacVicar and Dudek (1980) found only five (unidirectionally) connected pairs in 88 dual penetrations, and our experience has been similar (Miles and Wong, 1986b). On the other hand, such connections that do exist are powerful enough to enable a presynaptic action potential to cause a full postsynaptic action potential (Miles and Wong, 1983, 1986b; MacVicar and Dudek, 1980). We have found a monosynaptically connected pair of CA3 neurons such that a presynaptic burst in one cell evoked a postsynaptic burst in the other cell (Fig. 1). Using longitudinal (as opposed to the usual transverse) slices, Knowles and Schwartzkroin (1981) were unable to find excitatory interactions between pairs of CA1 cells. The fact that epileptic events occur in vivo in isolated CA1 islands suggests indirectly that CA1 excitatory interconnections may exist (Dichter et al., 1973). Furthermore, Benardo and Prince (1982) observed synchronized bursts in CA1 after local acetylcholine application, and Hablitz (1984) has recently observed picrotoxin-induced epileptiform activity in the isolated CA1 region. Therefore, the matter of recurrent excitatory connections remains open for CA1. The synaptic transmitter used at CA3 recurrent excitatory synapses is unknown, but the fact that epileptiform events are reversibly blocked by γ-d-glutamylglycine (see below) suggests that it may be an amino acid.

The well-known excitatory Schaffer collaterals (Lorente de No, 1934) run from CA3 cells to CA1 apical dendrites. They form a pathway that permits the spread of epileptic activity from CA3 to CA1 (see below).

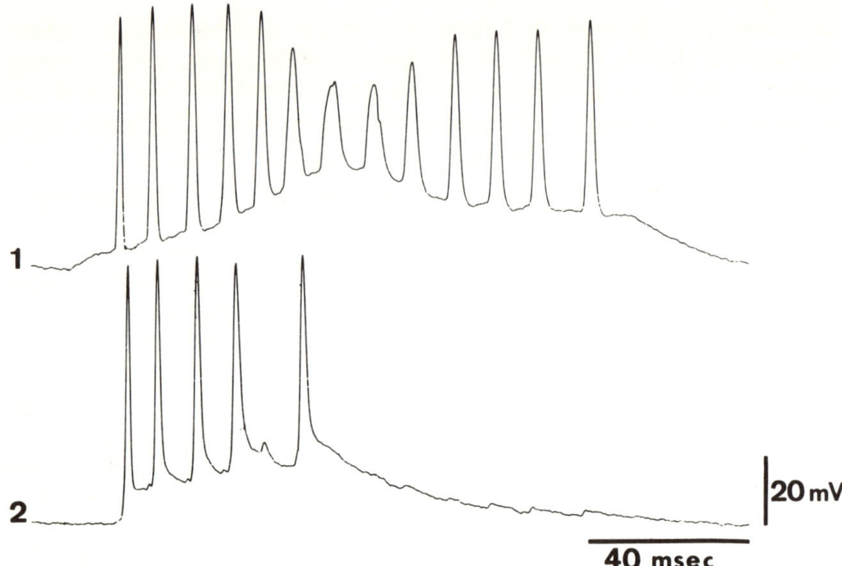

Fig. 1. Bursting in one hippocampal neuron can induce bursting in a connected neuron. These simultaneous intracellular recordings were made in the CA3 region of the hippocampal slice from a monosynaptically connected pair of cells. An 80-ms, 0.4-nA current injected into cell 1 induced a burst in cell 2. Note that the presynaptic burst is of longer duration than the postsynaptic burst, suggesting refractoriness in the postsynaptic cell or else block in the pathway from one cell to the other.

Local inhibitory circuits exist both within CA1 (Knowles and Schwartzkroin, 1981), within CA3 (Miles and Wong, 1984, 1986a), and in the connection from CA3 to CA1 [so-called feed-forward inhibition (Alger and Nicoll, 1982)]. We shall not discuss this inhibition in detail because the major component of this GABA-mediated inhibition is blocked by convulsant agents (Dingledine and Gjerstad, 1979; 1980; Miles et al. 1984; Lebeda et al., 1982; Wong and Prince, 1979). The functional significance of the recently described late IPSP, which is resistant to convulsant agents, is not yet known, although it is an active area of research.

2.3. Epileptic Events

2.3.1. Types of Events in the In Vitro Slice

There are two basic types of epileptic events in vitro, although one is really a special case of the other. The first type consists of the synchronized bursting of a population of pyramidal neurons. In such an event, a group of cells generates a burst during a given time interval whose duration is approximately the duration of a burst in an individual cell. Similar events occur in penicillin-induced epileptogenesis in vivo (Dichter and Spencer, 1969a, b; Matsumoto and Ajmone-Marsan, 1964a; Prince, 1968) and in human patients (Wyler et al., 1982). Synchronization in these latter cases is not as complete as in the in vitro case. Synchronized bursting also occurs in neocortical slices bathed in convulsant-containing solutions (Courtney and Prince, 1977; Gutnick et al., 1982). Electroencephalographic "interictal spikes," which occur in cortical regions of patients with focal seizure disorders (Currie et al., 1971), appear to have as their cellular basis precisely this type of synchronized bursting. We therefore refer to these events in vitro, by an abuse of terminology, as "interictal spikes."

The second type of in vitro epileptic event consists of the synchronized occurrence of a rapid series (15–22 Hz) of bursts. Each burst in the series occurs synchronously throughout the population. We call such events "synchronized afterdischarges." These afterdischarges probably correspond to tonic seizure discharges in epileptic patients (Brenner and Atkinson, 1982; Gastaut et al., 1963; Gastaut and Tassinari, 1975). These "epileptic recruiting rhythms" or "paroxysmal fast activities" consist of 10–25 Hz paroxysmal runs of widely distributed EEG or electrocorticographic waves. The reason we propose this correspondence between synchronized afterdischarges and tonic seizure discharges is that similar EEG waves are observed during penicillin-induced or electrically induced focal seizures (Matsumoto and Ajmone Marsan, 1964b; Sawa et al., 1968); during such events, cortical neurons depolarize in phase with the local field potential transients—hence, the cells must be depolarizing in phase with each other.

2.3.2. Experimental conditions Under Which Epileptic Events Are Observed

Interictal spikes occur when the slice is perfused with media containing GABA-blocking convulsant agents such as penicillin, bicuculline, and picrotoxin (Schwartzkroin and Prince, 1977; 1978; 1980; Wong and Traub, 1983; Miles et al., 1984). They are also observed under other bathing conditions that have more complex actions: kainic acid (Lothman et al., 1981) or high $[K^+]_0$ (Ogata et al., 1976; Rutecki and Johnston, 1983). Synchronized afterdischarges occur in healthy slices bathed in 0.1 m*M* picrotoxin (Miles et al., 1984; Hablitz, 1984), in rat hippocampal slices (from animals 9–19 d old), in penicillin (Swann and Brady, 1984), and in low chloride media (Ogata, 1978). Low Cl⁻ would be expected to abolish the major part of the GABA-dependent IPSP (Allen et al., 1977), as well as possibly increasing the input resistance of cells (by decreasing leak currents). Penicillin, bicuculline, and picrotoxin do not change the intrinsic properties of hippocampal neurons in the slice in any known way, although penicillin has been reported to augment Ca^{2+} currents of cultured neurons (Heyer et al., 1982). Other effects of penicillin, possibly relevant to epileptogenesis and synchronization, such as ectopic action potential generation (Gutnick and Prince, 1972; Noebels and Prince, 1978), have not yet been reported in the hippocampal slice.

2.4. Interictal Spikes

2.4.1. Experimental Phenomenology

Interictal spikes in vitro have several important features. They occur spontaneously and regularly, with interevent periods of several seconds and with the period remaining fairly constant for a given slice. Each event is associated with a so-called ringing field potential, several mV in amplitude, that reflects the synchronized population discharge (Fig. 2). This potential is comb-shaped, with an underlying positivity and superimposed fast transients. The detailed shape of this field potential reflects the degree of synchrony of individual action potentials in bursts generated by different cells; this in turn depends on the extracellular resistivity and the geometry of packing of the cells. All cells recorded from so far in our laboratory participated in each interictal spike. These events are initiated in the CA2-CA3 region (Wong and Traub, 1983). If the Schaffer collaterals are divided, epileptiform events continue in CA2-CA3, but not in CA1 (Schwartzkroin and Prince, 1978; Wong and Traub, 1983). Indeed, the isolated CA2-CA3 region can by itself generate synchronized bursts, so that the dentate gyrus is not required. The action of a convulsant on CA1 (i.e., disinhibition within CA1 itself) is required for propagation of interictal spikes from CA3 to CA1 (Mesher and Schwartzkroin, 1980). Within CA3, interictal spikes can be evoked by a variety of localized stimuli, either electric shocks to orthodromic (mossy fiber) or antidromic (Schaffer collateral or

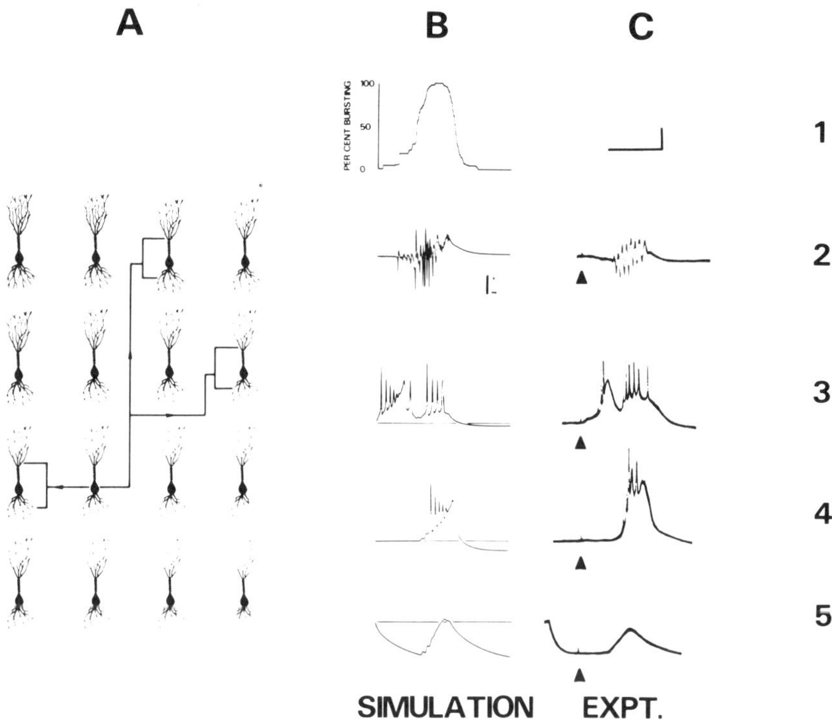

Fig. 2. Interictal spike in vitro. A. Schematic of randomly connected network of neurons used in simulations. Only some neurons and some connections are shown. All connections are excitatory. B. Simulation of synchronized burst in 100-cell model network, each cell having an average of five inputs from other cells. Stimulus is to four cells. B1. Number of cells bursting (i.e., with soma depolarized more than 20 mV). B2. Field potential. B3. Intracellular record of one of the stimulated cells, showing a double burst. B4. Intracellular potential of another cell, showing long-latency bursting. B5. The simulation was repeated with the cell of B4 hyperpolarized to reveal underlying EPSP. C. Synchronized bursts in penicillin-treated hippocampal slice, evoked by fimbrial stimuli (triangles) and recorded in the CA2 region. C2. Field potential. C3 and C4. Intracellular records showing double burst and long-latency burst, respectively. C5. Underlying EPSP revealed by hyperpolarization of cell of C4. Calibration: 50 ms in B, 60 ms in C, 25 mV for B3–B5, 4 mV for C2, and 20 mV for C3–C5 (B and C from Traub and Wong, 1982, reproduced with permission).

fimbria) pathways, or by ejection of droplets of hypertonic KCl from a small extracellular electrode (Wong and Traub, 1983). Remarkably, there is a latent period as long as 200 ms from an evoking stimulus to maximum field potential corresponding to the evoked epileptic event (Wong and Traub, 1983; Miles and Wong, 1983). Even more remarkably, as predicted by our model, intracellular stimulation of a single cell can influence the rhythm of a spontaneous series of interictal spikes (Miles and Wong, 1983). There is a ''population refractory period'' after an interictal spike, in the sense that the sooner one stimulates after a spontaneous event, the larger must be the stimulus to evoke another event. The rhythm of spontaneous interictal spikes is reset by an evoked event (Wong and Traub, 1983).

Further experiments have been performed that help to elucidate the mechanism of the interictal spike. When during interictal spikes CA3 cells are hyperpolarized enough to block action potentials (hyperpolarization to such extent being possible in a minority of cells), a large smooth depolarizing potential appears coincident with the population burst (Fig. 2) (Wong and Traub, 1983). This potential has the appearance of an EPSP. It is possible to reverse this potential by depolarizing the cell sufficiently, using a Cs^+-filled electrode (Johnston and Brown, 1981), further suggesting that it represents an EPSP. Voltage clamp experiments have lead to an estimate of 19–32 ns for the underlying synaptic conductance during a population burst, which is about 60–100 times the conductance of spontaneous EPSPs (Johnston and Brown, 1981) and comparable to the about 20 ns conductance produced by a (nonunitary) mossy fiber volley (Brown and Johnston, 1983). Thus, it appears that during an interictal spike, the population of cells bursts more or less simultaneously, each cell being triggered to burst by synaptic inputs from other cells. The conceptual problem, then, is to understand how this comes about.

A further question concerns how the interictal spike is terminated. Bursts that occur spontaneously or after current injection are followed by a long-duration hyperpolarization caused by a CA^{2+}-mediated K conductance (Hotson and Prince, 1980; Alger and Nicoll, 1980; Schwartzkroin and Stafstrom, 1980). The situation is more complicated during interictal spikes, however. Thus, cells bursting during penicillin-induced (but not picrotoxin-induced) interictal spikes have a Cl^--dependent hyperpolarization (Hablitz, 1981); this suggests that penicillin is a weaker blocker of the GABA-dependent Cl^- channel than is picrotoxin. Furthermore, the CA3 cellular hyperpolarization following penicillin-induced population bursts is insensitive to intracellular EGTA injection; thus, CA3 cells can apparently repolarize from this event without the Ca^{2+}-induced K^+ current. Blockade of $g_{K(Ca)}$ with 8-bromoadenosine 3',5'-cyclic monophosphate in CA1 cells during bicuculline-induced bursts does exert an effect—population bursts occur more frequently than before—but cells are still able to repolarize, and indeed the duration of bursts is not even increased (Newberry and Nicoll, 1984). We can summarize these results by saying that cells repolarize from a population burst because of (1) $g_{K(Ca)}$ (in part), (2) the delayed convulsant-resistant IPSP, which is probably K-dependent, and (3) a residual small GABA-dependent g_{Cl} (in the case of penicillin-induced bursts). The M-current (Halliwell and Adams, 1982) also must contribute to repolarization and the afterhyperpolarization, since acetylcholine prolongs CA1 population bursts induced by penicillin (Kriegstein et al., 1983). Inactivation of voltage-dependent intrinsic inward currents may also contribute to termination of spontaneous and epileptiform population bursts.

2.4.2. Interictal Spike Models

We have been able to synthesize many of these observations in the form of a computer model (Traub and Wong, 1982; 1983a) that makes transparent the underlying mechanism of the interictal spike. This model shows that three known features of the hippocampus suffice to explain synchronized population bursts: (1) intrinsic bursting (Wong and Prince, 1981), (2) recurrent excitation (MacVicar and Dudek, 1980), and (3) synaptic disinhibition, the established effect of many convulsant drugs (Dingledine and Gjerstad, 1979; 1980; Wong and Prince, 1979). The idea is that a burst in one cell—occurring spontaneously or after a stimulus—should be able to evoke bursts in follower cells if excitatory connections exist and are strong enough. In the presence of a normal amount of synaptic inhibition, the existing excitatory synaptic connections are not powerful enough to allow spread of bursting. Our model consists specifically of a collection (typically 100) ''neurons'' that are randomly interconnected with a low density (in agreement with experiment) of excitatory synapses: typically one cell averages 2.5–5.0 inputs. The agreement of this model with experiment is striking (Fig. 2).

The model shows that *both* network and intrinsic properties are essential for epileptiform events. Thus, the debate is rendered obsolete about whether the paroxysmal depolarization shift or ''PDS'' represents an intrinsic event or a ''giant EPSP.'' It is both.

That the EPSP underlying the PDS is large ("giant") simply reflects the synchronization of population bursting; one need not postulate that the effects of individual excitatory synapses are augmented — although they may be. It is further striking that a *low density* of excitatory synapses suffices for the production of an interictal spike. However, this situation resembles a chain reaction: as long as each cell has, on the average, more than one follower, the population discharge can grow. If the size of the population is 1000 cells and each cell has, say, two followers, then the connection probability is only 0.2%. The model explains the long latency from a localized stimulus to population discharge: This represents the initial stages of growth of the discharge from one or a few cells to the stage when enough cells are firing to produce a measurable field potential. A small degree of convulsant-resistant synaptic inhibition (not included in the present model) would also be expected to prolong the interictal spike latency (R. Traub, R. K. S. Wong, and R. Miles, unpublished data). The model predicts that when synaptic activity is blocked (e.g., with low Ca^{2+}, high Mg^{2+} media), even when the bursting threshold of individual cells is unchanged, then synchronized population discharges should not occur; this is precisely what happens.

The model has made two additional predictions that have since been verified: (1) excitatory synaptic connections should be powerful enough so that bursting in one cell can evoke a burst in at least some follower cells (Fig. 1); and (2) intracellular stimulation of at least some individual cells should initiate population discharges or influence their spontaneous rhythm (Miles and Wong, 1983). Since current sinks in both basilar and apical dendrites occur during the interictal spike (Swann et al., 1983), one would also guess that recurrent excitatory synapses make contact at both these locations, as is assumed in our model.

In assessing the validity of this model, we must try to consider both alternative explanations of the data, and criticisms of features of the model that seem inconsistent with experimental data. This will serve also to discuss other aspects of cellular synchronization.

(i) Could electrotonic junctions (MacVicar and Dudek, 1981) be responsible for convulsant-induced synchronized bursting? Dye-coupling between hippocampal neurons occurs in the form of small clusters rather than a large syncytium (Andrew et al., 1982), so that a localized stimulus can not engage the entire population through electrotonic junctions alone. Also, a low Ca^{2+}, high Mg^{2+} medium, which blocks synaptic transmission and prevents synchronized bursting, would not be expected to affect electrotonic junctions [which are, however, sensitive to changes in intracellular ion concentrations (Spray et al., 1984)]. However, electrotonic junctions can enhance synchronization produced through excitatory connections if these connections are strong enough (Traub and Wong, 1983b). If excitatory synaptic transmission is marginal, electrotonic junctions may serve to shunt synaptic currents and could actually *prevent* the development of synchronized bursts (Traub and Wong, 1983b).

(ii) Synchronized trains of action potentials (''field bursts'') occur in the hippocampus when synaptic transmission is blocked (Taylor and Dudek, 1982, 1984a, b; Jefferys and Haas, 1982; Yaari et al., 1983). This synchronization almost certainly results from ''field interactions'' between cells caused by currents flowing in the extracellular medium (Traub et al., 1985a). Could field interactions underlie the interictal spike? The phenomenology of field bursts is different than that of the interictal spike. Field bursts last longer (seconds rather than 100 ms), and the waveform of intracellular potential consists of a train of action potentials, rather than a burst riding on a depolarizing envelope followed by an afterhyperpolarization. Field bursts occur in situations in which synaptic transmission is blocked *and* individual neurons have increased excitability and are firing spontaneously. It can be shown that sufficiently excitable neurons packed together in a layer (as in the hippocampus) can entrain one another so that synchronization occurs (Traub et al., 1985a). During interictal spikes, in contrast, excitatory synaptic transmission is intact and the excitability of individual neurons is normal.

This is not to say that field interactions are unimportant for the interictal spike. During the syncrhonized population burst, cells are synaptically induced to fire time-overlapping trains of action potentials. Synchronization between cells of individual action potentials can occur through field interactions (Snow and Dudek, 1984). This phenomenon undoubtedly contributes to the ringing character of the epileptiform field potential (Traub et al., 1985a). In order to claim a *primary* character for field interactions in convulsant-induced interictal spikes, one would need to show that bursts in one or a few cells can induce bursting—through field interactions—in their neighbors. This seems unlikely, since cellular transmembrane depolarizations of only a few mV occur even when the entire population is already bursting (Richardson et al., 1984); thus, synchronization of the population should not develop after a localized stimulus through field effects alone. Field effects may promote spread of epileptiform activity around a central mass of already discharging cells.

(iii) Could changes in extracellular $[K^+]$ underlie convulsant-induced synchronization? In normal media to which a convulsant has been added, increases in $[K^+]_0$ appear to follow the onset of the interictal spike rather than precede it, as if $[K^+]_0$ increases were an epiphenomenon of population bursting rather than a cause (Pedley et al., 1976). The increased $[K^+]_0$ might serve to prolong the population burst, and also influences the frequency of interictal spiking (Oliver et al., 1978). In some cases, however, increased $[K^+]_0$ by itself allows for the appearance of synchronized bursts (Ogata et al., 1976; Rutecki and Johnston, 1983). The underlying mechanism for this is unknown. Our model predicts that in such cases the synaptic excitation–inhibition balance has been changed so that bursting can spread from cell to cell along excitatory synaptic pathways; this remains to be demonstrated experimentally.

One may also ask whether transient changes of $[K^+]$, in the narrow interstices between pyramidal cell bodies, might contribute to synchronization of individual action potentials. When a population of cells—having a laminated geometry as in the hippocampus—discharges synchronously, it induces a transmembrane depolarization in "passive cells"; intracellular recordings referenced to ground show, however, either no depolarization or a hyperpolarization (Taylor and Dudek, 1984a; Richardson et al., 1984). These membrane changes are readily explained by currents flowing in the extracellular medium and across cell membranes (Taylor and Dudek, 1984a; Traub et al., 1985a). It has been asserted, moreover, that localized increases in $[K^+]_0$ ought to cause depolarizations in ground-referenced intracellular recordings (Richardson et al., 1984) contrary to what is observed experimentally. A detailed analysis of intra- and extracellular ion flow with electrodiffusion equations, however, is necessary before one can be certain of this. At the present time, one can only say that synchronization of action potentials appears to be adequately explained by field potential interactions.

(iv) Interictal spikes occur in neocortex (Prince, 1968) and neocortical slices (Gutnick et al., 1982), where most neurons have limited or no ability to burst intrinsically (Connors et al., 1982); does this not contradict the model? The model does not require that all cells be capable of intrinsic bursting. Let us suppose (George and Connors, 1983) that a subnetwork of the neocortex exists, consisting of intrinsically bursting neurons interconnected by excitatory synapses. The model predicts that synchronized bursting would develop in this subnetwork—if there is disinhibition—and be projected to other neurons that might be (1) not intrinsically bursting, and/or (2) not themselves interconnected through excitatory synapses. The existence of such a subnetwork remains to be demonstrated directly, however. Epilepsy in the neocortex forces one to look at the underlying physics of our model in depth. A model of synchronization developing from a localized stimulus requires "amplification," so that, in effect, a small number of discharging cells can lead to a population discharge. There are several types of amplification: (1) intrinsic bursting, i.e., the tendency for a prolonged burst to develop after a brief stimulus, (2) axonal branching, so that bursting in one cell can excite two or more followers, and (3) an EPSP may outlast the stimulus that produces it. A diminution in one kind of amplification (e.g., intrinsic bursting) might be compensated for by an increase in the other two (axonal branching and EPSP duration). The mathematics of this has not been examined in detail.

(v) Backfiring, or ectopic action potential generation, has been observed in neocortical penicillin epileptogenic foci in vivo (Gutnick and Prince, 1972; Noebels and Prince, 1978). Could this not be an important synchronizing mechanism in the penicillin focus? This remains a theoretical possibility, difficult to assess since backfiring has not been demonstrated in the hippocampal slice. Even if backfiring occurs, it will be hard to judge its importance until the detailed "wiring pattern" of the CA3 region is known, as well as the precise loci at which ectopic action potentials are generated. Only then will it be possible to predict the effects on the neuronal system produced by a particular ectopic action potential.

(vi) The period of interictal spiking can be as long as 10 s or more, yet $g_{K(Ca)}$ lasts only 1 or 2 s; does this not imply that the model omits some important inhibitory process? The determinants of the period between interictal spikes remain unclear. It is interesting that blockade of $g_{K(Ca)}$ leads to a significant increase in the frequency of interictal spikes (Newbury and Nicoll, 1984), so it is possible that a small Ca^{2+}-mediated K^+ conductance persists beyond 2 s that is sufficient to regulate the period of population discharges. However, whether or not there are additional slow processes (electrogenic pump, and so on) *after* a population discharge is probably not germane to how the population discharge develops in the first place.

(vii) It has been shown that intracellular stimulation of about one out of three cells in CA3 can influence the rhythm of epileptiform population discharges. Yet in a random network of 1000 cells, where each cell has an average of 2.5 inputs, almost all cells should have this property (Traub et al., 1984). How can this discrepancy be explained? This discrepancy is probably an artifact of the slicing procedure, which selectively cuts axons of cells near the boundary of the slice more than it cuts axons of interior cells. The network in the slice is thus not statistically isotropic. It is also possible that only some cells send excitatory outputs to other cells, so that a subnetwork of the CA3 region is what generates synchronized bursts. This question awaits anatomical characterization of the excitatory recurrent collaterals in the hippocampus.

The above discussion should convince the reader that our model of the synchronization process is simple and yet still adequate to explain most of the experimental data. Nevertheless, many important experimental issues remain open.

2.5. Synchronized Afterdischarges

2.5.1. Experimental Phenomenology

This more complex form of epileptiform activity (Fig. 3) occurs in the presence of 0.1 mM picrotoxin (Miles et al., 1984; Hablitz, 1984) and sometimes in penicillin (Swann and Brady, 1984). As mentioned above, these afterdischarges are analogous to a brief tonic seizure. Each event lasts from about 200 to as long as 500 ms or more, and events recur spontaneously and periodically with a period of several seconds (Miles et al., 1984). Each cell (Fig. 3) produces a prolonged (100–150 ms) depolarizing ''primary burst,'' followed by a rhythmical series of ''secondary bursts'' or afterdischarges, with a period of 45–65 ms (15–22 Hz), finally terminating in a prolonged afterhyperpolarization (Miles et al., 1984). Both the primary and the secondary bursts are synchronized, as indicated by dual intracellular recordings, or by coincidence between each cellular event and the local field potential. Blockade of chemical synapses prevents the occurence of synchronized afterdischarges, even when individual cells can burst. An excitatory synaptic input, impinging from other cells, underlies both the primary burst and the secondary bursts; this has been shown by depolarizing cells with Cs$^+$ electrodes—past the EPSP reversal potential—during a synchronized afterdischarge (Miles et al., 1984).

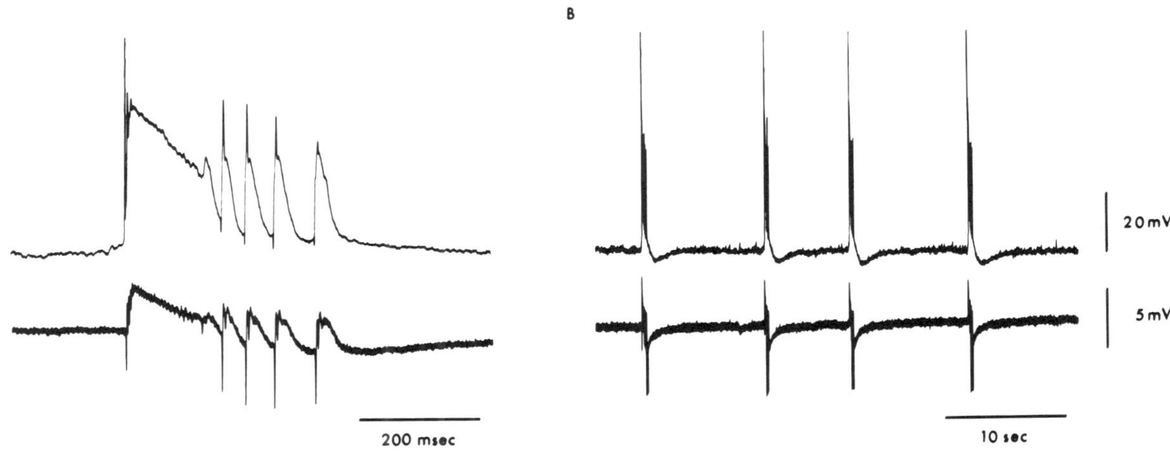

Fig. 3. Synchronized afterdischarges in the picrotoxin-treated hippocampal slice, CA3 region. Intracellular recordings above, simultaneous extracellular recordings below. Events are spontaneous. A single event (A) consists of a prolonged single burst, followed by a series (four in this case) of afterdischarges. These events recur, although not precisely periodically (B) (from Miles et al., 1984, reproduced with permission).

In summary, a synchronized afterdischarge shares many essential features with an interictal spike, the important difference being that an afterdischarge consists of a rapid series of bursts. It is reasonable to assume that the same principles (intrinsic bursting, mutual excitation, disinhibition) underlie both phenomena—especially since these principles are operative in both sets of experimental conditions in which these phenomena occur. One must ask, however, whether these principles are *sufficient* for the generation of afterdischarges. In particular, where does the 45–65 ms period of the afterdischarges come from? (It seems unlikely that the delayed convulsant-resistant IPSP is responsible, since this lasts for over 100 ms.)

2.5.2. Synchronized Afterdischarge Model

We have found that manipulation of system parameters (synaptic strength and connectivity) in our model of the interictal spike does not by itself lead to the generation of afterdischarges (Traub et al., 1984). If one postulates the existence of an additional form of neuronal refractoriness (which we have located in the communication between neurons, axons, and/or synapses), with an appropriate recovery time constant, then afterdischarges can be realistically simulated (Fig. 4) (Traub et al., 1984). This postulate is a reasonable one, in view of the known occurrence of intermittant axonal conduction and axonal block in several neuronal systems, both invertebrate (Smith, 1980) and vertebrate (Barron and Matthews, 1935; Chung et al., 1970; Grossman et al., 1979a, b; Kocsis et al., 1981; 1982; 1983). It remains to be demonstrated experimentally that axonal or synaptic conduction failure occurs in the hippocampus and that it has an appropriate recovery time course. Finch et al. (1983) have observed an example of such block in subicular cortex after slow repetitive stimulation. It is interesting to speculate that axonal branches in the hippocampus, passing through the densely packed pyramidal cell body layer, may be particularly susceptible to localized changes in $[K^+]$. Such changes would be expected to cause first increased and then decreased excitability (Kocsis et al., 1983; Malenka et al., 1981).

It has been shown experimentally that progressive blockade of excitatory synapses—with γ-d-glutamylglycine, or after switching to a low Ca^{2+}, high Mg^{2+} medium—leads to a progressive decrease in the number of secondary bursts, until only a single synchronized burst, an interictal spike, occurs. Eventually, after synaptic blockade, there is no synchronized activity (Miles et al., 1984). It is in

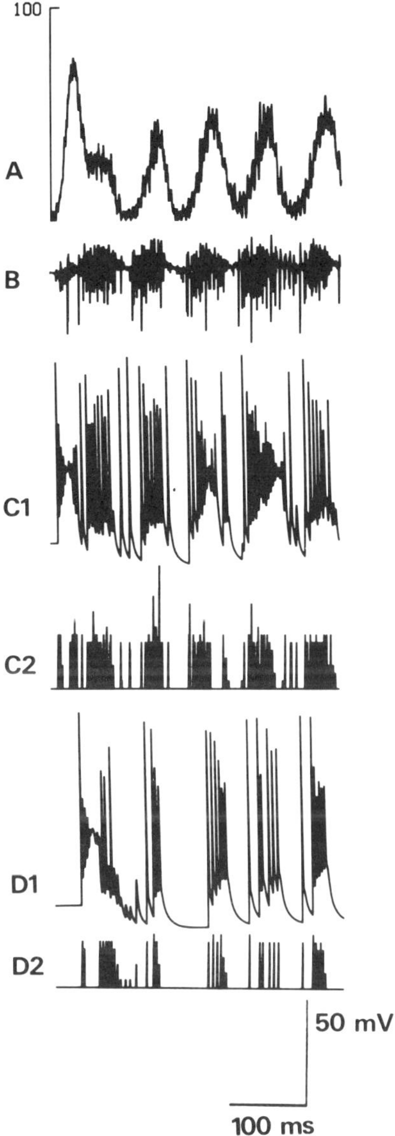

Fig. 4. Simulation of synchronized afterdischarges in 100-cell model, 2.5 inputs/cell on average, with ''axonal'' refractoriness. Stimulus is to four cells. A. Number of cells bursting (depolarized more than 20 mV). B. Field potential. C1 and D1. Intracellular records, showing afterdischarges. C2 and D2. Synaptic input to cells of C1 and D1, respectively, phasically oscillating as happens experimentally (from Traub et al., 1984, reproduced with permission).

this sense that an interictal spike is a special case of a synchronized afterdischarge. The progressive dropout of afterdischarges with synaptic blockade is reproduced by our model (Traub et al., 1984). This may explain why synchronized afterdischarges—as opposed to single synchronized bursts—are not routinely observed in the convulsant-perfused hippocampal slice. Afterdischarges can only occur (both experimentally and in the model) if excitatory synapses are "sufficiently strong." Any residual inhibition will have the effect of weakening the effectiveness of the excitatory connections, by diminishing the expected depolarization of follower cells produced by a burst in a given cell. Presumably, the quantitative extent of inhibitory blockade will vary with the choice of convulsant drug and its concentration.

3. Clinical Consequences of the Results on In Vitro Epilepsy

One may study epilepsy for many different reasons. One reason would be to learn about normal brain function from analysis of the abnormal activities of the brain. But a particularly compelling reason is to learn about the problems of real patients with clinical seizures. One must ask then whether the cellular events during convulsant-induced seizures in the slice are "the same" as during seizures in the living human. What do different clinical seizure types have in common at the cellular level? Why do clinical seizures occur when they do? One may ask many similar questions. Studies of human cortex, neurosurgically removed from patients with epilepsy (Prince and Wong, 1981; Schwartzkroin and Knowles, 1984), may provide some answers concerning the epileptogenic possibilities of abnormal brain, perhaps as compared with the properties of putatively normal brain, removed for example during aneurysm surgery; these studies do not, however, tell us what happens in an actual seizure. Extracellular unit recordings have been obtained during interictal events in human patients (Wyler et al., 1982), and gross depth and surface recordings have, of course, been obtained frequently during human seizures (Ajmone Marsan and O'Connor, 1973; Niedermeyer, 1982; Chatrian and Petersen, 1960), but these observations tell us little about the detailed cellular events that underlie a fit. To understand cellular events, one must do intracellular recording.

It is still a reasonable inference that the cellular events during convulsant-induced slice epileptiform events are similar to clinical seizure events. That interictal spikes represent synchronized (or at least partially synchronized) neuronal bursts has been shown in several different experimental models (Prince and Futamachi, 1968; Goldensohn and Purpura, 1963). Synchrony between neuronal bursting and focal electrocorticographic waves has been seen with penicillin and electrically induced seizures (Matsumoto and Ajmone Marsan, 1964a, b; Sawa et al., 1968). The relation between slice electrophysiology and other diverse seizure types (absence, myoclonic, akinetic, and so on) is of course problematic. Nevertheless, one can make some plausible speculations.

In the slice, GABA-mediated inhibition can be largely blocked by pharmacologic agents. Yet, in humans, there is always a background of inhibition, at least interictally. The interesting question, then, is: under what conditions might synaptic inhibition be reduced, so that recurrent excitatory forces—as described above—might come into play? Prince has discussed this question in similar terms (1983). *Chronically*, inhibition may be reduced by permanent loss of inhibitory neurons (as occurs after anoxic insults) (Sloper et al., 1980), or at least of GABA terminals (as in the alumina focus) (Ribak et al., 1979). *Acutely*, inhibition may become less effective simply because GABAergic effects are labile and can decrease during repetitive stimulation (Wong and Watkins, 1982; Ben-Ari et al., 1980). A superimposition of these effects in a cortical region could create conditions similar to those obtained in the convulsant-treated slice. Finally, it must be considered that some seizure types involve rather different mechanisms. One might imagine, for example, a population of neurons becoming so intensely depolarized through extrinsic influences (acute anoxia, proximity to a region of spreading depression) that the cells all fire repetitively even without the loss of synaptic inhibition. In a sense, this situation is still analogous to the one we have been discussing, since with neurons excitable enough, inhibition can be *present* yet virtually *ineffective*.

Acknowledgments

This article was supported in part by NIH grant NS18464 and a Klingenstein fellowship award (to R.K.S.W.). Some of the work described in this chapter was performed in collaboration with W. D. Knowles, F. E. Dudek, C. P. Taylor, and R. Numann.

References

Ajmone Marsan, C. and O'Connor, M.: *Electrocorticography. Handbook of Electroencephalography and Clinical Neurophysiology* vol. 10-C (A. Remond, ed.) pp. 3–49, Elsevier, Amsterdam, 1973.

Alger, B. E. and Nicoll, R. A.: Epileptiform burst afterhyperpolarization: Calcium-dependent potassium potential in hippocampal CA1 pyramidal cells. *Science* **210:** 1122–1124, 1980.

Alger, B. E. and Nicoll, R. A.: Feed-forward dendritic inhibition in rat hippocampal pyramidal cells studied *in vitro. J. Physiol.* **328:** 105–123, 1982.

Allen, G. I., Eccles, J., Nicoll, R. A., Oshima, T., and Rubia, F. J.: The ionic mechanisms concerned in generating the i.p.s.ps of hippocampal pyramidal cells. *Proc. Royal Soc. Lond. B* **198:** 363–384, 1977.

Andersen, P.: Organization of Hippocampal Neurons and Their Interconnections, In: *The Hippocampus*, vol. 1: *Sructure and Development* (R. L. Isaacson and K. H. Pribram, eds.) Plenum, New York, 1975.

Andersen, P., Bliss, T. V. P., and Skrede, K. K.: Lamellar organization of hippocampal excitatory pathways. *Exp. Brain Res.* **13:** 222–238, 1971.

Andrew, R. D., Taylor, C. P., Snow, R. W., and Dudek, F. E.: Coupling in rat hippocampal slices: Dye transfer between CA1 pyramidal cells. *Brain Res. Bull.* **8:** 211–222, 1982.

Barron, D. H. and Matthews, B. C.: Intermittent conduction in the spinal cord. *J. Physiol.* **85:** 73–103, 1935.

Benardo, L. S. and Prince, D. A.: Cholinergic excitation of mammalian hippocampal pyramidal cells. *Brain Res.* **249:** 315–333, 1982.

Ben-Ari, Y., Krnjevic, K., and Reinhardt, W.: Lability of synaptic inhibition of hippocampal pyramidal cells. *J. Physiol.* **298:** 36P–37P, 1980.

Brenner, R. P. and Atkinson, R.: Generalized paroxysmal fast activity: Electroencephalographic and clinical features. *Ann. Neurol.* **11:** 386–390, 1982.

Brown, D. A. and Griffith, W. H.: Persistent slow inward calcium current in voltage-clamped hippocampal neurones of the guinea-pig. *J. Physiol.* **337:** 303–320, 1983.

Brown, T. H. and Johnston, D.: Voltage-clamp analysis of mossy fiber s ynaptic input to hippocampal neurons. *J. Neurophysiol.* **50:** 487–507, 1983.

Brown, T. H., Fricke, R. A., and Perkel, D. A.: Passive electrical constants in three classes of hippocampal neurons. *J. Neurophysiol.* **46:** 812–827, 1981.

Brown, T. H., Wong, R. K. S.., and Prince, D. A.: Spontaneous miniature synaptic potentials in hippocampal neurons. *Brain Res.* **177:** 194–199, 1979.

Chatrian, G. E. and Petersen, M. C.: The convulsive patterns provoked by Indoklon, Metrazol, and electroshock: Some depth electrographic observations in human patients. *Electroenceph. Clin. Neurophysiol.* **12:** 715–725, 1960.

Chung, S.-H., Raymond, S. A., and Lettvin, J. Y. Multiple meaning in single visual units. *Brain Behav. Evol.* **3:** 72–101, 1970.

Clark, R. B. and Wong, R. K. S.: Three components of outward current in isolated mammalian cortical neurons. *Soc. Neurosci. Abs.* **9:** 601, 1983.

Connors, B. W., Gutnick, M. J., and Prince, D. A.: Electrophysiological properties of neocortical neurons in vitro. *J. Neurophysiol.* **48:** 1302–1320, 1982.

Courtney, K. R. and Prince, D. A.: Epileptogenesis in neocortical slices. *Brain Res.* **127:** 191–196, 1977.

Currie, S., Heathfield, K. W. G., Henson, R. A., and Scott, D. F.: Clinical course and prognosis of temporal lobe epilepsy. A survey of 666 patients. *Brain* **94:** 173–190, 1971.

Dichter, M. and Spencer, W. A.: Penicillin-induced interictal discharges from the cat hippocampus. I. Characteristics and topographical features. *J. Neurophysiol.* **32:** 649–662, 1969a.

Dichter, M. and Spencer, W. A.: Penicillin-induced interictal discharges from the cat hippocampus. II. Mechanisms underlying origin and restriction. *J. Neurophysiol.* **32:** 663–687, 1969b.

Dichter, M., Herman, C., and Selzer, M.: Penicillin epilepsy in isolated islands of hippocampus. *Electroenceph. Clin. Neurophysiol.* **34:** 631–638, 1973.

Dingledine, R. and Gjerstad, L.: Penicillin blocks hippocampal IPSPs, unmasking prolonged EPSPs. *Brain Res.* **168:** 205–209, 1979.

Dingledine, R. and Gjerstad, L.: Reduced inhibition during epileptiform activity in the *in vitro* hippocampal slice. *J. Physiol.* **305:** 297–313, 1980.

Finch, D. M. Nowlin, N. L., and Babb, T. L.: Demonstration of axonal projection of neurons in the rat hippocampus and subiculum by intracellular injection of HRP. *Brain Res.* **271:** 201–216, 1983.

Gastaut, H. and Tassinari, C. A.: Ictal Discharges in Different Types of Seizures, In: *Handbook of Electroencephalography and Clinical Neurophysiology* vol. 13, part A: *Epilepsies* (H. Gastaut and C. A. Tassinari, eds.) Elsevier, Amsterdam, 1975.

Gastaut, H., Roger, J., Ouachi, S., Timsit, M., and Broughton, R.: An electro-clinical study of generalized epileptic seizures of tonic expression. *Epilepsia* **4:** 15–44, 1963.

George, C. P. and Connors, B. W.: Initiation of synchronized bursting in neocortex. *Soc. Neurosci. Abs.* **9:** 396, 1983.

Goldensohn, E. S. and Purpura, D. P.: Intracellular potentials of cortical neurons during focal epileptogenic discharges. *Science* **139:** 840–842, 1963.

Grossman, Y., Parnas, I., and Spira, M. E.: Differential conduction block in branches of a bifurcating axon. *J. Physiol.* **295:** 283–305, 1979a.

Grossman, Y., Parnas, I., and Spira, M. E.: Ionic mechanisms involved in differential conduction of action potentials at high frequency in a branching axon. *J. Physiol.* **295:** 307–322, 1979b.

Gustafsson, B. and Wigstrom, H.: Shape of frequency-current curves in CA1 pyramidal cells in the hippocampus. *Brain Res.* **223:** 417–421, 1981.

Gustafsson, B., Galvan, M., Grafe, P., and Wigstrom, H.: A transient outward current in a mammalian central neurone blocked by 4-aminopyridine. *Nature* **299:** 252–254, 1982.

Gutnick, M. J. and Prince, D. A.: Thalamocortical relay neurons: Antidromic invasion of spikes from a cortical epileptogenic focus. *Science* **176:** 424–426, 1972.

Gutnick, M. J., Connors, B. W., and Prince, D. A.: Mechanisms of neocortical epileptogenesis in vitro. *J. Neurophysiol.* **48:** 1321–1335, 1982.

Hablitz, J. J.: Effects of intracellular injections of chloride and EGTA on postepileptiform burst hyperpolarizations in hippocampal neurons. *Neurosci. Lett.* **22:** 159–163, 1981.

Hablitz, J. J.: Picrotoxin-induced epileptiform activity in the hippocampus: Role of endogenous versus synaptic factors. *J. Neurophysiol.* **51:** 1011–1027, 1984.

Hablitz, J. J. and Johnston, D.: Endogenous nature of spontaneous bursting in hippocampal pyramidal neurons. *Cell. Molec. Neurobiol.* **1:** 325–334, 1981.

Halliwell, J. V. and Adams, P. R.: Voltage-clamp analysis of muscarinic excitation in hippocampal neurons. *Brain Res.* **250:** 71–92, 1982.

Heyer, E., Nowak, L. M., and MacDonald, R. L.: Membrane depolarization and prolongation of calcium-dependent action potentials of mouse neurons in cell culture by two convulsants: Bicuculline and penicillin. *Brain Res.* **232:** 41–56, 1982.

Hotson, J. R. and Prince, D. A.: A calcium-activated hyperpolarization follows repetitive firing in hippocampal neurons. *J. Neurophysiol.* **43:** 409–419, 1980.

Jackson, J. H.: On the Scientific and Empirical Investigation of Epileptics, In: *Selected Writings of John Hughlings Jackson* vol. I: *On Epilepsy and Epileptiform Convulsions* (J. Taylor, ed.), Hodder and Stoughton, London, 1931.

Jefferys, J. G. R. and Haas, H. L.: Synchronized bursting of CA1 hippocampal pyramidal cells in the absence of synaptic transmission. *Nature* **300:** 448–450, 1982.

Johnston, D.: Passive cable properties of hippocampal CA3 pyramidal neurons. *Cell. Molec. Neurobiol.* **1:** 41–55, 1981.

Johnston, D. and Brown, T. H.: Giant synaptic potential hypothesis for epileptiform activity. *Science* **211:** 294–297, 1981.

Johnston, D., Hablitz, J. J., and Wilson, W. A.: Voltage clamp discloses slow inward current in hippocampal burst-firing neurones. *Nature* **286:** 391–393, 1980.

Kandel, E. R. and Spencer, W. A.: Electrophysiology of hippocampal neurons. II. After-potentials and repetitive firing. *J. Neurophysiol.* **24:** 243–259, 1961.

Knowles, W. D. and Schwartzkroin, P. A.: Local circuit synaptic interactions in hippocampal brain slices. *J. Neurosci.* **1:** 318–322, 1981.

Kocsis, J. D., Malenka, R. C., and Waxman, S. G.: Enhanced parallel fiber frequency following after reduction of postsynaptic activity. *Brain Res.* **207:** 321–331, 1981.

Kocsis, J. D., Malenka, R. C., and Waxman, S. G.: Effects of extracellular potassium concentration on the excitability of the parallel fibres of the rat cerebellum. *J. Physiol.* **334:** 225–244, 1983.

Kocsis, J. D., Ruiz, J. A., and Cummins, K. L.: Modulation of axonal excitability mediated by surround electric activity: an intraaxonal study. *Exp. Brain Res.* **47:** 151–153, 1982.

Kriegstein, A. R., Suppes, T., and Prince, D. A.: Cholinergic enhancement of penicillin-induced epileptiform discharges in pyramidal neurons of the guinea pig hippocampus. *Brain Res.* **266:** 137–142, 1983.

Lebeda, F. J., Hablitz, J. J., and Johnston, D.: Antagonism of GABA-mediated responses by *d*-tubocurarine in hippocampal neurons. *J. Neurophysiol.* **48:** 622–632, 1982.

Lorente De No, R.: Studies on the structure of the cerebral cortex. II. Continuation of the study of the ammonic system. *J. Psychol. Neurol.* **46:** 113–177, 1934.

Lothman, E. W., Collins, R. C., and Ferrendelli, J. A.: Kainic acid-induced limbic seizures: Electrophysiologic studies. *Neurology* **31:** 806–812, 1981.

MacVicar, B. A. and Dudek, F. E.: Local synaptic circuits in rat hippocampus: Interactions between pyramidal cells. *Brain Res.* **184:** 220–223, 1980.

MacVicar, B. A. and Dudek, F. E.: Electrotonic coupling between pyramidal cells: A direct demonstration in rat hippocampal slices. *Science* **213:** 782–785, 1981.

Madison, D. V. and Nicoll, R. A.: Noradrenaline blocks accommodation of pyramidal cell discharge in the hippocampus. *Nature* **299:** 636–638, 1982.

Malenka, R. C., Kocsis, J. D., Ransom, B. R., and Waxman, S. G.: Modulation of parallel fiber excitability by postsynaptically mediated changes in extracellular potassium. *Science* **214:** 339–341, 1981.

Matsumoto, H. and Ajmone Marsan, C.: Cortical cellular phenomena in experimental epilepsy: Interictal manifestations. *Exp. Neurol.* **9:** 286–304, 1964a.

Matsumoto, H. and Ajmone Marsan, C.: Cortical cellular phenomena in experimental epilepsy: Ictal manifestations. *Exp. Neurol.* **9:** 305–326, 1964b.

Mesher, R. A. and Schwartzkroin, P. A.: Can CA3 epileptiform burst discharge induce bursting in normal CA1 hippocampal neurons? *Brain Res.* **183:** 472–476, 1980.

Miles, R. and Wong, R. K. S.: Single neurones can influence synchronized population discharge in the CA3 region of the guinea pig hippocampus. *Nature* **306:** 371–373, 1983.

Miles, R. and Wong, R. K. S.: Unitary inhibitory synaptic potentials in the guinea-pig hippocampus in vitro. *J. Physiol.* **356:** 97–113, 1984.

Miles, R. and Wong, R. K. S.: Excitatory synaptic connexions between guinea-pig CA3 hippocampal cells are revealed when synaptic inhibition is suppressed *in vitro. J. Physiol.* **372:** 14P, 1986a.

Miles, R. and Wong, R. K. S.: Excitatory synaptic interactions between CA3 neurones in the guinea-pig hippocampus. *J. Physiol.* **373:** 397–418, 1986b.

Miles, R., Wong, R. K. S., and Traub, R. D.: Synchronized afterdischarges in the hippocampus: Contribution of local synaptic interaction. *Neuroscience* **12:** 1179–1189, 1984.

Newberry, N. R. and Nicoll, R. A.: A bicuculline-resistant inhibitory post-synaptic potential in rat hippocampal pyramidal cells *in vitro. J. Physiol.* **348:** 239–254, 1984.

Niedermeyer, E.: Depth Electroencephalography, In: *Electroencephalography. Basic Principles, Clinical Applications and Related Fields* (E. Niedermeyer and F. Lopes da Silva, eds.) Urban and Schwarzenberg, Baltimore, 1982.

Noebels, J. L. and Prince, D. A.: Development of focal seizures in cerebral cortex: Role of axon terminal bursting. *J. Neurophysiol.* **41:** 1267–1281, 1978.

Numann, R., Wong, R. K. S., and Clark, R.: Electrophysiology of single dissociated cortical neurones. *Soc. Neurosci. Abs.* **8:** 413, 1982.

Ogata, N.: Possible explanation for interictal-ictal transition: Evolution of epileptiform activity in hippocampal slice by chloride depletion. *Experientia* **34:** 1035–1036, 1978.

Ogata, N., Hori, N., and Katsuda, N.: The correlation between extracellular potassium concentration and hippocampal epileptic activity in vitro. *Brain Res.* **110:** 371–375, 1976.

Oliver, A. P., Hoffer, B. J., and Wyatt, R. J.: Interaction of potassium and calcium in penicillin-induced interictal spike discharge in the hippocampal slice. *Exp. Neurol.* **62:** 510–520, 1978.

Pedley, T. A. and Traub, R. D.: Physiology of Epilepsy, In: *Scientific Basis of Clinical Neurology* (M. Swash and C. Kennard, eds.) Churchill Livingstone, London, 1985.

Pedley, T. A., Fisher, R. S., Futamachi, K., and Prince, D. A.: Regulation of extracellular potassium concentration in epileptogenesis. *Fed. Proc.* **35:** 1254–1259, 1976.

Prince, D. A.: The depolarization shift in ''epileptic'' neurons. *Exp. Neurol.* **21:** 467–485, 1968.

Prince, D. A.: Mechanisms of Epileptogenesis in Brain-Slice Model Systems, In: *Epilepsy* (A. A. Ward, Jr., J. K. Penry, and D. Purpura, eds.) Raven, New York, 1983.

Prince, D. A. and Futamachi, K. J.: Intracellular recordings in chronic focal epilepsy. *Brain Res.* **11:** 681–684, 1968.

Prince, D. A. and Wong, R. K. S.: Human epileptic neurons studied in vitro. *Brain Res.* **210:** 323–333, 1981.

Rall, W.: Theory of physiological properties of dendrites. *Ann. NY Acad. Sci.* **96:** 1071–1092, 1962.

Ribak, C. E., Harris, A. B., Vaughn, J. E., and Roberts, E.: Inhibitory, GABAergic nerve terminals decrease at sites of focal epilepsy. *Science* **205:** 211–214, 1979.

Richardson, T. L., Turner, R. W., and Miller, J. J.: Extracellular fields influence transmembrane potentials and synchronization of hippocampal neuronal activity. *Brain Res.* **294:** 255–262, 1984.

Rutecki, P. A. and Johnston, D.: Extracellular potassium controls the frequency of spontaneous interictal discharges in hippocampal slices. *Soc. Neurosci. Abs.* **9:** 396, 1983.

Sawa, M., Nakamura, K., and Naito, H.: Intracellular phenomena and spread of epileptic seizure discharges. *Electroenceph. Clin. Neurophysiol.* **24:** 146–154, 1968.

Schwartzkroin, P. A.: Characteristics of CA1 neurons recorded intracellularly in the hippocampal *in vitro* slice preparation. *Brain Res.* **85:** 423–436, 1975.

Schwartzkroin, P. A.: Further characteristics of hippocampal CA1 cells *in vitro. Brain Res.* **128:** 53–68, 1977.

Schwartzkroin, P. A.: Secondary range rhythmic spiking in hippocampal neurons. *Brain Res.* **149:** 247–250, 1978.

Schwartzkroin, P. A. and Knowles, W. D.: Intracellular study of human epileptic cortex: *In vitro* maintenance of epileptiform activity. *Science* **223:** 709–712, 1984.

Schwartzkroin, P. A. and Prince, D. A.: Penicillin-induced epileptiform activity in the hippocampal *in vitro* preparation. *Ann. Neurol.* **1:** 463–469, 1977.

Schwartzkroin, P. A. and Prince, D. A.: Cellular and field potential properties of epileptogenic hippocampal slices. *Brain Res.* **147:** 117–130, 1978.

Schwartzkroin, P. A. and Prince, D. A.: Changes in excitatory and inhibitory synaptic potentials leading to epileptogenic activity. *Brain Res.* **183:** 61–76, 1980.

Schwartzkroin, P. A. and Strafstrom, C. E.: Effects of EGTA on the calcium-activated afterhyperpolarization in hippocampal CA3 cells. *Science* **210:** 1125–1126, 1980.

Sloper, J. J., Johnson, P., and Powell, T. P. S.: Selective degeneration of interneurons in the motor cortex of infant monkeys following controlled hypoxia: A possible cause of epilepsy. *Brain Res.* **198:** 204–209, 1980.

Smith, D. O.: Mechanisms of action potential propagation failure at sites of axon branching in the crayfish. *J. Physiol.* **301:** 243–259, 1980.

Snow, R. W. and Dudek, F. E.: Electrical fields directly contribute to action potential synchronization during convulsant-induced epileptiform bursts. *Brain Res.* **323:** 114–118, 1984.

Snow, R. W., Taylor, C. P., and Dudek, F. E.: Electrophysiological and optical changes in slices of rat hippocampus during spreading depression. *J. Neurophysiol.* **50:** 561–572, 1983.

Spray, D. C., White, R. L., Campos De Carvalho, A., Harris, A. L., and Bennett, M. V. L.: Gating of gap junction channels. *Biophys. J.* **45:** 219–230, 1984.

Swann, J. W. and Brady, R. J.: Penicillin-induced epileptogenesis in immature rat CA3 hippocampal cells. *Dev. Brain Res.* **314:** 243–254, 1984.

Swann, J. W., Brady, R. J., Friedman, R. J., and Smith, E. J.: Penicillin-induced epileptiform discharges in CA3 hippocampal pyramidal cells: A current source density analysis. *Soc. Neurosci. Abs.* **9:** 395, 1983.

Taylor, C. P. and Dudek, F. E.: Synchronous neural afterdischarges in rat hippocampal slices without active chemical synapses. *Science* **218:** 810–812, 1982.

Taylor, C. P. and Dudek, F. E.: Excitation of hippocampal pyramidal cells by an electrical field effect. *J. Neurophysiol.* **52:** 126–142, 1984a.

Taylor, C. P. and Dudek, F. E.: Synchronization without active chemical synapses during hippocampal afterdischarges. *J. Neurophysiol.* **52:** 143–155, 1984b.

Traub, R. D.: Simulation of intrinsic bursting in CA3 hippocampal neurons. *Neuroscience* **7:** 1233–1242, 1982.

Traub, R. D. and Wong, R. K. S.: Synchronized burst discharge in disinhibited hippocampal slice. II. Model of cellular mechanism. *J. Neurophysiol.* **49:** 442–458, 1983a.

Traub, R. D. and Wong, R. K. S.: Synaptic mechanisms underlying interictal spike initiation in a hippocampal network. *Neurology* **33:** 257–266, 1983b.

Traub, R. D. and Wong, R. K. S.: Cellular mechanism of neuronal synchronization in epilepsy. *Science* **216:** 745–747, 1982.

Traub, R. D., Dudek, F. E., Taylor, C. P., and Knowles, W. D.: Simulation of hippocampal afterdischarges synchronized by electrical interactions. *Neuroscience* **14:** 1033–1038, 1985.

Traub, R. D., Dudek, F. E., Snow, R. W., and Knowles, W. D.: Computer simulations indicate that electrical field effects contribute to the shape of the epileptiform field potential. *Neuroscience* **15:** 947–958, 1985a.

Traub, R. D., Knowles, W. D., Miles, R., and Wong, R. K. S.: Synchronized afterdischarges in the hippocampus: Simulation studies of the cellular mechanism. *Neuroscience* **12:** 1191–1200, 1984.

Turner, D. A. and Schwartzkroin, P. A.: Electrical characteristics of dendrites and dendritic spines in intracellularly-stained CA3 and dentate neurons. *J. Neuroscience* **3:** 2381–2394, 1983.

Wong, R. K. S. and Prince, D. A.: Dendritic mechanisms underlying penicillin-induced epileptiform activity. *Science* **204:** 1228–1231, 1979.

Wong, R. K. S. and Prince, D. A.: Afterpotential generation in hippocampal pyramidal cells. *J. Neurophysiol.* **45:** 86–97, 1981.

Wong, R. K. S. and Traub, R. D.: The dendrites and somata of hippocampal pyramidal cells generate different action potential patterns. *Soc. Neurosci. Abs.* **8:** 412, 1982.

Wong, R. K. S. and Traub, R. D.: Synchronized burst discharge in the disinhibited hippocampal slice. I. Initiation in the CA2-CA3 region. *J. Neurophysiol.* **49:** 459–471, 1983.

Wong, R. K. S. and Watkins, D. J.: Cellular factors influencing GABA response in hippocampal pyramidal cells. *J. Neurophysiol.* **48:** 938–951, 1982.

Wong, R. K. S., Prince, D. A., and Basbaum, A. I.: Intradendritic recordings from hippocampal neurons. *Proc. Nat. Acad. Sci. USA* **76:** 986–990, 1979.

Wyler, A. R., Ojemann, G. A., and Ward, A. A., Jr.: Neurons in human epileptic cortex: Correlation between unit and EEG activity. *Ann. Neurol.* **11:** 301–308, 1982.

Yaari, Y., Konnerth, A., and Heinemann, U.: Spontaneous epileptiform activity of CA1 hippocampal neurons in low extracellular calcium solutions. *Exp. Brain Res.* **51:** 153–156, 1983.

Experimental Epilepsy Induced By Direct Topical Placement of Chemical Agents on the Cerebral Cortex

Charles R. Craig and Brenda K. Colasanti

1. Introduction

An almost endless variety of agents, when applied directly to the surface of the brain, are capable of producing convulsions. The most important compounds were last methodically reviewed by Ward (1972) and Prince (1972). Some of these agents produce convulsions that are immediate in onset, whereas with others there is a delay before the appearance of seizures. The convulsive episodes produced by several compounds last for only a few minutes, although the seizure states produced by others persist for the life of the animal. Some agents produce seizures in only certain species, whereas others produce convulsions in all species that have been examined. These variations should be taken into consideration when neurochemical factors responsible for the epileptogenic state are assessed, and they will be addressed in this chapter.

An important aspect in considering the role of neurotransmitter abnormalities in experimental seizure models is whether the epileptic states produced by local application of various substances to the brain have any relevance to human epilepsy. If there are similarities to human epilepsy, are the models produced by use of some epileptogenic materials more relevant to the study of the human disease than others? An attempt will be made to answer this question, and, in particular, to establish whether observed abnormalities in neurotransmitters in the experimental models may provide clues to neurochemical alterations existent in human epilepsy. One hallmark of human epilepsy is the intermittent and paroxysmal nature of the disorder. Only a limited number of the epileptogenic agents examined to date produce persistent intermittent seizure episodes in experimental animals.

In the ensuing discussion of the variety of agents that cause seizures when applied topically to the brain, the involved compounds have been arbitrarily separated into two groups: metals and nonmetals. Although the metals are relatively uniformly characterized chemically, the nonmetals include a diverse group of substances bearing no apparent chemical relationship.

2. Metals

Included among the metals and/or their salts capable of inducing a seizure state after topical application to the brain are aluminum (Kopeloff et al., 1942), cobalt and nickel (Kopeloff, 1960), tungsten (Blum and Liban, 1960), antimony, bismuth, cadmium, chromium, tin, titanium, iron, molybdenum, zirconium, mercury, vanadium, tungsten and tantalum (Chusid and Kopeloff, 1962), beryllium and lead (Chusid and Kopeloff, 1967), and zinc (Donaldson et al., 1971). The seizure-provoking properties of most of these metals have been examined only superficially and will not be considered further in this chapter. A reasonable amount of data has been gathered on the epileptogenic state induced by aluminum, cobalt, and iron.

Metals have a major inherent advantage over most other epileptogenic agents, in that following application to an area of the cortex, they remain at or close to that site, with only minimal diffusion to other areas of the brain. In addition, these metals normally are not actively transported by the blood, nor do they undergo metabolism.

2.1. Aluminum

Aluminum was the first metal demonstrated to be epileptogenic when applied directly to the cerebral cortex of monkeys (Kopeloff et al., 1942). This metal was initially used in the form of alumina cream, but it is currently employed as a commercially available aluminum hydroxide gel, which will be referred to in this chapter as alumina. Aluminum has also been shown to be epileptogenic in the cat (Mayman et al., 1965), and there is some evidence for its epileptogenicity in the dog (Aida, 1956; Essig, 1962) and in the rabbit (Harmony et al., 1968). This metal does not appear to cause a substantial epileptogenic response in the rat (Servit and Sterc, 1958; Craig and Colasanti, unpublished findings).

Ward (1972) thoroughly described the alumina model in the monkey and pointed out that it appears to offer a particularly good experimental model of epilepsy when applied to the sensorimotor cortex, and, more specifically, to the postcentral gyrus. The seizures are chronic, intermittent, and persistent. They may be more frequent at certain times of the day or night, and in certain female monkeys, the frequency of seizures appears to be greater around the time of estrus (Ward, 1972).

The time course for alumina-induced epilepsy in the monkey is different from that of other agents in that the seizure state is quite slow to develop. There is a long delay, usually 35–60 d, but sometimes as long as 8 mo, before clinical seizures are apparent (Ward, 1972). Once seizures appear, they continue to occur spontaneously throughout the lifetime of the animal. A major difference between this model and human epilepsy that might be significant is the finding that seizures can be quite easily elicited in monkeys by stress stimuli (Lockard and Barensten, 1967).

Morphologically, according to Ward (1972), loss of neurons and astrocytic gliosis is similar in alumina-induced epilepsy in the monkey and in focal human epilepsy. Furthermore, alterations in dendritic morphology, as evidenced by Golgi impregnation, are similar in both monkey alumina epilepsy and human epileptic foci.

In morphological studies in humans, there has been difficulty in discerning which changes are causal and which are the effects of long-term epilepsy. The elegant studies undertaken by Scheibel and coworkers (Scheibel et al., 1983) have provided clarification, but as Scheibel et al. (1983) point out, ''Until the etiology and pathogenesis of an altered state are fully understood, it may be difficult to determine whether observed tissue changes bear a causal relationship to that state.''

Although several months may be required for the development of alumina epilepsy in monkeys, the seizures are persistent. In contrast, convulsions develop in cats after about 35 d, but terminate by 50 d (Velasco et al., 1973a). Morphologically, light microscopy revealed the existence of cortical atrophy, disorganization of cortical layers, and neuronal depopulation in alumina-treated cats; however, many of these changes were also seen in control cats treated with silicon only (Velasco et al., 1973b). As the silicon-treated cats were not epileptic, the significance of these findings is questionable. A neuropathological study of alumina gel in cats was also conducted by Mayman et al. (1965), who observed a chronic type of inflammatory reaction after about 8 wk.

Stercova (1966) applied alumina cream to the motor cortex of the rat and followed histologically the development of the necrosis over a 2-yr period. She found nerve cell injury that persisted throughout the period of the study, but did not observe spontaneous seizures in any of the rats.

2.1.1. GABA

Insight into both morphological and neurochemical factors underlying epileptogenesis in the alumina model of epilepsy was provided by the recent studies of Ribak et al. (1979; 1982). This group used an immunocytochemical method for localizing glutamic acid decarboxylase (GAD), the gamma-aminobutyric acid (GABA) synthesizing enzyme, in brain slices from alumina gel-treated and control monkeys (Ribak et al., 1979). They noted a marked decrease of GAD-positive terminals in the immunocytochemical preparations from epileptic monkey cortex relative to nonepileptic controls. Ribak et al. (1982) subsequently conducted a careful quantitative ultrastructural analysis as a follow-up to their immunocytochemical studies of monkey neocortex following the injection of alumina gel. They made several observations: (1) The number of axosomatic symmetric synapses (assumed to be primarily inhibitory and GABAergic) within layer V pyramidal cells was decreased by 80% in the focus and by 50% in tissue adjacent to the focus. At the same time, the number of asymmetric synapses (assumed to be primarily excitatory) was reduced by only 25% in the focus and by 15% in adjacent tissue (para focus). They concluded that the large reduction in the number of symmetric synapses at the epileptic focus provides an indication that the previously observed loss of GABAergic terminals at sites of focal epilepsy is caused by terminal degeneration. The loss of such terminals could be the basis for seizure activity caused by a preferential decrease of inhibitory function at epileptic foci. The basis for the preferential loss of GABAergic terminals could be that these cells have higher metabolic activity than do excitatory cells (an increased number of mitochondria in GABA-ergic terminals supports this idea), and that they are, therefore, more susceptible to hypoxia. It was reasoned that alumina gel may act by compromising the cortical vasculature and may thereby cause hypoxia.

Bakay and Harris (1981) studied normal and alumina-epileptic monkey cortex using ligand binding techniques. The epileptic cortex demonstrated markedly reduced GABA receptor binding, as well as decreased GABA concentrations and reduced GAD activity. These authors also measured receptor binding for glutamate, muscarinic cholinergic, and both alpha and beta adrenergic receptors. They found significant decreases in receptor binding in the nongranulomatous area immediately adjacent to the lesion site for all four transmitters.

2.1.2. Taurine

In view of the interest in taurine as a possible anticonvulsant and a possible neurotransmitter, Mutani et al. (1974b) administered a 200 mg/kg dose of this amino acid to alumina-treated cats. He observed a decrease in seizure frequency.

2.1.3. Ions

Riedel et al. (1971) measured cation concentrations in the brains of alumina-treated cats and reported a decrease in potassium that was accompanied by an increase in sodium at the times of seizure activity. In another study, Harmony et al. (1968) studied $Na^+ - K^+$ ATPase in the foci between 6 and 15 d in rabbits following alumina cream placement, by which time there were already spikes in the EEG. However, these authors showed that aluminum in vitro was also capable of inhibiting $Na^+ - K^+$ ATPase. The inhibition of the enzyme might accordingly be a consequence of high concentration of the metal in the tissue and is probably not related to the seizure state.

2.2. Cobalt

Kopeloff (1960) originally demonstrated that pure cobalt was epileptogenic when applied to the cerebral cortex of the mouse. The epileptogenic effects of cobalt were subsequently confirmed in a variety of species, including rat (Dow et al., 1962), cat (Henjyoji and Dow, 1965), rabbit (Dimov, 1966), monkey (Chusid and Kopeloff, 1967), and gerbil (Payan, 1970). The majority of studies with cobalt have been carried out in the rat with the cat being the next most extensively employed species.

Henjyoji and Dow (1965) were the first to describe the seizure state produced in the cat by application of cobalt powder to the anterior sigmoid gyrus. Within 30 h, seizure activity was noted. The seizures, which were seen in 8 of 10 cats treated with cobalt, lasted only 3–4 d.

Colasanti et al. (1974) first demonstrated the time course for the development of convulsions in the cobalt model in the rat. They showed that very few seizures occur during the first 5 d after cobalt application, that there is a 7–10-d period of peak seizure activity, and that this is followed by a period of decreased seizure incidence. Few seizures are seen later than 2 wk after cobalt application. The number of seizures can be markedly increased by placement of cobalt on both the right and the left cerebral cortices (Craig et al., 1976). The time course of the seizures, however, is not altered. Ionic cobalt appears to be as effective in producing seizures as the pure metal (Willmore et al., 1975; Pekoe et al., 1978). Payan et al. (1965) observed that cobalt metal was epileptogenic in rats whether placed subdurally or extradurally.

2.2.1. Changes in Enzyme Levels

Brotchi et al. (1978) performed a histochemical study in which they examined dehydrogenases in "activated" astrocytes by using a photonic microscope to detect an insoluble, colored tetrazolium salt. They observed an increase in dehydrogenase activity in the area of the cobalt lesion, but not in the contralateral side. Based on observed seizure frequency, these authors concluded that the increased enzyme activity was not a consequence of the seizures. However, the authors observed the animals electrocorticographically for only 30 min each day and determined their degree of epileptogenicity on the basis of this limited amount of EEG and behavioral observation. In a subsequent paper, the same group also noted increases in phosphorylase and glycosyltransferases in "activated" astrocytes associated with the cobalt focus (Gerebtzoff et al., 1978). Singh (1980) studied lysosomal acid phosphatase in primary focus and in the homotopic area of the contralateral cortex in cobalt-treated rat brain. He observed a marked histochemically demonstrable increase in the activity of lysosomal acid phosphatase in both the primary and secondary focal areas.

2.2.2. Norepinephrine

Trottier et al. (1981) conducted a careful histochemical and biochemical study of the cortical noradrenergic system following administration of cobalt powder to rat cortex. Their results revealed a profound alteration in the noradrenergic system during the development of the cobalt focus. Histochemically, there was a modest decrease in the density of noradrenergic terminals at 2 d and a greater decrease at 8 and 14 d following cobalt. Thirty days after cobalt (when seizure activity had ceased), new adrenergic fibers had appeared and had invaded the perifocal area. The new fibers appeared randomly distributed and were not oriented parallel to the pial surface, as is typical in control rats. By 90 d fluorescence in the perifocal area was normal in appearance. Norepinephrine (N) concentration, likewise, was reduced by 2 d, with the greatest decrease occurring at 8 and 14 d. Although there was a tendency for cortical N levels to increase at the later time periods, they were still significantly lower than those of control animals not subjected to surgery. The only troublesome aspect of this study was an unexplained significant decrease in N in the sham-operated rats at 14 and 30 d.

In a subsequent document, Trottier et al. (1983) reported on the uptake and metabolism of N during cobalt epilepsy in the rat. They demonstrated a significant reduction in high-affinity uptake of N at 8 to 10 d after cortical cobalt application in tissue adjacent to the cobalt lesion. The high-affinity uptake remained depressed up to 40 d after cobalt, at which time it had returned to control levels. In the homotopic area of the contralateral cortex, a significant reduction in high-affinity N uptake occurred only at the 10-d period. N concentration and the concentration of its metabolite, 3,4-dihydroxyphenylethyleneglycol (DOPEG), were also measured. A 36% reduction in N concentration and a 19% reduction in DOPEG occurred at d 8–10; on the other hand, no significant changes were noted at either d 2 or 40.

2.2.3. Serotonin

The work of Trottier et al. (1983) also provided an analysis of the central serotonergic system in the cobalt model. As in the case of norepinephrine, a reduction in the uptake of serotonin (5-HT) occurred at d 8–10, but only at this time period and only in the ipsilateral cortex. Concentrations of 5-HT and of its major metabolite, 5-hydroxyindole acetic acid, were also measured. The only change in these levels was a significant reduction in 5-HT concentration at d 8–10.

2.2.4. Acetylcholine

Goldberg et al. (1972) measured choline acetyltransferase (CAT) and acetylcholinesterase (AChE) activities in tissue adjacent to the cobalt lesion exclusive of any visually necrotic tissue, as well as in tissue in the homotopic area of the contralateral cortex at d 2, 7, 21, and 35 after insertion of cobalt into the cortex. Both enzymes were significantly reduced at 7 d in the ipsilateral cortex, but had returned to control values by the later two time periods; no changes were observed in contralateral cortex. The changes in AChE and CAT were confirmed by Emson and Joseph (1975). In a follow-up study, Hoover et al. (1977) measured acetylcholine (ACh) levels in tissue adjacent to the lesion at d 3, 5, 7, 14, and 21 following cobalt or glass (control) implantation. They found a significant decrease in ACh on d 7 and 14 and a return to control levels by d 21. Hoover et al. (1977) also showed that the cholinesterase inhibitors, physostigmine and diisopropylfluorophosphate (DFP), were capable of preventing seizures as well as other EEG abnormalities in this model.

2.2.5. GABA

GABA has been clearly shown to be a major inhibitory neurotransmitter in the mammalian nervous system, with an extensive scientific literature relating GABA to many different types of seizure disorders. In the cobalt model, Van Gelder and Courtois (1972) initially reported a significant decrease in brain GABA in cobalt-treated cats. Similar changes were simultaneously observed in cobalt-epileptic mice (Van Gelder, 1972). Emson and Joseph (1975) subsequently demonstrated, in the rat, that GABA and its synthetic enzyme, GAD, were both significantly depressed in tissue surrounding the focus 4–10 d following cobalt implantation, with a return toward normal by d 24. GAD, but not GABA, was significantly reduced in the contralateral cortex as well. Ross and Craig (1981b) verified the above data for GABA and GAD in the ipsilateral cortex and demonstrated that these were not changes associated with cellular necrosis since copper, an element that produces extensive necrosis but no seizure activity, did not cause alterations in GABA or GAD at any time period examined.

Receptor ligand binding studies have clearly established the presence of GABA recognition sites on postsynaptic membranes (Enna and Snyder, 1975). Besides attaching to its own receptor, GABA can also interact at other sites, including transport sites and loci on enzymes concerned with its synthesis and degradation.

Balcar et al. (1978) studied kinetic parameters of high-affinity uptake of GABA into cobalt-induced cortical epileptogenic foci in rats and reported a decreased V_{max} for the uptake of GABA in the focal area. Ross and Craig (1981a) studied high-affinity postsynaptic GABA receptor binding characteristics (K_d and B_{max}) in crude synaptic membranes from rat brains treated with cobalt, copper, or glass. Tissue adjacent to the focus was collected 1 d (before seizures ordinarily are seen), 7 d (period of maximal seizures), and 21 d (period of minimal seizure activity) after surgery. No alteration in affinity or the number of postsynaptic binding sites was observed 1 d following cobalt implantation in comparison with values for animals treated with glass or copper. At 7 d, however, marked increases in postsynaptic GABA binding sites occurred in cobalt-treated rats. Again, there was no change in affinity for GABA binding. By 21 d, the number of GABA binding sites was still slightly elevated, although not significantly so.

The reason for an apparent increase in GABA receptor number is not clear. This change may well represent an adaptive process (such as that seen with the phenomenon of supersensitivity) as an attempt to compensate for the lowered levels of GABA, the reduced GAD activity, and the lowered set point of the high-affinity GABA transport mechanism resulting from cobalt implantation.

Another approach that has been used to demonstrate an alteration in the GABA system is Wolman's fluorescence procedure. Seidel et al. (1981) demonstrated a distinct decrease in GABA fluorescence in the pyramidal and granular layers of the hippocampus of the rat 21 d after cobalt administration to the sensorimotor cortex.

In a pharmacological study, Emson (1976) administered aminoxyacetic acid (AOAA), an inhibitor of gamma-aminobutyric aminotransferase (GABA-T), and valproic acid, an agent that has some GABA-T inhibitory action, among other effects, to cobalt-epileptic rats. He concluded that there was no simple relationship between the elevation of brain GABA and the anticonvulsant activity of AOAA. The dose of AOAA that produced the greatest increase in brain GABA levels (i.e., 10 mg/kg ip) was less effective as an anticonvulsant than smaller doses. Although valproic acid did cause an elevation in brain GABA levels, it was not effective as an anticonvulsant in this study.

2.2.6. Glutamic Acid

Just as GABA is recognized to be the major mammalian inhibitory neurotransmitter, there is now evidence that quantitatively glutamic acid may be the most important excitatory transmitter in the mammalian nervous system. Aspartic acid will be discussed with glutamic acid, since the effects of the two are virtually identical and they are interconvertible. These two amino acids do differ significantly in their relative distribution in different parts of the nervous system.

Van Gelder and Courtis (1972) reported significant decreases in both glutamic acid and aspartic acid in tissue adjacent to the focal area in cobalt-epileptic cats. Koyama (1972) confirmed these findings in the cat, and Van Gelder (1972) observed similar reductions in focal tissue from cobalt-epileptic mice. Craig and Hartman (1973) subsequently reported decreased concentrations of both glutamic acid and aspartic acid in cobalt-epileptic rats.

2.2.7. Taurine

Reduced levels of taurine have likewise been observed in focal tissue collected from cobalt-epileptic cats (Van Gelder and Courtois, 1972), mice (Van Gelder, 1972), and rats (Craig and Hartman, 1973). In a pharmacological study, Van Gelder (1972) administered taurine to both cobalt-epileptic mice and cats. He reported that repeated injections of this amino acid were capable of sharply reducing or abolishing seizure activity in both species. Interestingly, a return to normal levels of both GABA and glutamic acid, as well as taurine, paralleled the reduction in seizures. After administration of taurine to mice for 2 d, taurine, GABA, and glutamic acid levels had returned to normal by the 7th d. In cats, a complete recovery of GABA and glutamic acid was observed after taurine injections. In an extensive study on taurine in the cobalt-epileptic rat, Joseph and Emson (1976) documented a significant reduction in endogenous taurine in the area adjacent to the focus at 8 d, with subsequent return toward normal values at 24 and 30 d. These investigators were unable to detect any changes in the lowered levels of GABA or glutamic acid following oral administration of taurine. Chronic oral or intraventricular administration of taurine likewise failed to modify the development of spike activity in response to cobalt.

2.2.8. Other Amino Acids

Van Gelder and Courtois (1972) reported that increases in glutamine and glycine levels paralleled the reductions in the major inhibitory and excitatory amino acids in the brains of cobalt-epileptic cats. Koyama (1972) likewise detected increases in glycine in cobalt-treated cats and increases in serine and threonine as well. Van Gelder (1972) documented significant elevations in glycine, serine, and alanine in cobalt-epileptic mice, whereas Craig and Hartman (1973) reported on increases in alanine and lysine in the rat counterpart.

2.2.9. Lipids

Cenedella and Craig (1973) measured total free fatty acids, free cholesterol, esterified cholesterol, triglycerides, and phospholipids in cerebral cortex samples from control and cobalt-epileptic rats and reported that cobalt-induced epilepsy was associated with significant changes in cerebral cortical lipids in the area of the lesion, as well as in the nonnecrotic tissue adjacent to the lesion. The total lipid in the lesioned area decreased sharply as a result of reductions in free cholesterol and total phospholipids. The levels of cholesterol esters and triglycerides increased in the area of the lesion, and cholesterol esters were also increased in the adjacent tissue. In addition, there were decreases in the proportion of phosphatidylethanolamine in the phospholipids from the lesion site and adjacent tissue and decreases in the proportions of oleic, arachidonic, and nervonic acids (unsaturated acids), and an increase in the proportions of lignoceric acid in the phospholipids.

In a related study, Lunt and Grove (1975) measured unesterified fatty acids in an area of rat brain immediately surrounding the epileptic focus produced by cobalt powder. They reported an increase of about 80% in the content of unesterified fatty acids from tissue around the focal area. The predominant lipid was arachidonic acid. Bierkamper and Cenedella (1978) conducted a very extensive investigation on cerebral cortical cholesterol changes during cobalt-induced epilepsy in the rat. After implantation of cobalt (or other metals), measurements of cerebral cortical lipid concentrations were made in the lesion area and in the nonnecrotic tissue immediately adjacent to the lesion site. In rats treated with epileptogenic metals (i.e., cobalt and nickel), the cortical concentration of free cholesterol decreased in the adjacent and lesion site, although the concentration of cholesterol esters greatly increased in the adjacent area, in comparison with values for animals treated with copper and stainless steel. Differences in the fatty acid composition of cholesterol esters sampled from plasma and from the cobalt-lesioned cortex indicated that the accumulated cholesterol esters originated in the brain rather than from the plasma cholesterol ester pool. A time course study revealed that the changes in the cortical concentrations of free and esterified cholesterol precede the initial appearance of epileptiform activity as determined by electrocorticography. The conclusion of this study was that the reduction of free cholesterol levels probably reflects increased cholesterol metabolism in the cobalt-lesioned cortex. The postulated increased metabolism of cholesterol could not be explained by any changes in the activities of cholesterol ester synthetase and hydrolase(s).

Because cholesterol in the brain is localized almost exclusively in membranes, it is likely that the changes in brain sterols reflect losses in brain membranes. The resulting loss in membrane integrity could likely lead to the increases in excitability associated with epileptiform states.

2.2.10. Enkephalins

Colasanti et al. (1975) made the observation that "wet-dog shake" behavior was markedly enhanced in rats rendered epileptic by cobalt. Normally, wet-dog shakes are associated with narcotic withdrawal or precipitation of abstinence by naloxone. The relationship between endorphins and epilepsy is certainly not clear. It is well known that iv administration of the endogenous opioids, met-enkephalin and leu-enkephalin, is followed by the transient appearance of epileptiform activity in the EEG (Urca et al., 1977; Frenk et al., 1978). Because anticonvulsants specific for petit mal epilepsy antagonize the epileptogenic effect of leu-enkephalin, it has been postulated that enkephalins may be involved in the etiology of this particular seizure disorder.

2.2.11. Ions

Ionic disturbances are almost certainly involved in the manifestation of an epileptic state. Hunt and Craig (1973) reported an elevation of sodium and a modest decrease in potassium in tissue adjacent to the cobalt focus in rats. A marked increase in calcium in the focal area was also noted. Cobalt might partially exert its epileptic effect by blocking an active chloride extrusion mechanism (Sypert, 1982), since it has been postulated (Lux, 1971) that IPSPs are generated primarily by an increased postsynaptic membrane permeability to chloride ions.

2.3. Iron

There is some evidence that the epileptogenic state induced by salts of iron may have a different etiology than that produced by alumina or cobalt, although this has certainly not been established. It has been proposed that iron-induced epilepsy in rats may closely resemble posttraumatic epilepsy in humans (Willmore and Rubin, 1981; Sypert, 1982). In posttraumatic epilepsy, seizures may be a consequence of blood loss and the subsequent liberation of iron from hemoglobin and its sequestration into brain cells. Willmore and Rubin (1981) postulated that the iron salts could cause changes in membranes of subcellular organelles by causing peroxidation after the generation of superoxide radicals and/or peroxides. They tested this by administering two antiperoxidants, alpha-tocopherol and selenium, to ferrous chloride-treated rats. This treatment exerted a modest protective effect against seizures.

To date, no neurochemical studies have been conducted in animals rendered chronically epileptic by topical application of iron, although the anticonvulsant effects of marijuana cannabinoids have been examined (Turkanis and Karler, 1982). This is an area yet to be explored.

2.4. Zinc

Donaldson et al. (1971) reported that iv injection of ions of zinc, copper, iron, and manganese produced convulsions and that these same ions inhibited Na^+-K^+ ATPase from brain microsomes. Furthermore, the order of potency in producing convulsions was the same as inhibiting the enzyme: $Zn^{2+} > Cu^{2+} > Fe^{2+} > Mn^{2+}$. Barbeau and Donaldson (1974) proposed that inhibition of Na^+-K^+ ATPase, particularly in the hippocampus, may be related to epileptogenicity and that zinc, which is associated with cells in the mossy fiber layer of the hippocampus, may regulate the activity of the enzyme. Apparently, there has been little work subsequently undertaken with this proposal, although Pei et al. (1983) recently reported that the ic administration of zinc sulfate to rabbits produces seizures in virtually all the animals, but death as well in 90%.

3. Nonmetals

There are a large number of drugs and other chemical agents that reliably produce convulsions when applied directly to the brains of experimental animals. In some instances, the known pharmacological effects of these agents may provide clues as to the mechanisms that cause convulsions.

3.1. Penicillin G

Penicillin G is the most widely studied nonmetal epileptogenic agent. Stark et al. (1974) examined the time course for the development of an epileptogenic focus utilizing penicillin in immobilized, unanesthetized cats. This group reported the appearance of large-amplitude, single spikes within 2 min after the application of penicillin (15,000–60,000 U) to the cortex. The intensity of spiking increased for 1.5 h, after which time a progressive decrease in spiking was noted. Avoli (1980) noted a duration of epileptiform activity of between 3 and 5 h in the rat following bilateral application of penicillin to the cerebral hemispheres. Collins and Mehta (1978) reported a duration of seizures of slightly over 1 h following administration of penicillin to one side of the brain only.

It appears, therefore, that the duration of the epileptogenic response to topical penicillin is related to the time penicillin is actually at the site and that its action is terminated by its removal via the circulatory system. Presumably because the duration of the epileptogenic response to penicillin is too short to demonstrate morphological changes, no histochemical studies using this agent have been undertaken. Fischer and Langmeier (1980) observed that, within 15 min after application of penicillin, the number of synaptic vesicles in the homotopic area of the contralateral cortex in the rat was increased. There appeared to be more rounded vesicles in the penicillin-treated rats, but the involved neurotransmitters were not identified.

3.1.1. GABA

Schwartzkroin and Prince (1980) compared the effects on IPSPs and EPSP's of penicillin and bicuculline in the guinea pig in vitro hippocampus slice preparation. Both agents depressed IPSPs, but had no effects on EPSPs. These authors interpreted their results as indicating that both penicillin and bicuculline block cellular IPSPs, thereby allowing more distant excitatory events to trigger action potentials. This report is in agreement with a host of other studies, indicating that the principal epileptogenic action of penicillin is to produce a reduction of GABA-induced neuronal inhibition (Davidoff, 1972; Meyer and Prince, 1973; Van Duijn et al., 1973; Wong and Prince, 1979). Both penicillin and bicuculline have been reported to be specific antagonists of GABA (Curtis and Johnston, 1974). Moreover, recent studies indicate that penicillin exerts a nonspecific and reversible antagonism of several other inhibitory amino acids, including glycine and beta-alanine (Macon and King, 1979). Further evidence associating a GABA mechanism with the epileptogenicity of penicillin has been provided by the report of Collins and Mehta (1978). These investigators discovered that the GABA-T inhibitor AOAA was capable of markedly reducing the number of penicillin-induced spike discharges in the rat. They also reported that topical penicillin did not cause any change in the concentration of GABA *per se* in focal tissue. In the case of AOAA, however, there is no simple and direct relationship be-tween brain GABA concentration and anticonvulsant effect, since brain levels of GABA after AOAA treatment are higher after the anticonvulsant effect has dissipated.

3.1.2. Taurine

Mutani and coworkers (1974a,b) determined the effect of taurine on experimental epilepsy induced by penicillin and conjugated estrogens and strychnine. Although taurine exerted a marked anticonvulsant effect against convulsions induced by conjugated estrogens or strychnine, no protection against penicillin-induced seizures appeared until after 120 min. By this time, the epileptogenic effect of penicillin is markedly decreased.

3.1.3. Norepinephrine

Evidence has recently been forthcoming to suggest a role for noradrenergic pathways in the etiology of penicillin-induced epilepsy. Mueller and Dunwiddle (1983) determined the ability of a series of noradrenergic agonists to affect epileptiform activity in an in vitro rat hippocampal preparation superfused with penicillin and elevated potassium. Norepinephrine and several other alpha receptor agonists were all capable of decreasing interictal spike activity in this preparation, whereas the beta adrenoceptor agonists, 2-fluoro-norepinephrine and 1-isoproterenol increased interictal discharge rate. The alpha receptor antagonist phentolamine selectively blocked anticonvulsant response, whereas the beta antagonist, timolol, blocked the proconvulsant response.

Raabe et al. (1978) measured levels of the noradrenergic mediator cyclic-3′,5′-adenosine monophosphate (cAMP), and those of cyclic-3′,5′-guanosine monophosphate (cGMP) as well, in focal penicillin epilepsy in the cat. Although no changes were found in the case of cAMP, significant increases were found in cGMP during ictus, interictus, and postictus, with the greatest change occurring during the ictal stage.

In unrelated work, Roberts (1973) measured the incorporation of [5-^{3}H]-uridine into nuclear RNA of cells in various regions of the frog brain after topical application of penicillin. Significantly, [5-^{3}H]-uridine was found to exert an anticonvulsant effect in the frogs. Moreover, the [5-^{3}H]-uridine incorporation into nuclear RNA was increased twofold in both the focal area and the homotopic area of the contralateral cortex.

3.1.4. Acetylcholine

Unfortunately, the cholinergic system has been only cursorily examined in penicillin epilepsy. Using the guinea pig hippocampal slice preparation, Kriegstein et al. (1983) found that the focal application of ACh enhanced the multiphasic-evoked potential induced by penicillin, with an increase in the duration of the evoked reponse, as well as the number of spikes.

3.2. Convulsant Drugs

A number of convulsant drugs can produce seizures whether administered topically to the brain or parenterally. Among these are strychnine (a glycine antagonist), bicuculline (a GABA antagonist), picrotoxinin (a GABA antagonist), pentylenetetrazol (probably a GABA antagonist), and the convulsant hydrazides (which lower GABA by inhibiting GAD). Although these agents produce seizures, the seizures are not paroxysmal or intermittent. Instead, they are more or less continuous and either lead to death or complete recovery after termination of the drug effect by metabolism and/or excretion. They are not, therefore, considered true good models of epilepsy. Pharmacological studies suggest that they are acting by reducing normal inhibitory influences.

3.3. Ouabain

Ouabain provokes seizures when applied to the surface of the cerebral cortex (Lewin, 1970). The primary pharmacological property associated with this cardiac glycoside is its ability to inhibit cation transport and Na$^+$-K$^+$ ATPase.

3.4. Conjugated Estrogens

It has been recognized that fluctuations in sex hormones can alter seizure susceptibility (*see* Woodbury, 1969). Conjugated estrogens have been applied topically to the brain to produce an epileptogenic state that resembles absence epilepsy in that spike-slow wave discharges of about 2.5–3 cps predominate. As in the case of penicillin, the seizures persist for only a few hours (Marcus, 1972). Taurine markedly attenuates the seizure activity (Mutani et al., 1974a).

3.5. Kainic Acid

Kainic acid (KA), which is a rigid analog of glutamic acid, is a potent cytotoxic substance that has been widely employed as a lesioning agent to study neuronal cell loss. The administration of KA to the cortex results in a rather selective destruction of local neurons and spares terminals of axons projecting to the injection site from distant brain areas, as well as fibers of passage. The mechanisms underlying the neurotoxic effects of KA are not well understood, but appear to involve a sustained or repetitive depolarization and firing of neurons, perhaps by potentiating or mimicking the actions of endogenous excitatory amino acids (Sperk et al., 1983). Kainic acid applied to the hippocampus of the rat initiates a lethal status epilepticus. Nadler (1981) claims that the intraventricular injection of KA produces a behavioral seizure, lasting about 3–5 h, that occurs via brain damage similar to that in human temporal lobe epilepsy. A serious deficiency of the KA model is the fact that it does not produce spontaneous repetitive intermittent seizures (Nadler, 1981).

Kainic acid can also be administered systemically to produce long-lasting generalized tonic-clonic convulsions (Sperk et al., 1983). Again, this is an all-or-none event. An interesting phenomenon noted by Sperk et al. (1983) in this model in the rat is a marked increase in the incidence of wet-dog shakes, a phenomenon previously noted in the cobalt model (see above).

Several neurochemical changes have been noted after injection of kainic acid into the striatum. Sperk et al. (1981a) reported that 2 d after intrastriatal KA, the 5-HT metabolite 5HIAA was markedly increased and 5-HT itself was significantly decreased in the treated striatum. The dopamine (DA) metabolite, DOPAC, and DA turnover were both increased at the same time, although levels of DA itself were not altered. In a subsequent publication, Sperk et al. (1981b) reported a very-long-lasting (up to 10 wk) increase in the striatal histamine content after the intrastriatal injection of KA.

Although KA is not known to interact specifically with the GABA system, it appears that agents such as AOAA and gamma-vinyl GABA that increase GABA content in the CNS can antagonize KA-induced seizures (Nadler, 1981). Morphine enhances, and low doses of naloxone attenuate, KA-induced seizures (Fuller and Olney, 1979). In the latter study, KA was administered systemically.

4. Summary

In virtually every topically induced seizure model considered, alterations in the GABAergic system are seen. Regardless of whether the studies involved histochemical experiments, neuro-chemical measurements, or pharmacological manipulations, GABA-ergic mechanisms appear perturbed in all cases in which they have been examined. Experimental epilepsy induced by topical applica-tion of chemicals to the brain, regardless of the agent used to induce the seizure state or the particular species involved, is characterized by a decrease in brain GABA content, production, and utilization. Pharmacological techniques that lower brain GABA or antagonize its effects produce seizures, whereas techniques that elevate brain GABA, by and large, are anticonvulsants. This may indicate a preeminence of the GABA system in regulating the levels of brain excitability, or it may be a consequence of more studies being under-taken to study GABA since profound alterations in the cholinergic and noradrenergic systems, and perhaps the enkephalinergic system as well, appear in those instances in which they have been investigated. However, the volume of data implicating brain GABA in seizure mechanisms cannot be readily dismissed.

It is perhaps premature to consider one method for producing epilepsy to be superior to any other. The models that have been studied the most and that appear to provide reasonably consistent results include alumina-induced epilepsy in the monkey, cobalt-induced epilepsy in the rat, and the penicillin-induced seizure state in the cat or rat. Studies with the use of ferric salts in the rat are presently just emerging, and this model may prove to be a useful model of chronic experimental epilepsy.

References

Aida, S.: Experimental research on the function of the amygdaloid nuclei in psychomotor epilepsy. *Folia Psych. Neurol. Japonica* **10:** 181–207, 1956.

Avoli, M.: Electroencephalographic and pathophysiologic features of rat parenteral penicillin epilepsy. *Exp. Neurol.* **69:** 373–382, 1980.

Bakay, R. A. E. and Harris, A. B.: Neurotransmitter, receptor and biochemical changes in monkey cortical epileptic foci. *Brain Res.* **206:** 387–404, 1981.

Balcar, V. J., Pumain, R., Mark, J., Borg, J., and Mandel, P.: GABA-mediated inhibition in the epileptogenic focus, a process which may be involved in the mechanism of the cobalt-induced epilepsy. *Brain Res.* **154:** 182–185, 1978.

Barbeau, A. and Donaldson, J.: Zinc, taurine, and epilepsy. *Arch. Neurol.* **30:** 52–58, 1974.

Bierkamper, G. G. and Cenedella, R. J.: Cerebral cortical cholesterol changes in cobalt-induced epilepsy. *Epilepsia* **19:** 155–168, 1978.

Blum, B. and Liban, E.: Experimental basotemporal epilepsy in the cat. Discrete epileptogenic lesions produced in the hippocampus or amygdaloid by tungstic acid. *Neurology* **10:** 546–554, 1960.

Brotchi, J., Dresse, A., Scuvee-Moreau, J., and Gerebtzoff, M. A.: Histochemical study of cobalt-induced epilepsy. *Exp. Brain Res.* **32:** 459–469, 1978.

Cenedella, R. J. and Craig, C. R.: Changes in cerebral cortical lipids in cobalt-induced epilepsy. *J. Neurochem.* **21:** 743–748, 1973.

Chusid, J. G. and Kopeloff, L. M.: Epileptogenic effects of pure metals implanted in motor cortex of monkeys. *J. Applied Physiol.* **17,** 697–700, 1962.

Chusid, J. G. and Kopeloff, L. M.: Epileptogenic effects of metal powder implants in motor cortex of monkeys. *Inter. J. Neuropsychiat.* **3:** 24–28, 1967.

Colasanti, B. K., Hartman, E. R., and Craig, C. R.: Electrocorticogram and behavioral correlates during the development of chronic cobalt experimental epilepsy in the rat. *Epilepsia* **15:** 361–373, 1974.

Colasanti, B. K., Kosa, J. E., and Craig, C. R.: Appearance of wet dog shake behavior during cobalt experimental epilepsy in the rat and its suppression by reserpine. *Psychopharmacologia* **44:** 33–36, 1975.

Collins, R. C. and Mehta, S.: Effect of amino-oxyacetic acid (AOAA) on focal penicillin seizures. *Brain Res.* **157:** 311–320, 1978.

Craig, C. R. and Hartman, E. R.: Concentration of amino acids in the brain of cobalt-epileptic rat. *Epilepsia* **14:** 409–414, 1973.

Craig, C. R., Hoover, D. B., and Colasanti, B. K.: Appearance of epileptiform activity in the rat after bilateral cerebral implantation of cobalt. *Pharmacologist* **18:** 127, 1976.

Curtis, D. R. and Johnston, G. A. R.: Amino acid transmitters in the mammalian central nervous system. *Erg. Physiol.* **69:** 97–188, 1974.

Davidoff, R. S.: Penicillin and inhibition in the cat spinal cord. *Brain Res.* **45:** 638–642, 1972.

Dimov, S.: Changes in the Cerebral Bioelectric Activity of Rabbits Following Application of Cobalt to the Brain Cortex, In: *Comparative and Cellular Pathophysiology of Epilepsy* (Z. Servit, ed.) Experta Medica Foundation, Amsterdam, 1966.

Donaldson, J., St.-Pierre, T., Minnich, J., and Barbeau, A.: Seizures in rats associated with divalent cation inhibition of Na^+-K^+-ATPase. *Can. J. Biochem.* **49:** 1217–1224, 1971.

Dow, R. S., Fernandez-Guardiola, A., and Manni, E.: The production of cobalt experimental epilepsy in the rat. *Electroenceph. Clin. Neurophysiol.* **14:** 399–407, 1962.

Emson, P. C.: Effects of chronic treatment with amino-oxyacetic acid or sodium *n*-dipropylacetate on brain GABA levels and the development and regression of cobalt epileptic foci in rats. *J.Neurochem.* **27:** 1489–1494, 1976.

Emson, P. C. and Joseph, M. H.: Neurochemical and morphological changes during the development of cobalt-induced epilepsy in the rat. *Brain Res.* **93:** 91–110, 1975.

Enna, S. J. and Snyder, S. H.: Properties of gamma-aminobutyric acid (GABA) receptor binding in rat brain synaptic membrane fractions. *Brain Res.* **100:** 81–97, 1975.

Essig, C. F.: Focal convulsions during barbiturate abstinence in dogs with cerebro-cortical lesions. *Psychopharmacology* **3:** 432–437, 1962.

Fischer, J. and Langmeier, M.: Changes in the number, size and shape of synaptic vesicles in an experimental projected cortical epileptic focus in the rat. *Epilepsia* **21:** 571–585, 1980.

Frenk, H., Urca, G. and Liebeskind, J. C.: Epileptic properties of leucine- and methionine-enkephalin: Comparison with morphine and reversibility by naloxone. *Brain Res.* **147:** 327–337, 1978.

Fuller, T. A. and Olney, J. W.: Effects of morphine or naloxone on kainic acid neurotoxicity. *Life Sci.* **24:** 1793–1798, 1979.

Gerebtzoff, M. A., Brotchi, J., Duchesne, P. Y., and Dresse, A.: Phosphorylase et glycosyltransferases au niveau des astrocytes reactionnels. Étude Histochemique. *Comptes rendus des seances de la Societe de Biologie* **172:** 367–373, 1978.

Goldberg, A. M., Pollock, J. J., Hartman, E. R., and Craig, C. R.: Alterations in cholinergic enzymes during the development of cobalt-induced epilepsy in the rat. *Neuropharmacology* **11:** 253–259, 1972.

Harmony, T., Urba-Holmgren, R., Urbay, C. M., and Szara, S.: (Na-K)-ATPase activity in experimental epileptogenic foci. *Brain Res.* **11:** 672–680, 1968.

Henjyoji, E. Y. and Dow, R. S.: Cobalt-induced seizures in the cat. *Electroenceph. Clin. Neurophysiol.* **19:** 152–161, 1965.

Hoover, D. B., Craig, C. R., and Colasanti, B. K.: Cholinergic involvement in cobalt-induced epilepsy in the rat. *Exp. Brain Res.* **29:** 501–513, 1977.

Hunt, W. A. and Craig, C. R.: Alterations in cation levels and Na-K-ATPase activity in rat cerebral cortex during the development of cobalt-induced epilepsy. *J. Neurochem.* **20:** 559–567, 1973.

Joseph, M. H. and Emson, P. C.: Taurine and cobalt induced epilepsy in the rat: A biochemical and electrocorticographic study. *J. Neurochem.* **27:** 1495–1502, 1976.

Kopeloff, L. M.: Experimental epilepsy in the mouse. *Proc. Soc. Exp. Biol. Med.* **104:** 500–504, 1960.

Kopeloff, L. M., Barrera, S. E., and Kopeloff, N.: Recurrent convulsive seizures in animals produced by immunologic and chemical means. *Am. J. Psychiat.* **98:** 881–902, 1942.

Koyama, I.: Amino acids in the cobalt-induced epileptogenic and non-epileptogenic cat's cortex. *Can. J. Physiol. Pharmacol.* **50:** 740–752, 1972.

Kriegstein, A. R., Suppes, T., and Prince, D. A.: Cholinergic enhancement of penicillin-induced epileptiform discharges in pyramidal neurons of the guinea pig hippocampus. *Brain Res.* **266:** 137–142, 1983.

Lewin, E.: Epileptogenic foci induced with ouabain. *Electroenceph. Clin. Neurophysiol.* **29:** 402–403, 1970.

Lockard, J. S. and Barensten, R. I.: Behavioral experimental epilepsy in monkeys. I. Clinical seizure recording apparatus and initial data. *Electroenceph. Clin. Neurophysiol.* **22:** 482–486, 1967.

Lunt, G. G. and Grove, Y. K.: The effects of cobalt-induced epilepsy on the unesterified fatty acid content of rat cerebral cortex. *Biochem. Soc. Proc.* **3:** 701–702, 1975.

Lux, H. D.: Ammonium and chloride extrusion: hyperpolarizing synaptic inhibition in spinal motoneurons. *Science* **175:** 555–557, 1971.

Macon, J. B. and King, D. W.: Penicillin iontophoresis and the responses of somatosensory cortical neurons to amino acids. *Electroenceph. Clin. Neurophysiol.* **47:** 52–63, 1979.

Marcus, E. M.: Experimental Models of Petit Mal Epilepsy, In: *Experimental Models of Epilepsy — A Manual for the Laboratory Worker* (D. P. Purpura, J. K. Penry, D. Tower, D. M. Woodbury, and R. Walter, eds.) Raven, New York, 1972.

Mayman, C. I., Manlapaz, J. S., Ballantine, H. T., and Richardson, E. P.: A neuropathological study of experimental epileptogenic lesions in the cat. *J. Neuropath. Exp. Neurol.* **24:** 502–511, 1965.

Meyer, H. and Prince, D.: Convulsant action of penicillin: Effects on inhibitory mechanisms. *Brain Res.* **53:** 477–483, 1973.

Mueller, A. L. and Dunwiddle, T. V.: Anticonvulsant and proconvulsant actions of alpha and beta-noradrenergic agonists on epileptiform activity in rat hippocampus in vitro. *Epilepsia* **24:** 57–64, 1983.

Mutani, R., Bergamini, L., Fariello, R., and Delsedime, M.: Effects of taurine on cortical acute epileptic foci. *Brain Res.* **70:** 170–173, 1974a.

Mutani, R., Bergamini, L., Delsedime, M., and Durelli, L.: Effects of taurine in chronic experimental epilepsy. *Brain Res.* **79:** 330–332, 1974b.

Nadler, J.: Minireview: Kainic acid as a tool for the study of temporal lobe epilepsy. *Life Sci.* **29:** 2031–2042, 1981.

Payan, H. M.: Cobalt experimental epilepsy in gerbils. *Exp. Med. Surg.* **28:** 163–168, 1970.

Payan, H., Strebel, R., and Levine, S.: Epileptogenic effect of extradural and extracranial cobalt. *Nature* **208:** 792–793, 1965.

Pei, Y., Zhao, D., Huang, J., and Cao, L.: Zinc-induced seizures: A new experimental model of epilepsy. *Epilepsia* **24:** 160–176, 1983.

Pekoe, G. M., Craig, C. R., and Colasanti, B. K.: Epileptogenic effects of cobaltous chloride after prolonged intraventricular infusion in the rat. *Fed. Proc.* **37:** 2110, 1978.

Prince, D. A.: Topical Convulsant Drugs and Metabolic Antagonists, In: *Experimental Models of Epilepsy — A Manual for the Laboratory Worker* (D. P. Purpura, J. K. Penry, D. Tower, D. M. Woodbury, and R. Walter, eds.) Raven, New York, 1972.

Raabe, W., Nicol, S., Gumnit, R. J., and Goldberg, N. D.: Focal penicillin epilepsy increases cyclic GMP in cerebral cortex. *Brain Res.* **144:** 185–188, 1978.

Ribak, C. E., Bradburne, R. M., and Harris, A. B.: A preferential loss of GABAergic symmetric synapses in epileptic foci: A quantitative ultrastructural analysis of monkey neocortex. *J. Neuroscience* **2:** 1725–1735, 1982.

Ribak, C. E., Harris, A. B., Vaughn, J. E., and Roberts, E.: Inhibitory, GABAergic nerve terminals decrease at sites of focal epilepsy. *Science* **205:** 211–214, 1979.

Riedel, E., Kottler, A. H., and Wamsiedler, E.: Untersuchungen an Tieren mit experimentell erzeugter Epilepsie. Ueber Elektrolytverteilungen im Gehirn chronisch epileptischer Tiere. *Pharmakopsychiat. Neuropsychopharm.* **4:** 22–29, 1971.

Roberts, C. A.: Anticonvulsant effects of uridine: Comparative analysis of metrazol and penicillin induced foci. *Brain Res.* **55:** 291–308, 1973.

Ross, S. M. and Craig, C. R.: Studies on gamma-aminobutyric acid transport in cobalt experimental epilepsy in the rat. *J. Neurochem.* **36:** 1006–1011, 1981a.

Ross, S. M. and Craig, C. R.: γ-Aminobutyric acid concentration, L-glutamate l-decarboxylase activity, and properties of the γ-aminobutyric acid postsynaptic receptor in cobalt epilepsy in the rat. *J. Neurosci.* **1:** 1388–1396, 1981b.

Scheibel, A. B., Paul, L., and Fried, I.: Some Structural Substrates of the Epileptic State, In: *Basic Mechanisms of Neuronal Hyperexcitability* (H. H. Jasper and N. M. van Gelder, eds.) Alan R. Liss, New York, 1983.

Schwartzkroin, P. A. and Price, D. A.: Changes in excitatory and inhibitory synaptic potentials leading to epileptogenic activity. *Brain Res.* **183:** 61–78, 1980.

Seidel, J., Kastner, I., and Winkelmann, E.: A loss of GABAergic hippocampus innervation in rats with cobalt-induced epilepsy demonstrated by Wolman fluorescence method. *Acta Histochemica* **69:** 248–254, 1981.

Servit, Z. and Sterc, J.: Audiogenic epileptic seizures evoked in rats by artificial epileptogenic foci. *Nature* **181:** 1475–1476, 1958.

Singh, R.: Lysosomal acid phosphatase hyperactivity in cobalt epilepsy. *Indiun J. Exp. Biol.* **18:** 716–719, 1980.

Sperk, G., Berger, M., Hortnagl, H., and Hornykiewicz, O.: Kainic acid-induced changes of serotonin and dopamine metabolism in the striatum and substantia nigra of the rat. *Eur. J. Pharmacol.* **74:** 729–786, 1981a.

Sperk, G., Hortnagl, H., Reither, H., and Hornykiewicz, O.: Changes in histamine in the rat striatum following local injection of kainic acid. *Neuroscience* **6:** 2669–2675, 1981b.

Sperk, G., Lassmann, H., Baran, H., Kish, S. J., Seitelberger, F., and Hornykiewicz, O.: Kainic acid induced seizures: Neurochemical and histopathological changes. *Neuroscience* **10:** 1301–1316, 1983.

Stark, L. G., Edmonds, H. L., and Keesling, P.: Penicillin-induced epileptogenic foci. I. Time course and the anticonvulsant effects of diphenylhydantoin and diazepam. *Neuropharmacology* **13:** 261–267, 1974.

Stercova, A.: Dynamics of Neurohistopathological Changes in an Epileptogenic Focus Produced by Alumina Cream in the Rat, In: *Comparative and Cellular Pathophysiology of Epilepsy* (Z. Servit, ed.) Prague, Publishing House of Czechoslovak Academy of Sciences, Prague, 1966.

Sypert, G. W.: Metallic Salts and Epileptogenesis, In: *Physiology and Pharmacology of Epileptogenic Phenomena* (M. R. Klee, H. D. Lux, and E.-J. Speckmann, eds.) Raven, New York, 1982.

Trottier, S., Berger, B., Chauvel, P., Dedek, J., and Gay, M.: Alterations of the cortical noradrenergic system in chronic cobalt epileptogenic foci in the rat: A histofluorescent and biochemical study. *Neuroscience* **6**: 1069–1080, 1981.

Trottier, S., Claustre, Y., Carboche, J., Dedek, J., Chauvel, P., Nassif, S., and Scatton, B.: Alterations of noradrenaline and serotonin uptake and metabolism in chronic cobalt-induced epilepsy in the rat. *Brain Res.* **272**: 255–262, 1983.

Turkanis, S. A. and Karler, R.: Central excitatory properties of Δ9-tetrahydrocannabinol and its metabolites in iron-induced epileptic rats. *Neuropharmacology* **21**: 7–13, 1982.

Urca, G., Frenk, H., Liebeskind, J. C., and Taylor, A. N.: Morphine and enkephalin: Analgesic and epileptic properties. *Science* **197**: 83–86, 1977.

Van Duijn, H., Schwartzkroin, P. A., and Prince, D. A.: A possible action of penicillin on inhibitory processes in cat cortex. *Brain Res.* **53**: 470–476, 1973.

Van Gelder, N. M.: Antagonism by taurine of cobalt induced epilepsy in cat and mouse. *Brain Res.* **47**: 157–166, 1972.

Van Gelder, N. M. and Courtois, A.: Close correlations between changing content of specific amino acids in epileptogenic cortex of cats, and severity of epilepsy. *Brain Res.* **43**: 477–484, 1972.

Velasco, M., Velasco, F., Estrada-Villaneuva, F., and Olvera, A.: Alumina-cream-induced focal motor epilepsy in cats. I. Lesion size and temporal course. *Epilepsia* **14**: 3–14, 1973a.

Velasco, M., Velasco, F., Lozoya, X., Feria, A., and Gonzalez-Licea, A.: Alumina cream-induced focal motor epilepsy in cats. II. Thickness and cellularity of cerebral cortex adjacent to epileptogenic lesion. *Epilepsia* **14**: 15–27, 1973b.

Ward, A. A. Jr.: Topical Convulsant Metals, In: *Experimental Models of Epilepsy — A Manual for the Laboratory Worker* (D. P. Purpura, J. K. Penry, D. B. Tower, D. M. Woodbury, and R. D. Walter, eds.) Raven, New York, 1972.

Willmore, L. J., Fuller, P. M., Butler, A. B., and Bass, N. H.: Neuronal compartmentation of ionic cobalt in rat cerebral cortex during initiation of epileptiform activity. *Exp. Neurol.* **47**: 280–289, 1975.

Willmore, L. J. and Rubin, J. J.: Antiperoxidant pretreatment and iron-induced epileptiform discharges in the rat: EEG and histopathologic studies. *Neurology* **31**: 63–69, 1981.

Wong, R. K. S. and Prince, D. A.: Dendritic mechanism underlying penicillin induced epileptiform activity. *Science* **204**: 1228–1231, 1979.

Woodbury, D. M.: Role of pharmacological factors in the evaluation of anticonvulsant drugs. *Epilepsia* **10**: 121–144, 1969.

Seizures Induced
by Convulsant Drugs

Carl L. Faingold

1. Introduction

From the original insights of Jackson nearly a hundred years ago (1890) to the recent book by Lockard and Ward (1980), convulsive seizures have been viewed as a window into the functioning of the brain, the assumption being that pathophysiology yields considerable insight into normal physiology. Convulsant drugs have been a major tool over the years in trying to understand how and why the brain generates the extensive and dramatic electrical and behavioral events of generalized seizure. From Claude Bernard's pioneering work on the effects of strychnine on the spinal cord to Jeff Barker's investigations of convulsants in spinal cord neurons in culture (Barker et al., 1983), these chemical agents have been used as tools to aid in the understanding of the global events of epileptic disorders and as molecular probes of subsynaptic neuronal properties. Since the original discovery that physostigmine, the active ingredient in the calibar bean, was capable of prolonging the actions of a specific neurotransmitter, acetylcholine, the study of neurochemical actions of drugs with convulsant properties has developed into a dominant theme in investigations of the molecular bases of epilepsy. When Curtis and coworkers (1971) demonstrated that strychnine could specifically block the action of the inhibitory transmitter, glycine, which is a prominent neurotransmitter in the spinal cord (Werman et al., 1968; Young and Macdonald, 1983), this clearly represented an important mechanism subserving the ability of strychnine to produce the spinal seizures that Claude Bernard first observed so long before. Several useful reviews on convulsant drugs have been published previously (Hahn, 1960; Esplin and Zablocka-Esplin, 1969; Woodbury, 1980; Davidoff, 1983).

2. Neurotransmitter Specificity

Extensive supporting evidence for the ability of drugs with convulsant properties to affect the action of a specific neurotransmitter continues to be reported. Reports have appeared in recent years, however, that suggest that the convulsant drugs produce effects in addition to the neurotransmitter-specific actions in some CNS regions or neuronal systems that appear to be quite important to seizure induction. The effects of convulsant drugs are sometimes reported to be dose-dependent in that a transmitter-specific effect is observed at the lowest effective dose in some studies. However, effects on the action of multiple neurotransmitters and even on membrane properties not directly related to neurotransmitter action are also observed (Freeman, 1973; Pellmar and Wilson, 1977a; Barker and MacDonald, 1980; Rayport and Kandel, 1981; Heyer et al., 1982). Another factor in establishing the mechanisms of seizure induction that has not been considered sufficiently in many recent investigations is that neurons in different regions of the mammalian CNS are not affected to the same extent by convulsant drugs (Macdonald et al., 1979; Faingold and Stittsworth, 1980; *see* Faingold et al., 1985a for review). Thus, the neuroanatomy of the CNS must be prominently considered in attempting to understand the ability of convulsant drugs to produce seizures.

3. Mode of Drug Administration

The technical question about how the drug gets to the neuron must also be taken into account. Thus, systemic administration, iontophoretic application, or administration of the agent into the medium surrounding a neuron in vitro can each result in differential exposure of the cell body or cell processes of a neuron to the convulsant drug. These differences can affect the magnitude and even the nature of the effect observed (e.g., Faingold et al., 1984). This dichotomy may result from the finding that ionic conductance mechanisms that govern the excitability of the cell soma may differ from those that are predominant in cell processes even of the same neuron (Llinas and Sugimori, 1983). Ion conductance mechanisms, also referred to as ion channels or ionophores, are an important concept in neurotransmitter and convulsant drug actions, as discussed below. Each route of administration provides useful information, but administration to the whole animal in vivo produces the most relevant data on the mechanisms involved in production of seizures. Convulsant-induced seizures produce considerable changes in CNS transmitter levels (e.g., Nistri and Pepeu, 1974). These alterations may be caused by excessive neuronal firing during the seizure rather than mechanisms associated with seizure initiation with that specific convulsant drug. However, the finding that different patterns of change are seen with different convulsant drugs suggests that information on regional differences in convulsant effect may be gained.

4. Synaptic Mechanisms

Synaptic activity was viewed, until recently, as a simple balance between inhibitory and excitatory processes (Eccles, 1964), and it was believed that two general transmitter-related mechanisms could be involved in seizure induction by convulsants, excessive excitation and decreased inhibition (Esplin and Zablocka-Esplin, 1969). Although this was considered to be an either/or dichotomy, recent investigations suggest that many convulsant agents can exert both actions (Barker and MacDonald, 1980). A given convulsant drug may have a major effect on one mechanism, but other effects as well. An example of such a dual phenomenon is the ability of several convulsants to block the inhibitory action of gamma-aminobutyric acid (GABA) and also to increase the excitation caused by acetylcholine (ACh), by inhibiting its breakdown by acetylcholinesterase.

Much of the pioneering work on neuronal membrane properties was performed using the squid axon (Hodgkin and Huxley, 1952). In this well-studied case, rapid sodium and potassium ion conductance changes readily explain the majority of membrane properties underlying the excitability of this neuronal membrane. However, other mechanisms, including a persistent potassium current that is mediated by the entry of calcium into the cell, are reported to be involved in regulation of the excitability of other neural elements in CNS regions (Llinas and Sugimori, 1983). The relatively recent discovery that many of these voltage-sensitive currents are affected by neurotransmitters (Wilson and Wachtel, 1978;

Crill and Schwindt, 1983; 1986; Lewis et al., 1986) adds an additional complexity to our understanding of neurotransmitter action.

For example, the persistent calcium-mediated conductance change in neurons of the hippocampal slice is reported to be particularly important in control of repetitive firing mechanisms (Alger and Nicoll, 1980; Johnston et al., 1980; Hotson and Prince, 1980) and is affected by norepinephrine (Madison and Nicoll, 1982). Other voltage-sensitive currents have also been shown to be affected by neurotransmitters (Wilson and Wachtel, 1978; Halliwell and Adams, 1982; MacDonald et al., 1982). Convulsant drugs may produce effects on these currents through interactions with transmitter receptors or by directly affecting the conductance mechanisms or ion channels (Schwartzkroin and Pedley, 1979; Ticku and Maksay, 1983). Other recent complexities of transmitter action include the finding that specific receptors for different neurotransmitters, such as those for GABA and glycine, may use a common chloride ion channel (Barker and McBurney, 1979; Hamill et al., 1983). Significant differences in the kinetics of ion channel interactions may occur even among structural analogs of the same transmitter (Barker and Mathers, 1981).

5. Types of Preparation

The actions of convulsant drugs have been investigated in neurons from invertebrates to man, including many in vitro studies. The question arises as to what degree neurons in the goldfish, *Aplysia*, squid, or in in vitro may resemble those in vivo in the cat, rat, or man. Many studies have been carried out using neurons in tissue culture that were derived from embryonic mammalian sources. Although synapse formation does occur in vitro, considerable differences are observed in the degree of synaptic contact in culture from the degree that occurs in vivo (Macdonald et al., 1983). Quantitative differences of the effects of convulsant drugs on neurons cultured from different regions of mammalian CNS are also observed (Macdonald et al., 1979). In light of the different conductance mechanisms observed in different brain regions (Llinas and Sugimori, 1983), the effects of convulsants on neurons in vitro may involve a different combination of mechanisms from those involved in intact mammalian brain. Thus, the actions observed in these model systems may be highly instructive, but are not necessarily definitive of events occurring in mammalian neurons in vivo.

Another concern that must be remembered with in vitro studies is that the concentration of the agent should be comparable to known concentrations in brain that are associated with seizure induction in the intact animal. For example, in *Aplysia* neurons and spinal cord neurons in culture (Pellmar and Wilson, 1977a; Macdonald and Barker, 1979), the lowest concentrations of pentylenetetrazol required to see effects are 3–9 times the concentration in brain that produces seizures in the intact mouse (Yonekawa et al., 1980). However, the ability to use intracellular voltage and patch clamp techniques (Sakmann et al., 1983; Lewis et al., 1986) in invertebrate and cultured mammalian experiments in vitro renders molecular mechanisms open to examination that are not currently feasible in vivo where intracellular recording, let alone membrane clamping, during convulsant administration is an extremely difficult task (Tribble et al., 1983a,b).

6. Regional CNS Differences

In vitro approaches are invaluable and absolutely necessary to improve our understanding of synaptic and subsynaptic mechanisms. However, such studies are not sufficient to explain epileptogenesis in vivo, and examination of these mechanisms in larger neuronal assemblies must also be carried out. Although this is a difficult and laborious undertaking, some of the initial approaches, including investigations of comparative effects of convulsants on cultured neurons and slices derived from different brain regions (Macdonald et al., 1979; Prince, 1982) and comparative neuronal examinations in vivo (Faingold and Stittsworth, 1980; Faingold et al., 1983a; Pellegrini et al., 1983; *see* Faingold et al., 1985a for review), are yielding results that suggest that important differences in sensitivity to the effects of convulsants do occur. Differences in radiolabeled convulsant binding in different brain regions support the idea of regional selectivity of action (DeFeudis et al., 1978). Neurons in different brain regions appear to have important differences in conductance mechanisms (Llinas and Sugimori, 1983), which may help to explain the differences of convulsive drug action observed in the different brain regions in vitro and systemic studies in vivo (Prince, 1982; Faingold et al., 1985a). In addition, presynaptic neurotransmitter receptors, autoreceptors, or postsynaptic axon terminals may be present on neural elements in some brain regions but not in others (Mitchell and Martin, 1978; Noebels and Prince, 1978; Arbilla et al., 1979; Brennan et al., 1981; Frere et al., 1982). Although it has been suggested that ionic conductance mechanisms activated by specific neurotransmitters in the same brain region (cortex) of two different mammalian species may also be different (Marciani et al., 1980), it is not clear that the same cell types from the same cortical layers were examined in both studies. It is clear that sites of action within the CNS must be taken into account when evaluating the action of convulsants. Earlier studies, in fact, did consider regional actions, and lesion and ablation studies were widely done and must be considered (*see* Hahn, 1960 for review). However, many of the older ablation experiments involved rather large anatomical regions of the brain, and these studies could not take into account what has subsequently been learned about the functions of the specific nuclei in these large brain regions. More discrete lesions, such as those utlized by Garant and Gale (1983), need to be employed to better understand the important regional aspects of convulsant action in production of seizures. Research is in progress to evaluate the importance of certain prominent pathways in the brain for initiation and mediation of specific behavioral and EEG components of generalized seizure. In particular, the brainstem reticular formation (RF) (Penfield and Jasper, 1954; Faingold, 1985), the substantia nigra (Gale, 1985), and the corpus callosum, the major interhemispheric connecting pathway of the cerebral cortex (Pellegrini et al., 1979), are implicated in the initiation and control of generalized-onset seizures. These anatomical approaches are a useful framework from which to view the regional aspects of convulsant action.

A large number of studies indicate that convulsant application directly onto neurons in vivo and in vitro results in a neuronal firing pattern known as the paroxysmal depolarization shift (Ajmone Marsan, 1969; Prince, 1978). This phenomenon may have considerable practical importance because of the similarity of pattern to that observed in human focal epileptic neurons in vivo and in some types of neurons in vitro (Calvin et al., 1973; Prince and Wong, 1981). The paroxysmal depolarization shift has been suggested to involve the enhancement of EPSP mechanisms, but considerable controversy on this question exists (Ayala et al., 1973; Calvin, 1975; Schwartzkroin and Prince, 1980; Johnston and Brown, 1981; 1986). This phenomenon appears to be a general effect of convulsant drugs, since many very different chemicals including alumina gel can produce this pattern of neuronal discharge.

7. Use of Anesthetics

Many of the studies on convulsant drugs in vivo employ animals treated with anesthetics. This is a major problem because depressant anesthetics such as the barbiturates possess significant anticonvulsant properties. Nondepressant anesthetics such as ketamine and chloralose possess proconvulsant properties. In addition to producing epileptiform activity in the EEG, these agents are reported to produce enhanced sensory responses of the brain in a fashion similar to that seen with the convulsant drugs (*see* Table 1 and Fig. 1). In light of the different effects of these agents on various brain regions (e.g., Glass and Fromm, 1975), studies of convulsant drug action in the presence of these agents could potentially produce effects on different brain regions and to differing degrees than those that occur in the absence of pretreatment. For example, barbiturates may produce a relatively greater degree of effect on the brainstem reticular formation than in some other brain regions (Brazier, 1961; Bradley and Dray, 1973; Duggan et al., 1974; Bowery and Dray, 1978). When a convulsant is administered in the presence of a barbiturate anesthetic, the convulsant may not affect the brainstem reticular formation to the same extent it would normally. However, another brain region that is not as greatly affected by the barbiturate might participate to a greater extent than normal in seizure initiation. An example of this phenomenon is provided by the report of Chen et al. (1983), which shows that barbiturate administration blocks the ability of topical strychnine to enhance sensory responses in one type of cortical neuron, but not in an adjacent type of cortical neuron. The extensive redundancy of brain pathways would allow a different regional pattern of convulsant drug effects to culminate in a generalized seizure. Alterations of seizural phenomena are observed in the presence of anesthetics (Faingold and Berry, 1973; Noebels and Prince, 1978). Lesion and brain transection studies involving areas thought to be most sensitive to convulsant action have generally raised seizure threshold, but do not prevent seizure induction (Hahn, 1960; Velasco et al., 1976). These observations support the idea of alternative seizure pathways when the "preferred" or normal pathway is blocked pharmacologically or pathologically. Anticonvulsant drugs have also been utilized to gain evidence on the mechanism of action of convulsant drugs based on a transmitter-specific effect

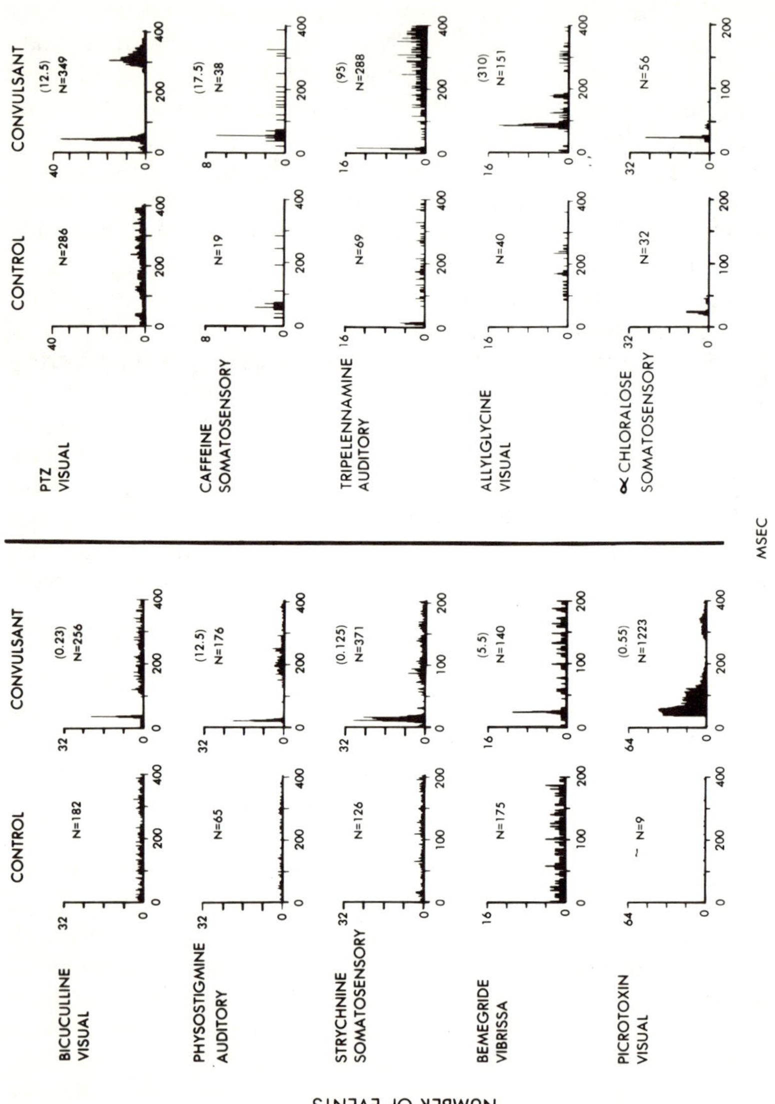
CONVULSANT
PTZ
VISUAL
(12.5)
N=349
40
0
200
400
CONTROL
N=286
40
0
200
400
CAFFEINE
SOMATOSENSORY
(17.5)
N=38
8
0
200
400
N=19
8
0
200
400
TRIPELENNAMINE
AUDITORY
(95)
N=288
16
0
200
400
N=69
16
0
200
400
ALLYLGLYCINE
VISUAL
(310)
N=151
16
0
200
400
N=40
16
0
200
400
CHLORALOSE
SOMATOSENSORY
N=56
32
0
100
200
N=32
32
0
100
200
MSEC
CONVULSANT
BICUCULLINE
VISUAL
(0.23)
N=256
32
0
200
400
CONTROL
N=182
32
0
200
400
PHYSOSTIGMINE
AUDITORY
(12.5)
N=176
32
0
200
400
N=65
32
0
200
400
STRYCHNINE
SOMATOSENSORY
(0.125)
N=371
32
0
100
200
N=126
32
0
100
200
BEMEGRIDE
VIBRISSA
(5.5)
N=140
16
0
100
200
N=175
16
0
100
200
PICROTOXIN
VISUAL
(0.55)
N=1223
64
0
200
400
N=9
64
0
200
400
NUMBER OF EVENTS

Fig. 1. Effects of gradual iv administration of 10 different agents with convulsant properties on the computer-summated responses to sensory stimuli of brainstem RF neurons of unanesthetized cat (poststimulus time histograms, 1 or 1/2 s stimulus presentation rate, 1 ms bin width, 25 or 50 stimulus presentations). The particular drug and sensory stimulus type is shown for each pair of histograms [*see* Faingold, (1978; 1980a) for description of analysis techniques and drug infusion parameters]. The amount of drug infused (mg/kg) is shown in the convulsant columns. Stimulus onset is at zero time. N is the total number of spikes in each histogram. Recovery of the response pattern to one resembling the control pattern after cessation of drug administration was observed, but is not shown.

Table 1
Effects of Convulsant Drugs on Sensory Responses of the CNS[a]

Convulsant drug	Evoked potential enhancement — Refs.	Neuronal response enhancement region — Refs.	
Aldrin	Joy, 1976		
Allylglycine	Kaplan and Williamson, 1978; Menini et al., 1980; Stutzmann et al., 1980	CTX, RF	Menini et al. (1981), Faingold (unpub.)
Bemegride	Rodin et al., 1966	RF, (Hp)	Faingold and Hoffmann (1981)
Bicuculline	Bigler, 1977a,b; Kaplan and Williamson, 1978	RF, (CTX)	Faingold et al. (1983a; 1984)[b]
Carbachol	Bharagava et al., 1978		
Chloralose	Winters, 1966; Glass and Fromm, 1975	RF	Faingold (unpub.)
Cobalt (topical)	Finch and Beatty, 1976		
Dieldrin	Joy, 1974a,b		
Dimethylhydrazine	Goff et al., 1970		
Flurothyl	A'Hearn, 1966		
Heptachlor	Joy, 1976		
Ketamine	Dafny and Rigor, 1978	CTX	Kayama and Iwama (1972)
Methionine sulfoximine	Hrebicek and Kolousek, 1968		

Drug	Reference	Site	Reference
Nicotine	Guha and Pradhan, 1976		
Oxotremorine	Bhargava et al., 1978		
Penicillin	Burchiel et al., 1976; Ebersole and Chatt, 1984; 1986	CTX, RF	Pellegrini et al. (1983), Macon and King (1979)[b]
Pentylenetetrazol	Gastaut and Hunter, 1950; Arduini and Arduini, 1954; Wenzel and Muller, 1971	RF, (AMG)	Faingold (1980a,b), Faingold et al. (1983b; 1984; 1985b)[b]
Phencyclidine	Shelburne and McLaurin, 1977		
Physostigmine	Bhargava et al., 1978	RF	Faingold (1978)
Picrotoxin	Bigler, 1977a,b; Kaplan and Williamson, 1978	RF	Faingold (unpub.)
Strychnine	Arduini and Arduini, 1954; Faingold, 1977	RF	Faingold (1980a), Faingold et al. (1984)[b]
Tetraethyl-pyrophosphate	A'Hearn, 1966		
Tripelennamine	Faingold, unpub.	RF	Faingold (unpub).

[a]Abbreviations: AMG, amygdala; CTX, cortex; Hp, hippocampus; RF, reticular formation; unpub., unpublished; parentheses ( ) indicate a lesser effect in comparative studies.
[b]Iontophoretic application.

of the anticonvulsant agent. This approach has certain pitfalls, including incongruities between the interactions of convulsants and anticonvulsants that are thought to affect the same transmitter (Rose, 1979). Certain barbiturates are reported to selectively enhance the action of GABA (Macdonald and Barker, 1979; Macdonald and McLean, 1982; 1986). However, the barbiturates used as anesthetics alter effects of other inhibitory as well as excitatory neurotransmitters and also affect ion conductance mechanisms in a variety of CNS regions and animal species (Bradley and Dray, 1973; Potashner et al., 1980; Macdonald and Barker, 1979; Lambert and Flatman, 1981; Ho and Harris, 1981; Cote and Wilson, 1980; Schulz and Macdonald, 1981; Heyer and Macdonald, 1982; Wachtel and Wilson, 1983). Therefore, the effect of the convulsants on neurotransmitter action is likely to be quantitatively different and may even be qualitatively different in the presence of anesthesia. Differences of anticonvulsant action on different brain regions, as well as evidence suggesting transmitter nonspecific actions of these agents, are reported (Stone and Javid, 1979).

8. Sensory-Induced Seizure Model

The anatomical loci responsible for seizure initiation following convulsant drug administration are extraordinarily difficult to identify because seizure onset is essentially impossible to predict accurately. An approach to overcoming this difficulty that has been used in neuronal studies in recent years is based on the observation that sensory stimuli can trigger generalized seizures following treatment with subthreshold levels of convulsant drugs. This method also permits the seizure initiation mechanism to be defined in a way that is not possible with convulsant administration alone, since the random and uncontrolled stimuli to which an animal is subjected by the external and internal milieu may actually play a role in initiating the seemingly ''spontaneous'' seizures. This approach also provides a model of human reflex epilepsy (Bickford and Klass, 1969). Studies since the early 1950's have shown that as the level of convulsant rises, the responses of the brain to sensory stimuli become extensively enhanced. The enhancement is quite evident at drug levels that are considerably below those that induce seizures and also well below the dose at which the sensory stimuli will trigger seizures (*see* Faingold, 1978; Faingold et al., 1985a for review). The 24 different convulsant drugs that produce this phenomenon of sensory response enhancement are shown in Table 1. Regional differences are observed in the degree of response enhancement produced by convulsant drugs. Responses in non-primary sensory regions of the brain are enhanced to a greater extent than primary sensory regions of the brain (Glass and Fromm, 1975; Faingold, 1977; Dafny and Rigor, 1978). The area most greatly enhanced and with the earliest onset of action is the brainstem reticular formation (Faingold, 1977; 1978). In addition, convulsant-induced generalized spontaneous preictal EEG spikes are also reported to occur in the RF before appearing in the cortex (Joy, 1974b).

However, sensory-evoked field potentials and EEGs that are generally recorded with large electrodes represent a population response generated, in part, by both inhibitory and excitatory post-synaptic potentials, which greatly complicates interpretation of neuronal events involved. Therefore, single-unit recording studies with microelectrodes are being carried out in order to establish the nature of the actual neuronal events involved in convulsant action on sensory responses and seizure initiation. A list of brain regions that have been examined using neuronal recordings with each of the convulsants is shown in Table 1 (last column). A summation of convulsant effects on RF neurons is shown in Fig. 1.

This review will examine the literature on convulsants with major actions on neurotransmitter activity with the above considerations about neurotransmitter specificity, anesthetic alterations of action, and brain site relative selectivity issues firmly in view. A generalized scheme for neuron and convulsion drug interaction for the convulsants discussed below is shown in Fig. 2. Additional general mechanisms of convulsant action that are not thought to involve neurotransmitter action have been described (Woodbury, 1980). Thus, seizures are induced by agents that alter brain metabolism (e.g., methionine sulfoximine) and agents that block active transport (e.g., ouabain), as well as inorganic compounds (e.g., ammonium chloride). Although these types of compounds are not thought to affect neurotransmitter actions, effects of certain of these agents on ionic conductance mechanisms may possibly play a role in the convulsant actions of these convulsants as well (Nicoll, 1978). These types of convulsants will not be discussed in this review. It is perhaps most important that the large number of investigators who have occasion to use convulsant drugs as tools to study the action of neurotransmitters be aware of the potential concerns about specificity of action and also test the specificity question in their own system. It cannot be assumed that previous work on the effects of these agents in other systems is definitive. Readers are re-

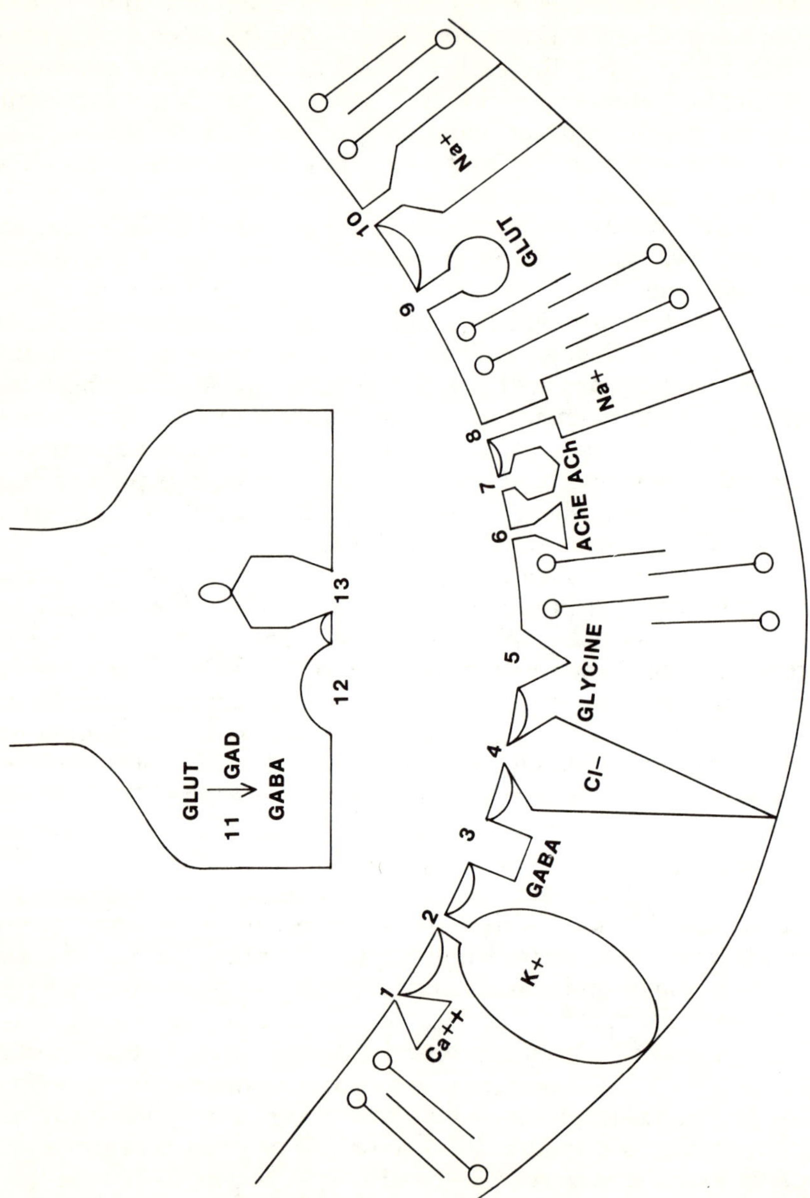

Na+
GLUT
Na+
ACh
AChE
GLYCINE
Cl-
GABA
K+
Ca++
GLUT
GAD
GABA
1
2
3
4
5
6
7
8
9
10
11
12
13

Fig. 2. Schematic representation of potential sites of convulsant action on neuronal membranes. Sites 1 and 2 are the voltage-sensitive calcium-mediated persistent potassium ionophore (ion channel) complex. Site 3 is the classical GABA$_A$ receptor site.[a] Site 4 is the chloride ionophore site that may be coupled to 3 and 5. Site 5 is the classical glycine receptor.[a] Site 6 is an acetylcholinesterase enzyme that is responsible for the breakdown of acetylcholine (ACh) and is located near the acetylcholine receptor (7). Site 7 is a classical acetylcholine (ACh) receptor (several receptor types known).[a] Site 8 is a sodium ionophore associated with ACh action. Site 9 is an excitant amino acid [e.g., glutamic acid (GLUT)] receptor (several receptor types known).[a] Site 10 is a sodium channel associated with the excitant amino acid receptor. Site 11 is the synthetic pathway for GABA synthesis from glutamic acid via glutamic acid decarboxylase (GAD). Site 12 is an unspecified presynaptic neurotransmitter receptor site (reported to exist for GABA and ACh). Site 13 is a hypothetical ionophore associated with 12. A common mode of convulsant action may involve blockade of the inhibitory action of GABA or glycine at its specific receptor site by a convulsant drug, resulting in reduced activation of site 4, the chloride ionophore. This would result in a reduced degree of hyperpolarization of the membrane and increased neuronal excitability. The convulsant drug may also interact with site 4, the chloride ionophore, and/or sites 1 and 2, the Ca^{2+}-mediated, persistent potassium ionophore, directly. Other convulsants act directly or indirectly on excitatory transmitter receptors (sites 7 and 9) to depolarize the neuronal membranes, which activates their own conductance mechanisms. The degree of selectivity of channels vs neurotransmitter receptors is not well established, and may be qualitatively or quantitatively different in different CNS regions. The final result of convulsant action is an increasing depolarization of the membrane that will lead to seizure initiation.

[a]The neurotransmitter receptors also interact with second messenger molecules, such as cyclic nucleotides or calmodulin, which affect metabolic processes in the cell that may also be important in seizures.

ferred to earlier reviews, particularly the reviews by Woodbury (1980) and Davidoff (1983), for recent useful expositions on the convulsant drug literature.

9. Strychnine

Strychnine is a prototype of the many convulsants that are reported to antagonize the actions of inhibitory transmitters. Strychnine is a naturally occurring, widely studied convulsant agent that has been used for its toxic properties as a rodenticide. Strychnine produces a spinal seizure with lowest drug doses. This "seizure" results in a 10–20 Hz regular discharge (see Fig. 3) in lower brainstem and cerebellum that is not associated with generalized convulsions (Johnson, 1955). The regular discharge in the brainstem of unanesthetized cats induced by iv strychnine can be blocked by spinal cord transection (*see* Fig. 3) (Faingold, 1980a). This high-frequency activity is preceded by sensory response enhancement in neurons of the medullary reticular formation (RF) of unanesthetized cats (Fig. 4C) (Faingold, 1977). The 10–20 Hz regular high-amplitude EEG is accompanied by bursts of action potentials in medullary RF neurons (Faingold, 1980a). (Fig. 4B) with a burst frequency of 10–20 Hz. In larger doses, a progression to a preconvulsive pattern is seen in the EEG, and at this point many mesencephalic RF neurons in the unanesthetized cat begin to exhibit sensory response enhancement (Fig. 1) (Faingold, 1977; Yokota et al., 1979; Faingold, 1980a). With increasing doses, a generalized EEG seizure followed by postictal depression is observed. The explanation of the various phases of strychnine's action requires detailed knowledge of the effects of strychnine on transmitter action and neuronal membrane properties. It was clearly established by Eccles and coworkers that strychnine exerts effects on spinal inhibition (Bradley et al., 1953). Strychnine application produces paroxysmal depolarization shifts and characteristic epileptiform firing patterns with topical administration in several species (Ajmone Marsan, 1969). It is reported that strychnine blocks postsynaptic glycine receptors in several CNS locations, particularly in the spinal cord (Curtis, 1969; Curtis et al., 1971) and medullary RF in the decerebrate cat and the lamprey (Hosli and Tebecis, 1970; Tebecis and DiMaria, 1972; Haas and Hosli, 1973; Martin, 1979). Strychnine is a relatively selective antagonist of the inhibition produced postsynaptically by glycine in various regions, including the cuneate

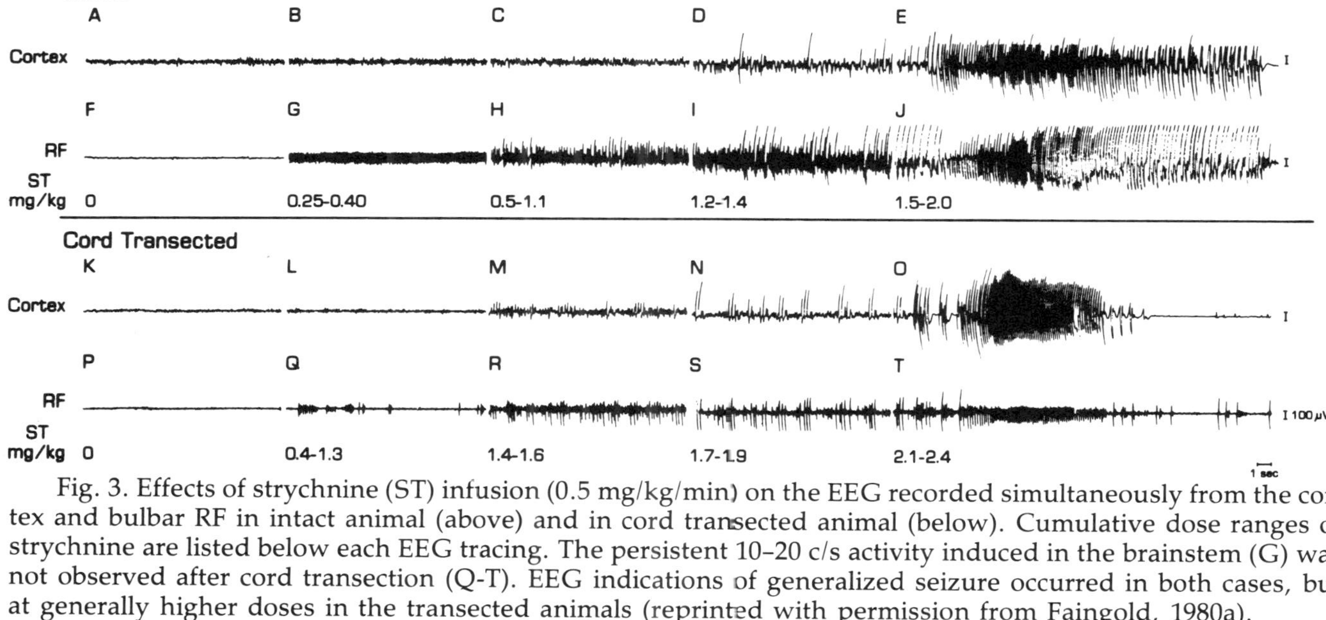

Fig. 3. Effects of strychnine (ST) infusion (0.5 mg/kg/min) on the EEG recorded simultaneously from the cortex and bulbar RF in intact animal (above) and in cord transected animal (below). Cumulative dose ranges of strychnine are listed below each EEG tracing. The persistent 10–20 c/s activity induced in the brainstem (G) was not observed after cord transection (Q-T). EEG indications of generalized seizure occurred in both cases, but at generally higher doses in the transected animals (reprinted with permission from Faingold, 1980a).

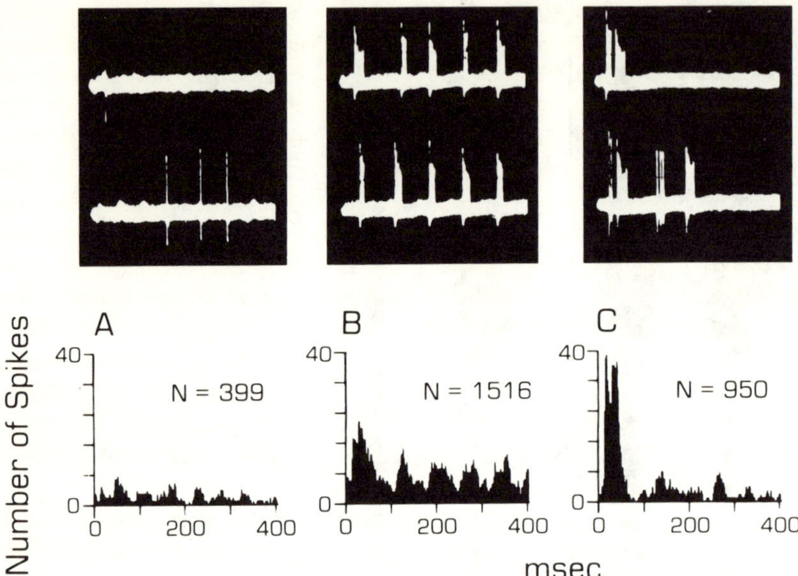

Fig. 4. The effects of strychnine administration (0.6 mg/kg) on the firing pattern and poststimulus time histograms (below) (*see* Fig. 1 for details) of a medullary RF neuron in response to somatosensory stimuli. Before drug administration (A), little consistent response to the stimulus is observed (neuronal maximum amplitude, 2 mV). Rhythmic bursting is observed in the neuron (B), which occurs at a burst frequency that mirrors the 10–20-Hz regular EEG discharge of the lower brainstem. The bursting pattern is usually preceded by and alternates with a pattern of greatly enhanced sensory responsiveness (C).

nucleus in the anesthetized cat (Hill et al., 1976) and the superior olivary nucleus in the anesthetized chinchilla (Moore and Caspary, 1983). The rate of activation of glycine-mediated ion channels is reported to be reduced when strychnine is applied onto mouse spinal cord neurons in culture (Barker et al., 1983). Strychnine blocks evoked inhibition as well as the action of iontophoretically applied glycine more frequently than it blocks the actions of GABA in certain CNS regions, including the cuneate nucleus in anesthetized cats (Hill et al., 1976; Simmonds, 1982). The distribution of radiolabeled strychnine binding in the CNS largely parallels the known distribution of high-affinity glycine binding in vitro in the rat (Young and Snyder, 1973; Zarbin et al., 1981), although strychnine-insensitive glycine binding sites are reported (DeFeudis et al., 1978). This idea of another type of glycine receptor that is not sensitive to strychnine is supported by the earlier report of a postsynaptic glycine receptor in the Renshaw cell of the spinal cord that is not readily blocked by strychnine in anesthetized cats (Ryall et al., 1972). Strychnine administration (iv) greatly reduces the amplitude of intracellularly recorded inhibitory postsynaptic potentials (IPSP) of neurons in glycine-mediated spinal cord pathways to 25% or less before spinal cord seizures developed in the anesthetized cat (Tribble et al., 1983b). These authors suggest that the finding that such a great degree of IPSP reduction is required does not support the blockade of presynaptic inhibition as the main cause of strychnine-induced spinal seizures (Tribble et al., 1983b). Presumed glycine receptors in frog and goldfish neurons are much less effectively or selectively antagonized by strychnine (Diamond et al., 1973; Barker et al., 1975a,b). Faber and Klee (1974) report a general blockade of IPSPs through blockade of chloride conductance changes in *Aplysia* neurons with strychnine. A reduction of threshold for inward current and increasing potassium inactivation in squid and lobster axons and *Aplysia* neurons with strychnine are also observed (Freeman, 1973; Klee et al., 1973; Shapiro et al., 1974). The action of strychnine on chloride conductance is also supported by the findings of Young and Snyder (1974), which report strychnine binding at a glycine-related chloride ionophore site in vitro in the rat. A model of the glycine receptor is suggested in which the strychnine binding site is closely associated with but separable from it, and a cooperative interaction between these sites is postulated (Davidoff, 1983). Strychnine is reported to have significant effects on calcium ion fluxes and to enhance transmission at a weak electrical synapse in *Aplysia* (Klee et al., 1978; Rayport and Kandel, 1981). Strychnine is also reported to lower response thresholds in rat cerebellar neurons in culture through a mechanism unrelated to the blockade of the action of glycine (Gahwiler, 1976). Strychnine is reported to enhance the action of acetylcholinesterase in vitro in rat brain tissue in low concentrations and to inhibit this enzyme at higher concentrations (Alid et al., 1974). This convulsant agent also blocks the actions of norepinephrine, 5-hydroxytryptamine, and GABA in cortical neurons of the anesthetized cat and spinal cord neurons of the decerebrate cat, and in higher doses strychnine blocks the action of GABA in the cuneate nucleus slice preparation and spinal cord (Phillis and York, 1967; Davidoff et al., 1969; Hill et al., 1973; Simmonds, 1978). The enhancement of the sensory responses of mesencephalic RF neurons in unanesthetized cat and rat by iontophoretically applied strychnine is reversed by application of either GABA or glycine (Faingold et al., 1984). Strychnine also increases GABA synthesis in anesthetized cat spinal cord by an unknown mechanism (Reynolds and Watkins, 1972). Benzodiazepines are reported to be very effective antagonists of strychnine seizures (Young et al., 1974). Although, it is reported that benzodiazepines generally act on an accessory receptor associated with the GABA receptor to enhance the action of GABA (Skolnick and Paul, 1982; Ticku, 1983), this may not occur in all regions of the CNS (Biggio et al., 1980). The ability of various benzodiazepine congeners to displace strychnine binding in vitro correlates well with their anticonvulsant potency in various species (Young et al., 1974), and the association of benzodiazepine binding with that of glycine in vitro is reported (Young et al., 1974). Strychnine is also reported to inhibit the interaction of GABA and benzodiazepine binding sites in vitro (Braestrup and Nielson, 1980). However, interactions of benzodiazepines and strychnine in anesthetized cat spinal cord and intact animal may not support this association in vivo (Curtis et al., 1976).

The bulk of experimental evidence suggests a relatively specific effect of low concentrations of strychnine at neuronal receptors for glycine in certain CNS regions, especially in the spinal cord, which could contribute significantly to the spinal cord seizures noted above. However, it is by no means clear that this is universally true in different brain regions associated with the generalized convulsive seizures observed with higher strychnine doses. Therefore, we must conclude that the other less neurotransmitter-specific effects of strychnine on membrane properties, ion conductance changes, and effects on the actions of other transmitters may also contribute to the induction of generalized convulsive seizures with this agent. A summation of reported actions of strychnine and the other agents in this review are shown in Table 2, with reference to the sites shown in Fig. 2.

10. Bicuculline

Bicuculline is a naturally occurring pthalide-isoquinoline alkaloid. When bicuculline is administered systemically to cats or rats, it produces a grand mal-like tonic-clonic generalized seizure with an EEG pattern similar to that of pentylenetetrazol (Welch and Henderson, 1934). Before a seizure-inducing dose level is achieved, sensory-evoked field potentials of the brain in intact and unanesthe-

Table 2
Sites and Mechanisms of Convulsant Drug Action[a]

Convulsant drug	Neurotransmitter	Synaptic actions[b]	Other effects	Most sensitive CNS regions
Strychnine	Glycine	5	a,b,c	Spinal cord (RF)
	(GABA, ACh, 5-HT, N)	(3,4,6,7,12)		
Bicuculline	GABA	3	a	RF (cortex)
	(Gly, ACh, 5-HT)	(1,2,4-7)		
Picrotoxin	GABA	4	a	
	(Gly, ACh, 5-HT)	(1,2,5,7,11)		
Penicillin	GABA	3	c	RF (cortex)
	(ACh, 5-HT, DA)	(1,2,4,7,12)		
PTZ		4	b,c	RF
	(GABA, Gly, ACh, DA, BZD)	(1,2,3,5,7)		
Bemegride	(GABA)	(3)		RF
Allylglycine	GABA	11		Cbell, brainstem
Cholinesterase inhibitors	ACh	6		RF
	(GABA)	(3)		
Kainic acid	Glut	9		Hp
	(GABA)	(3)		

[a]Abbreviations: ACh, acetylcholine; BZD, benzodiazepine; Cbell, cerebellum; DA, dopamine; GABA, γ-aminobutyric acid; Glut, glutamate; Gly, glycine; Hp, hippocampus; N, norepinephrine; PTZ, pentylenetetrazol; RF, reticular formation; 5-HT, 5-hydroxytryptamine. Key to other effects: a, threshold reduction; b, enhanced electrical synaptic transmisison; c, paroxysmal depolarization shift production; ( ), lesser effects.

[b]*See* Fig. 3 for references to sites of synaptic actions.

tized cats are enhanced (Kaplan and Williamson, 1978; Faingold et al., 1983a). Direct or iontophoretic application of bicuculline enhances visual responses of cortical neurons of anethetized cat, but also produces suppression of firing (Hill et al., 1973; Harris and Towe, 1976; Glass et al., 1980). The latter effect may not be related to actions on GABA receptors or may involve blockade of GABA autoreceptors because of high local concentrations of bicuculline reached with such applications (see below) (Arbilla et al., 1979; Napias et al., 1980). Recent studies indicate that the sensory responses of neurons in the medullary and mesencephalic RF of the unanesthetized cat are greatly enhanced by subconvulsant iv doses of bicuculline (*see* Fig. 1) (Faingold et al., 1983a). Cortical neurons in this animal also show response enhancement, but to a lesser degree and with a longer latency than RF neurons (*see* Fig. 5) (Faingold et al., 1983a). The particular cortical area examined in these studies is reported to have connections with the RF (Inubushi et al., 1978). Iontophoretic application of bicuculline enhances the response of RF neurons to sensory stimuli in unanesthetized cat (Faingold et al., 1984). This agent also enlarges somatosensory receptive fields of cortical neurons with certain types of sensory response patterns in anesthetized cats, although adjacent neurons with different sensory response patterns are not greatly affected (Hicks and Dykes, 1983). It is possible, but not proven, that RF enhancement following systemic bicuculline administration may mediate that in cortex. It appears that this cortical region does not directly mediate the RF enhancement, since RF neurons fire first so consistently after systemic administration of bicuculline (Faingold et al., 1983a).

Bicuculline is reported to block the action of GABA at the classical postsynaptic $GABA_A$ receptor in the cuneate nucleus slice and intact brain sites (Johnston, 1978; Simmonds, 1980). Blockade of the effect of GABA by bicuculline is also observed in mouse spinal cord and cortex neurons in culture (Macdonald and Barker, 1979; Nowak et al., 1982). This action is reported to be mediated by reduction of the rate of activation of GABA-activated ion channels (Barker et al., 1983). It is not settled whether the interaction of bicuculline with GABA receptors involves a competitive or noncompetitive mechanism (Curtis et al., 1970a,b; Shank et al., 1974; Enna et al., 1977; Schechter and Tranier, 1977; Nowak et al., 1982). Recent evidence suggests that a second type of GABA receptor ($GABA_B$) may occur presynaptically in certain regions of the CNS. This $GABA_B$ receptor is reported to be insensitive to bicuculline in usual concentrations (DeFeudis, 1983; Frere et al., 1982; Simmonds, 1983). In addition to its effect on $GABA_A$ receptors, iontophoretic application of bicuculline is reported to affect the actions of other putative neurotransmitters, including glycine (in anesthetized cat cuneate nucleus and spinal cord neurons), acetylcholine (in snail neurons), ethylenediamine (in cortex and globus pallidus neurons of anesthetized rats), and 5-hydroxytryptamine (on hypothalamic neurons in the anesthetized rat) (Dray, 1975; Hill et al., 1976; Krnjevic et al., 1977; Piggott et al., 1977; Mayer and Straughan, 1981; Perkins et al., 1981). In some cases the doses of bicuculline required to block certain of these transmitters are higher than those needed to block the effects of GABA (e.g., Krnjevic et al., 1977). Bicuculline is also reported to block acetylcholinesterase in vitro and to reduce the inhibitory effects of acetylcholine in *Aplysia* neurons and in medial septal neurons in anesthetized rats (Svenneby and Roberts, 1973; Miller and McLennan, 1974; Olsen et al., 1976; Yarowsky and Carpenter, 1978). Thus, this agent does not exert a highly selective GABA antagonism in many neuronal systems, including neurons in the cerebral cortex of the rat (Biscoe et al., 1972). The binding of GABA to synaptosomal membranes of rat brain in vitro is inhibited by bicuculline (Zukin et al., 1974). Bicuculline produces a reduction in IPSP amplitude in a GABA-mediated inhibitory pathway from the vestibular to the trochlear nerve in the anesthetized cat, but only by 10% or less before the onset of EEG seizures (Tribble et al., 1983a). The authors suggest that lower sensitivities to the actions of bicuculline of brain regions other than the brainstem may explain seizure induction. However, this study utilized barbiturate anesthesia, which may alter normal seizure mechanisms, as previously noted. Differences in the degree of effect of systemically administered convulsant drug on neurons from different brainstem nuclei in unanesthetized cats are seen with other convulsants (Faingold et al., 1983a), suggesting that the brainstem pathway that Tribble and coworkers (1983a) utilized is not necessarily representative of more sensitive brainstem neurons, such as those in the RF. Bicuculline administration is reported not to affect brain GABA levels (Loscher and Frey, 1977), although turnover of GABA is enhanced during bicuculline-induced seizures in anesthetized rats (Chapman and Evans, 1983).

Bicuculline is reported to exert direct effects on neuronal membranes involving mechanisms not unlike those noted previously with strychnine. Bicuculline affects potassium inactivation in the lobster axon (Freeman, 1973), which leads to an increased incidence

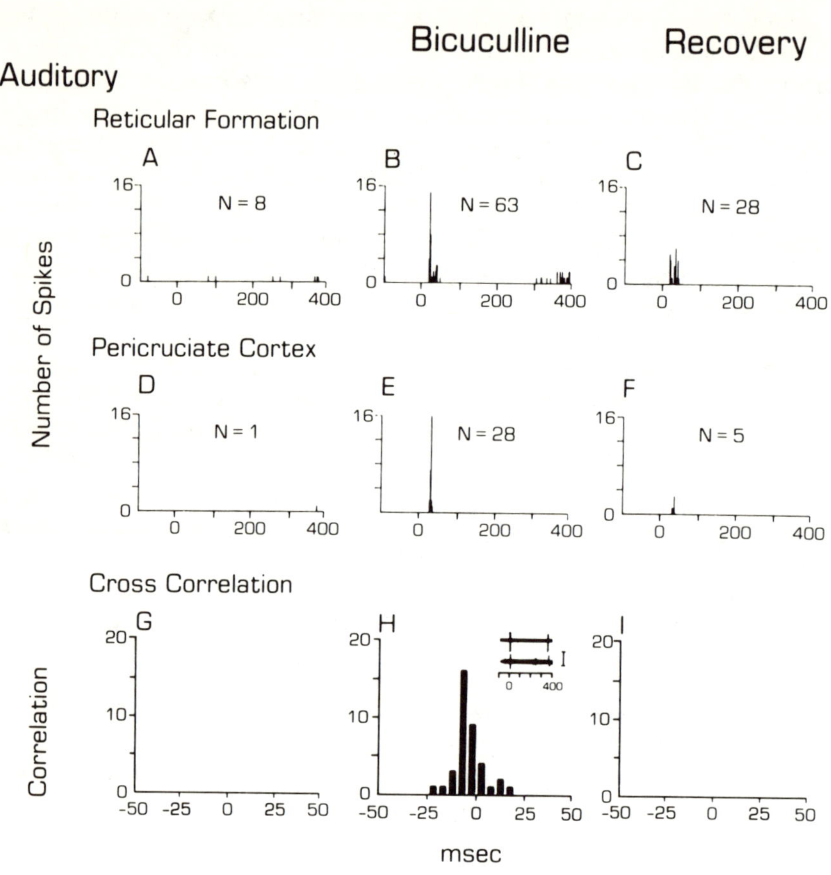

Fig. 5. Example of the effects of bicuculline (BIC) on the sensory responses and cross-correlation of simultaneously recorded neurons in mesencephalic RF and pericruciate cortex. Before BIC administration (in the first column), limited firing of each neuron wsa observed in response to auditory (A,D) or visual (J,M) stimuli, as shown in the histograms. No correlation (5 ms bin width) of firing was observed with either stimulus (G,P). Following BIC administration (0.08 mg/kg), both neurons became

Visual

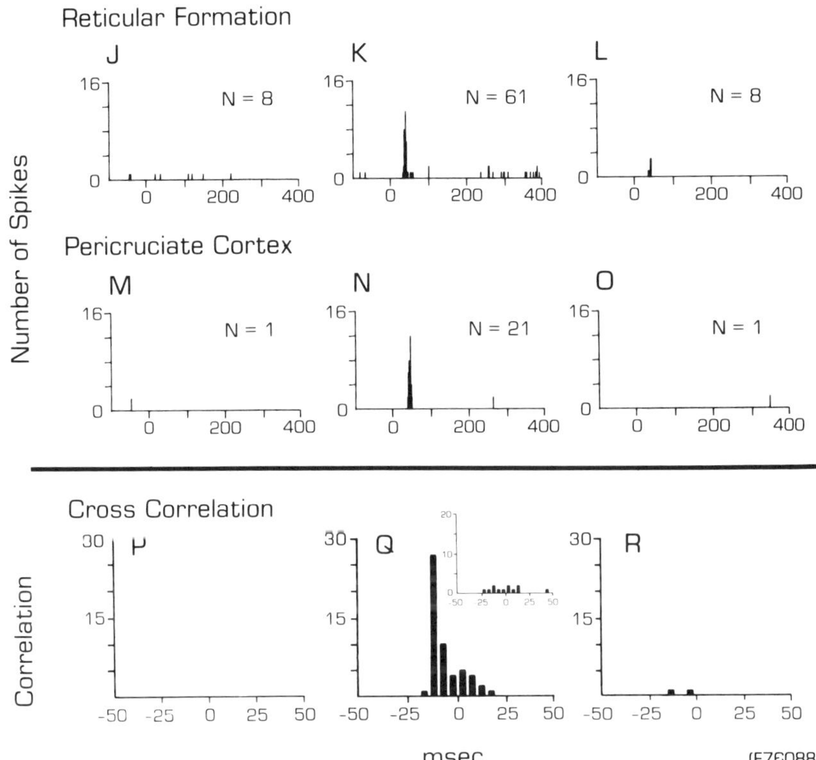

Fig. 5 *(continued)*
consistently responsive to auditory (B,E) and of firing to the auditory (H)
and visual (Q) stimuli was observed. At this time neither neuron was
responsive to the somatosensory stimuli and the correlogram did not show
a consistent peak (inset Q). The responses returned to predrug patterns
26 min after termination of drug administration. By convention, a negative
time value indicates that the RF neuron fired before the cortical neuron.
Calibration: 1 mV (reprinted with permission from Faingold et al., 1983a).

of multiple action potentials with this agent. In addition, bicuculline is also reported to reduce neuronal threshold in spinal cord neurons in culture by a mechanism that does not involve GABA receptor mediation (Barker et al., 1981). Bicuculline is also reported to block a calcium-mediated potassium conductance in mouse spinal cord neurons in culture by affecting nonsynaptic mechanisms, as well as antagonizing the action of GABA (Heyer et al., 1982). The auditory response threshold of RF neurons in the unanesthetized cat is also reduced with iv administration of this agent (Faingold et al., 1983a), which may not, however, be mediated by the same mechanism. In addition, it is reported from work in *Aplysia* that bicuculline will act at the chloride ionophore to block the chloride conductance increase caused by acetylcholine, as well as GABA, which both activate this mechanism in *Aplysia* neurons (Yarowsky and Carpenter, 1978). Thus, the blockade of GABA receptors appears to be a major effect of bicuculline. However, other important effects of bicuculline on the actions of other neurotransmitters and effects on ionic conductance mechanisms in different brain regions may also play an important role in seizures induced by this agent, as shown in Table 2.

11. Picrotoxin

Picrotoxin is a naturally occurring mixture of two dilactones. The primary active component is picrotoxinin. Picrotoxin is the original agent thought to be a specific antagonist of the action of GABA at its receptors in the CNS. Picrotoxin produces clonic seizures and generalized tonic-clonic seizures with tonic flexion and extension in several species. Lesion and local administration studies suggest that the brainstem is most sensitive to the convulsant action of this agent (Hahn, 1960). There is a peculiarity of onset that is different from that of the other convulsants discussed so far, in that there is a significant latency to seizure production even when the agent is given by the intracisternal route (Woodbury, 1980). It is suggested that this delayed onset may be caused by biotransformation or effects on the action of an enzyme (Loscher and Frey, 1977). This hypothesis is of interest in light of the constantly repetitive nature of the EEG seizures seen with picrotoxin administration when given by gradual iv administration, which is very similar to the constant repetitive pattern seen with allylglycine given by the same route. Allylglycine is reported to inhibit the action of glutamic acid decarboxylase (GAD) (Horton and Meldrum, 1973), an action also reported with picrotoxin (Loscher and Frey, 1977). Picrotoxin enhances sensory-evoked field potentials in anesthetized rats and behaving cats (Bigler, 1977a,b; Kaplan and Williamson, 1978), and RF neuronal responses to sensory stimuli in unanesthetized cat are also enhanced (Fig. 1). Picrotoxin is reported to block the synaptic action of GABA. Much of the original evidence for this action comes from invertebrate studies. Antagonism of the action of GABA is also seen in neurons of the cuneate nucleus (in the anesthetized cat and in vitro in the rat), olfactory cortex (slices from the guinea pig), cerebellum, and cortex (in vivo) (Galindo, 1969; Hill et al., 1973; Krnjevic, 1974; Simmonds, 1980; Scholfield, 1982; Simmonds, 1982). Picrotoxin is reported to reduce the rate of activation of GABA-activated ion channels in mouse spinal cord neurons in culture (Barker et al., 1983). The ability of picrotoxin to block the actions of GABA at many sites does not appear to be through direct blockade of GABA receptors (Simmonds, 1980; 1982), but rather through an action on a closely related but separate chloride conductance channel. This site is thought to subserve hyperpolarization that involves activation of the chloride ionophore, and blockade of the ionophore by picrotoxin reduces hyperpolarization and renders neuronal membranes more excitable (Takeuchi and Takeuchi, 1969; Fujimoto and Okabayashi, 1981; Squires and Saederup, 1982; Yarowsky and Carpenter, 1978; Simmonds, 1980). Thus picrotoxinin and derivatives do not alter GABA binding onto synaptic membranes in vitro (Ticku et al., 1978; Olsen et al., 1978; Zukin et al., 1974). A model of the picrotoxinin "receptor" site has been published (Klunk et al., 1983). This site is suggested to be an important locus of convulsant action (Squires et al., 1982). Picrotoxin also possesses other direct effects on membranes with some similarities to the convulsants already discussed, including effects on potassium conductance in lobster axons and membrane depolarization and reduction of neuronal response threshold with induction of repetitive action potentials in mouse neurons in culture (Freeman, 1973; Barker and MacDonald, 1980). Picrotoxin is also reported to block the actions of glycine (in neurons of the anesthetized cat cortex), serotonin (in ganglion neurons of the cat), and acetylcholine (in crab muscle and snail neurons) (DeGroat and Lalley, 1973; Hill et al., 1976; Piggott et al., 1977; Marder and Paupardin-Tritsch, 1980). Regional differences of picrotoxin-induced effects on GABA levels are observed within the brain (Saito and Tokunaga, 1967). Thus, much of the evidence supports a major effect of picrotoxin on chloride conductance channels associated with the GABA receptor, but this may not be the only important site involved in seizure production with this agent, as shown in Table 2. Actions of picrotoxin on other ion channels and on the action of other neurotransmitters in different CNS regions may also be important to seizure induction with this agent.

12. Penicillin

Penicillin is a widely utilized antibiotic agent that is epileptogenic. It is employed extensively in epilepsy research for its ability to produce focal seizures by topical application in vivo and in vitro. Systemic administration of penicillin results in generalized seizures with behavioral and EEG patterns that differ from the other drugs discussed thus far. Myoclonic jerking and 3/s spike and wave complex patterns are observed in the cat and rat (Sutton and Oldstone, 1969; Fariello, 1976) that have a striking resemblance to human *absence* (petit mal) seizures, although certain investigators characterize the pattern as multifocal (Fariello, 1976; Rodin et al., 1977). The spike and wave pattern in cats can be activated by photic stimulation, and sensory-evoked field potentials are reported to be enhanced by penicillin (Burchiel et al., 1976; Quesney, 1984). Cortical neuronal responses to sensory stimuli in the anesthetized cat are also enhanced with iontophoretic or topical application of penicillin (Macon and King, 1979; Ebersole and Chatt, 1981). Conditioned responses to sensory stimuli in behaving cats are blocked during the penicillin-induced spike and wave pattern in the cat (Taylor-Courval and Gloor, 1984). Evidence has been presented that this epilepsy model is driven from the cortex to the brainstem RF (Guberman and Gloor, 1974; Testa and Gloor, 1974; Avoli and Kostopoulos, 1983). However, studies utilizing simultaneous recordings in three areas in the unanesthetized cat indicate that a definite population of brainstem RF neurons exists that fire in association with the spike and wave complex and whose firing consistently precedes the firing of the cortical neurons (Pellegrini et al., 1983) in a manner similar to that observed with bicuculline (Faingold et al., 1983a).

Topical application of penicillin onto the cortex in several species and hippocampal slices in vitro results in the appearance of the paroxysmal depolarization shift (Ajmone Marsan, 1969; Prince, 1982). Neurons in the cortical slice preparation do not show this phenomenon (Gutnick et al., 1982). The reason for the differential effect of penicillin on cortical and hippocampal neurons may involve the finding that neurons in the deeper cortical layers (layer IV) are much more sensitive to the effects of this agent than the superficial cortical layers that are predominant in the slice (Chatt and Ebersole, 1982; Ebersole and Chatt, 1984). Abnormalities of dendritic morphology have been reported in neurons that exhibit the depolarization shift in the penicillin focus of anesthetized cats (Greenwood et al., 1983).

The mechanisms involved in penicillin action are suggested to involve blockade of the action of GABA. This effect is reported in neurons in the spinal cord, the cortex in anesthetized cat, cuneate nucleus, and hippocampus (in vitro) (Curtis et al., 1972; Hill et al., 1973; Dingledine and Gjerstad, 1980), but not universally (Krnjevic et al., 1977). In vitro studies suggest that penicillin selectively blocks GABA-mediated postsynaptic inhibition in mouse spinal cord neurons in culture (Macdonald and Barker, 1977). Penicillin blocks chloride conductance changes in response to GABA in crab muscle (Hochner et al., 1976) and release of GABA from the cortical slices (Cutler and Young, 1979). Seizure-inducing doses of penicillin do not block GABA-mediated IPSPs in Deiters nucleus neurons of anesthetized cats (Davenport et al., 1979). Penicillin is reported to block calcium-dependent persistent potassium conductance changes in spinal cord neurons in culture by a nonsynaptic action, as well as through blockade of the action of GABA (Heyer et al., 1982). Penicillin blocks IPSPs in hippocampus in vitro (Schwartzkroin and Prince, 1980) and blocks chloride conductance increases caused by several different neurotransmitters, including GABA, acetylcholine, dopamine, and serotonin in *Aplysia* (Wilson and Escueta, 1974; Pellmar and Wilson, 1977b). Penicillin also enhances the excitatory effects of acetylcholine in hippocampal slices and produces presynaptic enhancement at peripheral cholinergic synapses and in the hippocampal slice (Futamachi and Prince, 1975; Noebels and Prince, 1977; Kriegstein et al., 1983). An increased release of glutamate from cortical slices of the rat is reported with penicillin (de Boer et al., 1982). Nonsynaptic mechanisms and effects on calcium movement are also suggested to be involved in the production of the paroxysmal depolarization shift with penicillin (Schwartzkroin and Prince, 1980; Andersen and Gjerstad, 1981). Increases in extracellular calcium in spinal cord and especially cortex are reported prior to seizure initiation in anesthetized cat (Somjen, 1980). Thus, penicillin administration results in blockade of the effects of GABA, but effects on cholinergic mechanisms and membrane conductance mechanisms may also be involved in seizure production with this agent, as shown in Table 2.

13. Pentylenetetrazol

Pentylenetetrazol (PTZ)—also known as leptazol, cardiazol, and metrazol—is a widely studied convulsant. Administration of PTZ results in generalized tonic-clonic seizures in cats that can be initiated by sensory stimuli. In rat, forelimb and jaw clonus are observed first with threshold levels of this agent (Woodbury, 1980). The EEG changes induced by PTZ in several species involve interictal spikes of ever-increasing amplitude and eventually grand mal-like electrographic discharges in all regions of the brain simultaneously (Fig. 6) (Straw and Mitchell, 1967; Faingold and Berry, 1973; Faingold 1977). The ictal discharge is followed by postictal depression and gradual recovery to normal patterns. PTZ is reported to enhance sensory-evoked field potentials in cat, rat, and man (Gastaut and Hunter, 1950; Kimura, 1962; Bergamasco, 1966). Regional differences in onset and degree of evoked potential enhancement are observed, with the brainstem RF exhibiting the earliest onset and greatest degree of enhancement (Faingold, 1977). Multiunit activity in the mesencephalic RF in unanesthetized cats is also greatly enhanced by PTZ (Velasco et al., 1975). Mesencephalic lesions in unanesthetized cats raise seizure threshold (Velasco et al., 1976). PTZ produces sensory response enhancement in medullary and mesencephalic RF neurons in the vast majority of neurons examined in unanesthetized cats (Fig. 1) (Faingold and Caspary, 1977; Faingold, 1978; Faingold, 1980b). The degree of RF neuronal response enhancement greatly exceeds that in lateral geniculate and inferior colliculus neurons in unanesthetized cats (Faingold and Stittsworth, 1980; Faingold et al., 1983b). PTZ-induced visual afterdischarges are observed in lateral geniculate neurons in anesthetized rats (Bigler and Eidelberg, 1976). Electrical stimulation and cryoprobe blockade of brainstem auditory nuclei of unanesthetized cats suggest that auditory response enhancement with systemic PTZ involves a direct action on RF neurons (Faingold et al., 1983b), which has been confirmed using iontophoretic application onto RF neurons in unanesthetized cats and rats (Faingold et al., 1984). The mechanisms by which PTZ produces seizures are not well understood. PTZ is reported to block the action of GABA in mouse spinal

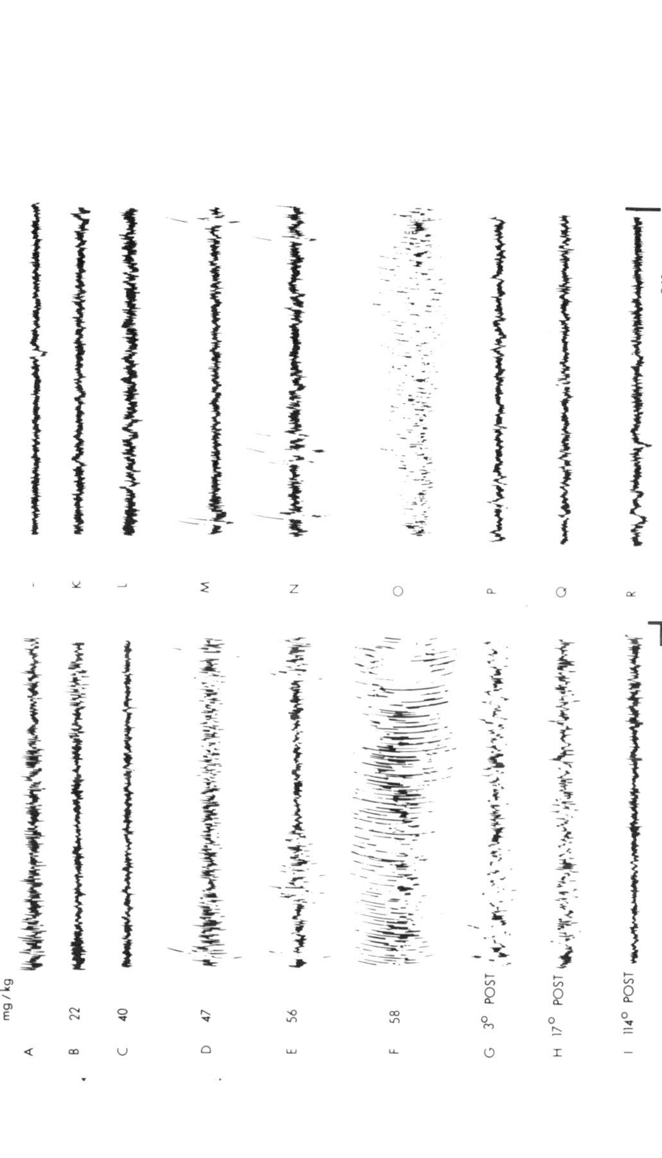

Fig. 6. Sequential concurrent EEG records from cortex and mesencephalic RF electrodes during the gradual infusion of pentylenetetrazol (1.25 mg/kg/min). (A) Predrug control activity; (B–C) Induction of desynchrony; (D) Appearance of sharp waves; (E) Clusters of sharp waves; (F) Seizure; (G–H) Postictal high-amplitude activity; (I) Recovery; RF changes; (J–R) Closely parallel cortical EEG activity. The cumulative dose of the drug is listed next to the activity seen with that dose. Calibrations 100 μV and 1 s. (reprinted with permission from Faingold, 1977).

cord neurons in culture, cortex neurons in anesthetized cats (in vivo), and cuneate nucleus neurons (in vitro) (Hill et al., 1973; Macdonald and Barker, 1977; Simmonds, 1978; 1982). These studies report that no changes in the action of glycine or glutamate are observed, although blockade of the inhibitory action of glutamate in snail neurons is reported (Piggott et al., 1977). PTZ is also reported to release glutamate from cortical slices of the rat (de Boer et al., 1982). PTZ application onto mouse spinal cord neurons in culture depresses membrane current responses to GABA, in part, through a reduction in the open-time of GABA-activated ion channels, although the latter mechanism does not appear to be the entire explanation (Barker et al., 1983). Repeated daily administration of PTZ in the dog causes GABA levels to fall, along with a decrease in seizure threshold and an increase in seizure severity (Losher, 1983). In *Aplysia* neurons, however, PTZ exerts a blocking effect on chloride-conductance increases produced by GABA, acetylcholine, dopamine, and glycine. PTZ is also reported to inhibit acetylcholinesterase in vitro (Mahon and Brink, 1970). Evidence supporting an interaction of PTZ with one type of benzo-diazepine receptor in various brain regions is reported based on structure–activity binding studies with tetrazole derivatives (Rehavi et al., 1982; Chweh et al., 1983), the ability of benzodiazepine agonists to block PTZ-induced seizures in mice, and the ability of certain benzodiazepine "inverse agonists" to enhance these seizures (Greksch et al., 1983; Jensen and Peterson, 1983). However, it is not clear that the inverse agonists are producing the proconvulsant effect by an action on benzodiazepine receptors (Boast et al., 1983). Inhibition of GAD activity is reported with PTZ in mouse brain, but at levels much higher than convulsant doses (Losher and Frey, 1977). PTZ also has other membrane effects unrelated to synaptic actions in snail neurons, including blockade of calcium-mediated potassium currents (Faugier-Grimaud, 1974; Mashimo and Sekiya, 1981). PTZ is reported to enhance transmission in a weak electrical synapse in *Aplysia* (Rayport and Kandel, 1981). PTZ application produces the paroxysmal depolarization shift in vivo in the cortex of the anesthetized cat and in snail neurons in vitro (Creutzfeldt et al., 1966; Speckmann and Caspers, 1973). This effect in snail neurons is blocked by several anticonvulsant drugs (Faugier-Grimaud, 1978). However, these findings do not correlate with the ability of these drugs to block PTZ-induced seizures in intact mammals (e.g., Faingold and Berry, 1973). A fall in extracellular free calcium and a rise in potassium with PTZ is reported in studies on the cerebral cortex of anesthetized cats (Heinemann et al., 1977), and an increase in intracellular calcium is reported in vitro in snail neurons during PTZ-induced bursting (Sugaya and Onozuka, 1978 a,b). Elevations of cyclic guanosine monophosphate are produced by PTZ in mice before seizure onset (Ferrendelli and Kinscherf, 1977).

Thus PTZ produces a variety of effects on neurotransmitter action and a general blockade of inhibition through effects on chloride conductance mechanisms, as shown in Table 2, with relative regional specificity for the brainstem RF. However, the exact importance of these mechanisms and sites of seizure induction by this agent remain to be examined.

14. Bemegride

Bemegride (megimide) is a synthetic compound with convulsant properties. It has had some clinical use as an analeptic and some experimental use as a diagnostic aid and model of epilepsy with similarities to pentylcnetetrazol (Hahn, 1960; Sellden, 1964; Rodin et al., 1971). The EEG changes induced by bemegride are quite similar to those produced by PTZ (*see* Fig. 6). Bemegride is reported to enhance sensory-evoked field potentials in both man and animals (Rodin et al., 1966; Kaplan and Williamson, 1978). Bemegride is reported to exert prominent and selective effects on high-frequency activity in the EEG of the brainstem RF of behaving cats (Rodin et al., 1971). Bemegride administration (iv) in subconvulsive doses enhances the responses of brainstem RF in both medullary and midbrain regions in unanesthetized cats (Fig. 1) (Faingold and Hoffman, 1981). The sensory response of less than half of the hippocampal neurons examined in unanesthetized cats are enhanced, but with an onset that is later in time and a latency that is greater than RF neurons (Faingold and Hoffman, 1981). During bemegride-induced EEG seizures in the unanesthetized cat the RF neurons fire in bursts leading to the generalized high-amplitude EEG, and these neurons fire just preceding each spike in the cortical EEG. Hippocampal neurons begin to fire in unison with cortical EEG spikes usually some time after the generalized EEG seizure begins (Faingold and Hoffmann, 1981).

The mechanisms involved in the action of bemegride are also suggested to involve blockade of the action of GABA, which is observed in guinea pig olfactory cortex in vitro and rat cuneate nucleus slice (Simmonds, 1978; Scholfield, 1982). Bemegride is reported to block primary afferent depolarization in the cuneate nucleus of decerebrate (and anesthetized cats) and anesthetized rats in a manner similar to that of other convulsants (Banna and Jabbur, 1970; Hayes et al., 1977). Paroxysmal activity induced by bemegride in leech ganglion neurons in vitro is similar to the paroxysmal depolarization shift seen with other convulsant drugs (Prichard, 1972; Kleinhaus, 1975). This effect may involve blockade of the action of GABA, which is reported in snail neuron and rat cuneate nucleus slice (Piggott et al., 1977; Simmonds, 1978). There is some evidence to suggest that bemegride blocks the action of acetylcholine and glutamate in snail neurons (Piggott et al., 1977). Thus the degree of contribution of the effects of bemegride on the actions of neurotransmitters to seizure induction is not established, despite the reports of actions on neurotransmitter and membrane mechanisms shown in Table 2.

15. Allylglycine

Allylglycine is a convulsant drug that in several species produces myoclonic and/or tonic seizures with a slow onset (Orlowski et al., 1977; Woodbury, 1980). The delayed onset is suggested to involve the conversion of this agent to a metabolite, 2-keto-4-pentenoic acid (Horton, 1980). The seizures involve repeated tonic-clonic convulsions (Hammad et al., 1983) accompanied by repetitive high-amplitude, high-frequency activity in the EEG, and the long onset is coupled with a long duration of recurrent seizures. Allyl-glycine administration results in exacerbation of genetic photo-myoclonic epilepsy in behaving baboons (Meldrum et al., 1979). Systemic administration of allylglycine produces enhancement of visual-evoked field potentials in cortex and brainstem RF in behaving normal and epileptic baboons and unanesthetized cats (Menini et al., 1980; Stutzmann et al., 1980; Faingold, unpublished). Enhanced visual responses of neurons in each of these structures subserves this enhancement (Menini et al., 1981) (*see* Fig. 1). Allyl-glycine inhibits neuronal firing in spinal cord neurons of anesthetized cats and in goldfish neurons (Curtis et al., 1970b; Roper, 1970). The mechanisms of convulsant action of this agent are reported to involve the reduction of GABA synthesis through inhibition of the synthetic enzyme, glutamic acid decarboxylase (GAD) (Alberici et al., 1969; Horton, 1980; Orlowski et al., 1977), possibly by reversible inactivation rather than competition with GAD (Fisher and Davies, 1974). A regional dichotomy of convulsant action between optical isomers of allylglycine is reported. The D and L isomers produce different seizure patterns in rodents (Meldrum, 1979). The D isomer induces irregular jerking, sudden jumps, and brief rotational seizures that differ from the forepaw myoclonus, wild running, and tonic seizure seen with the L isomer of allylglycine (Meldrum, 1979). GABA levels fall primarily in cerebellum, pons, and medulla in the rat after D allylglycine administration (Horton, 1980), but different regional patterns of GABA changes are observed with L allylglycine and other agents that inhibit GAD (Horton, 1980). The reason behind this regional selectivity may be caused by the regional differences in the distribution of D amino oxidase, which is involved in the activation of D allylglycine (Meldrum, 1979). Allylglycine is also a competitive inhibitor of cystathionase and blocks the synthesis of glutathione (Orlowski et al., 1977), which may be involved in the convulsant action of this agent. Benzodiazepines, barbiturates, phenytoin, valproate, and baclofen are reported to be effective anticonvulsants against allylglycine-induced seizures (Ashton and Wauquier, 1979). Thus allylglycine produces significant effects on GABA synthesis, but other effects may also contribute to seizure production with this agent.

16. Cholinesterase Inhibitors

Because fewer convulsant studies have been done to examine synaptic actions and CNS regional effects of cholinesterase inhibitors, they will be considered together. Some differences among individual agents are likely however. The agents that block the action of acetylcholinesterase, the enzyme responsible for the breakdown of acetylcholine (Brimijoin, 1983), also act as convulsant drugs. These agents include the long-lasting ''irreversible inhibitors,'' many of which are organophosphates (Ellin, 1982). The carbamates are reversible acetylcholinesterase inhibitors (ChEI) that include the tertiary amine derivatives, such as physostigmine, that readily penetrate into the brain. ChEI-induced seizures exhibit generalized spike bursts followed by high-voltage, high-frequency generalized spikes throughout the brain, accompanied by forelimb and facial clonus. Several ChEI agents have been shown to enhance sensory-evoked potentials (A'Hearn, 1966; Faingold, 1978), and several acetylcholine agonists are also reported to enhance sensory-evoked potentials (*see* Table 1). Physostigmine in subconvulsant doses also enhances the sensory responses of reticular formation neurons in the unanesthetized cat (Fig. 1). The mechanism by which ChEI agents produce seizures is generally attributed to increased action of the excitant transmitter acetylcholine (DuBois, 1963). Studies reporting that muscarinic receptor antagonists could abort the ChEI-induced seizures in the unrestrained rabbit have contributed to this idea (Johns and Himwich, 1950). However, benzodiazepines and other agents that have significant effects on the GABA system (Skolnick and Paul, 1982) effectively abolish ChEI-induced seizures (Lipp, 1973; Matin and Kar, 1973; Rump et al., 1973; Lundy and Magor, 1978; Lundy et al., 1978; Niemegeers et al., 1982; Honchar et al., 1983). Atropine, in doses that block the effects of excessive acetylcholine action in unrestrained rats, is reported not to block seizure induction with certain ChEI agents (Wecker et al., 1977; Lundy et al., 1978). The ability of muscarinic receptor-blocking agents to prevent ChEI-induced seizures is reported not to correlate with their potency as peripheral muscarinic blockers, and some anticonvulsant effects of these agents are observed against electroshock-induced seizures in mice and rats (Green et al., 1977). However, muscarinic blockers that penetrate well into the CNS consistently block physostigmine-induced seizures (Niemegeers et al., 1982). Increases in the number of dopamine and GABA receptors (in vivo but not in vitro), as well as increases in cyclic guanosine monophosphate in the cerebellum in vivo, are also observed after administration of certain irreversible ChEI agents in subconvulsant doses that accompany the decrease in acetylcholinesterase activity (Lundy and Magor, 1978; Sivam et al., 1983). These agents are also reported to increase 5-hydroxytryptamine levels and turnover in several regions of the rat brain (Prioux-Guyonneau et al., 1982). Other organophosphate agents that are structurally similar to the irreversible ChEI agents, but do not block acetylcholinesterase, also produce seizures and lower GABA levels in the brain of rodents (Kar and Matin, 1972; Bellet and Casida, 1973). Intravenous administration of certain of these compounds blocks the action of GABA and glycine on cuneate nucleus neurons in anesthetized rats (Bowery et al., 1976). Certain ChEI agents are also reported to affect calcium availability in rat neuromuscular junction and potassium conductance in electric fish (Abraham and Edery, 1978; Farquharson, 1978). Thus the majority of evidence suggests that ChEIs have a major effect on the action of acetylcholine, but actions on other transmitters and ion conductances may also play a role in seizure induction with these agents, as shown in Table 2.

17. Kainic Acid

Naturally occuring amino acids, such as glutamate, aspartate, and several derivatives, have significant convulsant properties. Although glutamate does not readily penetrate the blood–brain barrier in mature animals, its heterocyclic analog, kainic acid, which is derived from Japanese seaweed, penetrates well into the brain following systemic administration. Kainic acid induces "wet-dog shakes" followed by forelimb clonus and hindlimb extension in mice and rats (Olney et al., 1974; Sperk et al., 1983). EEG changes with kainic acid are reported to begin in the limbic system after a significant delay, with eventual spread to the rest of the brain in rats (Ben-Ari et al., 1981; Lothman et al., 1981). Kainic acid also produces a considerable amount of neurotoxicity (Olney, 1978; Coyle, 1983), at least in part, through its convulsive properties (Ben-Ari et al., 1979; 1981; Sperk et al., 1983). Considerable evidence is reported that supports excitant amino acids as putative excitatory transmitters in various regions of the nervous system, including the spinal cord and auditory system (Puil, 1983; Caspary et al., 1985; *see* Fonnum, 1984 for review). Evidence has been presented suggesting that excitant amino acid receptor subtypes exist, including one that is relatively selective for kainic acid (Davies and Watkins, 1979; Fagni et al., 1983). In addition to the action of kainic acid on postsynaptic receptors, this agent is also reported to release excitant amino acids from cortex and striatum slices through a presynaptic action (Collins et al., 1983; Potashner and Gerard, 1983). Excitant amino acid antagonists, benzodiazepines, barbiturates, and aminooxyacetic acid antagonize kainic acid-induced seizures in rats and mice (Stone and Javid, 1980; Fuller and Olney, 1981). Because the latter three anticonvulsants are thought to enhance the action of GABA and the ability of GABA agonists to block kainic acid seizures (Fariello et al., 1982), some investigators suggest that an effect of kainic acid on the action of GABA may also be involved in the convulsant effects of this compound (Fuller and Olney, 1981; Giorgi and Meek, 1984). Kainic acid produces a threshold reduction and enhancement of the input–output relationship of rat hippocampal neurons in vitro, as well as paroxysmal depolarization shifts like those observed with other convulsant agents in these neurons (Westbrook and Lothman, 1983). An important mechanism of these effects of kainic acid on neurons in the hippocampal slice is reported to involve presynaptic blockade of IPSPs without affecting postsynaptic responses to GABA (Fisher and Alger, 1984). Kainic acid is also reported to increase GABA receptor sensitivity in isolated frog spinal cord (Ono et al., 1983). Other studies suggest that the action of kainic acid on GABA may not be important to the convulsant actions of this agent (Kish et al., 1983). Kainic acid injections directly into the brain in rats and baboons are extensively utilized to provide a model of limbic epilepsy (Pisa et al., 1980; Ben-Ari et al., 1980; Cepeda et al., 1982). Some differences in the convulsant actions of glutamate and kainic acid are observed in that valproate, which is thought to enhance the action of GABA, effectively blocks the convulsant action of intracerebrally administered glutamate in mice (Stone and Javid, 1983), but valproate does not block seizures induced by sc kainic acid (Fuller and Olney, 1981). The neurotoxic action is associated with extensive changes in several neurotransmitter markers (Heggli and Malthe-Sorenssen, 1982; Heggli et al., 1981). Thus, neurotransmitter-specific actions on excitant amino acid receptors are a consistent finding with kainic acid. However, the role of actions of this agent on GABA and the actions of kainic acid to produce membrane depolarization, which do not appear to be mediated by neurotransmitter-specific receptors, may also contribute to seizure induction with this agent, as shown in Table 2.

18. Future Experiments

Although this review covers a large volume of literature on convulsant drugs, it represents only a selected sample of the vast quantity of information published on this subject. From the information reviewed, it is quite clear that convulsant drugs exert major and significant effects on the actions of neurotransmitters that have considerable import to the role of neurotransmitters in epilepsy. However, it is also clear that much more work, especially in unanesthetized mammalian brain in vivo, is required before we will achieve a comprehensive understanding of the neurotransmitter-related events that underlie initiation of seizures. In order to determine which CNS regions are most involved in seizure production in the intact organism, more comparative studies similar to that of Faingold and coworkers (1983a) are necessary. Examinations of intracellular events in vivo in other CNS regions, particularly in the RF, in a fashion similar to the studies of Tribble et al. (1983 a,b), are needed as well. Comparative studies on neurons in several brain regions of convulsant-induced changes of the effects of iontophoretically applied neurotransmitters in a fashion similar to the study of Macon and King (1979), but using systemic administration of convulsants, are also required before a truly definitive understanding of neurotransmitter mechanisms involved in seizure production by convulsant drugs can be achieved.

Acknowledgments

The author wishes to thank Donald Caspary, William Hoffmann, and James Stittsworth, Jr., for their assistance in performing experiments from this laboratory cited in this review, and Tammy Kissel and, especially, Carol Faingold for typing the manuscript. Thanks are also due to Brian Meldrum and Wilkie Wilson for critical comments on the manuscript. The author was supported by NIH grant NS 13849, Deafness Research Foundation, Southern Illinois University Central Research Committee and Pearson Family Foundation grant funds during the preparation of this review. The Division of Biomedical Illustration and Photography prepared the figures.

References

Abraham, Z. and Edery, H.: Cholinesterase inhibition as modifier of calcium availability at the rat diaphragm neuromuscular junction. *Israel J. Med. Soc.* **14:** 497, 1978.

A'Hearn, M. C.: Effects of cortical stimulants and cholinolytic agents on spontaneous and evoked potentials (doctoral dissertation). Madison, Wisconsin, University of Wisconsin, 1966.

Ajmone Marsan, C.: Acute Effects of Topical Epileptogenic Agents, In: *Basic Mechanisms of the Epilepsies* (H. H. Jasper, A. A. Ward, Jr., and A. Pope, eds.) Little Brown and Company, Boston, 1969.

Alberici, M., Rodriguez De Lores Arnaiz, G., and De Robertis, E.: Glutamic acid decarboxylase inhibition and ultrastructural changes by the convulsant drug allylglycine. *Biochem. Pharmacol.* **18:** 137–143, 1969.

Alger, B. E. and Nicoll, R. A.: Epileptiform burst after hyperpolarization: Calcium-dependent potassium potential in hippocampal CA 1 pyramidal cells. *Science* **210:** 1122–1124, 1980.

Alid, G., Valdes, L. F., and Orrego, F. J.: Strychnine as an anticholinesterase: In vitro studies with rat brain enzymes. *Experientia* **30:** 266–268, 1974.

Andersen, P. and Gjerstad, L.: Possible Mechanisms Underlying Epileptiform Discharges, In: *Brain Mechanisms of Perceptual Awareness and Purposeful Behavior* (O. Pompeiano and C. Ajmone Marsan, eds.) Raven, New York, 1981.

Arbilla, S., Kamal, L., and Langer, S. Z.: Presynaptic GABA autoreceptors on GABAergic nerve endings of the rat substantia nigra. *Eur. J. Pharmacol.* **57:** 211–217, 1979.

Arduini, A. and Arduini, M. G.: Effect of drugs and metabolic alterations on brain stem arousal mechanism. *J. Pharmacol. Exp. Ther.* **110:** 76–85, 1954.

Ashton, D. and Wauquier, A.: Effects of some anti-epileptic, neuroleptic and Gabaminergic drugs on convulsions induced by D, L-allylglycine. *Pharmacol. Biochem. Behav.* **11:** 221–226, 1979.

Avoli, M. and Kostopoulos, G.: Participation of corticothalamic cells in penicillin-induced generalized spike and wave discharges. *Brain Res.* **247:** 159–163, 1983.

Ayala, G. F., Dichter, M., Gumnit, R. J., Matsumoto, H., and Spencer, W. A.: Genesis of epileptic interictal spikes. New knowledge of cortical feedback systems suggests a neurophysiological explanation of brief paroxysms. *Brain Res.* **52:** 1–17, 1973.

Banna, N. R. and Jabbur, S. J.: The action of bemegride on presynaptic inhibition. *Neuropharmacology* **9:** 553–560, 1970.

Barker, J. L. and MacDonald, J. F.: Picrotoxin convulsions involve synaptic and nonsynaptic mechanisms on cultured mouse spinal neurons. *Science* **208:** 1054–1056, 1980.

Barker, J. L. and Mathers, D. A.: GABA analogues activate channels of different duration on cultured mouse spinal neurons. *Science* **212:** 358–361, 1981.

Barker, J. L. and McBurney, R. N.: GABA and glycine may share the same conductance channel on cultured mammalian neurones. *Nature* **277:** 234–236, 1979.

Barker, J. L., MacDonald, J. F., Mathers, D. A., McBurney, R. N., and Study, R. E.: Convulsant and Anticonvulsant Pharmacology of Cultured Mouse Spinal Neurons, In: *Neurotransmitters, Seizures, and Epilepsy* (P. L. Morselli, K. G. Lloyd, W. Loscher, B. Meldrum, E. H. Reynolds, eds.) Raven, New York, 1981.

Barker, J. L., McBurney, R. N., and Mathers, D. A.: Convulsant-induced depression of amino acid responses in cultured mouse spinal neurons studied under voltage clamp. *Br. J. Pharmacol.* **80:** 619–629, 1983.

Barker, J. L., Nicoll, R. A., and Padjen, A.: Studies on convulsants in the isolated frog spinal cord. II. Effects on root potentials. *J. Physiol.* **245:** 537–548, 1975a.

Barker, J. L., Nicoll, R. A., and Padjen, A.: Studies on convulsants in the isolated frog spinal cord. I. Antagonism of amino acid responses. *J. Physiol.* **245:** 521–536, 1975b.

Bellet, E. M. and Casida, J. E.: Bicyclic phosphorus esters: High toxicity without cholinesterase inhibition. *Science* **182:** 1135–1136, 1973.

Ben-Ari, Y., Tremblay, E., Ottersen, O. P., and Naquet, R.: Evidence suggesting secondary epileptogenic lesions after kainic acid: Pretreatment with diazepam reduces distant but not local brain damage. *Brain Res.* **165:** 362–365, 1979.

Ben-Ari, Y., Tremblay, E., Ottersen, O. P., and Meldrum, B. S.: The role of epileptic activity in hippocampal and "remote" cerebral lesions induced by kainic acid. *Brain Res.* **191:** 79–97, 1980.

Ben-Ari, Y., Tremblay, E., Riche, D., Ghilini, G., and Naquet, R.: Electrographic, clinical and pathological alterations following systemic administration of kainic acid, bicuculline or pentetrazole: Metabolic mapping using the deoxyglucose method with special reference to the pathology of epilepsy. *Neuroscience* **6:** 1361–1391, 1981.

Bergamasco, B.: Excitability cycle of the visual cortex in normal subjects during psychosensory rest and cardiazolic activation. *Brain Res.* **2:** 51–60, 1966.

Bhargava, V. K., Salamy, A., and Mckean, C. M.: Effects of cholinergic drugs on auditory evoked responses (CER) of rat cortex. *Neuropharmacology* **17:** 1009–1013, 1978.

Bickford, R. G. and Klass, D. W.: Sensory Precipitation and Reflex Mechanisms, In: *Basic Mechanisms of the Epilepsies* (H. H. Jasper, A. A. Ward, Jr., and A. Pope, eds.) Little Brown, Boston, 1969.

Biggio, G., Corda, M. G., De Montis, G., Stefanini, E., and Gessa, G. L.: Kainic acid differentiates GABA receptors from benzodiazepine receptors in the rat cerebellum. *Brain Res.* **193:** 589–593, 1980.

Bigler, E. D.: Comparison of effects of bicuculline, strychnine, and picrotoxin with those of pentylenetetrazol on photically evoked afterdischarges. *Epilepsia* **18:** 465–470, 1977a.

Bigler, E. D.: Neurophysiology, neuropharmacology and behavioral relationships of visual system evoked after-discharges: A review. *Bio. Behav. Rev.* **1:** 95–112, 1977b.

Bigler, E. D. and Eidelberg, E.: Principal cells in lateral geniculate: Effects of metrazol on capacity to after-discharge. *Brain Res. Bull.* **1:** 485–487, 1976.

Biscoe, T. J., Duggan, A. W., and Lodge, D.: Antagonism between bicuculline, strychnine, and picrotoxin and depressant amino-acids in rat nervous system. *Comp. Gen. Pharmacol.* **3:** 423–433, 1972.

Boast, C. A., Bernard, P. S., Barbaz, B. S., and Bergen, K. M.: The neuropharmacology of various diazepam antagonists. *Neuropharmacology* **22:** 1511–1521, 1983.

Bowery, N. G. and Dray, A.: Reversal of the action of amino acid antagonists by barbiturates and other hypnotic drugs. *Br. J. Pharmacol.* **63:** 197–215, 1978.

Bowery, N. G., Collins, J. F.,. and Hill, R. G.: Bicyclic phosphorus esters that are potent convulsants and GABA antagonists. *Nature* **261:** 601–603, 1976.

Bradley, P. B. and Dray, A.: Modification of the responses of brain stem neurones to transmitter substances by anaesthetic agents. *Br. J. Pharmacol.* **48:** 212–224, 1973.

Bradley, K., Easton, D. M., and Eccles, J. C.: An investigation of primary or direct inhibition. *J. Physiol.* **122:** 474–488, 1953.

Braestrup, C. and Nielsen, M.: Strychnine as a potent inhibitor of the brain GABA/benzodiazepine receptor complex. *Brain Res. Bull.* **5**(S2): 681–684, 1980.

Brazier, M. A. B.: Some effects of anaesthesia on the brain. *Br. J. Anaesth.* **33:** 194–204, 1961.

Brennan, M. J. W., Cantrill, R. C., and Krogsgaard-Larsen, P.: GABA Autoreceptors: Structure–Activity Relationships for Agonists, In: *GABA and Benzodiazepine Receptors* (E. Costa, G. DiChiara, and G. L. Gessa, eds.) Raven, New York, 1981.

Brimijoin, S.: Molecular forms of acetylcholinesterase in brain, nerve and muscle: Nature, localization and dynamics. *Prog. Neurobiol.* **21:** 291–322, 1983.

Burchiel, K. J., Myers, R. R., and Bickford, R. G.: Visual and auditory evoked responses during penicillin-induced generalized spike-and-wave activity in cats. *Epilepsia* **17:** 293–311, 1976.

Calvin, W. H.: Generation of spike trains in CNS neurons. *Brain Res.* **84:** 1–22, 1975.

Calvin, W. H., Ojemann, G. A., and Ward Jr., A. A.: Human cortical neurons in epileptogenic foci: Comparison of inter-ictal firing patterns to those of "epileptic" neurons in animals. *Electroencephal. Clin. Neurophysiol.* **34:** 337–351, 1973.

Caspary, D. M., Rybak, L. P., and Faingold, C. L.: The Effects of Inhibitory and Excitatory Amino-Acid Neurotransmitters on the response Properties of Brain Stem Auditory Neurons, In: *Auditory Biochemistry* (D. G. Drescher, ed.) Chas. C. Thomas, Springfield, Illinois, 1985.

Cepeda, C., Tanaka, T., Riche, D., and Naquet, R.: Limbic status epilepticus: Behaviour and sleep alterations after intra-amygdaloid kainic acid microinjections in *papio papio* baboons. *Electroenceph. Clin. Neurophysiol.* **54**: 603–613, 1982.

Chapman, A. G. and Evans, M. C.: Cortical GABA turnover during bicuculline seizures in rats. *J. Neurochem.* **41**: 886–889, 1983.

Chatt, A. B. and Ebersole, J. S.: Identification of penicillin diffusion at the onset of epileptogenesis in the cat striate cortex following differential laminar microinjeciton. *Brain Res.* **290**: 361–366, 1984.

Chatt, A. B. and Ebersole, J. S.: The laminar sensitivity of cat striate cortex to penicillin induced epileptogenesis. *Brain Res.* **241**: 382–387, 1982.

Chen, Z., Harding, G. W., and Towe, A. L.: Effect of strychnine on the cutaneous responsiveness of wide-field cerebral neurons after depression by pentobarbital. *Exp. Neurol.* **81**: 770–775, 1983.

Chweh, A. Y., Swinyard, E. A., and Wolf, H. H.: Pentylenetetrazol may discriminate between different types of benzodiazepine receptors. *J. Neurochem.* **41**: 830–833, 1983.

Collins, G. G. S., Anson, J., and Surtees, L.: Presynaptic kainate and N-methyl-D-aspartate receptors regulate excitatory amino acid release in the olfactory cortex. *Brain Res.* **265**: 157–159, 1983.

Cote, I. L. and Wilson, W. A.: Effects of barbiturates on inhibitory and excitatory responses to applied neurotransmitters in *Aplysia. J. Pharmacol. Exp. Therap.* **214**: 161–165, 1980.

Coyle, J. T.: Neurotoxic action of kainic acid. *J. Neurochem.* **41**: 1–11, 1983.

Creutzfeldt, O. D., Watanabe, S., and Lux, H. D.: Relations between EEG phenomena and potentials of single cortical cells. II. Spontaneous and convulsoid activity. *Electroenceph. Clin. Neurophysiol.* **20**: 19–37, 1966.

Crill, W. E. and Schwindt, P. C.: Role of Persistent Inward and Outward Membrane Currents in Epileptiform Bursting in Mammalian Neurons, In: *Basic Mechanisms of the Epilepsies* (A. V. Delgado-Escueta, A. A. Ward, Jr., D. M. Woodbury, and R. J. Porter, eds.) Raven, New York, 1986.

Crill, W. E. and Schwindt, P. C.: Active currents in mammalian central neurons. *Trends Neurosc.* **6**: 236–240, 1983.

Curtis, D. R.: The pharmacology of spinal postsynaptic inhibition. *Prog. Brain Res.* **31**: 171–189, 1969.

Curtis, D. R., Duggan, A. W., Felix, D., and Johnston, G. A. R.: GABA, bicuculline and central inhibition. *Nature New Bio.* **226**: 1222–1224, 1970a.

Curtis, D. R., Duggan, A. W., and Johnston, G. A. R.: The inactivation of extracellularly administered amino acids in the feline spinal cord. *Exp. Brain Res.* **10:** 447–462, 1970b.

Curtis, D. R., Duggan, A. W., and Johnston, G. A. R.: The specificity of strychnine as a glycine antagonist in the mammalian spinal cord. *Exp. Brain Res.* **12:** 547–565, 1971.

Curtis, D. R., Game, C. J. A., Johnston, G. A. R., McCulloch, R. M., and Maclachlan, R. M.: Convulsive action of penicillin. *Brain Res.* **43:** 242–245, 1972.

Curtis, D. R., Game, C. J. A., and Lodge, D.: Benzodiazepines and central glycine receptors. *Br. J. Pharmacol.* **56:** 307–311, 1976.

Cutler, R. W. P. and Young, J.: The effect of penicillin on the release of γ-aminobutyric acid from cerebral cortex slices. *Brain Res.* **170:** 157–163, 1979.

Dafny, N. and Rigor, B. M.: Neurophysiological approach as a tool to study effects of drugs on the central nervous system: Dose-effect of ketamine. *Exp. Neurol.* **59:** 275–285, 1978.

Davenport, J., Schwindt, P. C., and Crill, W. E.: Epileptogenic doses of penicillin do not reduce a monosynaptic GABA-mediated postsynaptic inhibition in the intact anesthetized cat. *Exp. Neurol.* **65:** 552–572, 1979.

Davidoff, R. A.: Studies of Neurotransmitter Actions (GABA, Glycine, and Convulsants), In: *Epilepsy* (A. A. Ward, Jr., J. K. Penry, and D. Purpura, eds.) Raven, New York, 1983.

Davidoff, R. A., Aprison, M. H., and Werman, R.: The effects of strychnine on the inhibition of interneurons by glycine and γ-aminobutyric acid. *Int. J. Neuropharmacol.* **8:** 191–194, 1969.

Davies, J. and Watkins, J. C.: Selective antagonism of amino acid-induced and synaptic excitation in the cat spinal cord. *J. Physiol.* **297:** 621–635, 1979.

deBoer, T., Stoff, J. C., and van Duijn, H.: The effects of convulsant and anticonvulsant drugs on the release of radiolabeled GABA, glutamate, noradrenaline, serotonin and acetylcholine from rat cortical slices. *Brain Res.* **253:** 153–160, 1982.

DeFeudis, F. V.: Do different populations of GABA-receptors exist in the vertebrate CNS? *Neurochem. Int.* **5:** 175–183, 1983.

DeFeudis, F. V., Orensanz Munoz, L. M., and Fando, J. L.: High-affinity glycine binding sites in rat CNS: Regional variation and strychnine sensitivity. *Gen. Pharmacol.* **9:** 171–176, 1978.

DeGroat, W. C. and Lalley, P. M.: Interaction between picrotoxin and 5-hydroxytryptamine in the superior cervical ganglion of the cat. *Br. J. Pharmacol.* **48:** 233–244, 1973.

Diamond, J., Roper, S., and Yasargil, G. M.: The membrane effects, and sensitivity to strychnine, of neural inhibition of the Mauthner cell, and its inhibition by glycine and GABA. *J. Physiol.* **232:** 87–111, 1973.

Dingledine, R. and Gjerstad, L.: Reduced inhibition during epileptiform activity in the *in vitro* hippocampal slice. *J. Physiol.* **305:** 297–313, 1980.

Dray, A.: Comparison of bicuculline methochloride with bicuculline and picrotoxin as antagonists of amino acid and monoamine depression of neurones in the rat brainstem. *Neuropharmacology* **14:** 887–891, 1975.

DuBois, K. P.: Toxicological Evaluation of the Anticholinesterase Agents, In: *Handbuch der Experimentellen Pharmakologie* vol. XV (G. B. Koelle, ed.) Springer, Heidelberg, 1963.

Duggan, A. W., Headley, P. M., and Lodge, D.: Acetylcholine-sensitive cells in the caudal medulla of the rat: Distribution, pharmacology and effects of pentobarbitone. *Br. J. Pharmacol.* **54:** 23–31, 1974.

Ebersole, J. S. and Chatt, A. B.: Toward a unified theory of focal penicillin epileptogenesis: An intracortical evoked potential investigation. *Epilepsia* **22:** 347–363, 1981.

Ebersole, J. S. and Chatt, A. B.: Laminar interactions during neocortical epileptogenesis. *Brain Res.* **298:** 253–271, 1984.

Ebersole, J. S. and Chatt, A. B.: Spread and Arrest of Seizures: The Importance of Layer 4 in Laminar Interactions During Neocortical Epileptogenesis, In: *Basic Mechanisms of the Epilepsies* (A. V. Delgado-Escueta, A. A. Ward, Jr., D. M. Woodbury, and R. J. Porter, eds.) Raven, New York, 1986.

Eccles, J. C.: *The Physiology of Synapses.* Springer-Verlag, Berlin, 1964.

Ellin, R. I.: Anomalies in theories and therapy of intoxication by potent organophosphorus anticholinesterase compounds. *Gen. Pharmacol.* **13:** 457–466, 1982.

Enna, S. J., Collins, J. F., and Snyder, S. H.: Stereospecificity and structure–activity requirements of GABA receptor binding in rat brain. *Brain Res.* **124:** 185–190, 1977.

Esplin, D. W. and Zablocka-Esplin, B.: Mechanisms of Action of Convulsants, In: *Basic Mechanisms of the Epilepsies* (H. H. Jasper, A. A. Ward, and A. Pope, eds.) Little, Brown, New York, 1969.

Faber, D. S. and Klee, M. R.: Strychnine interactions with acetylcholine, dopamine and serotonin receptors in *Aplysia* neurons. *Brain Res.* **65:** 109–126, 1974.

Fagni, L., Baudry, M., and Lynch, G.: Classification and properties of acidic amino acid receptors in hippocampus. I. Electrophysiological studies of an apparent desensitization and interactions with drugs which block transmission. *J. Neurosci.* **3:** 1538–1546, 1983.

Faingold, C. L.: Convulsant-induced enhancement of non-primary sensory evoked responses in reticular formation pathways. *Neuropharmacology* **16**: 73–81, 1977.

Faingold, C. L.: Brainstem reticular formation mechanisms subserving generalized seizures: Effects of convulsants and anticonvulsants on sensory-evoked responses. *Prog. Neuro-Psychopharmacol.* **2**: 401–422, 1978.

Faingold, C. L.: Strychnine effects on the sensory response patterns of reticular formation neurons. *Electroenceph. Clin. Neurophysiol.* **50**: 102–111, 1980a.

Faingold, C. L.: Enhancement of mesencephalic reticular neuronal responses to sensory stimuli with pentylenetetrazol. *Neuropharmacology* **19**: 53–62, 1980b.

Faingold, C. L.: Epilepsy: Brain-stem seizure mechanisms and drug action. An overview. *Fed. Proc.* **44**: 2412–2413, 1985.

Faingold, C. L. and Berry, C. A.: Quantitative evaluation of the pentylenetetrazol-anticonvulsant interaction on the EEG of the cat. *Eur. J. Pharmacol.* **24**: 381–388, 1973.

Faingold, C. L. and Caspary, D. M.: Changes in reticular formation unit response patterns associated with pentylenetetrazol-induced enhancement of sensory evoked responses. *Neuropharmacology* **16**: 143–147, 1977.

Faingold, C. L. and Hoffmann, W. E.: Effects of bemegride on the sensory responses of neurons in the hippocampus and brain stem reticular formation. *Electroenceph. Clin. Neurophysiol.* **52**: 316–327, 1981.

Faingold, C. L. and Stittsworth, Jr., J. D.: Comparative effects of pentylenetetrazol on the sensory responsiveness of lateral geniculate and reticular formation neurons. *Electroenceph. Clin. Neurophysiol.* **49**: 168–172, 1980.

Faingold, C. L., Hoffmann, W. E., and Caspary, D. M.: Bicuculline-induced enhancement of sensory responses and cross-correlations between reticular formation and cortical neurons. *Electroenceph. Clin. Neurophysiol.* **55**: 301–313, 1983a.

Faingold, C. L., Hoffmann, W. E., and Caspary, D. M.: On the site of pentylenetetrazol-induced enhancement of auditory responses of the reticular formation: Localized cooling and electrical stimulation studies. *Neuropharmacology* **22**: 961–970, 1983b.

Faingold, C. L., Hoffmann, W. E., and Caspary, D. M.: Effects of iontophoretic application of convulsants on the sensory responses of neurons in the brain-stem reticular formation. *Electroenceph. Clin. Neurophysiol.* **58**: 55–64, 1984.

Faingold, C. L., Hoffmann, W. E., and Caspary, D. M.: Mechanisms of sensory seizures: Brain-stem neuronal response changes and convulsant drugs. *Fed. Proc.* **44:** 2436–2441, 1985a.

Faingold, C. L., Hoffmann, W. E., and Caspary, D. M.: Comparative effects of convulsant drugs on the sensory responses of neurons in the amygdala and brainstem reticular formation. *Neuropharmacology* **24:** 1221–1230, 1985b.

Fariello, R. G.: Parenteral penicillin in rats: An experimental model of multifocal epilepsy. *Epilepsia* **17:** 217–222, 1976.

Fariello, R. G., Golden, G. T., and Pisa, M.: Homotaurine (3 amino-propanesulfonic acid; 3APS) protects from the convulsant and cytotoxic effect of systemically administered kainic acid. *Neurology* **32:** 241–245, 1982.

Farquharson, D.: Conductance changes associated with a soman-produced biphasic depolarization in electrophorus electroplax. *Biophys. J.* **21:** 54a, 1978.

Faugier-Grimaud, S.: Extrasynaptic mechanisms of Cardiazol-induced epileptiform activity of invertebrate neurons. *Brain Res.* **69:** 354–360, 1974.

Faugier-Grimaud, S.: Action of anticonvulsants on pentylenetetrazol-induced epileptiform activity on invertebrate neurones (*Helix Aspersa*). *Neuropharmacology* **17:** 905–918, 1978.

Ferrendelli, J. A. and Kinscherf, D. A.: Cyclic nucleotides in epileptic brain: Effects of pentylenetetrazol on regional cyclic AMP and cyclic GMP levels *in vivo. Epilepsia* **18:** 525–531, 1977.

Finch, D. M. and Beatty, J.: Visual evoked potentials during the development of a spiking cobalt focus in rat neocortex. *Electroenceph. Clin. Neurophysiol.* **41:** 137–152, 1976.

Fisher, R. S. and Alger, B. E.: Electrophysiological mechanisms of kainic acid-induced epileptiform activity in the rat hippocampal slice. *J. Neurosci.* **4:** 1312–1323, 1984.

Fisher, S. K. and Davies, W. E.: Some properties of guinea pig brain glutamate decarboxylase and its inhibition by the convulsant allylglycine (2-amino-4-pentenoic acid). *J. Neurochem.* **23:** 427–433, 1974.

Fonnum, F.: Glutamate: A neurotransmitter in mammalian brain. *J. Neurochem.* **42:** 1–11, 1984.

Freeman, A. R.: Electrophysiological analysis of the actions of strychnine, bicuculline and picrotoxin on the axonal membrane. *J. Neurobiol.* **4:** 567–582, 1973.

Frere, R. C., Macdonald, R. L., and Young, A. B.: GABA binding and bicuculline in spinal cord and cortical membranes from adult rat and from mouse neurons in cell culture. *Brain Res.* **244:** 145–154, 1982.

Fujimoto, M., and Okabayashi, T.: Effect of picrotoxin on benzodiazepine receptors and GABA receptors with reference to the effect of Cl⁻ ion. *Life Sci.* **28:** 895–901, 1981.

Fuller, T. A. and Olney, J. W.: Only certain anticonvulsants protect against kainate neurotoxicity. *Neurobehav. Tox. Teratol.* **3:** 355–361, 1981.

Futamachi, K. J. and Prince, D. A.: Effect of penicillin on an excitatory synapse. *Brain Res.* **100:** 589–597, 1975.

Gahwiler, B. H.: Spontaneous bioelectric activity of cultured Purkinje cells during exposure to glutamate, glycine, and strychnine. *J. Neurobiol.* **7:** 97–107, 1976.

Gale, K.: Mechanisms of seizure control mediated by γ-aminobutyric acid: Role of the substantia nigra. *Fed. Proc.* **44:** 2414–2424, 1985.

Galindo, A.: GABA-picrotoxin interaction in the mammalian central nervous system. *Brain Res.* **14:** 763–767, 1969.

Garant, D. S. and Gale, K.: Lesions of substantia nigra protect against experimentally induced seizures. *Brain Res.* **273:** 156–161, 1983.

Gastaut, H. and Hunter, J.: An experimental study of the mechanism of photic activation in idiopathic epilepsy. *Electroenceph. Clin. Neurophysiol.* **2:** 263–287, 1950.

Giorgi, O. and Meek, J. L.: γ-Aminobutyric acid turnover in rat striatum: Effects of glutamate and kainic acid. *J. Neurochem.* **42:** 215–220, 1984.

Glass, J. D. and Fromm, G. H.: Chloralose induced alteration of visually evoked response from specific and non-specific regions of cat neocortex. *Electroenceph. Clin. Neurophysiol.* **39:** 198–200, 1975.

Glass, J. D., Fromm, G. H., and Chattha, A. S.: Bicuculline and neuronal activity in motor cortex. *Electroenceph. Clin. Neurophysiol.* **48:** 16–24, 1980.

Goff, W. R., Allison, T., and Matsumiya, Y.: Effects of convulsive doses of 1, 1-dimethylhydrazine on somatic evoked responses in the cat. *Exp. Neurol.* **27:** 213–226, 1970.

Grecksch, G., de Carvalho, L. P., Venault, P., Chapouthier, G., and Rossier, J.: Convulsions induced by submaximal dose of pentylenetetrazol in mice are antagonized by the benzodiazepine antagonist Ro 15-1788. *Life Sci.* **32:** 2579–2584, 1983.

Green, D. M., Muir, A. W., Stratton, J. A., and Inch, T. D.: Dual mechanism of the antidotal action of atropine-like drugs in poisoning by organophosphorus anticholinesterases. *J. Pharm. Pharmacol.* **29:** 62–64, 1977.

Greenwood, R. S., Godar, S. E., and Winstead, K. K.: Focal penicillin seizures — motor activity and cellular physiology and morphology. *Brain Res. Bull.* **11:** 91–101, 1983.

Guberman, A. and Gloor, P.: Cholinergic drug studies of generalized penicillin epilepsy in the cat. *Brain Res.* **78:** 203–222, 1974.

Guha, D. and Pradhan, S. N.: Effects of nicotine on EEG and evoked potentials and their interactions with autonomic drugs. *Neuropharmacology* **15:** 225–232, 1976.

Gutnick, M. J., Connors, B. W., and Prince, D. A.: Mechanisms of neocortical epileptogenesis *in vitro. J. Neurophysiol.* **48:** 1321–1335, 1982.

Haas, H. L. and Hosli, L.: Strychnine and inhibition of bulbar reticular neurones. *Experientia* **29:** 542–544, 1973.

Hahn, F.: Analeptics. *Pharmacol. Rev.* **12:** 447–530, 1960.

Halliwell, J. V. and Adams, P. R.: Voltage-clamp analysis of muscarinic excitation in hippocampal neurons. *Brain Res.* **250:** 71–92, 1982.

Hamill, O. P., Bormann, J., and Sakmann, B.: Activation of multiple-conductance state chloride channels in spinal neurones by glycine and GABA. *Nature* **305:** 805–808, 1983.

Hammad, H. M., Al-Sayegh, A., Swanson, S., and Ebadi, M.: Dissociation between epileptic seizures induced by convulsant drugs and alteration in the concentrations of pyridoxal phosphate in rat brain regions. *Gen. Pharmacol.* **14:** 481–489, 1983.

Harris, F. A. and Towe, A. L.: Effects of topical bicuculline on primary evoked responses in pericruciate and precoronal cortex of the domestic cat. *Exp. Neurol.* **52:** 227–241, 1976.

Hayes, A. G., Gartside, I. B., and Straughan, D. W.: Effect of four convulsants on the time course of presynaptic inhibition and its relation to seizure activity. *Neuropharmacology* **16:** 725–730, 1977.

Heggli, D. E. and Malthe-Sorenssen, D.: Systemic injection of kainic acid: Effect on neurotransmitter markers in piriform cortex, amygdaloid complex and hippocampus and protection by cortical lesioning and anticonvulsants. *Neuroscience* **7:** 1257–1264, 1982.

Heggli, D. E., Aamodt, A., and Malthe-Sorenssen, D.: Kainic acid neurotoxicity: Effect of systemic injection on neurotransmitter markers in different brain regions. *Brain Res.* **230:** 253–262, 1981.

Heinemann, U., Lux, H. D., and Gutnick, M. J.: Extracellular free calcium and potassium during paroxysmal activity in the cerebral cortex of the cat. *Exp. Brain Res.* **27:** 237–243, 1977.

Heyer, E. J. and Macdonald, R. L.: Barbiturate reduction of calcium-dependent action potentials: Correlation with anesthetic action. *Brain Res.* **236:** 157–171, 1982.

Heyer, E. J., Nowak, L. M., and Macdonald, R. L.: Membrane depolarization and prolongation of calcium-dependent action potentials of mouse neurons in cell culture by two convulsants: Bicuculline and penicillin. *Brain Res.* **232:** 41–56, 1982.

Hicks, T. P. and Dykes, R. W.: Receptive field size for certain neurons in primary somatosensory cortex is determined by GABA-mediated intracortical inhibition. *Brain Res.* **274:** 160–164, 1983.

Hill, R. G., Simmonds, M. A., and Straughan, D. W.: A comparative study of some convulsant substances as γ-aminobutyric acid antagonists in the feline cerebral cortex. *Br. J. Pharmacol.* **49:** 37–51, 1973.

Hill, R. G., Simmonds, M. A., and Straughan, D. W.: Antagonism of γ-aminobutyric acid and glycine by convulsants in the cuneate nucleus of cat. *Br. J. Pharmacol.* **56:** 9–19, 1976.

Ho, I. K. and Harris, R. A.: Mechanism of action of barbiturates. *Ann. Rev. Pharmacol. Toxicol.* **21:** 83–111, 1981.

Hochner, B., Spira, M. E., and Werman, R.: Penicillin decreases chloride conductance in crustacean muscle: A model for the epileptic neuron. *Brain res.* **107:** 85–103, 1976.

Hodgkin, A. L. and Huxley, A. F.: The components of membrane conductance in the giant axon of *Loligo. J. Physiol.* **116:** 473–496, 1952.

Honchar, M. P., Olney, J. W., and Sherman, W. R.: Systemic cholinergic agents induce seizures and brain damage in lithium-treated rats. *Science* **220:** 323–325, 1983.

Horton, R. W.: GABA and seizures induced by inhibitors of glutamic acid decarboxylase. *Brain Res. Bull.* **5**(S2): 605–608, 1980.

Horton, R. W. and Meldrum, B. S.: Seizures induced by allylglycine, 3-mercaptopropionic acid and 4-deoxypyridoxine in mice and photosensitive baboons, and different modes of inhibition of cerebral glutamic acid decarboxylase. *Br. J. Pharmacol.* **49:** 52–63, 1973.

Hosli, L. and Tebecis, A. K.: Actions of amino acids and convulsants on bulbar reticular neurones. *Exp. Brain Res.* **11:** 111–127, 1970.

Hotson, J. R. and Prince, D. A.: A calcium-activated hyperpolarization follows repetitive firing in hippocampal neurons. *J. Neurophysiol.* **43:** 409–419, 1980.

Hrebicek, J. and Kolousek, J.: Preparoxysmal changes of spontaneous and evoked electrical activity of the cat brain after administration of methionine sulphoximine. *Epilepsia* **9:** 145–162, 1968.

Inubushi, S., Kobayashi, T., Oshima, T., and Torii, S.: Intracellular recordings from the motor cortex during EEG arousal in unanaesthetized brain preparations of the cat. *Jap. J. Physiol.* **28:** 669–688, 1978.

Jackson, J. H.: The Lumelian lectures on convulsive seizures. *Br. Med. J.* **1:** 765–771, 1890.

Jensen, L. H. and Petersen, E. N.: Bidirectional effects of benzodiazepine receptor ligands against picrotoxin- and pentylenetetrazol-induced seizures. *J. Neural Trans.* **58:** 183–191, 1983.

Johns, R. J. and Himwich, H. E.: A central action of some anti-histamines. Correction of forced circling movements and of seizure brain waves produced by the intracarotid injeciton of di-isopropyl fluorophosphate (DFP). *Am. J. Psychiat.* **107:** 367–372, 1950.

Johnson, B.: Strychnine paroxysms in brain stem. I. Anatomical distribution. *J. Neurophysiol.* **18:** 189–199, 1955.

Johnston, G. A. R.: Neuropharmacology of amino acid inhibitory transmitters. *Ann. Rev. Pharmacol. Toxicol.* **18:** 269–289, 1978.

Johnston, D. and Brown, T. H.: Control Theory Applied to Neural Networks Illuminates Synaptic Basis of Interictal Epileptiform Activity, In: *Basic Mechanisms of the Epilepsies.* (A. V. Delgado-Escueta, A. A. Ward, Jr., D. M. Woodbury, and R. J. Porter, eds.) Raven, New York, 1986.

Johnston, D. and Brown, T. H.: Giant synaptic potential hypothesis for epileptiform activity. *Science* **211:** 294–297, 1981.

Johnston, D., Hablitz, J. J., and Wilson, W. A.: Voltage clamp discloses slow inward current in hippocampal burst-firing neurones. *Nature* **286:** 391–393, 1980.

Joy, R. M.: Alteration of sensory and motor evoked responses by dieldrin. *Neuropharmacology* **13:** 93–110, 1974a.

Joy, R. M.: Temporal sequencing of dieldrin induced epileptiform spikes in the cat. *Proc. West. Pharmacol. Soc.* **17:** 82–86, 1974b.

Joy, R. M.: Convulsive properties of chlorinated hydrocarbon insecticides in the cat central nervous system. *Toxicol. Appl. Pharmacol.* **35:** 95–106, 1976.

Kaplan, B. J. and Williamson, P. D.: Electroencephalogram and somatosensory evoked potential changes after administration of six convulsant drugs. *Exp. Neurol.* **59:** 124–136, 1978.

Kar, P. P. and Matin, M. A.: Possible role of γ-aminobutyric acid in Paraoxon-induced convulsions. *J. Pharm. Pharmacol.* **24:** 996–997, 1972.

Kayama, Y. and Iwama, K.: The EEG, evoked potentials, and single-unit activity during ketamine anesthesia in cats. *Anesthesiology* **36:** 316–328, 1972.

Kimura, D.: Multiple response of visual cortex of the rat to photic stimulation. *Electroenceph. Clin. Neurophysiol.* **14:** 115–122, 1962.

Kish, S. J., Sperk, G., and Hornykiewicz, O.: Alterations in benzodiazepine and GABA receptor binding in rat brain following systemic injection of kainic acid. *Neuropharmacology* **22:** 1303–1309, 1983.

Klee, M. R., Faber, D. S., and Heiss, W.-D.: Strychnine- and pentylenetetrazol-induced changes of excitability in *Aplysia* neurons. *Science* **179:** 1133–1136, 1973.

Klee, M. R., Faber, D. S., and Hoyer, J.: Doublet Discharges and Bistable States Induced by Strychnine in a Neuronal Soma Membrane, In: *Abnormal Neuronal Discharges* (N. Chalazonitis and M. Boisson, eds.) Raven, New York, 1978.

Kleinhaus, A. L.: Electrophysiological actions of convulsants and anticonvulsants on neurons of the leech subesophageal ganglion. *Comp. Biochem. Physiol.* **52C:** 27–34, 1975.

Klunk, W. E., Kalman, B. L., Ferrendelli, J. A., and Covey, D. F.: Computer-assisted modeling of the picrotoxin and γ-butyrolactone receptor site. *Mol. Pharmacol.* **23:** 511–518, 1983.

Kostopoulos, G., Avoli, M., and Gloor, P.: Participation of cortical recurrent inhibition in the genesis of spike and wave discharges in feline generalized penicillin epilepsy. *Brain Res.* **267:** 101–112, 1983.

Kriegstein, A. R., Suppes, T., and Prince, D. A.: Cholinergic enhancement of penicillin-induced epileptiform discharges in pyramidal neurons of the guinea pig hippocampus. *Brain Res.* **266:** 137–142, 1983.

Krnjevic, K.: Chemical nature of synaptic transmission in vertebrates. *Physiol. Rev.* **54:** 418–540, 1974.

Krnjevic, K., Puil, E., and Werman, R.: Bicuculline, benzyl penicillin, and inhibitory amino acids in the spinal cord of the cat. *Can. J. Physiol. Pharmacol.* **55:** 670–680, 1977.

Lambert, J. D. C. and Flatman, J. A.: The interaction between barbiturate anaesthetics and excitatory amino acid responses on cat spinal neurones. *Neuropharmacology* **20:** 227–240, 1981.

Lewis, D. V., Huguenard, J. R., Anderson, W. W., and Wilson, W. A.: Membrane Currents Underlying Spike Bursting Pacemaker Activity and Frequency Adaptation in Invertebrates, In: *Basic Mechanisms of the Epilepsies.* (A. V. Delgado-Escueta, A. A. Ward, Jr., D. M. Woodbury, and R. J. Porter, eds.) Raven, New York, 1986.

Lipp, J. A.: Effect of benzodiazepine derivatives on soman-induced seizure activity and convulsions in the monkey. *Arch. Int. Pharmacodyn.* **202:** 244–251, 1973.

Llinas, R. R. and Sugimori, M.: Ionic Basis for the Electrophysiological Activity of Mammalian Neurons, In: *Epilepsy* (A. A. Ward, Jr., J. K. Penry, and D. Purpura, eds.) Raven, New York, 1983.

Lockard, J. S. and Ward, Jr., A. A.: *Epilepsy: A Window to Brain Mechanisms.* Raven, New York, 1980.

Loscher, W. and Frey, H. H.: Effect of convulsant and anticonvulsant agents on level and metabolism of γ-aminobutyric acid in mouse brain. *NS Arch. Pharmacol.* **296:** 263–269, 1977.

Loscher, W.: Alterations in CSF GABA levels and seizure susceptibility developing during repeated administration of pentetrazole in dogs. Effects of γ-acetylenic GABA, valproic acid and phenobarbital. *Neurochem. Intern.* **5:** 405–412, 1983.

Lothman, E. W., Collins, R. C., and Ferrendelli, J. A.: Kainic acid-induced limbic seizures: Electrophysiologic studies. *Neurology* **31:** 806–812, 1981.

Lundy, P. M. and Magor, G. F.: Cyclic GMP concentrations in cerebellum following organophosphate administration. *J. Pharm. Pharmacol.* **30:** 251–252, 1978.

Lundy, P. M., Magor, G., and Shaw, R. K.: Gamma aminobutyric acid metabolism in different areas of rat brain at the onset of soman-induced convulsions. *Arch. Int. Pharmacodyn.* **234:** 64–73, 1978.

MacDonald, J. F., Porietis, A. V., and Wojtowicz, J. M.: L-Aspartic acid induces a region of negative slope conductance in the current-voltage relationship of cultured spinal cord neurons. *Brain Res.* **237:** 248–253, 1982.

Macdonald, R. L. and Barker, J. L.: Penicillin and pentylenetetrazol selectively antagonize GABA-mediated postsynaptic inhibition of cultured mammalian neurons. *Neurology* **27:** 337, 1977.

Macdonald, R. L. and Barker, J. L.: Enhancement of GABA-mediated postsynaptic inhibition in cultured mammalian spinal cord neurons: A common mode of anticonvulsant action. *Brain Res.* **167:** 323–336, 1979.

Macdonald, R. L. and McLean, M. J.: Cellular bases of barbiturate and phenytoin anticonvulsant drug action. *Epilepsia* **23**(S1): S7–S18, 1982.

Macdonald, R. L. and McLean, M. J.: Anticonvulsant Drugs: Mechanisms of Action, In: *Basic Mechanisms of the Epilepsies* (A. V. Delgado-Escueta, A. A. Ward, Jr., D. M. Woodbury, and R. J. Porter, eds.) Raven, New York, 1986.

Macdonald, R. L., Pun, R. Y. K., Neale, E. A., and Nelson, P. G.: Synaptic interactions between mammalian central neurons in cell culture. I. Reversal potential for excitatory postsynaptic potentials. *J. Neurophysiol.* **49:** 1428–1441, 1983.

Macdonald, R. L., Young, A. B., and Nowak, L. M.: Bicuculline has different actions on mammalian spinal cord and forebrain neurons in primary dissociated cell cultures. *Soc. Neurosci. Abs.* **5:** 593, 1979.

Macon, J. B. and King, D. W.: Penicillin iontophoresis and the responses of somatosensory cortical neurons to amino acids. *Electroenceph. Clin. Neurophysiol.* **47:** 52–63, 1979.

Madison, D. V. and Nicoll, R. A.: Noradrenaline blocks accommodation of pyramidal cell discharge in the hippocampus. *Nature* **299:** 636–638, 1982.

Mahon, P. J. and Brink, J. J.: Inhibition of acetylcholinesterase *in vitro* by pentylenetetrazol. *J. Neurochem.* **17:** 949–953, 1970.

Marciani, M. G., Stanzione, P., Cherubini, E., and Bernardi, G.: Action mechanisms of γ-aminobutyric acid (GABA) and glycine on rat cortical neurons. *Neurosci. Lett.* **18:** 169–172, 1980.

Marder, E. and Paupardin-Tritsch, D.: Picrotoxin block of a depolarizing ACh response. *Brain Res.* **181:** 223–227, 1980.

Martin, R. J.: Glycine and GABA induced conductance changes in lamprey reticulospinal neurons and their antagonism by strychnine, thebaine, bicuculline and picrotoxin. *Comp. Biochem. Physiol.* **63C:** 109–115, 1979.

Mashimo, K. and Sekiya, Y.: The effects of pentylenetetrazol on voltage dependent and Ca-mediated potassium currents in the neurons of the snail, *Euhadra hickonis. Comp. Biochem. Physiol.* **69C:** 113–116, 1981.

Matin, M. A. and Kar, P. P.: Further studies on the role of γ-aminobutyric acid in paraoxon-induced convulsions. *Eur. J. Pharmacol.* **21:** 217–221, 1973.

Mayer, M. L. and Straughan, D. W.: Effects of 5-hydroxytryptamine on central neurones antagonized by bicuculline and picrotoxin. *Neuropharmacology* **20:** 347–350, 1981.

Meldrum, B.: *Convulsant Drugs, Anticonvulsants and GABA-Mediated Neuronal Inhibition: In GABA-Neurotransmitters.* (Alfred Benzon Symposium 12) (P. Krogsgaard-Larsen, J. Scheel-Kruger, and H. Kofod, eds.) Munksgaard, Copenhagen, 1979.

Meldrum, B. S., Menini, C., Naquet, R., Laurent, H., and Stutzmann, J. M.: Proconvulsant, convulsant and other actions of the D- and L-stereoisomers of allylglycine in the photosensitive baboon, *Papio papio. Electroenceph. Clin. Neurophysiol.* **47:** 383–395, 1979.

Menini, C., Stutzmann, J. M., Laurent, H., and Naquet, R.: Paroxysmal visual evoked potentials (PVEPs) in *Papio papio.* I. Morphological and topographical characteristics. Comparison with paroxysmal discharges. *Electroenceph. Clin. Neurophysiol.* **50:** 356–364, 1980.

Menini, C., Silva-Comte, C., Stutzmann, J. M., and Dimov, S.: Cortical unit activity during intermittent photic stimulation in *Papio papio.* Relationship with paroxysmal fronto-rolandic activity. *Electroenceph. Clin. Neurophysiol.* **52:** 42–49, 1981.

Miller, J. J. and McLennan, H.: The action of bicuculline upon acetylcholine-induced excitations of central neurones. *Neuropharmacology* **13:** 785–787, 1974.

Mitchell, P. R. and Martin, I. L.: Is GABA release modulated by presynaptic receptors? *Nature* **274:** 904–905, 1978.

Moore, M. J. and Caspary, D. M.: Strychnine blocks binaural inhibition in lateral superior olivary neurons. *J. Neurosci.* **3:** 237–242, 1983.

Napias, C., Bergman, M. O., Van Ness, P. C., Greenlee, D. V., and Olsen, R. W.: GABA binding in mammalian brain: Inhibition by endogenous GABA. *Life Sci.* **27:** 1001–1011, 1980.

Nicoll, R. A.: The blockade of GABA mediated responses in the frog spinal cord by ammonium ions and furosemide. *J. Physiol.* **283:** 121–132, 1978.

Niemegeers, C. J. E., Awouters, F., Lenaerts, F. M., Vermeire, J., and Janssen, P. A. J.: Prevention of physostigmine-induced lethality in rats. A pharmacological analysis. *Arch. Int. Pharmacodyn.* **259:** 153–165, 1982.

Nistri, A. and Pepeu, G.: Acetylcholine levels in the frog spinal cord following the administration of different convulsants. *Eur. J. Pharmacol.* **27:** 281–287, 1974.

Noebels, J. L. and Prince, D. A.: Presynaptic origin of penicillin after-discharges at mammalian nerve terminals. *Brain Res.* **138:** 59–74, 1977.

Noebels, J. L. and Prince, D. A.: Development of focal seizures in cerebral cortex: Role of axon terminal bursting. *J. Neurophysiol.* **41:** 1267–1281, 1978.

Nowak, L. M., Young, A. B., and Macdonald, R. L.: GABA and bicuculline actions on mouse spinal cord and cortical neurons in cell culture. *Brain Res.* **244:** 155–164, 1982.

Olney, J. W.: Neurotoxicity of Excitatory Amino Acids, In: *Kainic Acid as a Tool in Neurobiology* (E. G. McGeer, J. W. Olney, and P. L. McGeer, eds.) Raven, New York, 1978.

Olney, J. W., Rhee, V., and Ho, O. L.: Kainic acid: A powerful neurotoxic analogue of glutamate. *Brain Res.* **77:** 507–512, 1974.

Olsen, R. W., Greenlee, D., Van Ness, P., and Ticku, M. K.: Studies on the gamma-Aminobutyric Acid Receptor Ionophore Proteins in Mammalian Brain, In: *Amino Acids as Chemical Neurotransmitters* (F. Fonnum, ed.) Plenum, New York, 1978.

Olsen, R. W., Ban, M., and Miller, T.: Studies on the neuropharmacological activity of bicuculline and related compounds. *Brain Res.* **102:** 283–299, 1976.

Ono, H., Akahane, K., Fukuda, H., and Kudo, Y.: Supersensitivity of the GABA receptor in the frog spinal cord, as induced by kainic acid. *Comp. Biochem. Physiol.* **76C:** 231–236, 1983.

Orlowski, M., Reingold, D. F., and Stanley, M. E.: D- and L-stereoisomers of allylglycine: Convulsive action and inhibition of brain L-glutamate decarboxylase. *J. Neurochem.* **28:** 349–353, 1977.

Pellegrini, A., Ermani, M., Giaretta, D., Zanotto, L., Pasqui, L., and Testa, G.: Unitary activity of cortical thalamic and reticular neurons during spike and wave in feline generalized penicillin epilepsy. 15th Epilepsy International Symposium, Washington, DC, 1983.

Pellegrini, A., Musgrave, J., and Gloor, P.: Role of afferent input of sub-cortical origin in the genesis of bilaterally synchronous epileptic discharges of feline generalized penicillin epilepsy. *Exp. Neurol.* **64:** 155–173, 1979.

Pellmar, T. C. and Wilson, W. A.: Synaptic mechanism of pentylenetetrazole: Selectivity for chloride conductance. *Science* **197:** 912–914, 1977a.

Pellmar, T. C. and Wilson, W. A.: Penicillin effects on iontophoretic responses in *Aplysia californica. Brain Res.* **136:** 89–101, 1977b.

Penfield, W. and Jasper, H. H.: *Epilepsy and the Functional Anatomy of the Human Brain*, Little Brown, Boston, 1954.

Perkins, M. N., Bowery, N. G., Hill, D. R., and Stone, T. W.: Neuronal responses to ethylenediamine: Preferential blockade by bicuculline. *Neurosci. Lett.* **23:** 325–327, 1981.

Phillis, J. W. and York, D. H.: Strychnine block of neural and drug-induced inhibition in the cerebral cortex. *Nature* **216:** 922–923, 1967.

Piggott, S. M., Kerkut, G. A., and Walker, R. J.: The actions of picrotoxin, strychnine, bicuculline and other convulsants and antagonists on the responses to acetylcholine glutamic acid and gamma-aminobutyric acid on *Helix* neurones. *Comp. Biochem. Physiol.* **57C:** 107–116, 1977.

Pisa, M., Sanberg, P. R., Corcoran, M. E., and Fibiger, H. C.: Spontaneously recurrent seizures after intracerebral injections of kainic acid in rat: A possible model of human temporal lobe epilepsy. *Brain Res.* **200:** 481–487, 1980.

Potashner, S. J. and Gerard, D.: Kainate-enhanced release of D-[^{3}H]aspartate from cerebral cortex and striatum: Reversal by baclofen and pentobarbital. *J. Neurochem.* **40:** 1548–1557, 1983.

Potashner, S. J., Lake, N., Langlois, E. A., Plouffe, Jr., L., and Lecavalier, D.: Pentobarbital: Differential effects on the depolarization-induced release of excitatory and inhibitory amino acids from cerebral cortex slices. *Brain Res. Bull.* **5**(S2): 659–664, 1980.

Prichard, J. W.: Bemegride-induced paroxysmal discharges in leech ganglion. *Brain Res.* **45:** 594–598, 1972.

Prince, D. A.: Neurophysiology of epilepsy. *Ann. Rev. Neurosci.* **1:** 395–415, 1978.

Prince, D. A.: Epileptogenesis in Hippocampal and Neocortical Neurons, In: *Physiology and Pharmacology of Epileptogenic Phenomena* (M. R. Klee, H. D. Lux, and E. J. Speckmann, eds.) Raven, New York, 1982.

Prince, D. A. and Wong, R. K. S.: Human epileptic neurons studied in vitro. *Brain Res.* **210:** 323–333, 1981.

Prioux-Guyonneau, M., Coudray-Lucas, C., Coq, H. M., Cohen, Y., and Wepierre, J.: Modification of rat brain 5-hydroxytryptamine metabolism by sublethal doses of organophosphate agents. *Acta Pharmacol. et Toxicol.* **51:** 278–284, 1982.

Puil, E.: Actions and Interactions of *S*-Glutamate in the Spinal Cord, In: *Handbook of the Spinal Cord* vol. 1: *Pharmacology* (R. A. Davidoff, ed.) Marcel Dekker, New York, 1983.

Quesney, L. F.: Pathophysiology of generalized photosensitive epilepsy in the cat. *Epilepsia* **25:** 61–69, 1984.

Rayport, S. G. and Kandel, E. R.: Epileptogenic agents enhance transmission at an identified weak electrical synapse in *Aplysia. Science* **213:** 462–464, 1981.

Rehavi, M., Skolnick, P., and Paul, S. M.: Effects of tetrazole derivatives on [³H] diazepam binding in vitro: Correlation with convulsant potency. *Eur. J. Pharmacol.* **78:** 353–356, 1982.

Reynolds, A. P. and Watkins, J. C.: The effect of strychnine and of electrical stimulation on the labeling of γ-aminobutyric acid and other free amino acids from [U-¹⁴C] glucose in the spinal cord of the nembutalized rat. *Brain Res.* **36:** 343–351, 1972.

Rodin, E., Gonzalez, S., Caldwell, D., and Laginess, D.: Photic evoked response during induced epileptic seizures. *Epilepsia* **7:** 202–214, 1966.

Rodin, E., Kitano, H., Nagao, B., and Rodin, M.: The results of penicillin G administration on chronic unrestrained cats: Electrographic and behavioral observations. *Electroencephal. Clin. Neurophysiol.* **42:** 518–527, 1977.

Rodin, E., Onuma, T., Wasson, S., Porzak, J., and Rodin, M.: Neurophysiological mechanisms involved in grand mal seizures induced by metrazol and megimide. *Electroenceph. Clin. Neurophysiol.* **30:** 62–72, 1971.

Roper, S.: Inhibition of Mauthner cells by allylglycine. *Nature* **226:** 373–374, 1970.

Rose, D.: Facilitation of bicuculline- and picrotoxin-induced seizures by sodium valproate in rats. *Arch. Int. Pharmacodyn.* **239:** 78–85, 1979.

Rump, S., Grudzinska, E., and Edelwejn, Z.: Effects of diazepam on epileptiform patterns of bioelectrical activity of the rabbit's brain induced by fluostigmine. *Neuropharmacology* **12:** 813–817, 1973.

Ryall, R. W., Piercey, M. F., and Polosa, C.: Strychnine-resistant mutual inhibition of Renshaw cells. *Brain Res.* **41:** 119–129, 1972.

Saito, S. and Tokunaga, Y.: Some correlations between picrotoxin-induced seizures and γ-aminobutyric acid in animal brain. *J. Pharmacol. Exp. Ther.* **157:** 546–554, 1967.

Sakmann, B., Hamill, O. P., and Bormann, J.: Patch-clamp measurements of elementary chloride currents activated by the putative inhibitory transmitters GABA and glycine in mammalian spinal neurons. *J. Neur. Trans.* **(S)18:** 83–95, 1983.

Schechter, P. J. and Tranier, Y.: Effect of elevated brain GABA concentrations on the actions of bicuculline and picrotoxin in mice. *Psychopharmacology* **54:** 145–148, 1977.

Scholfield, C. N.: Antagonism of γ-aminobutyric acid and muscimol by picrotoxin, bicuculline, strychnine, bemegride, leptazol, *d-*

tubocurarine and theophylline in the isolated olfactory cortex. *NS Arch. Pharmacol.* **318:** 274–280, 1982.

Schulz, D. W. and Macdonald, R. L.: Barbiturate enhancement of GABA-mediated inhibition and activation of chloride ion conductance: Correlation with anticonvulsant and anesthetic actions. *Brain Res.* **209:** 177–188, 1981.

Schwartzkroin, P. A. and Pedley, T. A.: Slow depolarizing potentials in "epileptic" neurons. *Epilepsia* **20:** 267–277, 1979.

Schwartzkroin, P. A. and Prince, D. A.: Changes in excitatory and inhibitory synaptic potentials leading to epileptogenic activity. *Brain Res.* **183:** 61–76, 1980.

Sellden, U.: Repeated activation with Megimide in normal subjects: Electroencephalographic responses and threshold doses. *Electroenceph. Clin. Neurophysiol.* **17:** 1–10, 1964.

Shank, R. P., Pong, S. F., Freeman, A. R., and Graham, L. T.: Bicuculline and picrotoxin as antagonists of γ-aminobutyrate and neuromuscular inhibition in the lobster. *Brain Res.* **72:** 71–78, 1974.

Shapiro, B. I., Wang, C. M., and Narahashi, T.: Effects of strychnine on ionic conductances of squid axon membrane. *J. Pharmacol. Exp. Ther.* **188:** 66–76, 1974.

Shelburne, Jr., S. A. and McLaurin, R. L.: The effects of phencyclidine on visually evoked potentials of rhesus monkeys. *Electroenceph. Clin. Neurophysiol.* **43:** 95–98, 1977.

Simmonds, M. A.: Presynaptic actions of γ-aminobutyric acid and some antagonists in a slice preparation of cuneate nucleus. *Br. J. Pharmacol.* **63:** 495–502, 1978.

Simmonds, M. A.: Evidence that bicuculline and picrotoxin act at separate sites to antagonize γ-aminobutyric acid in rat cuneate nucleus. *Neuropharmacology* **19:** 39–45, 1980.

Simmonds, M. A.: Classification of some GABA antagonists with regard to site of action and potency in slices of rat cuneate nucleus. *Eur. J. Pharmacol.* **80:** 347–358, 1982.

Simmonds, M. A.: Multiple GABA receptors and associated regulatory sites. *Trends Neurosci.* **6:** 279–281, 1983.

Sivam, S. P., Norris, J. C., Lim, D. K., Hoskins, B., and Ho, I. K.: Effect of acute and chronic cholinesterase inhibition with diisopropylfluorophosphate on muscarinic, dopamine, and GABA receptors of the rat striatum. *J. Neurochem.* **40:** 1414–1422, 1983.

Skolnick, P. and Paul, S. M.: Benzodiazepine receptors in the central nervous system. *Intern. Rev. Neurobiol.* **23:** 103–140, 1982.

Somjen, G. G.: Stimulus-evoked and seizure-related responses of extracellular calcium activity in spinal cord compared to those in cerebral cortex. *J. Neurophysiol.* **44:** 617–632, 1980.

Speckmann, E. J. and Caspers, H.: Paroxysmal depolarization and changes in action potentials induced by pentylenetetrazol in isolated neurons of *Helix pomatia*. *Epilepsia* **14:** 397–408, 1973.

Sperk, G., Lassmann, H., Baran, H., Kish, S. J., Seitelberger, F., and Hornykiewicz, O.: Kainic acid induced seizures: Neurochemical and histopathological changes. *Neuroscience* **10:** 1301–1315, 1983.

Squires, R. F. and Saederup, E.: γ-Aminobutyric acid receptors modulate cation binding sites coupled to independent benzodiazepine, picrotoxin, and anion binding sites. *Mol. Pharmacol.* **22:** 327–334, 1982.

Squires, R. F., Casida, J. E., Richardson, M., and Saederup, E.: [^{35}S]*t*-Butylbicyclophosphorothionate binds with high affinity to brain-specific sites coupled to γ-aminobutyric acid-A and ion recognition sites. *Mol. Pharmacol.* **23:** 326–336, 1983.

Stone, W. E. and Javid, M. J.: Effects of anticonvulsants and glutamate antagonists on the convulsive action of kainic acid. *Arch. Int. Pharmacodyn.* **243:** 56–65, 1980.

Stone, W. E. and Javid, M. J.: Effects of anticonvulsants and other agents on seizures induced by intracerebral L-glutamate. *Brain Res.* **264:** 165–167, 1983.

Stone, W. E. and Javid, M. J.: Quantitative evaluation of the actions of anticonvulsants against different chemical convulsants. *Arch. Int. Pharmacodyn.* **240:** 66–78, 1979.

Straw, R. N. and Mitchell, C. L.: The effect of pentylenetetrazol on bioelectrical activity recorded from the cat brain. *Arch. Int. Pharmacodyn.* **168:** 456–466, 1967.

Stutzmann, J. M., Laurent, H., Valin, A., and Menini, C.: Paroxysmal visual evoked potentials (PVEPs) in *Papio papio*. II. Evidence for a facilitatory effect of intermittent photic stimulation. *Electroenceph. Clin. Neurophysiol.* **50:** 365–374, 1980.

Sugaya, E. and Onozuka, M.: Intracellular calcium: Its movement during pentylenetetrazole-induced bursting activity. *Science* **200:** 797–799, 1978a.

Sugaya, E. and Onozuka, M.: Intracellular calcium: Its release from granules during bursting activity in snail neurons. *Science* **202:** 1195–1197, 1978b.

Sutton, Jr., G. G. and Oldstone, M. B. A.: Evidence against pyridoxine deficiency as the mechanism of penicillin seizures. *Neurology* **19:** 859–864, 1969.

Svenneby, G. and Roberts, E.: Bicuculline and N-methylbicuculline-competitive inhibitors of brain acetylcholinesterase. *J. Neurochem.* **21:** 1025–1026, 1973.

Takeuchi, A. and Takeuchi, N.: A study of the action of picrotoxin on the inhibitory neuromuscular junction of the crayfish. *J. Physiol.* **205:** 377–391, 1969.

Taylor-Courval, D. and Gloor, P.: Behavioral alterations associated with generalized spike and wave discharges in the EEG of the cat. *Exp. Neurol.* **83:** 167–186, 1984.

Tebecis, A. K. and DiMaria, A.: Strychnine-sensitive inhibition in the medullary reticular formation: Evidence for glycine as an inhibitory transmitter. *Brain Res.* **40:** 373–383, 1972.

Ticku, M. K.: Benzodiazepine-GABA receptor-ionophore complex: Current concepts. *Neuropharmacology* **22:** 1459–1470, 1983.

Ticku, M. K. and Maksay, G.: Convulsant/depressant site of action at the allosteric benzodiazepine-GABA receptor-ionophore complex. *Life Sci.* **33:** 2363–2375, 1983.

Ticku, M. K., Ban, M., and Olsen, R. W.: Binding of [^{3}H] α-dihydropicrotoxinin, a γ-aminobutyric acid synaptic antagonist, to rat brain membranes. *Mol. Pharmacol.* **14:** 391–402, 1978.

Tribble, G. L., Schwindt, P. C., and Crill, W. E.: Reduction of postsynaptic inhibition tolerated before seizure initiation: Brain stem. *Exp. Neurol.* **80:** 304–320, 1983a.

Tribble, G. L., Schwindt, P. C., and Crill, W. E.: Reduction of postsynaptic inhibition tolerated before seizure initiation: Spinal cord. *Exp. Neurol.* **80:** 288–303, 1983b.

Tsumoto, T., Eckart, W., and Creutzfeldt, O. D.: Modification of orientation sensitivity of cat visual cortex neurons by removal of GABA-mediated inhibition. *Exp. Brain Res.* **34:** 351–363, 1979.

Velasco, F., Velasco, M., Estrada-Villaneuva, F., and Machado, J. P.: Specific and nonspecific multiple unit activities during the onset of pentylenetetrazol seizures. I. Intact animals. *Epilepsia* **16:** 207–214, 1975.

Velasco, F., Velasco, M., Maldona, H., and Estrada-Villanueva, F.: Specific and nonspecific multiple unit activities during the onset of pentylenetetrazol seizures. II. Acute lesions interrupting nonspecific system connections. *Epilepsia* **17:** 461–475, 1976.

Wachtel, R. E. and Wilson, Jr., W. A.: Barbiturate effects on acetylcholine-activated channels in *Aplysia neurons. Mol. Pharmacol.* **24:** 449–457, 1983.

Wecker, L., Mobley, P. L., and Dettbarn, W.-D.: Effects of atropine on paraoxon-induced alterations in brain acetylcholine. *Arch. Int. Pharmacodyn.* **227:** 69–75, 1977.

Welch, A. D. and Henderson, V. E.: A comparative study of hydrastine, bicuculline and adlumine. *J. Pharmacol. Exp. Ther.* **51:** 482–491, 1934.

Wenzel, J. and Muller, M.: Zurfunktionellen beziehung zwischen physiologischer und pentetrazol-induzierter rhythmischer aktivitat im hirnstronbild der freibeweglichen ratte. *Acta. Biol. Med. Germ.* **26:** 533–542, 1971.

Werman, R., Davidoff, R. A., and Aprison, M. H.: Inhibitory action of glycine on spinal neurons in the cat. *J. Neurophysiol.* **31:** 81–95, 1968.

Westbrook, G. L. and Lothman, E. W.: Cellular and synaptic basis of kainic acid-induced hippocampal epileptiform activity. *Brain Res.* **273:** 97–109, 1983.

Wilson, W. A. and Escueta, A. V.: Common synaptic effects of pentylene-tetrazol and penicillin. *Brain Res.* **72:** 168–171, 1974.

Wilson, W. A. and Wachtel, H.: Prolonged inhibition in burst firing neurons: Synaptic inactivation of the slow regenerative inward current. *Science* **202:** 772–775, 1978.

Winters, W. D.: Neuropharmacological studies and postulates on excitation and depression in the central nervous system. *Biol. Psychiat.* **9:** 313–345, 1966.

Woodbury, D. M.: Convulsant Drugs: Mechanisms of Action, In: *Antiepileptic Drugs: Mechanisms of Action* (G. H. Glaser, J. K. Penry, and D. M. Woodbury, eds.) Raven, New York, 1980.

Yarowsky, P. J. and Carpenter, D. O.: A comparison of similar ionic responses to γ-aminobutyric acid and acetylcholine. *J. Neurophysiol.* **41:** 531–541, 1978.

Yokota, T., Nishikawa, N., and Nishikawa, Y.: Effects of strychnine upon different classes of trigeminal subnucleus caudalis neurons. *Brain Res.* **168:** 430–434, 1979.

Yonekawa, W. D., Kupferberg, H. J., and Woodbury, D. M.: Relationship between pentylenetetrazol-induced seizures and brain pentylenetetrazol levels in mice. *J. Pharmacol. Exp. Ther.* **214:** 589–593, 1980.

Young, A. B. and Macdonald, R. L.: Glycine as a Spinal cord Neurotransmitter, In: *Handbook of the Spinal Cord vol. I. Pharmacology* (R. A. Davidoff, ed.) Marcel Dekker, New York, 1983.

Young, A. B. and Snyder, S. H.: Strychnine binding associated with glycine receptors of the central nervous system. *Proc. Natl. Acad. Sci. USA* **70:** 2832–2836, 1973.

Young, A. B. and Snyder, S. H.: The glycine synaptic receptor: Evidence that strychnine binding is associated with the ionic conductance mechanism. *Proc. Natl. Acad. Sci. USA* **71:** 4002–4005, 1974.

Young, A. B., Zukin, S. R., and Snyder, S. H.: Interaction of benzodiazepines with central nervous glycine receptors: Possible mechanism of action. *Proc. Natl. Acad. Sci. USA* **71:** 2246–2250, 1974.

Zarbin, M. A., Wamsley, J. K., and Kuhar, M. J.: Glycine receptor: Light microscopic autoradiographic localization with [³H]strychnine. *J. Neurosci.* **1**: 532–547, 1981.

Zukin, S. R., Young, A. B., and Snyder, S. H.: Gamma-aminobutyric acid binding to receptor sites in the rat central nervous system. *Proc. Natl. Acad. Sci. USA* **71**: 4802–4807, 1974.

The Role of Neurotransmitters in Electroshock Seizure Models

Ronald A. Browning

1. Introduction

The purpose of this report is to examine the evidence that specific neurotransmitters play a modulatory role in the susceptibility to and/or the expression of electrically induced experimental seizures. The fact that electrical stimulation of the brain can cause a seizure has been known for over a century (Fritsch and Hitzig, 1870; Albertoni, 1882).

Although advancements occurred in the 1920's and 1930's with regard to application of current to various parts of the head (Swinyard, 1972), the full potential of electroshock as an experimental model of epilepsy was not realized until Goodman and coworkers initiated their intensive studies in the 1940's. These investigators showed that the rodent and human nervous systems respond similarly to electrical stimulation of the brain and that the seizures elicited by electroshock are modified similarly in experimental animals and humans (*see* Swinyard, 1972, 1980 for detailed historical account). The studies of Goodman and coworkers (Toman et al., 1946; Swinyard, 1949; Swinyard et al., 1952) also established the electroshock method as a highly useful technique for screening new antiepileptic drugs, and their work led to its widespread use in numerous laboratories.

2. Electroshock Models

In their classical studies during the 1940's and 1950's, Goodman and coworkers (*see* Swinyard 1972 for references) developed four separate electrical models of experimental epilepsy that were subsequently used for testing antiepileptic drugs. These models included: (1) the minimal electroshock threshold test (referred to as ac-EST test), (2) the low-frequency electroshock threshold (referred to as the lf-EST or the 6 Hz-ES test), (3) the hyponatremic electroshock threshold test (referred to as HET), and (4) the maximal electroshock seizure test (referred to as the MES test). These models differ from one another in terms of the stimulus parameters employed (including the type of stimulation, stimulus frequency, and duration of the stimulus) and in the convulsive motor pattern produced. Another model that has contributed to our understanding of electroshock is the electrically induced spinal cord convulsion.

With regard to the present discussion, only the ac-EST, MES, and spinal cord convulsions need to be considered in any detail, since (except for two studies involving lf-EST) only these three tests (or slight modifications thereof) have been utilized to assess the role of neurotransmitters in seizures. Detailed descriptions of all electroshock seizure models can be found elsewhere (Swinyard, 1972; 1973).

2.1. The Minimal Electroshock Seizure Threshold (ac-EST Test)

The threshold for minimal electroshock seizures (ac-EST) is produced in mice or rats by stimulation with 60 Hz alternating current for 0.2 s using corneal electrodes. The minimum current necessary to produce 7–12 s of facial or face and forelimb clonus is determined by giving the animal small increments or decrements in current until the threshold is found. An apparatus similar to the one described by Woodbury and Davenport (1952) is generally used to deliver the stimulus. Using this apparatus, the current necessary to produce clonus varies between 6 and 9 mA for mice and between 20 and 36 mA for rats. The clonus produced by the ac-EST test is highly characteristic and involves clonus of the face and forelimbs usually accompanied by rearing similar to stage 5 kindled convulsions (Racine, 1972). As the current is increased, the clonus spreads to involve the whole body, with twisting and loss of posture. Further increments in the stimulating current (to about 10 mA above threshold) result in a change from clonus to explosive running, which may be interspersed with bouts of bouncing clonus involving symmetrical clonic activity in both fore- and hindlimbs. The latter seizure is similar to the running and clonus seen in minimal audiogenic seizures. The running, bouncing clonus is generally seen about 1–2 mA below the current necessary to elicit the tonic seizure (tonic flexion). Thus, using corneal electrodes and electroshock stimulation in rodents one can produce essentially three types of motor convulsions, depending on the magnitude of the current, (i.e., face and forelimb clonus, running with bouncing clonus, and tonus). Theoretically, one could study the effects of drugs or neurotransmitters on each of these motor patterns.

Some investigators employ ear-clip rather than corneal electrodes, and it is generally assumed or implied that seizures resulting from ear-clip and corneal stimulation are completely comparable. Recent studies in our laboratory, however, suggest that this is not the case. Indeed, the only overt clonus observed with transauricular stimulation in rats is the running or running bouncing clonus (Browning and Nelson, 1985), whereas the classical face and forelimb clonus with rearing could not be produced with ear-clip stimulation. Thus, as far as minimal electroshock seizures are concerned, the type of motor convulsion observed seems to depend on where the stimulating electrodes make contact with the skull. It seems likely, therefore, that the differences in motor pattern obtained from corneal and ear-clip stimulation are dependent on where the seizure discharge originates within the brain (see below).

2.2. The Maximal Electroshock Seizure (MES Test)

Behaviorally, maximal electroshock seizures do not differ from any other "maximal" seizure, whether produced by chemical convulsants or sound stimulation (Swinyard, 1973). Maximal electroshock seizures can be produced in mice, rats, rabbits, or cats by stimulation with 60 Hz alternating current for 0.2 s (Toman et al., 1946). Thus, the procedure and apparatus used to produce maximal electroshock seizures is exactly the same as that used to produce minimal electroshock seizures, except that the current intensity is 5–7 times greater than that necessary to elicit minimal electroshock seizures (clonus). In rats the "maximal seizure" consists of 3–4 s of tonic flexion where the body is ventroflexed, the forelimbs are extended, and the hindlimbs are flexed. This is followed by extension of the hindlegs that lasts 9–12 s. Finally there is a terminal clonus, which is essentially restricted to the hindlegs and amounts to a bicycle type kicking. The terminal clonus has recently been shown to be analagous to postdecapitation convulsions, and appears to be mediated by a discharge emanating from the spinal cord

(Fukuda et al., 1975a; Oishi et al., 1979). Like the postdecapitation convulsions, the terminal clonus also appears to be dependent on descending spinal noradrenergic neurons (Oishi et al., 1979).

In the MES test described by Toman et al. (1946), a supramaximal stimulus is used and the effect of drugs and/or neurotransmitters is evaluated on the components of the motor pattern. Seizure protection is said to have occurred if hindlimb extension (HLE) is abolished or markedly reduced in duration. On the other hand, seizure exacerbation is evidenced by a shortening of the flexor phase and a prolongation of the extensor phase. The most common endpoint used in the testing of antiepileptic drugs is the abolition of HLE. One problem with using rats is that 10–40% fail to consistently display HLE, and therefore must be pretested three or four times before use in an experiment (Swinyard, 1973; Buterbaugh, 1978).

A modification of the MES test that is rather widely used is the MES-threshold test, where one determines the minimum current (expressed as mA) needed to produce either tonic hindlimb extension (frequently used in mice) or the onset of tonus (i.e., tonic flexion). The tonic threshold is apparently more sensitive to transmitter manipulation than the supramaximal seizure, although both tests provide an examination of the tonic rather than the clonic seizure.

MES can also be produced by stimulating through ear-clip electrodes. Although the motor pattern is identical using ear-clip and corneal electrodes, less current is required to induce a maximal seizure using ear-clip stimulation in rats (Browning and Nelson, 1985). This is believed to be related to the path over which the current travels and the greater importance of the brainstem in the tonic seizure (see below). It seems possible that it would be easier to inhibit seizures induced by corneal stimulation than those induced by ear-clip stimulation. This should be kept in mind when comparing studies that have employed corneal electrodes with those that used ear-clip electrodes.

A modification of the MES test has been described by Bogue and Carrington (1953) that utilized a 7.5-mA (50-Hz) stimulus for up to 10 s (or until HLE occurs). This procedure has been employed by some investigators recently to evaluate the role of neurotransmitters in electroshock seizures and will be discussed later (*see* section 4.4.2). It should be pointed out that tonic seizures produced by prolonged stimulation may be more difficult to inhibit by drugs than those elicited by a short stimulus duration (Bogue and Carrington, 1953; Zablocka, 1963). Differences in stimulus duration may explain some of the disparity in the literature on neurotransmitters that will be considered later.

2.3. Electrically Induced Spinal Cord Seizures

Electrically induced spinal cord seizures may be considered another, albeit rather extreme, modification of the MES test, since this technique can be used to evaluate the effect of drugs on the maximal tonic pattern. In this technique, which was developed by Esplin and Freston (1960), the spinal cord is separated from the brainstem at the alanto-occipital junction, and the stimulating electrode, insulated except for the portion in contact with the spinal cord, is inserted into the cord from C_1 to C_4. The stimulus is different from MES in that dc square wave pulses are delivered by a Grass stimulator. The stimulus parameters usually employed are 1–2 ms pulses using 30–50 V. The frequency and duration of stimulation are varied. The severity of the seizure appears to increase as frequency becomes greater. At low frequencies (10/s) only running and clonus is seen. Whereas at 10–15/s the full MES pattern is observed. Even at higher frequencies, the stimulus must last for 3–4 s before the full pattern is obtained. Thus, the stimulus must last considerably longer than the 0.2 s used in MES, suggesting that the cord sustains seizure discharge less readily than the brain. The electrically induced spinal cord seizure can and has been used to localize the site where drugs and/or neurotransmitters act to modulate MES.

3. Hypothesis Concerning Anatomical Correlates of Electroshock-Induced Seizures

Although the neurophysiological mechanisms underlying the electroshock seizures are unknown, some attempt has been made to hypothetically link the motor convulsive pattern with anatomical locations within the CNS. Thus Woodbury and Esplin (1959) proposed that for electroshock to evoke a minimal seizure, it must discharge a substantial number of neurons over a finite period of time. This collection of neurons was called the ''oscillator'' in order to distinguish it from the stimulus that triggers the oscillator. Although the location of the oscillator within the brain was unknown, it was further hypothesized that it might be analogous to the centrencephalic system of Penfield and Jaspers (1954). It has been proposed that low-frequency EST activates the oscillator maximally, but there is no spread of discharge to other neurons (Swinyard, 1972; 1973). Minimal electroshock (ac-EST), on the other hand, was said to activate the oscillator and produce some spread to adjacent areas, whereas MES was thought to cause maximal activation of the oscillator and the spread of discharge over the entire brain. Using these various tests, it was believed that one could differentiate drugs that prevent discharge of the oscillator from those that prevent seizure spread (Swinyard, 1973). With regard to the maximal electroshock seizure, Zablocka and Esplin (1964) have suggested that the maximal seizure pattern is dependent on three participating mechanisms: (1) a sustained discharge of the oscillator, (2) the spread of discharge throughout the entire brain, and (3) the integration of supraspinal discharge at the spinal level to produce the stereotype MES motor pattern. Drugs that display anticonvulsant activity by abolishing HLE in the MES test are generally believed to act by preventing the spread of seizure discharge.

Although experimental support for the first two mechanisms suggested by Zablocka and Esplin (1964) to account for the MES pattern is lacking, there is support for the third mechanism. Indeed, Esplin and Freston (1960), employing the spinal cord seizures described above, showed that the ultimate determinants of the MES pattern are located in the spinal cord, and that the motor pattern seen following supramaximal brain stimulation is the natural response of the spinal cord to intense supraspinal discharge.

Based on studies carried out since 1959 it is now possible to be somewhat more specific concerning the anatomical correlates of electroshock seizures. The results of lesion (Browning et al., 1981a, b; Browning and Engel, 1984) and stimulation (Kreindler et al., 1958; Bergmann et al., 1963; Chui and Burnham, 1982) studies in rats and rabbits suggest that the tonic components of maximal seizures result from a discharge that emanates from the brainstem, and involves the brainstem reticular formation. Indeed, a growing body of evidence obtained from lesion, stimulation, and pharmacological studies (*see* Browning, 1985 for review) supports the notion that the face and forelimb clonus observed in the ac-EST test is driven by a discharge emanating from somewhere in the forebrain. On the other hand, the running bouncing clonus resulting from ear-clip stimulation, as well as all of the tonic components of MES, are believed to emanate from a discharge in the brainstem (Browning and Nelson, 1985; Browning et al., 1985). Thus, in order to be consistent with the current literature, it is necessary to modify the hypothesis advanced by Woodbury and Esplin (1959). Accordingly, maximal seizures (i.e., tonic) occur not when the discharge spreads throughout the entire brain, but when the relevant brainstem system is activated (i.e., exhibits paroxysmal discharge that is independent of continued input from other brain areas). The existing evidence is also consistent with the proposal that there are at least two systems in the brain that drive the spinal cord through the seizure: (1) a forebrain system responsible for the clonus of minimal electroshock, and (2) a brainstem system that when partially activated produces running-bouncing clonus and, when maximally activated, produces tonic convulsions (Browning, 1985). Alternatively, one could view the transition from clonus to tonus as a spread of discharge from the forebrain to the brainstem, and an initiation of the brainstem seizure. In the latter case, the hypothesis of Woodbury and Esplin (1959) was not actually wrong, but was potentially misleading because of its vagueness. In the case of the MES test, however, it seems likely that the stimulus activates both the forebrain and brainstem simultaneously, and the brainstem overrides the forebrain in gaining control of the spinal cord (Kreindler et al., 1958; Tanaka and Mishima, 1953).

4. Neurotransmitters and Electrically Induced Seizures

Inasmuch as this literature has been addressed previously (Maynert et al., 1975) in a comprehensive review that covered all seizure models, it is not our purpose to simply recapitulate what was stated before and update what has occurred since 1975, although some of this is unavoidable. Rather we would like to readdress the literature in light of the aforementioned hypothesis by focusing on the role of specific neurotransmitters in (1) minimal electroshock (clonus), (2) maximal electroshock (tonus), and (3) spinal cord seizures. Using this approach it is hoped that we will not only be able to evaluate the importance of neurotransmitters in electrically induced seizures in general, but that we might also gain some insight concerning the anatomical region in which these transmitters act to modify seizures. However, identification of the site of action of a given transmitter will ultimately come only when we can selectively manipulate it in discrete brain regions or at specific synapses. Some success has been realized in this area through selective lesioning of transmitter-containing neurons, and/or through microinjection of the drugs into specific brain nuclei, and these studies will be discussed in connection with the appropriate transmitter.

Since most of the evidence for the involvement of neurotransmitters in electroshock seizures is based on the use of drugs believed to alter the concentration of neurotransmitters at their receptors, it is necessary to raise all the caveats discussed previously (Maynert et al., 1975). The inability to measure the concentration of these chemicals in the synaptic cleft, and the lack of complete drug specificity, require certain assumptions on the part of the investigator that sometimes preclude firm conclusions. Nevertheless, when studies employing diverse chemicals in several different laboratories provide harmonious findings concerning the role of a given neurotransmitter, we can be fairly confident in our conclusions.

In their review in 1975, Maynert and coworkers summarized the available literature and interpreted the data in terms of increased or decreased availability of the neurotransmitters at receptor sites. Although this approach entails several assumptions, it still seems valid and will therefore be adopted here. In considering treatments that decrease the availability of neurotransmitters at receptor sites, in the past we have operated under the belief that the use of lesions or selective neurotoxins to destroy specific neurotransmitter-containing neurons provided more direct and easily interpretable evidence than the pharmacological studies. However, in view of the recent evidence that two or more transmitters may coexist in the same neuron (Hokfelt et al., 1980), it appears that interpretation of lesion studies may be less straightforward than previously believed.

4.1. Acetylcholine

Previous reviews on acetylcholine (ACh) and seizures (Maynert, 1969; Maynert et al., 1975) clearly supported the notion that this transmitter is a facilitator of seizures, and may itself be a convulsant. This conclusion comes largely from studies showing that direct application of ACh to the brain results in seizures (Feldberg, 1945; Feldberg and Sherwood, 1954; Haley and McCormick, 1957; Guerrero-Figueroa et al., 1964). In examining the literature on ACh and comparing it with other biogenic amine neurotransmitters, one is struck by the relative dearth of pharmacological studies designed to determine the importance of ACh in the various seizure models, and electroshock is no exception. This dearth of information undoubtedly reflects the relative paucity of pharmacological tools that have been available for manipulating ACh in the CNS and the lack of convenient techniques for assaying brain ACh. It appears that recent advances may obviate both of these drawbacks in the near future.

In focusing on the role of ACh in electroshock-induced seizures, two things become apparent: (1) it is a facilitator of some types of seizures, and (2) it exerts anticonvulsant activity in some seizure models.

4.1.1. Acetylcholine in Minimal Electroshock Seizures

Only two laboratories have examined the effect of increased ACh on minimal electroshock, and one of these employed cholinergic agonists. In both studies the findings are consistent with seizure facilitation. Zablocka and Esplin (1963) found that pilocarpine, a lipophilic muscarinic agonist capable of reaching the brain, caused a marked reduction in the threshold for clonic seizures produced by minimal electroshock in both rats and mice. In both of these species the threshold-lowering effect of pilocarpine was blocked by pretreatment with atropine. This facilitating effect of pilocarpine on ac-EST is consistent with the fact that this drug produces clonic seizures in rats with a convulsant dose-50 of 105 mg/kg. Pilocarpine also appears to lower the clonic threshold for pentylenetetrazol (PTZ)-induced seizures in rats and mice (Zablocka and Esplin, 1963).

The other study concerning the role of ACh in minimal electroshock is more difficult to interpret because much of it was devoted to examining the interaction between Ro 4-1284 (a drug that depletes catecholamines and 5-HT) and cholinergic agonists and/or antagonists. In this study, rats were stimulated with 26 mA through corneal electrodes, which normally causes only clonus (Hatoum, 1974). Physostigmine, in a dose that alone had no effect, caused an enhancement of the seizure-facilitating action of Ro 4-1284, whereas scopolamine (a muscarinic antagonist) had the opposite effect. In these studies seizure facilitation was measured as an increase in the incidence of tonic convulsions. Whether this is caused by an increase in the severity of clonus, which at higher intensities triggers tonus, or a facilitation of the tonic seizure as well, is not clear. Nevertheless, ACh given intracerebroventricularly (icv) to rats pretreated only with physostigmine also increased the incidence of tonic seizures, suggesting that ACh enhances the severity of minimal electroshock seizures even in the absence of Ro 4-1284

(Hatoum, 1974). These findings are consistent with other findings that show that cholinergic agonists given icv can trigger forebrain seizure discharge (Snead, 1983), and when injected into the amygdala (Wasterlain and Jonec, 1983) or lateral ventricle repeatedly (Snead, 1983) can function as a kindling agent. Studies directed at examining the effects of decreased ACh receptor stimulation on electroshock-induced clonus are either inadequate or have not been done. Thus, based on the relatively few studies that have been done, it appears that ACh is a facilitator of electroshock-induced clonus.

4.1.2. Acetylcholine in Maximal Electroshock Seizures

In contrast to the rather consistent seizure-facilitating effects of ACh on minimal electroshock, the effects of ACh on MES appear to be either inhibitory, weakly facilitating, or absent depending on the study and the species employed.

In rats the same dose of pilocarpine that lowered the ac-EST was found to abolish HLE (Zablocka, 1963; Zablocka and Esplin, 1963), indicating a protective effect against tonic seizures. The fact that the same dose of pilocarpine (40 mg/kg) facilitates clonic seizures and inhibits tonic seizures induced by electroshock emphasizes further the likelihood that clonic and tonic seizures are regulated by separate systems in the brain. Moreover, it is conceivable that separate forebrain (clonic) and brainstem (tonic) effects of ACh may be mediated by distinct subclasses of muscarinic receptors (M_1 and M_2), which seem to be differentially distributed between forebrain and brainstem (Birdsall and Hulme, 1983).

Stimulation of muscarinic receptors with pilocarpine was found to produce little effect on MES in mice (Zablocka and Esplin, 1963), suggesting a species difference in the effect of cholinergic agonists on electroshock-induced tonus. On the other hand, some investigators examining the MES threshold in mice have observed anticonvulsant effects of cholinergic agonists, although they were not always in agreement. Arecoline, another muscarinic agonist, was found to protect mice against HLE at 1 mg/kg doses, when threshold currents were employed (Lazarova and Roussinov, 1977). Chen et al. (1968) examined the effect of several different cholinergic agonists on the threshold for electroshock-induced HLE and found that arecoline and pilocarpine alone lowered the threshold slightly at low doses, whereas pilocarpine elevated it at higher doses (16 mg/kg) and oxotremorine (another central cholinergic agonist) produced seizure protection at almost all doses. Moreover, arecoline, pilocarpine, oxotremorine, and physostigmine were all found to potentiate the anticonvulsant effects of methazolamide in mice (Chen et al., 1968). These findings show that mice receiving cholinergic agonists may be protected against tonic electroshock-induced seizures under certain conditions. Even when seizure facilitation was observed in mice treated with cholinergic agonists, the effect was small compared to the seizure-facilitating effects of reserpine, and the effect was not dose dependent. The hypothesis that increased ACh receptor stimulation is antagonistic to electroshock-induced tonic seizures is further supported by the assertion that strains of mice with higher levels of brain ACh appear to be more resistant to MES, as evidenced by a longer latency and a shorter duration of extension, than mice with lower levels (Karczmar, 1974).

Although studies examining the effects of increased ACh receptor stimulation indicate a cholinergic-mediated protection against electroshock-induced tonic seizures, the effects of reduced cholinergic receptor stimulation on MES are not in harmony with this idea. Indeed, atropine alone has been found to reduce HLE in rats (Zablocka, 1963), and other anticholinergic drugs were shown to protect mice against maximal electroshock-induced HLE (Millichap et al., 1968). The reason for this discrepancy with atropine and other antimuscarinic agents is not known, but may be related to independent anticonvulsant effects or to effects on other neurotransmitters, such as catecholamines and serotonin (Karczmar, 1974).

Overall it appears that cholinergic neurons in the brain may facilitate minimal electroshock-induced clonus, but either inhibit or have no effect on the tonic component of MES. However, additional studies are needed to elucidate more precisely the role of ACh in clonic and tonic convulsions.

4.2. Norepinephrine (NE)

Examination of the literature leaves no doubt that norepinephrine (NE) attenuates seizures in almost every seizure model available (*see* reviews: Maynert, 1969; Maynert et al., 1975; Jobe, 1981; Kalichman, 1982; Burley and Ferrendelli, 1984). The seizure-inhibitory effect of NE appears to be most marked in genetic models of epilepsy where the animal has an enhanced propensity for seizures (Jobe and Laird, 1981), and is somewhat less obvious when one is attempting to examine its role in normal animals receiving supramaximal seizure-evoking stimuli. Indeed, from the information now available amounting to 30 yr of investigation, one cannot escape the conclusion that NE and γ-aminobutyric acid (GABA) are more influential than any other neurotransmitters in modulating seizure susceptibility and seizure intensity, and the electroshock model is no exception.

The first indication that NE might be involved in seizures was provided by studies employing the electroshock model (Chen et al., 1954; Prokop et al., 1959). The drugs employed in these early studies (reserpine and MAO inhibitors) altered dopamine and serotonin in concert with NE, allowing for no conclusions about which of these biogenic amines was responsible for the effects. Nevertheless, it opened the door for subsequent investigations, which have been able to delineate a specific and important role for NE. Examination of the role of NE in each of the electroshock models described above (including electrically induced spinal cord seizures) can provide important information on the role of NE in seizures of varying severity (clonic vs tonic) and may provide some insight concerning which regions of the CNS are involved in the seizure-attenuating effect of NE.

4.2.1. Norepinephrine in Minimal Electroshock Seizures

Low-frequency (6-Hz) electroshock seizures, which cause maximal activation of the oscillator without spread of discharge (Swinyard, 1973), have not been examined in connection with neurotransmitters, with two exceptions: (1) Using extremely high doses of reserpine (8 mg/kg) Chen et al. (1954) reported a facilitation of low-frequency electroshock seizures in mice 4.5 h after treatment, and (2) Buterbaugh and London (1977) assessed the effect of reserpine (2.5 mg/kg), α-methyl-p-tyrosine, and p-chlorophenylalanine on low-frequency EST and found it was not affected by monoamine depletion of any kind. In view of the marked seizure-facilitating effects of reserpine on all other seizure models, it is probably safe to conclude that NE (as well as dopamine and 5-HT) does not play a very significant role in low-frequency EST.

Because there are few studies that have assessed the importance of NE in minimal electroshock seizures (ac-EST), they will not be categorized under headings of "increased" or "decreased" neurotransmitter. Attempts to assess the effect of icv-administered NE (16 μg) on ac-EST were confounded by the hypothermic effects of NE, which in turn caused seizure facilitation (Browning and

Maynert, 1978b). However, when body temperature was controlled, NE (16 μg) failed to influence the electroshock-induced clonic seizure. On the other hand, Stull et al. (1977) found that NE-administered icv could attenuate minimal electroshock seizures (26 mA) that had been exacerbated by pretreatment with Ro 4-1284 (a reserpine-like drug). Thus, the minimal clonic seizure was first converted to a tonic seizure by monoamine depletion, and NE was administered to reduce seizure intensity. However, the dose of NE required to attenuate the seizure was between 125 and 208 μg. In the same study Stull et al. (1977) also found that dopamine given icv could also attenuate the Ro 4-1284-induced enhancement of minimal electroshock seizures, but much higher doses were needed (1200 μg). Other studies concerned with the role of NE in minimal electroshock seizures do not readily distinguish between the relative importance of NE and dopamine. For example, in mice whose ac-EST has been lowered by pretreatment with reserpine, administration of the catecholamine precursor, L-DOPA, was found to antagonize the seizure-facilitating effect of reserpine, suggesting that catecholamines were involved in the effect of reserpine (Azzaro et al., 1972). Similar results were found when 5-hydroxytryptophan (5-HTP, a serotonin precursor) was administered, implicating 5-HT as well as NE and dopamine. Subsequent studies using the same experimental design in mice showed that restoration of the ac-EST to normal after reserpine was prevented if dopamine β-hydroxylase was inhibited, to prevent NE synthesis (Wenger et al., 1973). Inasmuch as selective inhibition of 5-HT synthesis by p-chlorophenylalanine also prevented a return to normal susceptibility, the authors concluded that both NE and 5-HT were important in regulating minimal electroshock thresholds in mice. Other studies using a similar protocol in rats suggested that repletion of NE and/or dopamine could provide protection from the seizure-facilitating effects of reserpine in a minimal electroshock test, but that repletion of 5-HT alone could not (Jobe et al., 1974).

A decrease in the ac-EST has also been reported in rats following depletion of catecholamines with α-methyl-p-tyrosine (Buterbaugh and London, 1977), or following destruction of catecholamine neurons with 6-hydroxydopamine (Browning and Maynert, 1978a). Thus, it seems clear that depletion of catecholamines (NE and dopamine) facilitate electroshock-induced clonus. It also seems likely that NE can attentuate clonus produced by minimal electroshock, but the effect of dopamine is more equivocal.

4.2.2. *Norepinephrine and Maximal Electroshock Seizures*

4.2.2.1. INCREASE IN NE SYNAPTIC CONCENTRATION. A variety of pharmacological manipulations designed to increase catecholamines in the synaptic cleft have been shown to protect rats and mice from electrically induced tonic seizures, but none of these can attribute the effect exclusively to NE. For example, attenuation of electrically induced tonic seizures has been achieved with amphetamine (Rudzik and Johnson, 1970), tricyclic antidepressants (Lange et al., 1976; Chen et al., 1968), L-DOPA (Kilian and Frey, 1973), or monoamine oxidase inhibitors (Prokop et al., 1959).

4.2.2.2. DECREASE IN NE SYNAPTIC CONCENTRATION. Studies examining the effects of catecholamine depletion on maximal electroshock provide more conclusive evidence that NE plays an important role in modulating electrically induced tonic seizures. Inhibition of catecholamine synthesis with α-methyl-p-tyrosine (AMPT) has been shown to reduce the threshold for electrically induced tonic convulsions in both rats and mice (Kilian and Frey, 1973; Buterbaugh and London, 1977), although Buterbaugh (1978) has found that catecholamine depletion does not affect the MES threshold in rats. Kilian and Frey (1973) concluded that the seizure-facilitating effects of AMPT could be attributed to NE, since the electroshock tonic threshold was also depressed by drugs that block NE synthesis and spare dopamine. Moreover, the tonic threshold was reduced by an alpha-adrenergic antagonist (phentolamine), but was unaffected by a dopamine receptor antagonist (haloperidol). Basically alpha-adrenoceptor blockers have been shown to facilitate MES in mice, whereas beta-adrenoceptor antagonists appear to have an anticonvulsant effect that seems to be unrelated to the blockade of adrenergic receptors. Owing to the fact that this topic has not changed significantly since it was reviewed previously, the reader is referred to the review by Maynert et al. (1975) for more detail.

Selective destruction of noradrenergic and/or dopaminergic neurons with 6-hydroxydopamine (6-OHDA) or discretely placed electrolytic lesions further substantiate the importance of NE in electrically induced tonic convulsions. Indeed, since the original observation over a decade ago (Browning and Maynert, 1970) that 6-OHDA facilitates electrically and chemically induced seizures, this compound has been shown to affect convulsions in essentially every seizure model examined.

Intracerebral administration of 6-OHDA by the intraventricular or intracisternal routes has been shown to prolong the duration of hindlimb extension in rats subjected to MES (Browning and

Maynert, 1978a; Simonton, 1978), indicating an increase in the severity of the tonic seizure (Tedeschi et al., 1956). Similar results have been obtained in mice following the icv administration of 6-OHDA (Fukuda et al., 1975a). It appears that postictal recovery from MES is also markedly prolonged following 6-OHDA in both rats (Browning and Maynert, 1978a) and mice (Fukuda et al., 1975b). Other investigators have shown that the threshold for electroshock-induced HLE produced by ear-clip stimulation for 6 s is lowered by icv administration of 6-OHDA in rats (Quattrone and Samanin, 1977; Quattrone et al., 1978; Crunelli et al., 1981). Moreover, the tonic seizure-facilitating effects of 6-OHDA in rats appear to be mediated exclusively by the depletion of NE rather than dopamine, since no changes were found in rats whose noradrenergic neurons were protected from the neurotoxic action of 6-OHDA by desmethylimipramine (Quattrone et al., 1978; Crunelli et al., 1981) or protriptyline (Simonton, 1978). 6-Hydroxydopamine has also been shown to reduce the effectiveness of several anticonvulsants such as phenobarbital, phenytoin, acetazolamide, and carbamazepine in rats (Quattrone et al., 1978; Browning and Simonton, 1978).

4.2.2.3. LOCALIZATION OF NE EFFECTS ON MES. Studies designed to produce regional CNS depletion of NE with 6-OHDA have provided some information on where NE may be acting to alter the electrically induced tonic seizure. Based on the hypothesis discussed above, we would predict that brainstem and spinal cord NE would be more important in tonic seizures. Selective depletion of spinal cord NE by intraspinal injections of 6-OHDA was found to prolong the duration of HLE in mice receiving maximal electroshock stimulation through corneal electrodes (Oishi et al., 1979). This finding is in agreement with the effect of 6-OHDA on spinal cord convulsions (see below) and suggests that the exacerbation of the tonic seizure pattern by NE depletion is, at least in part, produced at the level of the spinal cord. It has also been found that the clonus (hindlimb kicking) that follows the tonic extensor phase of MES is inhibited by depletion of spinal cord NE, as are postdecapitation convulsions (Fukuda et al., 1975a; Oishi et al., 1979). At least some of the descending NE neurons involved in postdecapitation convulsions appear to arise from the locus ceruleus (Suenaga et al., 1979). Therefore, it seems likely that the last phase of maximal electroshock seizures (clonus) is driven from the spinal cord.

Depletion of forebrain NE by microinjections of 6-OHDA into the ascending noradrenergic pathways was reported to increase the duration of electroshock seizures produced through ear-clip stimulation with currents of 10, 15, and 22.5 mA (Mason and Corcoran, 1979). These findings are difficult to interpret because the authors did not provide a description of which components of the seizure were prolonged. However, there was no effect on the incidence of tonic extension and no effect on the latency of tonus. Thus, it seems likely that this lesion increased either the clonus or the postictal phase. Interpretations might have been easier if they had employed MES or ac-EST to produce the seizure rather than the technique they used. It is interesting to note that destruction of the nigrostriatal dopaminergic neurons with 6-OHDA had no effect on electroshock seizures produced with ear-clip electrodes (Mason and Corcoran, 1979).

Electrolytic destruction of the locus ceruleus has recently been found to lower the threshold for tonic seizures induced by electroshock (Oishi and Suenaga, 1982), without affecting the duration of tonic extension. Although it suggests an important role for the NE-containing neurons of the locus ceruleus in tonic seizures, it does not clarify the site of action of NE, since depletion was observed from the cortex to the spinal cord.

4.2.3. *Norepinephrine in Electrically-Induced Spinal Cord Seizures*

As described above, stimulation of the cervical spinal cord following its separation from the brain produces a seizure identical to MES. Thus, this model is ideal for determining the role of spinal cord neurotransmitters in the maximal electroshock pattern. Jobe et al. (1978) treated rats with 6-OHDA icv and examined spinal cord convulsions and NE levels 14 d later. The tonic extensor component and the total duration of the seizure was found to be significantly prolonged, whereas the tonic flexor component was unaffected. This seizure-enhancing effect of 6-OHDA on spinal cord convulsions could be attributed to NE depletion because it was absent in rats whose NE neurons were protected from 6-OHDA by pretreatment with desipramine. These findings are consonant with the increase in the duration of tonic extension observed in mice subjected to MES following intraspinal injections of 6-OHDA (see above). The effect of 6-OHDA on spinal cord seizures is also in agreement with other studies that examined the effects of pharmacological manipulations of NE on spinal cord seizures (Jobe and Ray, 1980a,b).

Based on the foregoing, it appears that the attenuating effect of NE on the MES pattern may occur through noradrenergic neurons in the spinal cord. On the other hand, it seems likely that the noradrenergic neurons influencing the tonic seizure threshold are located in the brainstem. It should be noted, however, that although the MES threshold is reduced in mice following NE depletion (Kilian and Frey, 1973), this threshold in rats seems to be unaffected by similar treatments (Buterbaugh, 1978).

4.3. Dopamine

As indicated in the previous section, similarities between NE and dopamine in terms of their chemical structure, biosynthesis, metabolism, storage, and release, make it difficult to distinguish the importance of one over the other when pharmacological manipulations are involved. Nevertheless, several manipulations such as selective lesioning of the NE- or dopamine-containing neurons, inhibition of dopamine β-hydroxylase to selectively decrease NE synthesis, the use of 6-OHDA following pretreatment with noradrenergic uptake inhibitors, or the use of some selective agonists and antagonists has allowed some success in distinguishing between these two catecholamines. However, many of these studies have already been cited in connection with NE, and we will therefore limit our remarks in this section to summarizing statements.

4.3.1. Dopamine in Minimal Electroshock Seizures

Although the icv administration of dopamine failed to alter the ac-EST in rats when body temperature was controlled (Browning and Maynert, 1978b), larger doses (1200 μg) were found to reduce seizure intensity in rats whose seizures had been facilitated by Ro 4-1284 (Stull et al., 1973; 1977). Apomorphine (a dopamine agonist) was also found to antagonize the seizure facilitating effects of Ro 4-1284, whereas pimozide (a dopamine antagonist) blocked the effects of both dopamine and apomorphine (Stull et al., 1973). In the latter studies, however, it is difficult to know whether the effect of dopamine was exerted on the clonic or tonic seizure systems (assuming they are different) because the minimal electroshock current (26 mA) administered caused mostly tonic seizures in the Ro 4-1284-treated rats, and dopamine was found to abolish tonus. Using a similar protocol in a separate study, Jobe et al. (1974) also concluded that repletion of dopamine after its removal by Ro 4-1284 could reduce the severity of seizures caused by minimal electroshock. On the other hand, repletion of dopamine following reserpine was found to be unrelated to the restoration of the ac-EST in mice (Wenger et al., 1973). Thus, the role of dopamine in minimal electroshock remains equivocal.

4.3.2. Dopamine in Maximal Electroshock Seizures

The effect of the ic administration of dopamine on MES has not been examined. However, information on the role of dopamine in MES is provided by pharmacological manipulations, some of which have been described in the section on NE.

Stimulation of dopamine receptors with apomorphine (1–10 mg/kg) has generally failed to provide protection against MES in mice (McKenzie and Soroko, 1972; Kleinrok et al., 1978), although there is one conflicting report using identical procedures (Gyorgy, 1983). Other investigators have also concluded that dopamine does not play an important role in regulating MES threshold in mice based on the differential effects of tyrosine hydroxylase and dopamine β-hydroxylase inhibitors (Kilian and Frey, 1973; Rudzik and Johnson, 1970).

An increase in the threshold for electrically induced tonic seizures was observed in rabbits following an increase in brain dopamine levels by treatment with iproniazid, reserpine, and L-DOPA (De Schaepdryver et al., 1962). More recently, the protection of young chicks against MES produced by amphetamines has been attributed to dopamine release, since it was potentiated by FLA-63 (a dopamine β-hydroxylase inhibitor) and antagonized by pimozide (Osuide et al., 1983).

Studies that have employed 6-OHDA to assess the role of catecholamines in MES in the rat have uniformly found that selective depletion of dopamine has no effect on the tonic seizure (Mason and Corcoran, 1979; Quattrone et al., 1978; Crunelli et al., 1981). The latter findings conflict with the observation of McKenzie and Soroko (1972) showing a dose-dependent inhibition of electroshock-induced hindlimb extension in rats treated with apomorphine. However, the possibility that apomorphine antagonized MES through an effect on other transmitter systems cannot be excluded, since it is known to increase 5-hydroxytryptamine levels (Lee and Geyer, 1984) and GABA turnover (Hokfelt et al., 1976).

4.4. 5-Hydroxytryptamine (5-HT)

Although the literature is replete with studies suggesting that 5-hydroxytryptamine (5-HT, serotonin) has seizure-attenuating effects, some investigators have failed to confirm this (*see* reviews by Maynert et al., 1975; Browning et al., 1978). A reexamination of the literature suggests that one reason for the disparity in findings is that 5-HT may affect clonic and tonic seizures differently. Moreover, 5-HT appears to also affect electroshock and pentylenetetrazol (PTZ) seizures differently. Some of the discrepancies in the literature must inevitably be attributed to improper attention to the fact that *p*-chlorophenylalanine (PCPA) is not entirely specific for 5-HT neurons.

4.4.1. 5-HT in Minimal Electroshock Seizures

There are relatively few reports concerning the role of 5-HT on minimal electroshock-induced clonus, and these are limited to an examination of 5-HT depletion or depletion followed by repletion. The studies uniformly indicate that 5-HT has no effect on minimal electroshock-induced seizures in rats. The ac-EST was unaffected following depletion of 5-HT with PCPA (Buterbaugh, 1978) or destruction of serotonergic neurons with 5,7-dihydroxytryptamine (Browning et al., 1978), despite the fact that both of these treatments exacerbated MES (see below). Jobe et al. (1974) also concluded that 5-HT did not play a role in seizures induced by minimal electroshock stimulation (26 mA), since repletion of 5-HT stores by 5-HTP following reserpine, failed to antagonize the seizure-facilitating effects of reserpine.

Although there are only two studies on the role of 5-HT in electroshock-induced clonus in mice, the findings are essentially consistent. Buterbaugh and London (1977) reported a 17% reduction in the ac-EST following treatment of mice with PCPA. It should be noted, however, that the dose of PCPA used by Buterbaugh and London (1977) also caused a 20–25% reduction in brain catecholamines. In another study using mice, Wenger et al. (1973) concluded that 5-HT was important in maintaining a normal ac-EST, because PCPA prevented a normal restoration of the seizure threshold after it had been lowered with reserpine.

The studies in rats seem more conclusive and suggest that 5-HT does not play a role in minimal electroshock seizures in this species. The studies in mice are weakened by the possibility that the effects may be attributed to catecholamines as well as 5-HT. Nevertheless, those studies raise the possibility that there are species differences concerning the effects of 5-HT on ac-EST.

4.4.2. 5-HT in Maximal Electroshock Seizures

A large number of concordant studies provide evidence that 5-HT is an attenuator of MES in both rats and mice. For example, elevations in brain 5-HT produced by the combination of 5-HTP (a 5-HT precursor) and a MAO inhibitor have been found to elevate the threshold for tonic seizures in mice (Chen et al., 1968; Kilian and Frey, 1973) and rats (Prokop et al., 1959). More recently, Buterbaugh (1978) reported that drugs that facilitate 5-HT neurotransmission, such as 5-HTP (high doses), fenfluramine (a 5-HT releaser), fluoxetine (a 5-HT reuptake blocker), and 5-methoxydimethyltryptamine (a 5-HT agonist), can be shown to block the HLE component of MES in rats. Moreover, Buus Lassen (1978) has obtained similar results in mice given high doses of 5-HTP alone, or low doses of 5-HTP in combination with a 5-HT uptake inhibitor. One exception to these harmonious findings is the work of Crunelli et al. (1979), who failed to observe any effect of enhancing 5-HT neurotransmission on electrically induced tonic seizure thresholds in rats. The latter investigator, however, employed ear-clip electrodes and the continuous stimulation technique of Bogue and Carrington (1953) described earlier, which might account for the disparity in findings (see below).

Reductions in brain 5-HT, or interference with 5-HT neurotransmission, has consistently been shown to lower the threshold for electrically induced tonic seizures in mice (Chen et al., 1968; Gray and Rau, 1971; Kilian and Frey, 1973; Buterbaugh and London, 1977) and rats (Buterbaugh, 1978; Waller and Buterbaugh, 1983). Buterbaugh (1977; 1978) has shown that depletion of brain 5-HT converts naturally occurring nonextensor rats (those that fail to exhibit HLE when subjected to MES) to extensor rats and also lowers the threshold for tonic seizures.

Several laboratories have studied the effect of selective destruction of serotonergic neurons (with either 5,7-DHT or surgical lesioning) on maximal electroshock seizures. Some of the latter studies provide insight into the anatomical localization of serotonin's seizure-attenuating effect (*see* section 4.4.3). Intracerebroventricular administration of 5,7-DHT (200 μg) was found to shorten the latency to HLE and prolong the duration of HLE in rats subjected to MES (Browning et al., 1978). More recently 5,7-DHT and 6-OHDA have been used in neonatal rats to study the role of brain monoaminergic systems in the development of the electroshock tonic convulsive pattern (London et al., 1982; Waller and Buterbaugh, 1983). These studies show that rats treated with 5,7-DHT on postnatal d 3 display a lower MES threshold on d 21 that is still present on d 30. On the other hand, treatment with 6-OHDA on postnatal d 1 and 2 resulted in a lower MES threshold that disappeared by the third postnatal week. These investigators have concluded that although catechol-amines seem to regulate the tonic threshold prior to 3 wk of age, there is an apparent transition from catecholaminergic to serotonergic regulation of the tonic electroshock threshold during the third postnatal week. It would appear that after the third week and even in adulthood (Buterbaugh, 1978), 5-HT is much more important than catecholamines in regulating the MES threshold. Rats treated with 5,7-DHT were also found to be resistant to the antiextensor effects of some anticonvulsant drugs in the MES test (Browning et al., 1978).

In contrast, Crunelli et al. (1979; 1981) have concluded that 5-HT plays no role in electrically induced tonic convulsions or in the action of anticonvulsants on these seizures, based on their findings with icv-administered 5,7-DHT (150 or 200 μg). The only obvious difference between the studies of Crunelli et al. (1979; 1981) and those of other investigators showing an effect of 5,7-DHT (Browning et al., 1978; Waller and Buterbaugh, 1983) is the type of stimulation employed. Crunelli et al. (1979; 1981) have used a modification of the method of Bogue and Carrington (1953), in which seizures are induced by continuous stimulation (for up to 6 s) through ear-clip electrodes. As indicated earlier, it is believed that the current path and therefore the structures activated are different, depending on whether ear-clip or corneal electrodes are used (Browning and Nelson, 1985). It is conceivable that with the prolonged stimulation used by Crunelli et al. (1979) the current spreads to the spinal cord and activates it directly. However, other studies will be needed before this discrepancy can be adequately explained.

4.4.3. Localization of 5-HT Action
in Maximal Electroshock Seizures

Although widespread CNS depletion of 5-HT by the icv administration of 5,7-DHT or the systemic administration of PCPA has been shown to facilitate MES, mechanical lesions of the ascending serotonergic neurons (in the dorsal and median raphe nuclei) were found to have no effect on MES or on the suppression of MES by anticonvulsant drugs (Browning et al., 1978). Other investigators have made similar observations while comparing the effects of raphe lesions and PCPA on tonic seizures (Crunelli et al., 1979; Quattrone et al., 1978). Thus, several laboratories have now provided evidence that forebrain 5-HT is not involved in suppressing electrically induced tonic seizures. Moreover, it appears that spinal cord 5-HT also plays no role in electrically induced tonic seizures. Depletion of CNS 5-HT with 5,7-DHT, or manipulation of 5-HT with other drugs, was found to have no effect on electrically induced spinal cord convulsions (Jobe et al., 1978; Buterbaugh, 1978). Similarly, intrathecal injections of 5,7-DHT to selectively deplete 5-HT in the spinal cord without affecting the brain was found to have no effect on MES in rats (Browning and Engel, 1984). It seems likely, therefore, that serotonergic neurons in the brainstem or cerebellum are more important than those in the forebrain or spinal cord in regulating electrically induced tonic seizures.

4.5. Gamma-Aminobutyric Acid (GABA)

Even before GABA was shown to be the major inhibitory neurotransmitter of the brain, it was recognized that this compound had potentially important influences on seizures. Among the first to recognize this were Killam and Bain (1957), who suggested that the decrease in GABA caused by semicarbizide might be responsible for the convulsions caused by this compound. Since then it has been found that essentially anything that interferes with neurotransmission at GABAergic synapses results in convulsions. Thus, drugs that inhibit glutamic acid decarboxylase (GAD, the enzyme that catalyzes the formation of GABA from glutamic acid) or block GABA receptors are convulsants (Meldrum, 1975). However, the demonstration by several laboratories (Baxter and Roberts, 1960; Maynert and Kaji, 1962; Kuriyama et al., 1966) of a lack of correlation between steady-state GABA levels and cerebral excitability (i.e., convulsant or anticonvulsant activity) led to controversy and confusion with regard to the role of GABA in convulsions. This lack of correlation has now been largely accounted for by the realization that steady-state GABA levels may not reflect the amount of GABA reaching the receptor. For example, it has been demonstrated that cerebral excitability is correlated with the concentration of GABA in nerve terminals (Wood et al., 1980; Gale and Iadarola, 1980), or with newly synthesized GABA that apparently represents the GABA pool of greatest physiological significance (Tapia, 1980). Since GAD is the enzyme involved in the synthesis of GABA utilized for release, the activity of this enzyme is more likely to reflect functional activity at GABAergic synapses than is the steady-state GABA level of the brain (Wood, 1975).

Inasmuch as a reduction in GABA levels or inhibition of GABA synthesis inevitably leads to seizures, the influence of GABA on experimental models of epilepsy can only be studied using treatments that increase GABAergic neurotransmission. The anticonvulsant properties of GABA and GABAmimetic drugs are potentially of great clinical interest, and have been studied in a wide variety of experimentally induced seizures (many of which are discussed elsewhere in this volume). The literature on the role of GABA in convulsions up through 1974 has been reviewed by several authors (Meldrum, 1975; Maynert et al., 1975; Wood, 1975). However, prior to 1975 few studies had been performed to examine the relationship between GABA and electroshock seizures because of the dearth of pharmacological tools available for manipulating GABA. In the last 8 yr, however, a number of new compounds have become available for enhancing GABAergic neurotransmission (Enna and Maggi, 1979), and many of these have recently been used in connection with electroshock-induced seizures.

4.5.1. GABA in Minimal Electroshock Seizures

Studies on the role of GABA in electrically induced clonic seizures are all but absent. Indeed, we found only two reports. Whereas the systemic administration of GABA was found to have no effect, the icv administration of 10–80 μg was found to suppress clonus induced by minimal electroshock or pentylenetetrazol in mice (Gulati and Stanton, 1960). However, these same doses had no effect on the MES pattern in mice. In another study Maynert and Kaji (1962) found that elevating brain GABA levels with hydrazine or hydoxylamine had no effect on clonic seizures induced by electroshock. A more recent study by Wood et al. (1980) may explain the lack of anticonvulsant activity observed with hydrazine. It appears that hydrazine increases whole brain GABA without increasing synaptosomal (nerve ending) GABA, primarily because it is also a rather strong inhibitor of GAD.

4.5.2. GABA in Maximal Electroshock Seizures

Three classes of pharmacological agents are now available for facilitating GABAergic neurotransmission. These include: (1) GABA-transaminase (GABA-T) inhibitors, (2) GABA-reuptake inhibitors, and (3) directly acting GABA-agonists. All of these have now been studied in connection with MES.

4.5.2.1. GABA-T-INHIBITORS ON MES. Inhibition of GABA: α-Ketaglutarate transaminase (GABA-T), the enzyme that catalyzes the conversion of GABA to succinic semialdehyde has long been used to increase GABA levels in brain. One difficulty with the use of GABA-T inhibitors is that they are rarely specific for GABA-T. For example, many of the well-known GABA-T inhibitors, such as aminooxyacetic acid (AOAA), are carbonyl trapping agents and therefore inhibit all pyridoxal-dependent enzymes, including GAD to some extent. Aminooxyacetic acid, however, apparently inhibits GABA-T to a much greater extent than GAD, and therefore can be used to facilitate GABAergic neurotransmission. More recently a new group of GABA-T inhibitors that inhibit the enzyme irreversibly has been developed. These are referred to as catalytic enzyme inhibitors since they are inactive when administered, but when acted on by GABA-T form highly reactive intermediates that then bind to the enzyme irreversibly (Enna and Maggi, 1979). Two of these irreversible GABA-T inhibitors, γ-vinyl GABA, and γ-acetylenic GABA, have been widely used to study experimentally induced seizures.

Kuriyama et al. (1966) found that AOAA inhibits HLE in mice subjected to MES with peak seizure protection occurring at 1.5 h after AOAA administration. However, the increase in brain GABA did not peak until 6 h after AOAA administration, by which time the effect on seizures had disappeared. Subsequent studies by Wood and Peesker (1975; 1976) have suggested that AOAA may inhibit MES in mice at 1.5 h after injection through a nonGABAergic mechanism. This drug also inhibits seizures produced by convulsants through a GABA-dependent mechanism that peaks 6 h after administration.

In contrast to Kuriyama et al. (1966), Loscher and Frey (1978) found an increase in the MES threshold in mice 6 hours after the administration of AOAA. The latter investigators also reported a good correlation between the time course for changes in GABA content and the elevation of MES threshold. Although the reason for the disparity in findings between Kuriyama et al. (1966) and Loscher and Frey (1978) is not known, it may be related to the fact that Loscher and Frey measured the MES threshold, which is apparently more easily influenced by GABA (Worms and Lloyd, 1981; Loscher, 1982) than is the MES pattern.

Others have found that several GABA-T inhibitors, such as γ-vinyl GABA, γ-acetylenic GABA, gabaculine, ethanolamine-O-sulfate, and valproic acid are capable of elevating the threshold for MES in mice (*see* Schecter et al., 1977; Gale and Iadarola, 1980; Loscher, 1981; 1982 for details on specific agents). Moreover, Gale and Iadarola (1980) found that γ-vinyl GABA blocks HLE in rats subjected to MES with a time course that exactly parallels the increase in nerve terminal GABA, rather than the time course for the increased total tissue GABA content. A similar correlation between the increase in nerve terminal GABA and the suppression of HLE was also reported for valproic acid and AOAA (Gale and Iadarola, 1980). Loscher (1981), on the other hand, failed to observe a good correlation between the increase in synaptosomal (nerve ending) GABA and elevations in the MES threshold in mice, although protection from PTZ-induced seizures was found to correlate well with synaptosomal GABA concentration. Perhaps species differences can account for this apparent discrepancy between Loscher (1981) and Gale and Iadarola (1980). Nevertheless, it is clear that GABA-T inhibitors exhibit significant anticonvulsant action in the maximal electroshock test. In comparing effectiveness of GABA-T inhibitors for elevating the MES threshold in mice, it appears that AOAA is most effective, followed by GABAculine and γ-acetylenic GABA (Loscher, 1982). The rather weak effects of γ-vinyl GABA (high doses required) and ethanolamine-O-sulfate are apparently related to their poor penetration of the blood–brain barrier.

The question of whether or not spinal cord GABA plays a role in the GABA-mediated suppression of electroshock-induced convulsions was addressed by Jobe and coworkers (1980) using GABA-T inhibitors. These investigators found that although both γ-vinyl GABA and γ-acetylenic GABA effectively elevated spinal cord GABA levels, γ-acetylenic GABA produced anticonvulsant effects, but γ-vinyl GABA did not. Moreover, the anticonvulsant effect of γ-acetylenic GABA failed to correlate temporally with the increase in spinal cord GABA. It was suggested, therefore, that spinal cord GABAergic neurons do not play an important role in regulating electrically induced seizures. However, it will be necessary to study the effect of spinal cord GABA manipulations in conjunction with minimal and maximal electroshock and to correlate these with changes in nerve ending GABA before definitive conclusions are reached regarding the role of spinal cord GABA.

4.5.2.2. GABA REUPTAKE INHIBITORS. Inhibition of GABA uptake would be expected to block the primary mechanism for terminating the action of GABA in the synaptic cleft and thereby prolonging its action. In their 1975 review, Maynert et al. concluded that the search for new anticonvulsant drugs should be focused on the development of GABA uptake inhibitors. Although a number of fairly selective GABA uptake inhibitors have been identified from in vitro studies, these were found to be highly polar compounds and unable to penetrate the brain after peripheral administration. This difficulty was overcome by Frey et al. (1979) using the methyl and ethyl esters of several uptake inhibitors (nipecotic acid, *cis*-4-hydroxynipecotic acid, and guavacine). These esters can apparently enter the brain, where they are hydrolyzed to the free uptake inhibitors. Frey et al. (1979) have tested these drugs on the MES and PTZ thresholds in mice. The ethyl ester of (−)-nipecotic acid (a potent inhibitor of high affinity GABA uptake) was found to significantly elevate the MES and PTZ thresholds in mice. The methyl ester of (±)-nipecotic acid also elevated the MES threshold in mice, but it was less potent than (−)-nipecotic acid. The methyl ester of (±)-*cis*-4-hydroxynipecotic acid elevated only the MES threshold, whereas guavacine methyl ester and (+)-nipecotic acid failed to affect the MES threshold, but elevated the PTZ threshold.

One of the major problems with the use of esters of GABA uptake inhibitors is that they all caused pronounced symptoms of peripheral cholinergic stimulation, including tremor, salivation, lacrimation, and defecation (Frey et al., 1979). Furthermore, the cholinergic stimulation produced by both esters of guavacine and the methyl ester of (±)-nipecotic acid was so strong that the anticonvulsant testing had to be performed after pretreatment with atropine. In the case of the other drugs, the signs of cholinergic stimulation had disappeared before seizure testing was performed. Although these findings suggest that GABA uptake inhibitors afford protection against MES in mice, interpretation of these results are obviously confounded by their effects on cholinergic systems and the knowledge that ACh may also suppress tonic seizures (see above).

4.5.2.3. GABA-AGONISTS. The third approach to facilitating GABA neurotransmission is to administer GABA receptor agonists that can directly activate the GABA–receptor complex. Although GABA itself cannot readily penetrate the CNS, the lipophilic cetyl derivative of GABA (cetyl GABA) does reach the brain, where it is hydrolyzed to increase free GABA levels. Cetyl GABA, along with several other compounds that have been shown to function as GABA agonists, such as muscimol, THIP (Enna and Maggi, 1979), and progabide (Lloyd et al., 1982) have now been tested as anticonvulsants in the MES and PTZ seizure models (Frey and Loscher, 1980; Worms and Lloyd, 1981; Loscher, 1982; Worms et al., 1982). Except for THIP, all of these agonists have been found to raise the MES threshold as well as the PTZ threshold in mice (Loscher 1982; Worms et al., 1982; Worms and Lloyd, 1981) and to protect rats or mice from a wide variety of other seizures (Worms and Lloyd, 1981; Worms et al., 1982). THIP was, however, effective against PTZ seizures (Worms and Lloyd, 1981; Loscher, 1982). Whether any of these agonists can be developed as effective antiepileptic agents remains to be determined. One drawback is their pronounced neurotoxicity. Indeed, in a comparative study conducted by Loscher (1982), only progabide and cetyl GABA were found to exhibit a margin of safety similar to valproic acid, a currently used antiepileptic drug.

It would appear from the above evidence that enhancing GABA neurotransmission in the CNS does elevate the MES threshold in mice. However, it has been suggested that the modification of the MES pattern following supramaximal electroshock is more resistant to modification by GABAergic drugs than is the MES threshold (Worms and Lloyd, 1981; Loscher, 1982). Nevertheless, it is clear that GABA-T inhibitors can also protect rats against supramaximal electroshock-induced seizures (Gale and Iadarola, 1980). In comparing the effects of GABA-T inhibitors, GABA uptake blockers, and GABA agonists in their ability to protect against electroshock and PTZ seizures, Loscher (1982) has concluded that although all of them can elevate the convulsive thresholds, except for progabide and ($\pm$)-*cis*-nipecotic acid methyl ester, they are more potent against PTZ-induced convulsions than against those produced by electroshock. However, at this point we must conclude that GABA seems able to suppress both clonic and tonic seizures (perhaps equally well).

4.5.3. Site of Anticonvulsant Effect of GABA

Of all the neurotransmitters discussed in this chapter and probably in this book, GABA is the only one for which a site of action has been identified in the brain. Iadarola and Gale (1982) found that microinjections of γ-vinyl GABA or muscimol into the substantia nigra abolish HLE in the MES test and also reduce the intensity of convulsions produced by bicuculline and PTZ. Microinjections of γ-vinyl GABA into the forebrain or other areas of the brainstem failed to suppress MES unless sufficient time was allowed for it to diffuse to the substantia nigra (Iadarola and Gale, 1982).

Microinjections of GABA agonists or GABA-T-inhibitors into the substantia nigra have also been shown to suppress amygdala-kindled seizures (Le Gal La Salle, et al., 1983; McNamara et al., 1983). In addition, intranigral injections of muscimol appear to protect adult rats against flurothyl seizures, although facilitating these seizures in 15-d-old rat pups (Moshe and Albala, 1984). The studies examining the effect of intranigral GABAmimetic drugs basically support studies involving the systemic administration of GABAergic drugs and suggest that GABA affects both clonic and tonic seizures. Moreover, these studies suggest that the anticonvulsant action of GABAergic drugs is exerted within the substantia nigra. Although the effect of intranigral injections of GABAmimetic drugs on minimal electroshock seizures has not been examined, it seems likely that these kinds of seizures would also be inhibited by such injections.

4.6. Opiate Peptides

Although we may expect to see a great deal of new information concerning peptide neurotransmitters and seizures in the future, only the opiate-like peptides (enkephalins and endorphins) have thus far been extensively investigated in connection with seizures in general and in relation to electroshock seizures in particular. Although some may question a true neurotransmitter role for the enkephalins and endorphins, few would deny that these neuropeptides function as neuromodulators (Krieger, 1983).

Except for the opiate antagonists (naloxone, naltrexone), all of the pharmacological treatments used to study the role of opiate peptides in seizures have involved the administration of agonists (morphine, β-endorphins, D-ala-met-enkephalinamide) to facilitate neurotransmission. This is because there are as yet no known methods for lowering the levels of endogenous opiates. Indeed, in comparison with the biogenic amine neurotransmitters, the techniques for manipulating the synaptic concentration of opiate peptides are quite limited. Nevertheless, it is clear that these peptides and morphine have rather powerful effects on seizures.

4.6.1. Opiates in Generalized Seizures

Morphine and the endogenous opiates have clearly been shown to be facilitators of seizures under some conditions, and inhibitors of seizures under others. Indeed, the proconvulsant and anticonvulsant action of morphine and endogenous opiates has recently been reviewed by Frenk (1983), who provides some order to what otherwise appears to be a totally confusing field. Frenk, who has worked extensively in this field, concludes that there are three proconvulsant systems in the CNS on which opiates act, one of which utilizes μ receptors, one of which utilizes δ receptors, and one of which does not involve specific receptors. In addition, Frenk (1983) suggests that there is an anticonvulsant system activated by opiates, which utilizes μ receptors. However, the existence of all these systems is highly speculative, and may represent an overly complicated interpretation of the available data.

If we examine the available literature in light of the model we proposed earlier (i.e., distinct forebrain and brainstem seizure-driving systems), an interesting pattern emerges. Opiates, like muscarinic agonists (see above), appear to facilitate and/or induce forebrain seizures, while inhibiting the brainstem seizure-facilitating system. The evidence that opiates enhance seizure activity in the forebrain of rats is now overwhelming. Intracerebroventricular injections of morphine or opiate peptides in low doses have uniformly been shown to produce electrographic seizure activity in limbic structures (Urca et al., 1977; Henriksen et al., 1978; Frenk et al., 1978; *see* Frenk, 1983 for additional references). Whereas these seizures are not associated with generalized convulsive (motor) activity, they are frequently accompanied by myoclonic twitches and wet dog shakes (Frenk, 1983). Many investigators have implicated the hippocampus as the site where seizure activity is initiated following icv administration of opiates, but epileptiform activity has also been observed after injections of morphine or enkephalins into the thalamus, amygdala, and caudate nucleus (Frenk, 1983). Limbic seizures following the icv administration of opiates appear to be mediated by opiate receptors, since they are blocked by naloxone. The specific opiate receptor involved in seizure facilitation has not been identified, and there is some evidence that different opiate receptor subtypes mediate different seizure patterns (Snead and Bearden, 1982). Although icv administration of morphine usually fails to produce generalized convulsive behavior after single doses, this compound can function as a kindling agent when given repeatedly (Snead, 1983), resulting in fully kindled motor convulsions. Kindling has also been achieved with microinjections of methionine enkephalin into the posterior amygdala or injections of β-endorphin into the posterior amygdala or ventral hippocampus (Cain and Corcoran, 1984). Moreover, the seizures produced by icv administration of morphine are similar to those produced by low doses of icv carbachol in the rat (Snead, 1983), again emphasizing the similarity with the forebrain seizures elicited by cholinergic agonists. Some differences were noted between the seizures produced by icv morphine and icv leucine-enkephalin, suggesting that different opiate receptor subtypes may be involved in the action of these two compounds (Snead, 1983). High (toxic) doses of morphine (300 mg/kg, ip) are also known to cause generalized convulsions, but this appears to be a nonspecific effect, and seems to be mediated through an action on the cortex and spinal cord (Frenk et al., 1984a, b).

Morphine has also been reported to display proconvulsant effects when subconvulsant doses are given in conjunction with various chemical convulsants such as PTZ. In the mouse, morphine is clearly a facilitator of PTZ-induced convulsions (Mannino and Wolf, 1974). However, in the rat several studies suggested that morphine and opiate peptides produced anticonvulsant effects on PTZ and flurothyl seizures (Adler et al., 1976; Cowan et al., 1981; Frenk, 1983). At first glance this would appear to be incompatible with the hypothesis that opiates facilitate forebrain (clonic) seizures, since both PTZ and flurothyl seizures appear to begin with clonus and have been categorized in our model as forebrain-type seizures at their onset (Browning, 1985). However, in all studies in which opiates were reported as anticonvulsant, the investigators measured the onset latency to the motor convulsion following the administration of flurothyl or PTZ. In some studies in which additional information about the seizure is provided, it appears that although the onset is prolonged, the seizure itself is more severe, and the incidence of death is increased (Urca and Frenk, 1980; Wallenstein, 1983). Indeed, Wallenstein (1983) found that although morphine increased the onset latency for PTZ seizures in rats, it facilitated these seizures by: (1) causing a greater number of seizures, (2) causing a greater percentage of deaths from seizures, and (3) lowering the dose of PTZ needed to produce generalized convulsive seizures. So that even in the rat, morphine seems to facilitate PTZ seizures, although they take longer to begin. The reason for the delayed onset is not known, but it is possible that opiates decrease blood flow to the brain structures involved in seizure initiation, therefore reducing the rate at which the convulsant reaches these areas. It is also possible that morphine enhances seizure discharge in limbic structures, while raising the threshold for paroxysmal discharge in those systems required for motor expression. However, once this threshold is reached and discharge begins, the convulsions are more severe because of the enhanced activity in the limbic forebrain. Nevertheless, these findings show that even in the rat opiates facilitate chemically induced forebrain (clonic) seizures. Morphine has also been found to facilitate convulsions induced by GABA antagonists in rats (Foote and Gale, 1983). On the other hand, the studies on the maximal electroshock seizure reveal that opiates have a clear-cut anticonvulsant effect.

4.6.2. Opiates in Electroshock Induced Seizures

Few laboratories have examined the effect of opiates on electroshock seizures, and all but one of these looked at MES exclusively. Based on the foregoing discussion, we would predict that morphine facilitates clonic seizures induced by minimal electroshock, but it was found to have no effect in this seizure test over a wide range of doses (Tripathi and Singh, 1978). However, there is complete concordance from four different laboratories showing that systemically administered morphine abolishes tonic HLE in a naloxone reversible fashion in mice (Chen et al., 1968) and rats (Tripathi and Singh, 1978; Berman and Adler, 1981; Foote and Gale, 1983) subjected to MES. More recently, icv-administered morphine, β-endorphin, or (D-ala-D-Leu) enkephalin were found to inhibit MES in mice (Puglisi-Allegra et al., 1984). These findings, taken with the evidence that opiates induce seizure discharge in forebrain structures, are consistent with the notion that the forebrain and brainstem seizure-driving systems are distinct from one another. Furthermore it appears that morphine may suppress tonic seizures by an action in the brainstem, since microinjections of morphine and enkephalins into the ventral midbrain tegmentum (in the region of the substantia nigra) were found to confer protection against MES (Garant et al., 1982).

5. Summary

Electroshock has long been employed as a technique for producing experimental seizures in animals. It is now clear that many different types of seizures can be elicited with electroshock, depending on the frequency and intensity of the current employed, as well as the location of the electrodes on the head. Using corneal electrodes with the standard 60-Hz stimulus for 0.2 s, one can produce any one of three motor patterns, depending on the magnitude of the current employed. Low currents (20–30 mA in the rat) produce a clonus that resembles kindled seizures, and we have proposed that these represent a paroxysmal discharge emanating from forebrain structures. Slightly higher currents (35–45 mA) produce either running or bouncing clonus, which differs from clonus produced by lower currents. Finally, at still higher currents the tonic threshold is reached and tonic flexion with or without HLE occurs.

The bouncing clonus and tonus are believed to originate from structures in the brainstem. Most convulsive patterns can also be elicited by direct electrical stimulation of the spinal cord, indicating this structure is the ultimate determinant of the motor pattern. These observations suggest a role for the spinal cord in all electroshock seizures. When ear-clip electrodes are used in place of corneal electrodes, it is possible to produce only the brainstem and/or spinal cord type of seizures. The low-frequency electroshock seizures (produced with 6-Hz stimuli) probably represent a paroxysmal discharge in forebrain limbic structures without generalized convulsive seizures, much like one sees following icv administration of opiates. However, the role of neurotransmitters in the low-frequency EST test has not been well studied.

We have reviewed the literature dealing with the importance of neurotransmitters in electroshock-induced seizures and attempted to categorize and interpret it in light of the above hypothesis concerning the different neural substrates underlying a particular type of electroshock seizure (*see* Table 1). Thus, neurotransmitters influencing minimal electroshock seizures induced by corneal electrodes presumably are acting on the forebrain seizure-driving system, whereas those that alter MES are believed to exert their predominant effect on the brainstem.

Acetylcholine appears to have a proconvulsant effect on the forebrain-driven clonic seizure (ac-EST), while having anticonvulsant effects on tonic (MES, brainstem) seizures in rats. This may, however, be species-specific, since poor anticonvulsant effects were observed in mice. Based on a wide variety of investigations that have examined the role of norepinephrine (NE) in seizures, it appears that NE is inhibitory to both forebrain (ac-EST) and brainstem (MES) seizures. Moreover, NE is also antagonistic to spinal cord seizures. Thus, NE appears to exert anticonvulsant effects at all levels of the neuroaxis.

At the present time we must conclude that the role of dopamine in electroshock seizures is controversial, and may be species-dependent. Although dopamine has not been well studied in connection with electroshock-induced clonus, several investigators have assessed its importance in electrically induced tonic seizures (MES). It appears that dopamine may suppress MES in rabbits and chickens, although having little or no effect in rats and mice. Any conclusions concerning rats and mice, however, are weakened by reports of conflicting observations. On the other hand, the lack of consistent observations with diverse pharmacological approaches

TABLE 1
Effect of Neurotransmitters on Electrically Induced Seizures[a]

Transmitters	lf-EST	ac-EST	MES	Spinal cord seizures
Acetylcholine	Not studied	Facilitates seizure	Inhibits in rats; Little effect in mice	Not studied
Norepinephrine	(AMPT, no effect)	Inhibits	Inhibits	Inhibits
Dopamine		Inhibits(?)[b]	No effect(?)	No effect
5-Hydroxytryptamine	(PCPA, No effect)	No effect in rats; Inhibition in mice(?)	Inhibits	No effect
GABA	Not studied	Inhibition(?)	Inhibits	No effect(?)
Opiate peptides	Not studied	No effect(?)	Inhibits	Not studied

[a]Abbreviations: AMPT, α-methyl-p-tyrosine; PCPA, parachlorophenylalanine; lf-EST, low-frequency electroshock threshold ac-EST, minimal electroshock threshold; MES, maximal electroshock threshold.

[b](?), Indicates that this has not been well studied and available evidence is inconclusive.

indicates that dopamine does not play an important role in electroshock-induced seizures in rodents.

5-Hydroxytryptamine (5-HT) does not affect the ac-EST in rats, but may exert a weak anticonvulsant effect in mice, although more studies should be conducted in mice to discern whether this is real. In contrast to its lack of effect on ac-EST, 5-HT has uniformly been shown to exert an anticonvulsant effect on MES in both rats and mice. Thus, 5-HT like ACh appears to exert differential effects on the ac-EST (forebrain seizures) and MES (brainstem seizures). Based on these findings and studies on spinal cord seizures, it appears that the anticonvulsant effect of 5-HT is localized to the brainstem and/or cerebellum.

GABA appears to be anticonvulsant in all electroshock models, although it has not been widely studied in connection with the EST test or electrically induced spinal cord seizures. Studies employing three classes of pharmacological agents designed to elevate the synaptic concentration of GABA consistently showed that GABA elevates the MES threshold in mice, although it appears to be more effective in elevating the threshold than in altering the MES seizure pattern. Studies in rats suggest that GABA can also attenuate the MES pattern in this species. GABA represents the only neurotransmitter for which a site of action for attenuating electroshock seizures has been localized to a specific nucleus. Several studies now implicate the substantia nigra as a site where GABA acts to inhibit MES as well as other kinds of seizures. Although excitatory amino acid neurotransmitters undoubtedly play a role in seizures, their role in electroshock seizures has not been sufficiently studied to warrant a discussion in this review.

Opiate peptides, like ACh and 5-HT, appear to exert differential effects on forebrain (ac-EST)- and brainstem (MES)-seizure systems. Morphine and opiate peptides clearly produce seizures in forebrain (limbic) structures when administered icv. Therefore, one would predict that they should lower the ac-EST. Although only one laboratory has looked at this, no effect on EST was found. Nevertheless, opiates do facilitate clonic convulsions produced by chemical agents and, like ACh, can themselves function as kindling agents. On the other hand, morphine has consistently been found to abolish HLE in the MES test in rats. Moreover, there is some evidence that this anticonvulsant effect is mediated by an action in the ventral midbrain tegmentum. A summary of the neurotransmitter effects on different electroshock models is provided in Table 1.

References

Adler, M. W., Lin, C. H., Keinath, S. H., Braverman, S. and Geller, E. B.: Anticonvulsant action of acute morphine administration in rats. *J. Pharmacol. Exp. Ther.* **198:** 655–660, 1976.

Albertoni, P.: Untersuchung uberdie Wirkuing einiger Arzneimittel auf die Erregbarkeit des Grosshirns nebst Beitragen zur Therapie der Epilepsie. *Arch. Exp. Pathol. Pharmacol.* **15:** 248–288, 1882.

Azzaro, A. J., Wenger, G. P., Craig, C. R., and Stitzel, R. E.: Reserpine-induced alterations in brain amines and their relationship to changes in the incidence of minimal electroshock seizures in mice. *J. Pharmacol. Exp. Ther.* **180:** 558–568, 1972.

Baxter, C. F. and Roberts, E.: Demonstration of thiosemicarbazide-induced convulsions in rats with elevated brain levels of γ-aminobutyric acid. *Proc. Soc. Exp. Biol. Med.* **104:** 426–427, 1960.

Bergmann, F., Costin, A., and Gutman, J.: A low threshold convulsive area in the rabbit's mesencephalon. *Electroencephalogr. Clin. Neurophysiol.* **15:** 683–690, 1963.

Berman, E. F. and Adler, M. W.: Effect of opioids and β-endorphin on maximal electroconvulsive (MES) seizures in rats. *Fed. Proc.* **40:** 281, 1981.

Birdsall, N. J. M. and Hulme, E. C.: Muscarinic receptor subclasses. *Trends Pharmacol. Sci.* **4:** 459–463, 1983.

Bogue, J. Y. and Carrington, H. C.: The evaluation of "mysoline" — a new anticonvulsant drug. *Br. J. Pharmacol.* **8:** 230–236, 1953.

Browning, R. A.: Role of the brainstem reticular formation in tonic-clonic seizure: Lesion and pharmacological studies. *Fed. Proc.* **44**(8): 2425–2431, 1985.

Browning, R. A. and Engel, S.: Role of brainstem and spinal cord serotonin (5-HT) in maximal electroshock (MES)-induced seizures. *Fed. Proc.* **43:** 569, 1984.

Browning, R. A. and Maynert, E. W.: Increased seizure susceptibility in 6-hydroxydopamine treated rats. *Fed. Proc.* **29:** 966, 1970.

Browning, R. A. and Maynert, E. W.: Effect of intracisternal 6-hydroxydopamine on seizure susceptibility in rats. *Eur. J. Pharmacol.* **50:** 97–101, 1978a.

Browning, R. A. and Maynert, E. W.: Effects of intraventricularly administered monoamines on seizure susceptibility and body temperature in rats. *Neuropharmacology* **17:** 649–653, 1978b.

Browning, R. A. and Nelson, D. K.: Variation in threshold and pattern of electroshock-induced seizures in rats depending on site of stimulation. *Life Sci.* **37:** 2205–2211, 1985.

Browning, R. A. and Simonton, R. L.: Antagonism of the anticonvulsant action of phenytoin, phenobarbital and acetazolamide by 6-hydroxy-dopamine. *Life Sci.* **22:** 1921–1930, 1978.

Browning, R. A., Nelson, D. K., Mogharreben, N., Jobe, P. C., and Laird, H. E.: Effect of midbrain and pontine tegmental lesions on audiogenic seizures in genetically epilepsy-prone rats. *Epilepsia* **26**(2): 175–183, 1985.

Browning, R. A., Hoffman, W. E., and Simonton, R. L.: Changes in seizure susceptibility after intracerebral treatment with 5,7-dihydroxytryptamine: Role of serotonergic neurons. *Ann. NY Acad. Sci.* **305:** 437–456, 1978.

Browning, R. A., Turner, F. J., Simonton, R. L., and Bundman, M. C.: Effect of midbrain and pontine tegmental lesions on the maximal electroshock seizure pattern in rats. *Epilepsia* **22:** 583–594, 1981a.

Browning, R. A., Simonton, R. L., and Turner, F. J.: Antagonism of experimentally-induced tonic seizures following a lesion of the midbrain tegmentum. *Epilepsia* **22:** 595–601, 1981b.

Burley, E. S. and Ferrendelli, J. A.: Regulatory effects of neurotransmitters on electroshock and pentylenetetrazol seizures. *Fed. Proc.* **43:** 2521–2524, 1984.

Buus Lassen, J.: Potent and long-lasting potentiation of two 5-hydroxy-tryptophan-induced effects in mice by three selective 5-HT uptake inhibitors. *Eur. J. Pharmacol.* **47:** 351–358, 1978.

Buterbaugh, G. G.: A role for central serotonergic systems in the pattern and intensity of the convulsive response of rats to electroshock. *Neuropharmacology* **16:** 707–719, 1977.

Buterbaugh, G. G.: Effect of drugs modifying central serotonergic function on the response of extensor and nonextensor rats to maximal electroshock. *Life Sci.* **23:** 293–404, 1978.

Buterbaugh, G. G. and London, E. D.: The relationship between magnitude of electroshock stimulation and the effect of digitoxigenin, pentylenetetrazol and brain monoamine reduction on electroshock convulsive thresholds. *Neuropharmacology* **16:** 617–623, 1977.

Cain, D. P. and Corcoran, M. E.: Intracerebral β-endorphin, met-enkephalin and morphine: Kindling of seizures and handling-induced potentiation of epileptiform effects. *Life Sci.* **34:** 2535–2542, 1984.

Chen, G., Ensor, C. R., and Bohner, B.: A facilitation of reserpine action on the central nervous system. *Proc. Soc. Exp. Biol. Med.* **86:** 507–510, 1954.

Chen, G., Ensor, C. R., and Bohner, B.: Studies of drug effects on electrically-induced extensor seizures and clinical implications. *Arch. Int. Pharmacodyn.* **172:** 183–218, 1968.

Chiu, P. and Burnham, W. M.: The effects of anticonvulsant drugs on convulsions triggered by direct stimulation of the brainstem. *Neuropharmacology* **21:** 355–359, 1982.

Cowan, A., Tortella, F. C., and Adler, M. W.: A comparison of the anticonvulsant effects of two systematically active enkephalin analogues in rats. *Eur. J. Pharmacol.* **71:** 117–121, 1981.

Crunelli, V., Bernasconi, S., and Samanin, R.: Evidence against serotonin involvement in the tonic component of electrically-induced convulsions and in carbamazepine anticonvulsant activity. *Psychopharmacology* **66:** 79–85, 1979.

Crunelli, V., Cervo, L., and Samanin, R.: Evidence for a Preferential Role of Central Noradrenergic Neurons in Electrically-Induced Convulsions and Activity of Various Anticonvulsants in the Rat, In: *Neurotransmitters, Seizures and Epilepsy* (P. L. Morselli, K. G. Lloyd, W. Loscher, B. Meldrum, and E. H. Reynolds, eds.) Raven, New York, 1981.

DeShaepdryver, A. F., Piette, Y., and Delaunois, A. L.: Brain amines and electroshock threshold. *Arch. Int. Pharmacodyn.* **140:** 350–367, 1962.

Enna, S. J. and Maggi, A.: Minireview: Biochemical pharmacology of gabaergic agonists. *Life Sci.* **24:** 1727–1738, 1979.

Esplin, D. W. and Freston, J. W.: Physiological and pharmacological analysis of spinal cord convulsions. *J. Pharmacol. Exp. Ther.* **130:** 68–80, 1960.

Feldberg, W.: Present views on the mode of acetylcholine in the central nervous system. *Physiol. Rev.* **25:** 596–642, 1945.

Feldberg, W. and Sherwood, S. I.: Injections of drugs into the lateral ventricle of the cat. *J. Physiol.* **123:** 148–167, 1954.

Foote, F. and Gale, K.: Morphine potentiates seizures induced by GABA antagonists and attenuates seizures induced by electroshock in the rat. *Eur. J. Pharmacol.* **95:** 259–264, 1983.

Frenk, H.: Pro- and anticonvulsant actions of morphine and the endogenous opioids: Involvement and interactions of multiple opiate and non-opiate systems. *Brain Res. Rev.* **6:** 197–210, 1983.

Frenk, H. Watkins, L. R., and Mayer, D. J.: Differential behavioral effects induced by intrathecal microinjections of opiates: Comparison of convulsive and cataleptic effects produced by morphine, methadone and D-ala-methionine-enkephalinamide. *Brain Res.* **299:** 31–42, 1984a.

Frenk, H., Watkins, L. R., Miller, J., and Mayer, D. J.: Nonspecific convulsions are induced by morphine but not D-ala-methionine-enkephalinamide at cortical sites. *Brain Res.* **299:** 51–59, 1984b.

Frenk, H., Urca, G., and Leibeskind, J. C.: Epileptic properties of leucine and methionine-enkephalin: Comparison with morphine and reversibility by naloxone. *Brain Res.* 327–337, 1978.

Frey, H. H. and Loscher, W.: Cetyl GABA: Effect on convulsant thresholds in mice and acute toxicity. *Neuropharmacology* **19:** 217–220, 1980.

Frey, H. H., Popp, C., and Loscher, W.: Influence of inhibitors of the high affinity GABA uptake on seizure thresholds in mice. *Neuropharmacology* **18:** 581–590, 1979.

Fritsch, G. and Hitzig, E.: Ueber die electrische Erregbarkeit des Grosshirns. *Arch. Anat. Physiol. Wissenschaftlichen Medizine.* **37:** 300–332, 1870.

Fukuda, T., Araki, Y., and Suenaga, N.: Inhibitory effects of 6-hydroxydopamine on the clonic convulsions induced by electroshock and decapitation. *Neuropharmacology* **14:** 579–583, 1975a.

Fukuda, T., Araki, Y., Takisita, S., and Koga, T.: Correlation between brain monoamine levels and postictal coma following electroshock. *Arch. Int. Parmacodyn.* **213:** 58–63, 1975b.

Gale, K. and Iadarola, M. J.: Seizure protection and increased nerveterminal GABA: Delayed effects of GABA transaminase inhibition. *Science* **208:** 288–291, 1980.

Garant, D. S., Iadarola, M. J., and Gale, K.: Pharmacologic manipulations of peptide-mediated transmission in rat substantia nigra: Anticonvulsant effects. *Soc. Neurosci. Abst.* **8:** 986, 1982.

Gray, W. D. and Rauh, C. E.: The relation between monoamines in brain and the anticonvulsant action of inhibitors of carbonic anhydrase. *J. Pharmacol. Exp. Ther.* **177:** 206–218, 1971.

Guerrero-Figueroa, R., Verster, F., DeBarros, A., and Heath, R. G.: Cholinergic mechanisms in subcortical mirror focus and effects of topical application of γ-aminobutyric acid and acetylcholine. *Epilepsia* **5:** 140–155, 1964.

Gulati, O. D. and Stanton, H. C.: Some effects on the central nervous system of gamma-amino-*N*-butyric acid (GABA) and certain related amino acids administered systemically and intracerebrally to mice. *J. Pharmacol. Exp. Ther.* **129:** 178–185, 1960.

Gyorgy, L.: Role of dopaminergic and gaba-ergic interactions in seizure susceptibility. *Arch. Int. Pharmacodyn.* **241:** 280–286, 1983.

Haley, T. J. and McCormick, W. G.: Pharmacological effects produced by intracerebral injection of drugs in the conscious mouse. *Br. J. Pharmacol.* **12:** 12–15, 1957.

Hatoum, H. T.: The role of acetylcholine in electroshock seizure in the rat (Masters thesis). Shreveport, Louisiana, Northeast Louisiana University, 1974.

Henriksen, S. J., Bloom, F. E., McCoy, F., Ling, N., and Guillemin, R.: Beta-Endorphin induces nonconvulsive limbic seizures. *Proc. Natl. Acad. Sci. USA* **75:** 5221–5225, 1978.

Hokfelt, T., Ljungdahl, A., Perez De La Mora, M., and Fuxe, K.: Further evidence that apomorphine increases GABA turnover in the DA cell body rich and DA nerve terminal areas of brain. *Neurosci. Lett.* **2:** 239–242, 1976.

Hokfelt, T., Lundberg, J. M., Schultzberg, M., Johansson, O., Ljungdahl, A., and Rehfeld, J.: Co-existence of Peptides and Putative Transmitters in Neurons, In: *Neural Peptides and Neuronal Communications* (E. Costa and M. Trabucchi, eds.) Raven, New York, 1980.

Iadarola, M. J. and Gale, K.: Substantia nigra: Site of anticonvulsant activity mediated by γ-aminobutyric acid. *Science* **218:** 1237–1240, 1982.

Jobe, P. C.: Pharmacology of Audiogenic Seizures, In: *Pharmacology of Hearing: Experimental and Clinical Bases* (R. P. Brown and E. A. Daigneault, eds.) Wiley Interscience, New York, 1981.

Jobe, P. C. and Laird, H. E.: Neurotransmitter abnormalities as determinants of seizure susceptibility and intensity in the genetic models of epilepsy. *Biochem. Pharmacol.* **30:** 3137–3144, 1981.

Jobe, P. C. and Ray, T.: The effects of reserpine iproniazid and L-Dopa on electrically-induced spinal cord seizures. *Res. Comm. Chem. Pathol. Pharmacol.* **29:** 417–427, 1980a.

Jobe, P. C. and Ray, T.: Effects of norepinephrine and 5-hydroxytryptamine reuptake inhibitors on electrically-induced spinal cord seizures in rats. *Res. Comm. Chem. Pathol. Pharmacol.* **30:** 185–188, 1980b.

Jobe, P. C., Graham, L. T., Ray, T. B., and Mims, M. E.: Effects of γ-acetylenic GABA and γ-vinyl GABA on electrically-induced spinal cord convulsions and on spinal cord GABA concentration. *Life Sci.* **27:** 2005–2011, 1980.

Jobe, P. C., Stull, R. E., and Geiger, P. F.: The relative significance of norepinephrine, dopamine and 5-hydroxytryptamine in electroshock seizure in the rat. *Neuropharmacology* **13:** 961–968, 1974.

Jobe, P. C., Geiger, P. F., Ray, T. B., Woods, T. W., and Mims, M. E.: The relative significance of spinal cord norepinephrine and 5-hydroxytryptamine in electrically-induced seizure in the rat. *Neuropharmacology* **17:** 185–190, 1978.

Kalichman, M. W.: Neurochemical correlates of the kindling model of epilepsy. *Neurosci. Biobehav. Rev.* **6:** 165–181, 1982.

Karczmar, A.: Brain Acetylcholine and Seizures, In: *Psychobiology of Convulsive Therapy* (M. Fink, S. Kety, J. M. McGaugh, and T. A. Williams, eds.) Winston and Sons, Washington, DC, 1974.

Kilian, M. and Frey, H. H.: Central monoamines and convulsive thresholds in mice and rats. *Neuropharmacology* **12:** 681–692, 1973.

Killam, K. F. and Bain, J. A.: Convulsant hydrazides. I. *In vitro* and *in vivo* inhibition of vitamin B$_6$ by convulsant hydrazides. *J. Pharmacol. Exp. Ther.* **119:** 255–262, 1957.

Kleinrok, F., Czuczwar, S., Wojak, A., and Prezegalinski, E.: Brain dopamine and seizure susceptibility in mice. *Pol. J. Pharmacol. Pharm.* **30:** 513–519, 1978.

Kreindler, A., Zuckermann, E., Steriade, M., and Chimion, J.: Electroclinical features of convulsions induced by stimulation of brainstem. *J. Neurophysiol.* **21:** 430–436, 1958.

Krieger, D. T.: Brain peptides: What, Where and Why? *Science* **222:** 975–985, 1983.

Kuriyama, K., Roberts, E., and Rubinstein, M. K.: Elevation of γ-aminobutyric acid in brain with amino-oxyacetic acid and susceptibility to convulsive seizures in mice: A quantitative re-evaluation. *Biochem. Pharmacol.* **15:** 221–236, 1966.

Lange, S. C., Julien, R. M., and Fowler, G. W.: Biphasic effects of imipramine in experimental models of epilepsy. *Epilepsia* **17:** 183–196, 1976.

Lazarova, M. and Roussinov, K.: On some relationships between gamma-aminobutyric acid (GABA) and the cholinergic mechanisms in electric convulsions. *Acta Physiol. Pharmacol. Bulg.* **3:** 37–43, 1977.

Lee, E. H. Y. and Geyer, M. A.: Indirect effects of apomorphine on serotoninergic neurons in rats. *Neuroscience* **11:** 437–442, 1984.

Le Gal La Salle, G., Kaijima, M., and Felblum, S.: Abortive amygdaloid kindled seizures following microinjections of γ-vinyl-gaba in the vicinity of the substantia nigra. *Neurosci. Lett.* **36:** 69–74, 1983.

Lloyd, K. G., Arbilla, S., Beaumont, K., Briley, M., DeMontis, G., Scatton, B., Langer, S. Z., and Bartholini, G.: γ-Aminobutyric acid (GABA) receptor stimulation. II. Specifity of progabide (SL 76002) and SL 75102 for the GABA receptor. *J. Pharmacol. Exp. Ther.* **220:** 672–677, 1982.

London, E. D., Waller, S. B., Vocci, F. J., and Buterbaugh, G. G.: Age-dependent reduction in maximum electroshock convulsive threshold associated with decreased concentrations of brain monoamines. *Pharmacol. Biochem. Behav.* **16:** 441–447, 1982.

Loscher, W.: Relationship between drug-induced changes in seizure threshold and the GABA content of brain and brain nerve endings. *NS Arch. Pharmacol.* **317:** 131–134, 1981.

Loscher, W.: Comparative assay of anticonvulsant and toxic potencies of sixteen gaba-mimetic drugs. *Neuropharmacology* **21**: 803–810, 1982.

Loscher, W. and Frey, H. H.: Aminooxyacetic acid: Correlation between biochemical effects, anticonvulsant action and toxicity in mice. *Biochem. Pharmacol.* **27**: 103–108, 1978.

Mannino, R. A. and Wolf, H. H.: Opiate receptor phenomenon: Proconvulsant action of morphine in the mouse. *Life Sci.* **15**: 2089–2096, 1974.

Mason, S. T. and Corcoran, M. E.: Depletion of brain noradrenaline, but not dopamine, by intracerebral 6-hydroxydopamine potentiates convulsions induced by electroshock. *J. Pharm. Pharmacol.* **31**: 209–211, 1979.

Maynert, E. W.: The role of biochemical and neurohumoral factors in the laboratory evaluation of antiepileptic drugs. *Epilepsia* **10**: 145–162, 1969.

Maynert, E. W. and Kaji, A. K.: On the relationship of brain γ-aminobutyric acid to convulsions. *J. Pharmacol. Exp. Ther.* **137**: 114–121, 1962.

Maynert, E. W., Marczynski, T. J., and Browning, R. A.: The role of neurotransmitters in the epilepsies. *Adv. Neurology* **13**: 79–147, 1975.

McKenzie, G. M. and Soroko, F. E.: The effects of apomorphine, (+)-amphetamine and L-Dopa on maximal electroshock convulsions — a comparative study in the rat and mouse. *J. Pharm. Pharmacol.* **24**: 696–701, 1972.

McNamara, J. O., Rigsbee, L. C., and Galloway, M. T.: Evidence that substantia nigra is crucial to neural network of kindled seizures. *Eur. J. Pharmacol.* **86**: 485–486, 1983.

Meldrum, B. S.: Epilepsy and γ-aminobutyric acid-mediated inhibition. *Int. Rev. Neurobiol.* **17**: 1–36, 1975.

Millichap, J. G., Pitchford, G. L., and Millichap, M. G.: Anticonvulsant activity of antiparkinsonism agents. *Proc. Soc. Exp. Biol. Med.* **127**: 1187–1190, 1968.

Moshe, S. L. and Albala, B. J.: Nigral muscimol infusions facilitate the development of seizures in immature rats. *Dev. Brain Res.* **13**: 305–308, 1984.

Oishi, R. and Suenaga, N.: The role of the locus coeruleus in regulation of seizure susceptibility in rats. *Japan. J. Pharmacol.* **32**: 1075–1081, 1982.

Oishi, R., Suenaga, N., Hidaka, T., and Fukuda, T.: Inhibitory effect of intraspinal injection of 6-hydroxydopamine on the clonic convulsion in maximal electroshock seizure. *Brain Res.* **169**: 189–193, 1979.

Osuide, G., Wambebe, C., and Ngur, D.: Studies on the pharmacology of *d*-amphetamine on maximal electroconvulsive seizure in young chicks. *Psychopharmacology* **81**: 119–121, 1983.

Penfield, W. and Jaspers, H.: *Epilepsy and the Functional Anatomy of the Human Brain*. Little, Brown, Boston, 1954.

Prokop, D. J., Shore, P. A., and Brodie, B. B.: Anticonvulsant properties of monoamine oxidase inhibitors. *Ann. NY Acad. Sci.* **80**: 643–651, 1959.

Puglisi-Allegra, S., Castellano, C., Csanyl, V., Doka, A., and Oliverio, A.: Opioid antagonism of electroshock-induced seizures. *Pharmacol. Biochem. Behav.* **20**: 767–769, 1984.

Quattrone, A. and Samanin, R.: Decreased anticonvulsant activity of carba-mazepine in 6-hydroxydopamine-treated rats. *Eur. J. Pharmacol.* **41**: 333–336, 1977.

Quattrone, A., Crunelli, V., and Samanin, R.: Seizure susceptibility and anticonvulsant activity of carbamazepine, diphenylhydantoin and phenobarbital in rats with selective depletions of brain monoamines. *Neuropharmacology* **17**: 643–647, 1978.

Racine, R. J.: Modification of seizure activity by electrical stimulation II. Motor seizures. *Electroenceph. Clin. Neurophysiol.* **32**: 281–294, 1972.

Rudzik, A. D. and Johnson, G. A.: Effect of Amphetamine and Amphetamine Analogs on Convulsive Thresholds, In: *International Symposium on Amphetamines and Related Compounds* (E. Costa and S. Garattini, eds.) Raven, New York, 1970.

Schecter, P. J., Tranier, Y., Jung, M. J., and Sjoerdsma, A.: Antiseizure activity of gamma-acetylenic gamma-aminobutyric acid: A catalytic irreversible inhibitor of gamma-aminobutyric acid transaminase. *J. Pharmacol. Exp. Ther.* **201**: 606–612, 1977.

Simonton, S. L.: The relative importance of norepinephrine and dopamine as inhibitors of electroshock-induced seizures (Masters thesis). Carbondale, Illinois, Southern Illinois University, 1978.

Snead III, O. C.: Seizure induced by carbachol, morphine and leucine-enkephalin: A comparison. *Ann. Neurol.* **13**: 445–451, 1983.

Snead III, O. C. and Bearden, L. J.: The epileptogenic spectrum of opiate agonists. *Neuropharmacology* **21**: 1137–1144, 1982.

Stull, R. E., Jobe, P. C., Geiger, P. F., and Ferguson, G. G.: Effects of dopamine receptor stimulation and blockade on Ro 4-1284-induced enhancement of electroshock seizure. *J. Pharm. Pharmac.* **25**: 842–844, 1973.

Stull, R. E., Jobe, P. C., and Geiger, P. F.: Brain areas involved in the catecholamine mediated regulation of electroshock seizure intensity. *J. Pharm. Pharmac.* **29**: 8–11, 1977.

Suenaga, N., Oishi, R., and Fukuda, T.: The role of the locus coeruleus in decapitation convulsions of rats. *Brain Res.* **177**: 83–93, 1979.

Swinyard, E. A.: Laboratory assay of clinically effective antiepileptic drugs. *J. Am. Pharm. Assoc.* **38:** 201–204, 1949.

Swinyard, E. A.: Electrically Induced Convulsions, In: *Experimental Models of Epilepsy* (D. P. Purpura, J. K. Penry, D. M. Woodbury, D. B. Tower, and R. D. Walter, eds.) Raven, New York, 1972.

Swinyard, E. A.: Assay of Antiepileptic Drug Activity in Experimental Animals: Standard Tests, In: *International Encyclopedia of Pharmacology and Therapeutics* sect. 19, vol. 1 (J. Mercier, ed.) Pergamon, New York, 1973.

Swinyard, E. A.: Introduction: History of the Antiepileptic Drugs, In: *Antiepileptic Drugs: Mechanisms of Action* (G. H. Glaser, J. K. Penry, and D. M. Woodbury, eds.) Raven, New Y ork, 1980.

Swinyard, E. A., Brown, W. C., and Goodman, L. S.: Comparative assays of antiepileptic drugs in mice and rats. *J. Pharmacol. Exp. Ther.* **106:** 319–330, 1952.

Tanaka, K. and Mishima, O.: The localization of the center dealing with the tonic extensor seizure of electroshock. *Jap. J. Pharmacol.* **3:** 6–9, 1953.

Tapia, R.: Convulsions and the Functions of Gabaergic Synapses, In. *Neurochemistry and Clinical Neurology*, Alan R. Liss, New York, 1980.

Tedeschi, D. H., Swinyard, E. A., and Goodman, L. S.: Effects of variations in stimulus intensity on maximal electroshock seizure pattern, recovery time and anticonvulsant potency of phenobarbital in mice. *J. Pharmacol. Exp. Ther.* **116:** 107–113, 1956.

Toman, J. E. P., Swinyard, E. A., and Goodman, L. S.: Properties of maximal seizures and their alteration by anticonvulsant drugs and other agents. *J. Neurophysiol.* **9:** 231–240, 1946.

Tripathi, K. D. and Singh, R. B.: Effect of morphine on electroshock seizures in rats. *Ind. J. Exp. Bio.* **16:** 992–993, 1978.

Urca, G. and Frenk, H.: Pro- and anticonvulsant action of morphine in rats. *Pharmacol. Biochem. Behav.* **13:** 343–348, 1980.

Urca, G., Frenk, H., Liebeskind, J. C., and Taylor, A. N.: Morphine and enkephalin: Analgesic and epileptic properties. *Science* **197:** 83–86, 1977.

Wallenstein, M.: Effect of Morphine pretreatment on pentylenetetrazol-induced seizures in the rat. *Neuropharmacology* **22:** 1187–1192, 1983.

Waller, S. B. and Buterbaugh, G. G.: Tonic convulsive thresholds and responses during the postnatal development of rats administered 6-hydroxydopamine or 5,7-dihydroxytryptamine within three days following birth. *Pharmacol. Biochem. Behav.* **19:** 973–978, 1983.

Wasterlain, C. G. and Jonec, V.: Chemical kindling by muscarinic amygdaloid stimulation in the rat. *Brain Res.* **271:** 311–323, 1983.

Wenger, G. R., Stitzel, R. E., and Craig, C. R.: The role of biogenic amines in the reserpine-induced alteration of minimal electroshock seizure thresholds in the mouse. *Neuropharmacology* **12:** 693–703, 1973.

Wood, J. D.: The role of γ-aminobutyric acid in the mechanism of seizures. *Prog. Neurobiol.* **5:** 77–95, 1975.

Wood, J. D. and Peesker, S. J.: The anticonvulsant action of GABA-elevating agents: A re-evaluation. *J. Neurochem.* **25:** 277–282, 1975.

Wood, J. D. and Peesker, S. J.: A dual mechanism for the anticonvulsant action of aminooxyacetic acid. *Can. J. Physiol. Pharmacol.* **54:** 534–540, 1976.

Wood, J. D., Russell, M. P., and Kurylo, E.: The γ-aminobutyrate content of nerve endings (synaptosomes) in mice after intramuscular injection of γ-aminobutyrate-elevating agents: A possible role in anticonvulsant activity. *J. Neurochem.* **35:** 125–130, 1980.

Woodbury, D. M. and Esplin, D. W.: Neuropharmacology and neurochemistry of anticonvulsant drugs. *Res. Publ. Assoc. Nerv. Ment. Dis.* **37:** 24–56, 1959.

Woodbury, L. A. and Davenport, V. D.: Design and use of a new electroshock seizure apparatus and analysis of factors altering seizure thresholds and pattern. *Arch. Int. Pharmacodyn.* **92:** 97–107, 1952.

Worms, P. and Lloyd, K. G.: Functional Alterations of GABA Synapses in Relation to Seizures, In: *Neurotransmitters, Seizures and Epilepsy* (P. L. Morselli, K. G. Lloyd, W. Loscher, B. Meldrum, and E. H. Reynolds, eds.) Raven, New York, 1981.

Worms, P., Depoorteve, H., Durand, A., Morselli, P. L., Lloyd, K. G., and Bartholini, G.: γ-Aminobutyric acid (GABA) receptor stimulation. I. Neuropharmacological profiles of progabide (SL 76002) and SL75102, with emphasis on their anticonvulsant spectra. *J. Pharmacol. Exp. Ther.* **220:** 660–670, 1982.

Zablocka, B.: Effects of autonomic agents, alone and in combination with antiepileptic drugs, on electroshock seizures in rats. *Arch. Int. Pharmacodyn.* **142:** 533–538, 1963.

Zablocka, B. and Esplin, D. W.: Central excitatory and depressant effects of pilocarpine in rats and mice. *J. Pharmacol. Exp. Ther.* **140:** 162–169, 1963.

Zablocka, B. and Esplin, D. W.: Role of seizure spread in determining maximal convulsion pattern in rats. *Arch. Int. Pharmacodyn.* **147:** 525–542, 1964.

Neurotransmitters in Human Epilepsy

Mitchell J. Kresch, Bennett A. Shaywitz,
Sally E. Shaywitz, George M. Anderson,
James L. Leckman, and Donald J. Cohen

1. Introduction

Recently, it has become increasingly evident that a relationship
exists between monoaminergic neurotransmitters and seizure disorders. Other chapters in this volume have examined such findings
at the preclinical level in animal studies. This chapter is designed
to provide evidence linking seizures to disturbances in neurotransmitter function in the human. We provide, first, an overview of
the classification and clinical presentation of seizure disorders. We
next briefly review the pharmacological background of the neurotransmitter systems and discuss the strategies employed in human
investigations. Studies detailing the relationship between seizures
and each of the principal neurotransmitters are then described.

2. Classification of Human Epilepsy

At the outset, the reader should be aware of the terminology
employed. The word "epilepsy" is derived from a Greek word
meaning "to seize upon," and the terms seizure disorder and epilepsy are used interchangeably.

Epilepsy may be caused by a variety of insults to the nervous
system, including infection (such as encephalitis or meningitis);
trauma, resulting in intracranial bleeding; metabolic derangements,
such as hyperammonemia or amino aciduria; degenerative CNS

disease; cerebrovascular disease; or congenital anomalies. In the majority of cases, however, no etiology is apparent and we refer to the clinical problem as idiopathic epilepsy. Current nosology recognizes two principal kinds of seizures—generalized and partial (Wright et al., 1982).

2.1. Generalized Epilepsy

Generalized seizures occur when neuronal discharges involve widespread areas of both cerebral hemispheres. This is reflected in clinical changes indicative of bilateral hemisphere involvement and an electroencephalographic (EEG) pattern of bilateral discharges. Four types of generalized seizures are described below: absence, atypical absence, myoclonic, and generalized.

Absence seizures, sometimes referred to as petit mal seizures, are characterized by impairment of consciousness with or without convulsive movements. Classically, these absence seizures are associated with a momentary lapse of consciousness, during which the individual blankly stares ahead. Electroencephalographically, these are manifest as bilateral, synchronous 3–3.5-Hz bursts of spike and wave activity in both hemispheres. Occasionally, these spells have an associated mild clonic or tonic component. Some absence seizures, on the other hand, may have an atonic component, in which the patient abruptly loses all muscle tone, falling to the ground. Automatisms, repetitive stereotyped behavior, such as pacing back and forth or alternately sitting and standing, may occur with absence seizures (Wright et al., 1982).

Atypical absence seizures occur in patients who may suffer from mental retardation and developmental delay. These patients, most often children, may exhibit lapses in consciousness with features of complex partial as well as generalized tonic-clonic seizures. In addition, these spells may be associated with akinetic or atonic episodes. Atypical absence seizures are manifest electroencephalographically as bilateral, irregular 2–3-Hz bursts of activity, and this combination of clinical and EEG findings is referred to as the Lennox-Gastaut variant (Wright et al., 1982).

Myoclonic seizures are sudden, rapid, massive muscle contractions resulting in flailing of an extremity or, more classically, flexion of the body at the hips (jackknife or salaam seizures). During infancy, such seizures are known as infantile spasms and may occur hundreds of times in a 24-h period. When they are accompanied by a chaotic EEG pattern, termed "hypsarrythmia," they are nearly always associated with very poor neurological outcome.

Generalized tonic-clonic seizures, also called grand mal epilepsy, result in rhythmic contraction of muscle groups leading to convulsive movements. Occasionally, only one component is present, resulting in tonic convulsions (stiffening of muscle groups only) or clonic convulsions (rhythmic jerking movements).

2.2. Partial (Focal) Epilepsy

Partial or focal seizures (Wright et al., 1982) occur when initial neuronal discharges are localized in one area of a cerebral hemisphere, as evidenced by clinical or electroencephalographic changes.

Simple partial seizures involve motor or sensory cortical areas and are characterized by either focal convulsive movements or hallucinations (e.g., auditory, visual, olfactory), without impairment of consciousness. Psychic symptoms, such as dysmnesia (feeling of deja vu), may also occur.

When focal seizures are associated with impairment of consciousness, they are referred to as complex partial seizures. Simple partial seizures may evolve into complex partial seizures with or without automatisms, or complex partial seizures may occur at the onset with or without automatisms. Because they may originate in the temporal lobes, complex partial seizures are often described as temporal lobe, or psychomotor, epilepsy. Simple or complex partial seizures may evolve into generalized tonic-clonic seizures.

Given the diversity of clinical and electrophysiologic manifestations of seizures, it is not unreasonable to suggest that a variety of etiologies and pathogenetic mechanisms are responsible. If derangements in neurotransmitters influence seizures one would not be surprised to find that different neurotransmitters are involved in particular seizure types. Thus it is important to characterize seizures as precisely as possible in any study designed to examine the relationship between particular neurotransmitters and seizures.

3. Neurotransmitters

3.1. Indoleamines

The indoleamine, serotonin (5-hydroxytryptamine or 5-HT), is derived from the amino acid tryptophan. Synthesis proceeds via the enzyme tryptophan hydroxylase, the rate-limiting enzyme, yielding 5-hydroxytryptophan, 5-HTP. A nonspecific enzymatic reaction using *l*-aromatic amino acid decarboxylase catalyzes the decarboxylation of 5-HTP to serotonin (5-HT). After release from the presynaptic storage vesicles, serotonin is catabolized by monoamine oxidase to 5-hydroxyindole-acetic-acid (5-HIAA) (Bowers, 1970). Serotonin concentration in the CNS may be increased by providing the precursor 5-HTP, which readily passes through the blood–brain barrier, or by inhibiting the catabolism of 5-HT by administering an inhibitor of monoamine oxidase.

Histoflourescence studies indicate that almost all serotonergic neurons in the brain are located in the midline raphe nuclei within the reticular formation of the caudal mesencephalon, pons, and medulla (Morgane and Stern, 1974). Fibers from the raphe nuclei project via the medial longitudinal fasciculus and the medial forebrain bundle to the suprachiasmatic and periventricular nuclei of the hypothalamus, the hippocampus, the amygdala, the lateral geniculate nucleus of the thalamus, the cerebral cortex, and the spinal cord (Aghajanian et al., 1969; Jouvet, 1972; Morgane and Stern, 1974).

The raphe nuclei are felt to be involved in the generation of slow wave sleep (initial stages of sleep) (Jouvet, 1967; 1969; 1972). In addition, it has been suggested that although the more rostral raphe nuclei may regulate slow wave sleep, the more caudal raphe nuclei may be involved in priming the structures and mechanisms of the deeper stage of sleep, rapid eye movement (REM) sleep (Jouvet, 1969; 1972).

3.2. Catecholamines

Dopamine is synthesized from the precursor amino acids phenylalanine and/or tyrosine. Phenylalanine is hydroxylated to tyrosine, which, in turn, is hydroxylated by the rate-limiting enzyme, tyrosine hydroxylase (E.C. 1.14.16.2) to form dihydroxyphenylalanine (DOPA). DOPA is then converted by DOPA decarboxylase (E.C. 4.1.1.28) into dopamine. Dopamine, itself, inhibits tyrosine hydroxylase through negative feedback. Norepinephrine is formed from dopamine via the enzyme dopamine beta-hydroxylase (E.C. 1.14.17.1) (Moskowitz and Wurtman, 1975; Wurtman et al., 1974). The catecholamines are synthesized and stored in presynaptic vesicles and released during neural activity. Some of the active catecholamine undergoes reuptake into presynaptic vesicles. Catabolism of the remaining catecholamines proceeds via two principal enzymes, catechol-*O*-methyltransferase (COMT, E.C. 2.1.1.6) and monoamine oxidase (MAO, E.C. 1.4.3.4). The final product of dopamine metabolism is homovanillic acid, HVA, whereas the principal metabolite of norepinephrine in the brain appears to be 3-methoxy-4-hydroxyphenyl-glycol, MHPG. Since catabolism of the catecholoamines occurs both presynaptically, as well as after release, measurement of metabolites can be a poor index of synaptic activity (Moskowitz and Wurtman, 1975; Moir et al., 1970).

Dopaminergic cell bodies are located in substantia nigra, interpeduncular nucleus, and hypothalamus, and send projections via the nigro-neostriatal tract, limbic, and hypothalamic projections. Dopaminergic systems appear to be involved in both extrapyramidal functions as well as arousal (Bertler and Rosengran, 1959; Moskowitz and Wurtman, 1975).

Cells of origin for noradrenergic projections are located in the locus coeruleus on the floor of the fourth ventricle in the pontine tegmentum (Morgane and Stern, 1974; Walter and Eccleston, 1973). The locus coeruleus sends projections via the dorsal reticular formation to the mesencephalic tegmentum; some of these fibers course ventromedially to terminate in the medial hypothalamus, whereas others join the medial forebrain bundle and terminate in the lateral hypothalamus, amygdala, and limbic system (Loizou, 1969a; b). In addition, cerulocerebellar projections have been demonstrated (Moskowitz and Wurtman, 1975). The locus ceruleus has been implicated in the physiology of a variety of neuropsychiatric disorders, such as anxiety, affective disorders, and panic disorders, as well as normal physiologic processes such as sleep (Jouvet, 1972).

3.3. Ontogeny of Monoamines

Evidence from both human and animal studies suggests that serotonin levels in the CSF or brain increase from birth to early childhood (Shaywitz et al., 1980b). This increase has been correlated with an increased number of axon terminals and synaptic vesicles in rats (Loizou, 1969c; Loizou and Salt, 1970). Thus, increasing monoamine levels in the developing brain may reflect synaptogenesis and the development of neural connections in the brain.

In contrast, studies in both humans and animals have documented that brain and CSF concentrations of catecholamines and their metabolites decrease with advancing age after infancy. Such data suggest that the turnover of all three monoamines is increased in early life, but then decreases with maturation. Adult levels are probably attained in adolescence (Shaywitz et al., 1978; 1980b).

3.4. Amino Acids

Several amino acids act as putative neurotransmitters. Glutamate is an excitatory neurotransmitter that causes increased sodium and potassium permeability leading to depolarization (Maynert et al., 1975). Gamma-aminobutyric acid (GABA), on the other hand, is an inhibitory neurotransmitter formed through the decarboxylation of glutamic acid by the enzyme glutamic acid decarboxylase (E.C. 4.1.1.15) (Wood, 1975). GABA causes hyperpolarization of the postsynaptic membrane as a result of an increase in chloride permeability (Wood, 1975; Maynert et al., 1975).

Glutamate is the most abundant amino acid neurotransmitter in the brain, with most of its content in the cerebral cortex (Maynert et al., 1975). GABAergic neurons are found in the globus pallidus, which sends projections to the substantia nigra. GABAergic neurons are also located in the hypothalamus and the cerebellar Purkinje cells (Maynert et al., 1975). These systems may be involved in mechanisms of EEG activation and arousal, as well as extrapyramidal function (Maynert et al., 1975).

4. Methodology

There are several methodological approaches to the study of the relationship between neurotransmitters and epilepsy in humans. Except for autopsy determination of brain neurotransmitter content, most indices of function are obtained by examining concentrations of neurotransmitters or their metabolites in urine, blood, or CSF (Moskowitz and Wurtman, 1975; Moir et al., 1970).

The most commonly used strategy uses measurement of the CSF levels of the acid metabolites of the neurotransmitters and is based on the belief that the CSF concentration of their metabolites provides information about the metabolic turnover of the parent amines in brain (Moskowitz and Wurtman, 1975). An additional assumption is that the concentration of CSF metabolites is derived from the brain rather than spinal cord, and that lumbar CSF concentrations reflect levels of the parent amines in the brain. Support for these assumptions includes studies of ventriculospinal CSF gradients in man as well as animal studies (Moskowitz and Wurtman, 1975; Moir et al., 1970).

A more recent strategy involves the use of probenecid to assess turnover of monoamines in brain. It is based on evidence that CSF concentrations of a metabolite reflect the concentration in adjacent CNS tissue; active transport of the metabolites out of the CSF appears to occur near the fourth ventricle, and this is blocked by probenecid (Moir et al., 1970). Using probenecid to prevent or minimize efflux of these metabolites from the CSF helps to reduce the ventriculospinal CSF concentration gradients and provides an estimate of the synthesis, storage, release, reuptake, and metabolism of the parent amines, i.e., turnover. Variability in dosage and action of probenecid can be circumvented by utilizing the ratio of concentrations of metabolite to probenecid in the CSF (Shaywitz et al., 1975; 1980a).

There are at least six indices for detecting whether a neurological disease involves abnormalities in neurotransmitters (Moskowitz and Wurtman, 1975). First, abnormalities in brain neurotransmitter content may be revealed. Second, changes in turnover of neurotransmitters as reflected in changes in CSF concentration of metabolites may be documented. Third and fourth, blood and urinary abnormalities in the concentration of neurotransmitter metabolites may exist in association with the disease under study. The fifth criterion is that pharmacologic manipulation of neurotransmitters be associated with the disease. Finally, animal models relating changes in neurotransmitters to the disease should be developed.

5. Serotonin and Epilepsy

Mohan et al. (1982) examined a heterogeneous group of patients with seizure disorders, but were unable to detect any difference in CSF concentrations of 5-HIAA (or HVA) between untreated epileptics and control patients. Because their study group was composed of patients with many different types of epilepsy, a relationship between serotonin turnover and a specific type of epilepsy could have been masked. Furthermore, the failure to document changes in the concentration of 5-HIAA in lumbar CSF does not necessarily exclude the possibility that serotonin may play a role in the genesis of seizures. Thus, alterations in metabolites may still occur locally in serotonergic neurons and synapses near a seizure focus.

Decreased 5-HIAA concentrations in both lumbar and ventricular CSF were reported in epileptics compared to controls (Garelis and Sourkes, 1973). Chadwick et al. (1975a) did not find any differences in lumbar CSF concentrations of 5-HIAA between a heterogeneous group of epileptic patients compared to controls. Elevated 5-HIAA levels in lumbar CSF were, however, found in patients with high concentrations of anticonvulsant levels in serum. In addition, significant correlations were noted between serum levels of anticonvulsants, and the number of anticonvulsants used, and CSF concentrations of 5-HIAA. Here too a methodologic problem of concern was the inclusion of a heterogeneous patient group with a variety of seizure disorders. Their findings of an association between high concentrations of 5-HIAA in CSF and elevated serum levels of anticonvulsants suggest that serotonergic mechanisms may be related to anticonvulsant effects, and it is intriguing to speculate that anticonvulsant intoxication may be related to disturbances in serotonergic activity in the brain.

In theory, turnover rather than simply endogenous concentrations of the metabolite should provide better characterization of monoaminergic function in clinical disorders, and in practice, turnover may be approximated using the probenecid loading technique. Employing this methodology, Shaywitz et al. (1975) determined lumbar CSF concentrations of 5-HIAA in children with seizure disorders and controls. Results indicated decreased 5-HIAA concentrations in epileptic children, suggesting decreased turnover of serotonin in the CNS. As in the previous studies reviewed, this was a heterogeneous group of children with many types of seizure disorders, and whether one type of seizure may be related to decreased serotonin turnover could not be elucidated. Nevertheless, these studies suggest that reduced turnover of serotonin, but not necessarily a reduced serotonin brain content, may be related to epilepsy.

Pharmacologic manipulations that tend to decrease endogenous serotonin have been associated with increased susceptibility to seizures. Reserpine, which depletes serotonin as well as catecholamines in the brain, decreases the seizure threshold (Kobayashi and Mori, 1977). Conversely, monoamine oxidase inhibitors, which nonselectively increase postsynaptic levels of serotonin and catecholamines, seem to raise the seizure threshold and protect one from seizures (Kobayashi and Mori, 1977). These pharmacologic manipulations are nonselective and it is difficult to separate the effects on indoleamines from catecholamines in relation to seizures.

One particular group of seizure disorders, myoclonic epilepsy and intention myoclonus, has been well studied in relation to serotonin turnover. Intention (action) myoclonus is an involuntary arrhythmic jerking related to rapid contraction of a particular muscle group. This myoclonic movement is exacerbated by volitional motor activity and by emotional, tactile, auditory, and visual stimuli. One major cause of intention myoclonus is anoxic brain damage.

Chadwick et al. (1975b) noted that some patients with postanoxic intention myoclonus exhibited low 5-HIAA concentrations in lumbar CSF, and these patients were shown to improve when treated with 5-HTP and/or clonazepam. Interestingly, in animal studies clonazepam, a benzodiazepine anticonvulsant, has been reported to increase serotonin and 5-HIAA in both CSF and brain (Chadwick et al., 1975b).

Van Woert and Sethy (1975) reported decreased CSF 5-HIAA in a very small group of patients with postanoxic intention myoclonus using the probenecid load technique. After treatment with 5-HTP and carbidopa, patients noted marked improvement in symptoms and had elevated 5-HIAA levels in CSF, as compared to pretreatment levels. In a later study, Van Woert et al. (1977) reported improvement of symptoms in 12 out of 13 patients with postanoxic intention myoclonus treated with 5-HTP and carbidopa. In this study, pretreatment levels of 5-HIAA in lumbar CSF were reduced compared to controls, but increased 14-fold with treatment (Van Woert et al., 1977).

It is reasonable to postulate that anoxic brain damage may destroy neurons in the serotonergic raphe nuclei, and such a deficiency in brain serotonin may be causally related to postanoxic intention myoclonus. Serotonin is an inhibitory neurotransmitter, and ablation of the raphe nuclei in the midbrain may decrease these inhibitory impulses. The resultant decreased inhibitory control may lead to stimulus-sensitive hyperexcitability and/or a decrease in seizure threshold. Destruction of serotonergic neurons may result in a hyperexcitable, stimulus-sensitive central nervous system leading to symptoms of myoclonic seizures.

6. Dopamine and Epilepsy

Papeschi et al. (1972) reported markedly reduced levels of HVA (and 5-HIAA) in ventricular CSF of patients with a variety of neurological diseases, including temporal lobe epilepsy. Since the dopaminergic nigroneostriatal pathway is involved in extrapyramidal function and those movements have some similarity to psychomotor activity, the above data suggest a role for dopaminergic neural transmission in decreased susceptibility to complex partial seizures.

In a study noted above using probenecid loading, Shaywitz et al. (1975) found reduced CSF HVA as well as 5-HIAA in children with epilepsy. These results suggest decreased turnover of both dopamine and serotonin in the CNS of children with epilepsy. Other studies have also shown decreased concentrations of 5-HIAA and HVA in both lumbar and ventricular CSF using the probenecid technique (Garelis and Sourkes, 1973; Van Woert and Sethy, 1975; Van Woert et al., 1977).

Reflex epilepsy refers to the induction of seizures utilizing various sensory stimuli. A particular strain of mice, for example, are genetically endowed with susceptibility to seizures induced by sound stimuli. In the mouse model, this susceptibility appears to resolve in the first 12 wk of life, and the decrease in seizure susceptibility appears to be associated with an increase in brain dopamine and norepinephrine synthesis and turnover (Shaywitz et al., 1978).

Photosensitive seizures are the most studied form of reflex epilepsy in humans. Quesney et al. (1980) reported that the dopamine receptor agonist, apomorphine, prevented photically induced seizures in 9 of 11 patients with photosensitive epilepsy. The effects began 15 min after sc injection, and lasted about 45 min without changes in visually evoked potentials. A follow-up study by this group indicated that the opiate antagonist, naloxone, failed to block the anticonvulsant effect of apomorphine, suggesting that the effect of apomorphine in this type of epilepsy is mediated by dopaminergic rather than opiate receptors (Quesney et al., 1981). Apomorphine had no effect on spontaneous (nonreflex) seizures (Quesney et al., 1980; 1981).

Studies have shown that microiontophoretic application of dopamine to subcortical and neocortical neurons results in inhibition (Phillis and Tebecis, 1969). Photic stimulation in the cat results in decreased dopamine release in the visual cortex (Reader et al., 1976), and chlorpromazine, a dopamine receptor blocking agent, enhances photosensitive seizures (Lamprecht, 1977; Trimble, 1977). Quesney et al. (1981) have hypothesized that photic stimulation increases thalamocortical activity, resulting in decreased release of dopamine. As a consequence, cortical inhibition is decreased and seizure activity may occur in susceptible individuals. Apomorphine, a dopamine receptor antagonist, reverses such effects.

Pharmacologic evidence also supports a role of dopaminergic neurotransmission in epilepsy. Neuroleptics, such as chlorpromazine and haloperidol, effect dopamine receptor blockade and result in electroencephalographic evidence of seizure activity (Lamprecht, 1977; Trimble, 1977). In contrast, amphetamine increases dopaminergic transmission, decreases seizure activity, and improves EEG patterns (Kobayashi and Mori, 1977; Lamprecht, 1977; Trimble, 1977). Although apomorphine reduces susceptibility to photosensitive seizures, Marrosu et al. (1983) noted that apomorphine resulted in activation of the EEG in patients with partial seizures, but had no effect on the EEG in patients with nonreflex generalized seizures. It is likely that such paradoxical findings are related to dose effects. Thus, high doses may stimulate postsynaptic dopamine receptors, resulting in motor activity and sterotypy, whereas low doses of apomorphine affect a presynaptic dopamine autoreceptor, resulting in decreased dopamine synthesis and activity with subsequent decrease in motor activity. Thus, although high doses of apomorphine may increase dopaminergic activity and the seizure threshold, low doses of apomorphine may enhance seizures by decreasing dopaminergic activity.

Further evidence for dopaminergic influences on seizures comes from studies on the relationship between epilepsy and schizophrenia. Clinical experience suggests that as schizophrenia worsens, susceptibility to seizures decreases and, conversely, seizures seem to improve schizophrenic symptomatology. In addition, there is a very low incidence of seizure disorders among patients with schizophrenia compared with the general population. Dopaminergic neurotransmission appears to be the unifying concept behind this relationship (Trimble, 1977).

Pharmacologic studies also support this relationship. Thus dopaminergic agonists, such as amphetamines and L-DOPA, may exacerbate schizophrenic symptoms and reduce seizures, whereas dopamine receptor antagonists, such as phenothiazines, used in the treatment of schizophrenia, may increase susceptibility to seizures (Kobayashi and Mori, 1977; Lamprecht, 1977; Trimble, 1977). Drugs that deplete brain monoamines, such as reserpine, result in increased seizure susceptibility and decreased schizophrenic symptoms (Kobayashi and Mori, 1977; Lamprecht, 1977; Trimble, 1977). Anticonvulsant therapy tends to increase dopaminergic receptor stimulation, but is associated with worsening schizophrenia (Kobayashi and Mori, 1977). Lamprecht's hypothesis suggests that decreasing the inhibitory feedback resulting in increased dopamine release, or increasing the set point (threshold) for firing of the postsynaptic cell in a central dopaminergic synapse, results in an increased number of dopamine receptors occupied (Lamprecht, 1977); and may be related to schizophrenia. In contrast, increased inhibitory feedback or a lower threshold for postsynaptic action potential generation causes a decreased number of dopamine receptors to be occupied, which may be related to epilepsy (Lamprecht, 1977).

7. Norepinephrine and Epilepsy

The relationship between norepinephrine and epilepsy is supported by two lines of investigation. The first is pharmacological. Thus, administration of tricyclic antidepressants, which inhibit reuptake and enhance synaptic concentrations of norepinephrine, is associated with decreased susceptibility to seizures (Kobayashi and Mori, 1977). The second is based upon findings from cerebellar stimulation. Good evidence indicates that cerebellar stimulation not only suppresses hippocampal and amygdala seizures, but inhibits electroshock and chemically induced seizures as well. Wood et al. (1977a) found that in a heterogeneous group of patients with epilepsy, stimulation of implanted *cerebellar* electrodes resulted in improvement in seizures as well as elevated norepinephrine levels in CSF. Patients with *cerebral* cortical electrodes who underwent chronic stimulation did not improve clinically and had depressed CSF norepinephrine concentrations (Wood et al., 1977b).

It appears that catecholamines (norepinephrine and dopamine) decrease neuronal activity and susceptibility to epileptogenic stimuli. The data from the cerebellar stimulation studies suggest that a noradrenergic mechanism may be involved in seizure suppression and is consonant with the results of studies in animals, which have shown that decreased norepinephrine synthesis and reduced norepinephrine concentration in the CNS are associated with decreased seizure thresholds and increased susceptibility to seizures. Other studies indicate that iv or ic injection of norepinephrine exerts anticonvulsant action (Wood et al., 1977a).

Noradrenergic pathways from the locus ceruleus in the pons project to the cerebellar cortex, cerebral cortex, mesencephalon, and hippocampus, although some of the noradrenergic projections to the cerebellum synapse onto Purkinje cells. Wood et al. (1977a, b) postulate that cerebellar stimulation results in antidromic activation of the locus ceruleus. Increased locus ceruleus activity may cause increased release of norepinephrine in cerebral cortex, reticular formation, and hippocampus, which inhibits neuronal firing and seizure activity. Slow frequency cerebellar stimulation may activate noradrenergic inhibitory synapses on cerebellar Purkinje cells. This in turn may decrease Purkinje cell output, mediated by the inhibitory neurotransmitter gamma-aminobutyric acid (GABA). Decreased GABA levels in CSF are associated with the spread of induced seizure activity (Wood et al., 1977a; b). Thus it is hypothetically possible for increased noradrenergic activity to result in either an increase or decrease in seizure threshold.

8. Amino Acids and Epilepsy

Although animal studies indicate that a deficiency of GABA at the synaptic cleft plays a role in traumatic "chemical" epilepsy (Wood, 1975; Ribak et al., 1979) and in electrically and audiogenically induced seizures (Maynert et al., 1975), the evidence for GABA in the etiology of seizures in humans is inconclusive. Whereas one study demonstrated decreased GABA levels in cerebral cortex of epileptic patients (Van Gelder et al., 1972), other studies have found normal GABA levels in the actual epileptic foci of patients with focal epilepsy (McGeer et al., 1971; Perry et al., 1975). A study of a heterogeneous group of seizure patients using lumbar CSF measurements as an indirect measure of brain neurotransmitter levels found an association between low GABA concentrations and generalized tonic-clonic (grand mal) and complex partial (psychomotor) seizures (Wood et al., 1979; 1980). This study, however, did not control for anticonvulsant treatments. The fact that amino acid neurotransmitters also function in energy metabolism confounds interpretation of both CSF and brain concentrations in all of these studies. Reduced CSF concentrations of GABA are also found in other neurologic disorders, making this a nonspecific finding (Wood et al., 1980).

One can only conclude that decreased GABA levels may be associated with generalized tonic-clonic seizures and complex partial seizures. It is possible that cerebellar stimulation and, hence, Purkinje cell firing and release of GABA may result in the inhibition of cerebral seizure foci. CSF GABA levels are decreased, however, in patients with intractable seizures undergoing chronic cerebellar stimulation (Wood et al., 1980). Alternatively, Purkinje cell loss with resultant decreased GABA levels may result in generalized tonic-clonic, myoclonic, and complex partial seizures (Cooper et al., 1976).

9. Effects of Anticonvulsant Drugs on Neurotransmitters

Anticonvulsant medications may act via central monoaminergic mechanisms. The benzodiazepine, clonazepam, used in the treatment of myoclonic seizures and intention myoclonus has been shown to elevate serotonin and its metabolites in both CSF and animal brains (Chadwick et al., 1975b). Increased serum levels of diphenylhydantoin (DPH) and phenobarbital have been correlated with increased concentrations of HVA and 5-HIAA in lumbar CSF, and marked elevation in serotonin metabolites in the CSF are seen in patients with toxic levels of DPH (Chadwick et al., 1975a; 1977). It is possible that the signs and symptoms of anticonvulsant intoxication may be secondary to elevated serotonin levels or turnover.

10. Summary

Evidence supports a relationship between decreased levels of monoamines and various types of epilepsy. Deficiencies in serotonin turnover appear to be most closely related to myoclonic seizures and postanoxic intention myoclonus, perhaps related to dysfunction of the serotonergic raphe nuclei. Interestingly, the serotonergic raphe system appears to be involved in the regulation of slow wave sleep, a stage during which myoclonic jerking movements are not uncommon. Excess of serotonin or increased turnover of serotonin may be related to the manifestations of anticonvulsant intoxication.

Dopaminergic neuronal systems appear to play a major role in decreasing susceptibility to seizures. Deficiencies in dopamine turnover are seen in temporal lobe and reflex epilepsy. In addition, the antagonistic relationship between epilepsy and schizophrenia supports a dopaminergic mechanism in the pathophysiology of both disorders.

Noradrenergic turnover appears to be decreased in patients with certain seizure disorders. Chronic cerebellar stimulation both decreases seizure susceptibility and increases norepinephrine turnover.

The evidence for the role of GABA in human epilepsy is controversial. Decreased levels of GABA in the synaptic cleft may be associated with generalized tonic-clonic and complex partial epilepsy.

Thus evidence suggests that dysfunction, or hypofunction, of the dopaminergic nigrostriatal extrapyramidal systems and the noradrenergic locus ceruleus may be associated with seizures in general, and temporal lobe (complex partial) or reflex epilepsy in particular. Further study of neurotransmitter turnover in particular nuclei and neural systems in the brain in relation to specific types of seizure disorders may help to elucidate more exact pathophysiologic associations between neurotransmitters and human epilepsy.

References

Aghajanian, G. K., Bloom, F. E., and Sheard, M. H.: Electron microscopy of degeneration within the serotonin pathway of rat brain. *Brain Res.* **13:** 266–273, 1969.

Bertler, A. and Rosengren, E.: Occurrence and distribution of dopamine in brain and other tissues. *Experientia* **15:** 10–11, 1959.

Bowers, M. B.: 5-Hydroxyindoleacetic acid in the brain and CSF of the rabbit following administration of drugs affecting 5-hydroxytryptamine. *J. Neurochem.* **17:** 827–828, 1970.

Chadwick, D., Jenner, P., and Reynolds, E. H.: Amines, anticonvulsants and epilepsy. *Lancet* **i:** 473–476, 1975a.

Chadwick, D., Harris, R., Jenner, P., Reynolds, E. H., and Marsden, C. D.: Manipulation of brain serotonin in the treatment of myoclonus. *Lancet* **ii:** 434–435, 1975b.

Chadwick, D., Jenner, P., and Reynolds, E. H.: Serotonin metabolism in human epilepsy: The influence of anticonvulsant drugs. *Ann. Neurol.* **i:** 218–224, 1977.

Cooper, I. S., Amin, J., Riklan, M., Waltz, J. M., and Poon, T. P.: Chronic cerebellar stimulation in epilepsy. *Arch. Neurol.* **33:** 559–570, 1976.

Garelis, E. and Sourkes, T. L.: Factors affecting monoamine metabolite concentrations in the cerebrospinal fluid. *Neurology* **23(4):** 410, 1973.

Jouvet, M.: Neurophysiology of the states of sleep. *Physiol. Rev.* **47:** 117–177, 1967.

Jouvet, M.: Biogenic amines and the states of sleep. *Science* **163:** 32–41, 1969.

Jouvet, M.: The Role of Monoamines and Acetylcholine-Containing Neurons in the Regulation of the Sleep-Waking Cycle, In: *Ergebnisse der Physiol.* vol. 64, Springer-Verlag, New York, 1972.

Kobayashi, K. and Mori, A.: Brain monoamines in seizure mechanism (review). *Folia Psychiat. Neurolo. Japonica* **31(3):** 483–489, 1977.

Lamprecht, F.: Epilepsy and schizophrenia: A neurochemical bridge. *J. Neural Trans.* **40:** 159–170, 1977.

Loizou, L. A.: Rostral projections of noradrenaline-containing neurones in the lower brain stem. *J. Anat.* **104:** 593, 1969a.

Loizou, L. A.: Projections of the nucleus locus coeruleus in the albino rat. *Brain Res.* **15:** 563–566, 1969b.

Loizou, L. A.: The development of monoamine-containing neurones in the brain of the albino rat. *J. Anat.* **104:** 588, 1969c.

Loizou, L. A. and Salt, P.: Regional changes in monamines of the rat brain during postnatal development. *Brain Res.* **20:** 467–470, 1970.

Marrosu, F., DelZompo, M., and Corsini, G. U.: The role of dopamine in human epilepsy: Effect of apomorphine. *Prog. Clin. Biol. Res.* **124:** 95–104, 1983.

Maynert, E. W., Marczynski, T. J., and Browning, R. A.: The role of the neurotransmitters in the epilepsies. *Adv. Neurol.* **13:** 79–147, 1975.

McGeer, P.. L., McGeer, E. G., and Wada, J. A.: Glutamic acid decarboxylase in Parkinson's disease and epilepsy. *Neurology* **21:** 1000–1007, 1971.

Mohan, A., Nag, D., Misra, R. N., Gujrati, V. R., Shanker, K., Doval, D. C., Saxena, R. C., and Bhargava, K. P.: Serotonergic and dopaminergic metabolites in cerebrospinal fluid of epileptics. *Pharmazie* **37(11):** 803, 1982.

Moir, A. T. B., Ashcroft, G. W., Crawford, T. B., Eccleston, D., and Guldberg, H. C.: Cerebral metabolites in cerebrospinal fluid as a biochemical approach to the brain. *Brain* **93:** 357–368, 1970.

Morgane, P. J. and Stern, W. C.: Chemical Anatomy of Brain Circuits in Relation to Sleep and Wakefulness, In: *Advances in Sleep Research* vol. 1 (E. Weitzman, ed.), Spectrum, New York, 1974.

Moskowitz, M. A. and Wurtman, R. J.: Catecholamines and neurologic diseases (part I). *N. Eng. J. Med.* **293(6):** 274–280, 1975.

Papeschi, R., Molina-Negro, P., Sourkes, T. L., and Erba, G.: The concentration of homovanillic and 5-hydroxyindoleacetic acids in ventricular and lumbar CSF. *Neurology* **22:** 1151–1159, 1972.

Perry, T. L., Hansen, S., Kennedy, J., Wada, J. A., and Thompson, G. B.: Amino acids in human epileptogenic foci. *Arch. Neurol.* **32:** 752–754, 1975.

Phillis, J. W. and Tebecis, A. K.: The responses of thalamic neurons to iontophoretically applied monoamines. *J. Physiol.* **193:** 715–745, 1969.

Quesney, L. F., Andermann, F., Lal, S., and Prelevic, S.: Transient abolition of generalized photosensitive epileptic discharge in humans by apomorphne, a dopamine-receptor agonist. *Neurology* **30:** 1169–1174, 1980.

Quesney, L. F., Andermann, F., and Gloor, P.: Dopaminergic mechanism in generalized photosensitive epilepsy. *Neurology* **31:** 1542–1544, 1981.

Reader, T. A., Champlain, J., and Jasper, H.: Catecholamines released from cerebral cortex in the cat: Decrease during sensory stimulation. *Brain Res.* **111:** 95–108, 1976.

Ribak, C. E., Harris, A. B., Vaughn, J. E., and Roberts, E.: Inhibitory GABAergic nerve terminals decrease at sites of focal epilepsy. *Science* **205:** 211–214, 1979.

Shaywitz, B. A., Cohen, D. J. and Bowers, M. B.: Reduced cerebrospinal fluid 5-hydroxyindoleacetic acid and homovanillic acid in children with epilepsy. *Neurology* **25:** 72–79, 1975.

Shaywitz, B. A., Yager, R. D., and Gordon, J. W.: Ontogeny of brain catecholamine turnover and susceptibility to audiogenic seizures in DBA/2J mice. *Dev. Psychobio.* **11(3):** 243–250, 1978.

Shaywitz, B. A., Cohen, D. J., and Bowers, M. B.: Cerebrospinal Fluid Monoamine Metabolites in Neurological Disorders of Childhood, In: *Neurobiology of Cerebrospinal Fluid* vol. I (J. H. Wood, ed.) Plenum, New York, 1980a.

Shaywitz, B. A., Cohen, D. J., Leckman, J. F., Young, J. G., and Bowers, M. B.: Ontogeny of dopamine and serotonin metabolites in the cerebrospinal fluid of children with neurological disorders. *Dev. Med. Child Neurol.* **22:** 748–754, 1980b.

Trimble, M.: The relationship between epilepsy and schizophrenia: A biochemical hypothesis. *Bio. Psychiat.* **12(2):** 299–304, 1977.

Van Gelder, N. M., Sherwin, A. L., and Rasmussen, T.: Amino acid content of epileptogenic human brain. *Brain Res.* **40:** 385–393, 1972.

Van Woert, M. H. and Sethy, V. H.: Therapy of intention myoclonus with L-5-hydroxytryptophan and a peripheral decarboxylase inhibitor, MK 486. *Neurology* **25:** 135–140, 1975.

Van Woert, M. H., Rosenbaum, D., Howieson, J., and Bowers, M. B.: Long-term therapy of myoclonus and other neurologic disorders with L-5-hydroxytryptophan and carpidopa. *N. Eng. J. Med.* **296:** 70–75, 1977.

Walter, D. S. and Eccleston, D.: Increase of noradrenaline metabolism following electrical stimulation of the locus coeruleus in the rat. *J. Neurochem.* **21:** 281–289, 1973.

Wood, J. D.: The Role of Gamma Aminobutyric Acid in the Mechanism of Seizures, In: *Progress in Neurobiology* (G. A. Kerkut and J. W. Phillis, eds.) Pergamon, New York, 1975.

Wood, J. H., Ziegler, M. G., Lake, C. R., Sode, J., Brooks, B. R., and Van Buren, J. M.: Elevations in cerebrospinal fluid norepinephrine during unilateral and bilateral cerebellar stimulation in man. *Neurosurgery* **1(3)**: 260–264, 1977a.

Wood, J. H., Lake, C. R., Ziegler, M. G., Sode, J., Brooks, R. R., and Van Buren, J. M.: Cerebrospinal fluid norepinephrine alterations during electrical stimulation of cerebellar and cerebral surfaces in epileptic patients. *Neurology* **27**: 716–724, 1977b.

Wood, J. H., Hare, T. A., Glaeser, B. S., Ballenger, J. C., and Post, R. M.: Low cerebrospinal fluid gamma aminobutyric acid content in seizure patients. *Neurology* **29**: 1203–1208, 1979.

Wood, J. H., Hare, T. A., Glaeser, R. S., Brooks, B. R., Ballenger, J. C., and Post, R. M.: Cerebrospinal fluid GABA variations with seizure type and cerebellar stimulation in man. *Brain Res. Bull.* **5(2)**: 747–753, 1980.

Wright, F. S., Dreifuss, F. E., Wolcott, G. J., Swaiman, K. F., Low, N. L., Freeman, J. M., and Nelson, K. B.: Seizure Disorders, In: *The Practice of Pediatric Neurology* 2nd ed., (K. F. Swaiman and F. S. Wright, eds.) C. V. Mosby, St. Louis, 1982.

Wurtman, R. J., Larin, F., Mostafapour, S., and Fernstrom, J. D.: Brain catechol synthesis: Control by brain tyrosine concentration. *Science* **185**: 183–184, 1974.

Neurotransmitter Systems and the Epilepsy Models

Distinguishing Features and Unifying Principles

Phillip C. Jobe and Hugh E. Laird II

1. Introduction

A small body of data offers provocative views of the neuro-chemical etiology of epilepsy. A substantial portion of this information points toward specific neurotransmitter systems as being important components of the seizure-regulating mechanisms. Diverse epilepsy models are being used to elucidate the influences of both excitatory and inhibitory neurotransmitters. Much of the information and many concepts have not been systematically utilized in the attempt to develop new antiepileptic drugs. Yet the need for new approaches is apparent. Jones and Woodbury (1982) have noted that our worldwide attempts to develop new medications for prevention and management of these neurological disorders have not produced substantial results for the past few decades.

Several dogmatic concepts may have prevented broader-based progress. First, too much emphasis may have been placed on the concept that epilepsy in animals is fundamentally different than that in humans. Misconceptions about pharmacological evidence may have contributed to these potential errors. For example, some drugs are known to be convulsant in humans, but anticonvulsant in animals. Such findings have been interpreted as substantiating the concept of a fundamental difference in the two types of epilepsy. The convulsant/anticonvulsant effects of the tricyclic and tetracyclic antidepressant drugs provide an example of a dogma built with pharmacological tools. These agents have been viewed as placing a patient at risk for convulsions. Consequently, few people have viewed them as potential antiepileptic agents. Such conclusions have been reached despite demonstrations in animal models of epilepsy that the antidepressant drugs are potent anticonvulsant agents. Interestingly, the convulsant effects of the antidepressant drugs are also detectable in animals. Indeed the evidence suggests that nothing more than a dose–response relationship governs whether an antidepressant drug is anticonvulsant or convulsant. Studies in the genetically epilepsy-prone rat have revealed that the convulsant dose$_{50}$ of antidepressants is 200–900% of the anticonvulsant dose$_{50}$ (Jobe et al., 1984). Reports relative to human seizures also suggest that overdoses are associated with seizures. Although not widely known, a few studies in humans have suggested that lower doses of these drugs produce antiepileptic effects (Fromm et al., 1972, 1978; Ojemann et al., 1983).

Another dogma may have been constructed by investigations of known anticonvulsant drugs used in treating epilepsy. The assumption in the investigations is that the mechanism of action of the clinically useful anticonvulsant drugs may be instructive relative to the etiological abnormalities of the epilepsies. Yet, despite the intensive nature of such experimentation, little progress has been made toward the goal of improved therapy. As would be expected, such approaches have produced new drugs that affect epilepsy in a similar manner as do the old drugs.

Another dogmatic concept may be derived from our interpretation of the clinical symptomatology of the epilepsies. Presently, the epilepsies are divided into four different types of syndromes: localized, generalized, undetermined as to localized or generalized, and special syndromes (e.g., febrile convulsions, isolated unprovoked epileptic events, and so on) (Commission on Classification and Terminology of the International League Against Epilepsy, 1985). The clinical subdivisions within these broad categories are often considered to be composed of relatively homogeneous disease entities. Individuals within the subdivisions display somewhat similar electroencephalographic activity, overt convulsions (or other behaviors), and responses to anticonvulsant drugs. Whether certain patients within these clinical subdivisions are characterized by different types of neurochemical defects is largely unknown. It seems quite possible that people who exhibit similar electroencephalographic activity, convulsions, and responses to anticonvulsant drugs may have fundamentally divergent neurochemical causes of their neurological dysfunction. If such differences truly exist, neurochemical identification of the divergent subgroups will be essential as a foundation for developing rationally based treatments.

Neurochemical distinctions between the DBA/2J mice and the genetically epilepsy-prone rats may prove to be an instructive resource relative to current dogmas predicted on clinical views of the epilepsies. Clinically, both of these animals have acoustic epilepsy, both exhibit generalized convulsions, and both respond favorably to the anticonvulsant drugs such as phenytoin (Dailey and Jobe, 1985; Loscher and Meldrum, 1984). Yet, the neurochemical etiology of epilepsy in the two models does not appear to be the same. Several lines of evidence suggest that dopaminergic deficits are determinants of seizure predisposition in DBA/2J mice (*see* Chapman and Meldrum, 1986). In contrast, both pharmacological and pathophysiological data provide an increasingly persuasive basis for rejecting dopaminergic systems as determinants of seizure activity in the genetically epilepsy-prone rats (Laird and Jobe, 1986). Yet, in both the epileptic mice and rats, phenytoin is effective as an anticonvulsant. Even in normal, nonepileptic animals, phenytoin has the capacity to act as an anticonvulsant. Thus, the anticonvulsant effect of phenytoin does not require the presence of seizure predisposition. Moreover, the effectiveness of this drug is manifest in at least two types of animals with neurochemically divergent epilepsies. Thus, phenytoin has a broadly based anticonvulsant effect rather than a specific antiepileptic effect. If these views are correct, continued investigation of phenytoin may only set the stage for the emergence of new anticonvulsants that also fail to produce specific antiepileptic effects.

Therefore the assumption that phenytoin represents an antiepileptic drug because it is effective in several diagnostic categories of clinical epilepsy may lead us falsely to the conclusion that its cellular and neurochemical mechanisms of action also represent antiepileptic rather than nonspecific anticonvulsant effects. Indeed, most of the current clinically useful antiepileptic drugs are similar to phenytoin. They are effective anticonvulsants in nonepileptic subjects. We believe that a new direction is needed in antiepileptic drug research. In this new approach, we should establish protocols whereby drugs with specific antiepileptic properties can be identified. Medicinal agents identified through this new process can be anticipated to result in improved therapy in the epilepsies.

This view of antiepileptic drug specificity represents an essential departure from the intellectual constructs of the past. Commonly the term specificity has been used to denote a high ratio between the sedative dose of a drug and the anticonvulsant dose of the same drug (Krall et al., 1978a, b; Consroe et al., 1980). According to this line of reasoning, an anticonvulsant drug without specific anti-epileptic properties would suppress seizures only in doses sufficient to cause sedation. Also according to previous logic, an antiepileptic drug has been imputed to have specificity if the ratio between the neurotoxic dose and the anticonvulsant dose were high. Antiepileptic specificity has also been assumed to be an inherent property of an anticonvulsant drug if, upon chronic dosing, tolerance developed to its sedative, but not its seizure-suppressing, properties.

According to our new definition, antiepileptic drug specificity is regarded as a special propensity of a pharmacologic agent to produce an anticonvulsant effect in animals with a documented seizure predisposition (i.e., epilepsy). Does the candidate antiepileptic drug cause an anticonvulsant effect in epileptic animals, but not in normal animals subjected to supramaximal electroshock? Alternatively, does the candidate medicinal agent produce an anticonvulsant effect in both types of animals, but with lower concentrations required in the brains of the epileptic than in the normal subject?

Conceptually, the enhanced sensitivity of the epileptic subject could occur through several mechanisms. Consider the possibility that the seizure-prone state of an animal might be caused by a deficiency in the availability of anticonvulsant neurotransmitter. Partial compensation for this state could result from up-regulation of post-synaptic receptors, which in turn causes an exaggerated response to an exogenously administered agonist. If such an agonist were tested as a candidate anticonvulsant drug in the presence of these extra receptors, the epileptic animal would be expected to exhibit seizure suppression in response to doses that would not be sufficiently high to cause an anticonvulsant effect in nonepileptic controls.

In contrast to the idea that antiepileptic specificity might be manifested by an exaggerated anticonvulsant response, a diminished response could also be a means for identification of possible specificity. As an example, consider the idea that a seizure-prone state of an animal might be caused by a genetically determined decrement in the density of functional anticonvulsant postsynaptic receptors. In such a situation, normal synaptic concentrations of the anticonvulsant neurotransmitter would be too low to support a normal level of transmission. Also, because of the deficiency in receptor density, exogenously administered agonists would be pharmacologically weak. Compared to the situation in nonepileptic animals, larger doses would be required to cause sufficient activation of the postsynaptic anticonvulsant receptors to result in the anticipated effect.

From a clinical perspective, antiepileptic drug specificity characterized by enhanced rather than reduced seizure suppression would be more attractive. Reduced suppression would require that higher doses of the antiepileptic drug be employed therapeutically. Higher doses would probably result in more severe toxicities. Thus, in a rationally designed antiepileptic drug testing program, the primary focus would be in developing drugs with antiepileptic specificity characterized by enhanced seizure suppression.

These concepts support the idea that identification of the roles of the neurotransmitter systems in the determination of epileptic states will have a profound impact on the clinical treatment of these disorders. Progress leading to improved therapy will be facilitated by identification of the dogmas that prevent the development of new approaches for identification of future antiepileptic medications. Much of the neurochemical information needed to make initial revisions in our approach to the development of antiepileptic drugs is already available. Integration of these diverse data into a useful conceptual matrix is imperative. Toward this goal the remaining sections of this chapter are designed to critically examine the roles of neurotransmitters in several diverse models of epilepsy. Our objective is to identify differences and similarities in the roles of these neurotransmitter systems as revealed by studies using the different models. Distinguishing characteristics and unifying principles relative to these roles and their relationship to the treatment of epilepsy are a point of major focus.

2. Summary of the Implicated Neurotransmitter Systems

2.1. GABAergic Systems

GABAergic terminal fields apparently have the capacity to protect the central nervous system against excessive neuronal activity, including that underlying seizures. The evidence supporting this concept is convincing. GABAergic agents invariably have the capacity to suppress seizure activity in most models of epilepsy. These observations are summarized in Table 1.

TABLE 1
GABAergic Role in Seizure Regulation[a]

Model	Suppress seizures	Defect reported	Proposed roles	
			Physiological	Pathophysiological
Epileptic mice	Yes	Decreased receptor density; Increased receptor affinity	Protects against excessive neuronal activity	Innate abnormalities present in opposing directions. Etiological significance remains unclear
GEPRs	Yes	Increased receptor density and neuronal number	Protects against excessive neuronal activity	Innate increments partly compensate for seizure condition. Alternatively, increments may reflect disinhibition.
Epileptic gerbil	Yes	Increased neuronal number; Decreased GABA levels, receptor density	Protects against excessive neuronal activity	Innate abnormalities present in opposing directions. Etiological significance remains unclear
Epileptic chickens	Yes	Increased GABA levels	Protects against excessive neuronal activity	Innate increment partly compensates for seizure condition
Epileptic humans	Probably	Lack of sufficient documentation	Protects against excessive neuronal activity	Innate abnormalities not elucidated
Kindled seizures	Yes	Increased release, receptor density	Protects against excessive neuronal activity	Innate increments partly compensate for seizure condition

Electroshock seizures	Yes	Decreased uptake; Increased binding	Protects against excessive neuronal activity	Increments occur as a consequence of the seizures
Bicuculline seizures	Yes	Blockade of receptors; Increased GABA turnover	Protects against excessive neuronal activity	Receptor blockade thought to cause seizures
Topical cobalt seizures	Perhaps	Decreased GABA levels, GAD[b] activity, uptake; Increased receptor density	Protects against excessive neuronal activity	Induced presynaptic decrement cause of seizures; Induced postsynaptic increment reflects incomplete compensation for lack of synaptic GABA
In vitro slices	Yes	Drugs that block receptors cause synchronized bursting; Decreased sensitivity to GABA after high-frequency stimulation	Protects against excessive neuronal activity	GABAergic decrements may be partially responsible for synchronized bursting

[a]Citations: Epileptic mice: Chapman and Meldrum, 1986; GEPRs (genetically epilepsy-prone rats): Laird and Jobe, 1986; Chapman et al., 1986; epileptic gerbil: Lomax et al., 1986; Loscher, 1985, Loscher et al., 1983, Peterson et al., 1984; epileptic chickens: Johnson and Tuchek, 1986; epileptic humans: Kresh et al., 1986; kindled seizures: McNamara et al., 1986; electroshock seizures: Browning, 1986; Essman and Essman, 1980; bicuculline seizures: Faingold, 1986; topical cobalt seizures: Craig and Colasanti, 1986; in vivo seizures: Traub et al., 1986; Krnjevic, 1982; Prince, 1983; Wong, 1982.

[b]GAD, glutamic acid decarboxylase.

Evidence for this concept is perhaps somewhat weaker for seizures produced by topical cobalt than for other models. Moreover, such evidence for human epilepsy is also less than persuasive. Nevertheless, when considered within the context of the other epilepsy models, these informational deficiencies may be no more than a reflection of the incomplete nature of the studies for cobalt seizures and human epilepsy. We suspect that when these two forms of seizures have been more fully scrutinized, the data will be as persuasive as for the other models.

The overwhelming number and diversity of observations showing that GABAergic agents protect against excessive neuronal activity in general and seizure episodes in particular have led to the hypothesis that GABAergic deficiencies are etiologically significant factors for epilepsy. This concept derives a large amount of additional support from experiments showing that drugs that reduce GABAergic transmission also cause seizures and/or convulsions in intact animals (see Faingold, 1986). Additionally, these drugs provoke electrophysiological manifestations of epileptic events in in vitro slice preparations (*see* Gjerstad et al., 1982; Wong, 1982; Traub et al., 1986). Although, these agents may also interact with the nervous system through other mechanisms (Faingold 1986), GABAergic antagonism is an important element in their capacity to provoke excessive neuronal activity.

Even a site for GABAergic influences on seizure processes has been localized to a specific nucleus. Several investigations implicate the substantia nigra as at least one of the sites where GABAergic neurons acts to inhibit electroshock seizures (*see* Browning, 1986). Another site, the deep prepiriform cortex, has been identified as a site in which generalized convulsions can be elicited by a focal injection of bicuculline in low doses (Piredda et al., 1985). This site may be selectively sensitive to GABA antagonists or it may be a substrate for the initiation of seizure activity by a variety of chemoconvulsants. A more complete characterization of this site is needed.

Despite the persuasive nature of the arguments that GABAergic transmission protects against the appearance of seizures, we believe that it is premature to generalize this concept to the pathophysiology of epilepsy. If the concept were valid, one might expect that among the diverse models of epilepsy, evidence for innate GABAergic deficits would be overwhelming. However, as set forth in Table 1, such documentation is minimal.

Indeed in the genetically epilepsy-prone rat, the epileptic gerbil, epileptic chickens, and kindled rodents, the data indicate that elevated rather than reduced GABAergic activity may be characteristic of seizure predisposition. In our view, these observations are consistent with the concept that, in animals that develop a neuronal state characterized by seizure predisposition, GABAergic systems may respond to the underlying pathology by a compensatory increase in their level of inhibition. According to this concept, some forms of epilepsy will be characterized by GABAergic increments. In clinically detectable cases, the degree of GABAergic compensation would be insufficient to completely offset the neurochemical causes of the epilepsy. Antiepileptic drug therapy with a GABAergic agent would be effective in these individuals by additionally augmenting seizure-suppressing transmission. An antiepileptic effect would become evident when drug-induced inhibitory control became sufficient to reduce seizure expression below the pretreatment level.

Some overtly normal individuals might in reality have the neurochemical antecedents of epilepsy. Yet, their clinical seizures would be prevented by a sufficient level of GABAergic compensation. Future experimental protocols will detect such complexities, if appropriately designed. In the past, Jobe et al. (1981) identified the presence of determinants of seizure predisposition in animals that were overtly normal.

An alternative hypothesis relative to the GABAergic increments has also been proposed. Roberts and Ribak (1986) suggest that the abnormally high number of GABAergic neurons present in the genetically epilepsy-prone rat and the epileptic gerbil are neuroanatomically arranged so that disinhibition occurs. An excessive number of GABAergic terminals may be forming synapses with other GABAergic neurons. As a result, GABAergic cells inhibit other GABAergic neurons so that disinhibition occurs. According to this line of reasoning, the seizure-prone states of the epileptic gerbil and genetically epilepsy-prone rat are at least partially caused by the resultant overall decrease in GABAergic inhibition of excitatory activity.

2.2. Other Inhibitory Amino Acid Systems

Glycine and taurine have been minimally studied for possible roles in epilepsy. Only in strychnine-induced seizures does a role for glycine appear to be a realistic probability. In Faingold's (1986) view, most evidence supports the concept that low doses of strychnine cause seizures by blocking postsynaptic glycine receptors, especially in the spinal cord. Whether glycinergic antagonism in other brain regions is also responsible for the generalized seizures observed with high doses of strychnine remains undetermined. Indeed, Faingold (1986) has concluded that effects on other nonglycinergic transmitter systems, as well as ''. . . other less neurotransmitter-specific effects of strychnine on membrane properties, ion conductance changes . . . may also contribute to the induction of generalized convulsive seizures with this agent.''

Abnormally low taurine concentrations in focal tissue collected from cobalt-epileptic cats, rats, and mice have been detected (*see* Craig and Colasanti, 1986). Exogenously administered taurine reportedly exerts an anticonvulsant effect in cobalt-epileptic mice and cats, but not rats.

Taurine may cause a specific antiepileptic effect in the genetically epilepsy-prone rat. Laird and Huxtable (1978) have shown that 200 mmol of this amino acid injected into the inferior colliculi raises the electroshock seizure threshold in the genetically epilepsy-prone rat, but not in nonepileptic controls.

Pathophysiological studies of taurine in the brains of genetically epilepsy-prone rats reveal a deficient high-affinity taurine transport system for this amino acid when comparisons are made to nonepileptic controls (Bonhaus and Huxtable, 1984). This observation coupled with the epilepsy-specific anticonvulsant effect of taurine prompted Bonhaus and Huxtable (1984) to suggest that a defect in high-affinity taurine transport in the brain contributes to seizure predisposition in the genetically epilepsy-prone rat.

If taurine is an endogenous antiepileptic substance, in what manner does it produce its effect? Taurine does not appear to function as a neurotransmitter (Bonhaus et al., 1983). It has not been shown to be stored in nerve endings and released into synapses in response to a nerve action potential. Does an alternative mechanism exist whereby endogenous taurine functions as an anticonvulsant substance? Evidence suggests that taurine is mainly dissolved in the cytosol of the neuron (Bonhaus et al., 1983). Its concentration in the cytosol may be an important determinant of its anticonvulsant activity. In the heart and brain, intracellular taurine is believed to participate in the regulation of calcium fluxes and transmitter release (Kuriyama et al., 1983). Perhaps one of these modulatory functions of taurine is ultimately translated into an anticonvulsant effect.

2.3. Excitatory Amino Acid Systems

As pointed out by Faingold (1986) and Craig and Colasanti (1986), the naturally occurring excitatory amino acids, glutamate and aspartate, have the capacity to cause seizures. Moreover, other compounds that have the capacity to stimulate excitatory amino acid receptors in the central nervous system also have convulsant properties. For example, kainic acid, an analog of glutamic acid, has the capacity to produce convulsions.

In epileptic mice (DBA/2J), excitatory amino acid antagonists suppress audiogenic seizures (*see* Chapman and Meldrum, 1986). Although information relative to endogenous defects in excitatory amino acid systems is not documented for epileptic mice, information on this subject has now become available for the genetically epilepsy-prone rat. Some observations suggest that excitatory amino acid transmission may be abnormally high (*see* Laird and Jobe, 1986). These animals are characterized by elevated density of glutamate binding sites, with no abnormality in binding affinity. Potassium-stimulated release of glutamate is higher than in nonepileptic controls.

In contrast, other neurochemical indices suggest that seizure-experienced genetically epilepsy-prone rats may be characterized by deficiencies at synapses that utilize glutamate or aspartate as transmitters. Chapman et al. (1986) reported that, during interictal periods, the concentrations of aspartate are abnormally low in the striatum, substantia nigra, and inferior colliculus of the genetically epilepsy-prone rat. Glutamate concentrations are abnormally low in the hippocampus and striatum. Whether turnover rates of these neurotransmitters are abnormal has not been reported.

The observations of Chapman et al. (1986) were made using genetically epilepsy-prone rats that had been exposed to a seizure-provoking stimulus 1 or more d before the animals were sacrificed for aspartate determinations. Also using seizure-experienced animals, Huxtable et al. (1982) measured aspartate and glutamate concentrations in genetically epilepsy-prone rats that were seizure-experienced in that they had been exposed to an acoustic stimulus to provoke convulsions prior to neurochemical measurements. Under these conditions, no abnormalities were found in aspartate concentrations.

Because of the experimental designs utilized, neither the abnormalities reported by Chapman et al. (1986) nor the normalities reported by Huxtable et al. (1982) can be confidently attributed to the seizure-naive state of the genetically epilepsy-prone rat. Both teams of investigators utilized animals that had a known history of seizure episodes. Data are needed now to determine whether an aspartate and/or glutamate deficit exists in the brains of genetically epilepsy-prone rats that are naive to seizures.

Chapman and coworkers (1986) have also demonstrated that aspartate and glutamate concentrations tend to rise in response to a seizure. Accordingly, in genetically epilepsy-prone rats that have previously sustained audiogenic seizures, aspartate and glutamate levels are higher during a subsequent ictal state than during a subsequent interictal period.

These observations of Chapman and associates (1986) lead to two hypotheses: (1) the interictal state of the genetically epilepsy-prone rat is characterized by a presynaptic deficit in activity of aspartate and glutamate neurons, and (2) audiogenic seizures may occur partially in response to a temporary increase in excitatory amino acid activity. Direct evidence relative to aspartergic and glutaminergic indices in animals that are naive to seizures has not been reported.

Pharmacologically, excitatory amino acid antagonists have been reported to block audiogenic seizure susceptibility in the genetically epilepsy-prone rat (Faingold et al., 1984). In normal rats the injection of an excitatory amino acid agonist elicits audiogenic seizure susceptibility as well as spontaneous wild running episodes (Meldrum et al., 1985). These observations reinforce the concept that audiogenic seizures may occur partially in response to an increment in excitatory amino acid transmission. Moreover, the pharmacological data suggests that the seizure-prone state of the genetically epilepsy-prone rat may be partially caused by excessive excitatory transmission. Unfortunately, pathophysiological investigations have not revealed whether the abnormality is one of excessive or deficient excitatory transmission. Intuitively, one would expect that indices of the interaction between the excitatory neurotransmitter and its postsynaptic receptors would be excessive in epileptic brains. Such an excessive level of excitatory transmission would be expected even if the direct cause of seizure predisposition were a deficient level of inhibitory activity. Diminished inhibition would allow an excessive level of excitation. Such increased excitation would be supported by an augmented rate of action potentials and an elevated release of the excitatory neurotransmitters. The excessive synaptic concentration of neurotransmitter would reinitiate the excessive activity at the postsynaptic membrane on the next neuron.

Such a condition of excessive excitatory transmission would expectedly be present during a seizure. The observations of Chapman et al. (1986) support this idea. However, the conditions that underlie seizure predisposition must be present in the seizure-naive state, as well as in the interictal period in seizure-experienced animals. It is during these conditions or intervals that excitatory transmission might not appear excessive. A deficient inhibitory system might actually be sufficient to prevent most commonly encountered excitatory stimuli from producing an excessive level of pervasive activity. According to this line of reasoning, the inhibitory defect would become functionally important in response to unusually high demands. If this concept were true, excitatory transmission would not be excessively high during the interictal period. The abnormally low inhibitory activity would be sufficient to contain usual excitatory transmission, but insufficient to contain excitatory events of an excessive degree.

These concepts underscore the possibility that epileptic animals might not be characterized by an excessive degree of excitatory transmission, except perhaps during a seizure itself. Such a state of interictal normality would be expected, especially if the primary epileptic defect were not within the excitatory neurons. If excitatory neurons were directly responsible for seizure predisposition, an elevated index of such transmission would be quite probable during the seizure-naive state or the interictal state. If inhibitory neurons were directly responsible for the seizure predisposition, whereas excitatory neurons were participating in a manner secondary to the primary defect, an excitatory increment might well be undetectable during these nonictal periods.

Studies in the epileptic gerbil provide little or no evidence that excitatory amino acid transmission plays an important role in the seizure-prone state of these animals (Loscher, 1985). In this regard, the glutamate receptor antagonist, 2-amino-7-phosphonoheptanoic acid (2-APH) blocked major seizures in epileptic gerbils only in doses that impaired righting reflexes. Another glutamic acid receptor antagonist, 2-amino-5-phosphonopentanoic acid (2-APP), appeared to be less active than 2-APH. No comparisons were made to nonepileptic gerbils so that the question of epileptic specificity remains undetermined.

2.4. Cholinergic Systems

Cholinergic abnormalities may be associated with the seizure-prone state of various types of genetically epileptic mice such as DBA/2J, EP, and A2G (see Chapman and Meldrum, 1986). In our view, the existing pathophysiological data are consistent in supporting the concept that an abnormally high level of central cholinergic activity exists in EP mice. However, evidence of a similar abnormality in this system in the DBA/2J and A2G mice is subject to equivocal interpretation. Some of the measured neurochemical indices suggest an increased cholinergic increment in these two types of epileptic mice, whereas one index suggests a decrement.

Documentation that a defect in cholinergic systems in the genetically epilepsy-prone rat is responsible for the seizure-prone state of these animals has not been forth coming (*see* Laird and Jobe, 1986). The limited studies that have been performed to date provide evidence that seizure-naive animals of this epileptic strain have normal indices of cholinergic function. Nevertheless, these seemingly normal cholinergic neurons may participate in the seizure process. Undoubtedly, normal neurons can be recruited as carriers of excessive discharges of seizure episodes. In support of this view, audiogenic seizure-induced cholinergic abnormalities have been detected in the genetically epilepsy-prone rat (*see* Laird and Jobe, 1986).

2.5. Noradrenergic Systems

Noradrenergic neurons are capable of producing marked influences on seizure activity. Table 2 summarizes the noradrenergic potential for seizure regulation. References to the literature are indicated in the footnotes of this table. In some models of epilepsy, exogenously administered noradrenergic agonists have the capacity to suppress seizure activity, whereas drug-induced noradrenergic decrements have the reverse effect. Evidence for a noradrenergic role in regulating seizure activity is substantial and straightforward for the genetically epilepsy-prone rat. Although less persuasive, some data also support the concept that drug-induced noradrenergic activity suppresses seizures in epileptic humans, kindled rodents, and the electroshock and topical cobalt models.

Are these models that exhibit an anticonvulsant response to noradrenergic drugs also characterized by noradrenergic abnormalities? The information summarized in Table 2 suggests that the answer is yes. Accordingly, functional noradrenergic deficits appear to exist in the genetically epilepsy-prone rats, kindled rodents, and topical cobalt models. All three of these models are forms of chronic seizure expression. Human epilepsy is also a chronic disorder. Whether noradrenergic deficits characterize epileptic humans is not known.

In some other epilepsy models, noradrenergic influences seem to be opposite to those present in the genetically epilepsy-prone rats, kindled rodents, and rats treated with topical cobalt. For example, drug-induced noradrenergic increments may exacerbate seizures in the epileptic mice (DBA/2J and tottering) and in the epileptic gerbil. Tottering epileptic mice are especially interesting because non-drug-treated subjects have an increment in the number of central noradrenergic terminals. When the responses to pharmacological agents are considered in concert with innate noradrenergic increments, it is reasonable to speculate that elevated noradrenergic activity may participate in the epileptic state of the tottering mice.

Neurochemical localization of seizure-regulating noradrenergic neurons will contribute importantly to an understanding of epileptic processes. Evidence for identifying brain areas important in regulating audiogenic seizures in the genetically epilepsy-prone rat (*see* Laird and Jobe, 1986; Jobe et al., 1986) and in the tottering mouse (*see* Chapman and Meldrum, 1986) has emerged. Also, there are some data relative to other epilepsy models, such as kindled rodents (*see* McNamara et al., 1986) and electroshock seizures (*see* Browning, 1986).

In summary, drug-induced modifications of noradrenergic activity appears to influence seizure activity in numerous models of epilepsy. In a few of these models, abnormalities in noradrenergic transmission may be determinants of seizure predisposition or resistance.

2.6. Other Neurotransmitter Systems

Dopaminergic, serotonergic, and opioid systems also appear to participate in seizure processes in some models of epilepsy. Presently, it appears that dopaminergic systems may regulate seizures in epileptic (DBA/2J) mice (*see* Chapman and Meldrum, 1986). In contrast, the dopaminergic system does not appear to influence seizures in the genetically epilepsy-prone rats (*see* Laird and Jobe, 1986). Opioid peptide neurons may eventually be shown to participate as determinants of the epileptic diathesis in the epileptic gerbil (Lomax et al., 1986). The possibility that opioid systems may

TABLE 2
Noradrenergic Role in Seizure Regulation[a,b]

Model	Suppress seizures	Defect reported	Proposed roles	
			Physiological	Pathophysiological
Epileptic mice, DBA/2J	Doubtful, increments may worsen seizures	Decreased alpha-1 receptor density; NE levels are variable; Lower COMT activity; Elevated NE synthesis rate	Participates in regulation of neuronal activity	Drug data suggests that noradrenergic increments may exacerbate seizures; Innate abnormalities present in opposing directions. Etiologic role remains unclear
Epileptic mice, Tottering	Increments may cause seizures	Increased number of NE terminals	Participates in regulation of neuronal activity	Innate increments may participate as a cause of seizures
GEPRs	Yes	Decreased NE levels, turnover rate, tyrosine hydroxylase activity, alpha-1 receptor density, facilitation of GABAergic inhibition, neuronal number	Participates in regulation of neuronal activity	Innate decrements in NE activity act as partial causes of the epileptic state

Epileptic gerbils	Doubtful, decrement may suppress seizures	None	Participates in regulation of neuronal activity	Not known
Epileptic chickens	No known relationship	Increased NE levels; Decreased turnover rate	Participates in regulation of neuronal activity	Not known
Epileptic humans	Perhaps	Not known	Participates in regulation of neuronal activity	Not known
Kindled seizures	Yes, drug-induced decrements facilitate development of fully kindled state. Repletion of stores reduces seizure severity. Decrements increase seizure severity in fully kindled animals	Long-lasting abnormality not adequately documented. Abnormalities that have been detected may have occurred as consequences of seizure activity	Participates in regulation of neuronal activity	Some nervous systems may have deficient noradrenergic activity and thereby kindle in response to stimuli that otherwise would not have resulted in kindled seizures
Electroshock seizures	icv NE attenuates tonic extensor convulsions and	Increased tyrosine hydroxylase activity, NE synthesis and	Participates in regulation of neuronal activity	Noradrenergic increments may occur as a consequence of seizures.

(continued)

TABLE 2 *(continued)*
Noradrenergic Role in Seizure Regulation[a,b]

Model	Suppress seizures	Defect reported	Proposed roles	
			Physiological	Pathophysiological
	perhaps general-ized clonus	turnover, V_{max} for uptake; Decreased high affinity NE uptake		These increments provide increased protection against additional seizures
Bicuculline seizures	Evidence suggests absence of NE activity in regulating these seizures	Increased tyrosine hydroxylase activity, after bicuculline seizures; Decreased NE levels after bicuculline seizures	Participates in regulation of neuronal activity	Noradrenergic abnormalities associated with bicuculline seizures in normal animals may not play an etiological role. The abnormalities may represent a response to seizures rather than a partial cause
Topical cobalt seizures	NE precursor may suppress discharges	Decresed NE terminal density in perifocal area. Sprouting of	Participates in regulation of neuronal activity	Induced NE decrements may act as partial cause of epileptic state.

NE fibers occurs
with disappearance
of epileptic state;
Increased beta-recep-
tor density and
tyrosine hydroxylase
activity, plus super-
sensitivity to applied
NE

Whether postsynaptic
receptor upregulation
completely compen-
sates for presynaptic
NE defect is not fully
resolved

[a]Citations: DBA/2J epileptic mice: Chapman and Meldrum, 1986; Dailey and Jobe, 1984, Nyquist-Battie et al., 1986; totter-ing mice: Chapman and Meldrum, 1986; GEPRs (genetically epilepsy-prone rats): Laird and Jobe, 1986; epileptic gerbils: Cox and Lomax, 1976; Lomax et al., 1986; epileptic chickens: Johnson and Tuchek, 1986; human epilepsy: Kresh et al., 1986; kindled seizures: McNamara et al., 1986; electroshock seizures: Hendley, 1976; Masserano et al., 1981; Schildkraut and Draskoczy, 1974; bicuculline seizures: Faingold, 1986; Calderini et al., 1978; Gale et al., 1983; Ingvar, et al., 1983; topical cobalt: Craig and Colasanti, 1986; Trottier et al., 1981; Bregman et al., 1985; Clayton and Emson, 1975; Schuvee-Moreau et al., 1977.

[b]Abbreviations: icv, intracerebroventricular; COMT, catechol-*O*-methyltransferase; NE, norepinephrine.

regulate seizures in other models is intriguing (*see* Browning, 1986; McNamara et al., 1986).

3. Unifying Principles and the Emerging Conceptual Matrix

Can we identify unifying principles for the roles of the neurotransmitters in epilepsy? Does evidence suggest that a single neurotransmitter abnormality is responsible for epilepsy? Is a certain combination of neurotransmitter abnormalities causative? Perhaps some neurotransmitter abnormalities initiate seizures, whereas others may be responsible for seizure predisposition without the capacity to initiate.

The data summarized in this chapter suggest that any one neurotransmitter system may play a diverse role in epilepsy. Moreover, a variety of transmitter systems has been associated with the seizure state within single models and across models.

Yet within this diversity some unifying principles are emerging. Although experiments of the future may alter many of the insights that now appear to be valid, we believe that some instructive value is inherent in setting forth principles predicated upon our current views.

3.1. Emerging Principles

GABAergic and noradrenergic indices are abnormal within a broad section of epilepsy models. Aberrant GABAergic activity is present in genetically epileptic mice, rats, gerbils, chickens, and in the kindling, electroshock, bicuculline, topical cobalt, and in vitro slice models of epilepsy. Noradrenergic abnormalities are present in most of these same models, with the possible exception of the epileptic gerbil.

The role of a particular transmitter system may be different from one model of epilepsy to another. In one case, the GABAergic decrements may be a partial cause of seizures in the bicuculline, topical cobalt, and in vitro slice models. In the second case, the GABAergic increments may compensate partially for seizure predisposition in the genetically epilepsy-prone rats, epileptic chickens, and kindled rodents. In the last case, GABAergic increments may be a cause of seizure predisposition through a process of disinhibition.

In either of the first two cases, GABAergic neurons are exerting anticonvulsant influences on the brain. Yet because the influence is functionally inadequate, seizures are initiated in the first case and seizure predisposition remains apparent in the second case. In the third instance, a proliferation and derangement of GABAergic synapses causes excessive inhibition of other GABAergic neurons, thereby reducing the net inhibitory influence on excitatory activity.

A great diversity of roles also seems apparent for noradrenergic neurons. Noradrenergic increments appear to suppress seizures in the genetically epilepsy-prone rat, kindled rodents, and animals exposed to electroshock and topical cobalt. In contrast, noradrenergic increments may exert a proconvulsant influence in epileptic mice and epileptic gerbils.

Abnormalities in some neurotransmitter systems appear capable of initiating seizure activity, whereas abnormalities in others appear to cause seizure predisposition without precipitating seizure episodes. Accordingly, drug-induced decrements in GABAergic transmission appear to initiate seizures in nervous systems that are neurologically normal before administration of the anti-GABAergic agent. In contrast, drug-induced noradrenergic decrements do not appear to initiate seizures. Convulsions do not seem to occur as a result of noradrenergic decrements alone. However, drug-induced noradrenergic decrements do lower the electroshock seizure thresholds in rats. Moreover, in the genetically epilepsy-prone rats, innate noradrenergic decrements do lower the electroshock seizure thresholds in rats. Moreover, in the genetically epilepsy-prone rats, innate noradrenergic decrements appear to be one of the determinants susceptibility and seizure severity. Certain types of seizures occur in these animals that do not occur in neurologically normal subjects. These unusual states of seizure susceptibility appear to be partially dependent on noradrenergic defects. Also, some genetically epilepsy-prone rats are characterized by more severe seizures than other types of such rats. This difference in severity appears to be partially dependent on noradrenergic defects.

Seizure predisposition that is dependent on one set of neurochemical factors may include exaggerated sensitivity to some but not all seizure-initiating modalities. For example, in the genetically epilepsy-prone rat, which is characterized by a certain set of neurochemical aberrations, seizure predisposition includes susceptibility to seizures initiated endogenously (spontaneous seizures) or by audiogenic or hyperthermic stimuli (*see* Laird and Jobe, 1986). Other manifestations of predisposition in this model include an exaggerated sensitivity to seizures produced by morphine administered intracerebroventricularly, bicuculline injected directly into the inferior colliculus, or the barbiturate withdrawal syndrome, as well as by electroconvulsive shock. In contrast, seizure predisposition in the genetically epilepsy-prone rat does not appear to include responses to strychnine, benzodiazepines administered rapidly by the intravenous route, tricyclic or tetracyclic antidepressant drugs, or systemically administered bicuculline.

Seizure predisposition that is dependent on one set of neurochemical factors may include exaggerated sensitivity or resistance to some but not all anticonvulsant modalities. These abnormal responses that occur because of underlying neurochemical abnormalities could form a basis for detecting antiepileptic specificity. For future development of clinically useful medications, it may be helpful to regard antiepileptic drug specificity as a special propensity of a pharmacological agent to produce an anticonvulsant effect in animals with a documented seizure predisposition.

In line with this type of reasoning, abnormal anticonvulsant responses would be expected for noradrenergic agonists in the genetically epilepsy-prone rat and in the epileptic tottering mouse, dopaminergic agonists in the genetically epileptic DBA/2 mice, and GABAergic agonists in the topical cobalt model. These possibilities require careful experimental scrutiny. The probability that taurine causes a specific antiepileptic effect in the genetically epilepsy-prone rat is of special interest. Some experimental verification has already been obtained. A small amount of this substance injected into the inferior colliculi raises the electroshock seizure threshold in the genetically epilepsy-prone rats, but not in nonepileptic controls. Moreover, the genetically epilepsy-prone rat has a documented abnormality in high-affinity taurine transport.

The anticonvulsant mechanism of action of a drug may not provide insights into the neuronal abnormalities that cause seizure predisposition. In almost every model of epilepsy, agents that augment GABAergic activity have the capacity to produce anticonvulsant effects. The evidence for this GABAergic phenomenon is substantial. Yet strong support is lacking for the concept that GABAergic abnormalities are responsible for seizure predisposition. Indeed, GABAergic abnormalities in epileptic conditions may partially compensate for seizure predisposition produced through other neurochemical mechanisms.

Epilepsy in different patients may be caused by different underlying neurochemical causes. Clinical distinctions between different forms of epilepsy may be inadequate for distinguishing between the different neurochemical causes of epilepsy. Some forms of clinical epilepsy may appear diagnostically similar but have divergent neurochemical causes. Also, some forms of human epilepsy may appear clinically divergent, yet have similar neurochemical causes. Evidence from animal models of epilepsy supports these concepts.

4. Future Directions

The search for the neurochemical bases of the epilepsies needs a better degree of balance. The search has heretofore focused too sharply on GABAergic mechanisms and on the processes of seizure initiation. Yet the epilepsies may result from a diverse array of neurochemical derangements, only one of which may be GABAergic. Moreover, the process by which seizures are initiated is only one of a group of neurobiological factors that probably account for the epilepsies. What are the processes that determine whether a brain will become epileptic in response to an initiating event? Some brains do not exhibit seizures despite the presence of localized or perhaps generalized triggering events. Neurologically normal nervous systems are resistant to seizures. What are the neurobiological factors that cause seizure predisposition? Neuronal abnormalities that allow responses to the triggering events should be subjected to a more intensive investigation.

The neuroanatomical localization of neurotransmitter systems that may function as determinants of epileptic states will provide a substantial improvement in the conceptual understanding of seizure disorders. Useful information has been generated regarding the sites in which GABAergic regulation may be prominent in the normal nervous system. Whether these sites are mechanistically responsible for seizure predisposition in any of the genetic models of epilepsy or in kindling is unknown. Other studies have suggested sites that may be relevant for GABAergic participation in seizure processes in the epileptic gerbil and for noradrenergic and serotonergic participation in the genetically epilepsy-prone rat.

Another important future goal is to identify and develop treatments that offer antiepileptic drug specificity rather than anticonvulsant specificity. This process must use animal models of epilepsy that are capable of separating these two types of drug effects. Moreover, the process will be facilitated by relying upon information developed in studies of neurochemical etiologies of the various epilepsies in humans and other animals.

Future investigations with human epilepsy should include improved paradigms for elucidating neurochemical abnormalities among epileptic patients and for identifying seizure disorders with different neurochemical etiologies. Nuclear magnetic reasonance techniques may be helpful for identifying at least some of the neurotransmitter defects that underlie the various human epilepsies. The power of this technique for the quantitative spectrographic analyses of in vitro neurochemistry is only now beginning to be appreciated by the biomedical community. Also, can the coexistence of one or more of the epilepsies with other neurological diseases provide hints for distinguishing between the different forms of the epilepsies? As a matter of speculation, we might ask whether the etiology of epilepsies coupled with major affective disorder are different from the neurochemical etiologies of epilepsies that occur in the absence of neurological or behavioral disorders.

References

Bonhaus, D. W. and Huxtable, R. J.: Seizure-susceptibility and decreased taurine transport in the genetically epileptic rat. *Neurochem. Int.* **6:** 365–368, 1984.

Bonhaus, D. W., Laird, H., Mimaki, T., Yamamura, H. I., and Huxtable, R. J.: Possible Bases for the Anticonvulsant Action of Taurine. In: *Sulfur Amino Acids: Biochemical and Clinical Aspects* (K. Kuriyama, R. J. Huxtable, and H. Iwata, eds.) A. R. Liss, New York, 1983.

Bregman, B., Le Saux, F., Trottier, S., Chauvel, P., and Maurin, Y.: Chronic cobalt-induced epilepsy: Noradrenaline iontophoresis and adrenoceptor binding studies in the rat cerebral cortex. *J. Neural. Trans.* **63:** 109–118, 1985.

Browning, R. A.: The Role of Neurotransmitters in Electroshock Seizures Models. In: *Neurotransmitters and Epilepsy* (P. C. Jobe and H. E. Laird, II, eds.) Humana, Clifton, New Jersey (in press), 1986.

Calderini, G., Carlsson, A., and Nordstrom, C. H.: Monoamine metabolism during bicuculline-induced epileptic seizures in the rat. *Brain Res.* **157**(2): 295–302, 1978.

Chapman, A. G. and Meldrum, B. S.: Epilepsy Prone Mice: Genetically-Determined Sound-Induced Seizures. In: *Neurotransmitters and Epilepsy* (P. C. Jobe and H. E. Laird, II, eds.) Humana, Clifton, New Jersey (in press), 1986.

Chapman, A. G., Faingold, C. L., Hart, G. P., Bowker, H. M., and Meldrum, B. S.: Brain regional amino acid levels in seizure susceptible rats: Changes related to sound-induced seizures. *Neurochem. Int.* (in press), 1986.

Clayton, P. R. and Emson, P. C.: Changes in monoamine-related enzymes in cobalt-induced epilepsy. *Biochem. Soc. Trans.* **3:** 261–263, 1975.

Commission on Classification and Terminology of the International League Against Epilepsy: Proposal for Classification of Epilepsies and Epileptic Syndromes. *Epilepsia* **26**(2): 268–278, 1985.

Consroe, P., Kudray, K., and Schmitz, R.: Acute and chronic antiepileptic drug effects in audiogenic seizure-susceptible rats. *Exp. Neurol.* **70:** 626–637, 1980.

Cox, B. and Lomax, P.: Brain amines and spontaneous epileptic seizures in the mongolian gerbil. *Pharmacol. Biochem. Behav.* **4:** 263–267, 1976.

Craig, C. R. and Colasanti, B. K.: Experimental Epilepsy Induced by Direct Topical Placement of Chemical Agents on the Cerebral Cortex. In: *Neurotransmitters and Epilepsy* (P. C. Jobe and H. E. Laird, II, eds.) Humana, Clifton, New Jersey (in press), 1986.

Dailey, J. W. and Jobe, P. C.: Effect of increments in the concentration of dopamine in the central nervous system of audiogenic seizures in DBA/2J mice. *Neuropharmacology* **23**(9): 1019–1024, 1984.

Dailey, J. W. and Jobe, P. C.: Anticonvulsant drugs and the genetically epilepsy-prone rat. *Fed. Proc.* **44:** 2640–2644, 1985.

Essman, E. J. and Essman, W. B.: Synaptosomal GABA uptake and receptor binding effects of a convulsion. *Brain Res. Bull.* **5:** 209–211, 1980.

Faingold, C. L.: Seizures Induced by Convulsant Drugs. In: *Neurotransmitters and Epilepsy* (P. C. Jobe and H. E. Laird, II, eds.) Humana, Clifton, New Jersey (in press), 1986.

Faingold, C. L., Meldrum, B. S., and Millan, M. H.: Blockade of audiogenic seizure susceptibility by focal injection into the inferior colliculus of an excitant amino acid antagonist. *Br. J. Pharmacol.* **84:** 95P, 1984.

Fromm, G. H., Amores, C. Y., and Thies, W.: Imipramine in epilepsy. *Arch Neurol.* **27:** 198–204, 1972.

Fromm, G. H., Wessel, H. B., Glass, J. D., Alvin, J. D., and Van Horn, G.: Imipramine in absence and myoclonic-astatic seizures. *Neurology* **28:** 953–957, 1978.

Gale, K., Iadarola, M. J., and Casu, M.: Anticonvulsant action of GABA-ergic drugs: Cellular and anatomical localization of their site of action *in vivo*. *Prog. Clin. Biol. Res.* **124:** 39–62, 1983.

Gjerstad, L., Dingeldine, R., Andersen, P., and Langmoen, I. A.: Reduction of Two Types of GABAergic Inhibition During Epileptiform Activity in Hippocampal Pyramidal Cells. In: *Physiology and Pharmacology of Epileptogenic Phenomena* (M. R. Klee, H. D. Lux, and E-J. Speckmann, eds.) Raven, New York, 1982.

Hendley, E. D.: Electroconvulsive shock and norepinephrine uptake kinetics in the rat brain. *Psychopharmacol. Comm.* **2:** 17–25, 1976.

Huxtable, R. J., Laird, H., Bonhaus, D., and Thies, A. C.: Correlations between amino acid concentrations in brains of seizure-susceptible and seizure-resistant rats. *Neurochem. Int.* **4:** 73–78, 1982.

Ingvar, M., Lindvall, O., Folbergrova, J., and Siesjo, B. K.: Influence of lesion of the noradrenergic locus coeruleus system on cerebral metabolic response to bicuculline-induced seizure. *Brain Res.* **264**(2): 225–231, 1983.

Jobe, P. C., Brown, R. D., and Dailey, J. W.: Effect of Ro 4-1284 on audiogenic seizure susceptibility and intensity in epilepsy-prone rats. *Life Sci.* **28:** 2031–2038, 1981.

Jobe, P. C., Reigel, C. E., Jr., Mishra, P. K., and Dailey J. W.: Neurotransmitter Abnormalities as Determinants of Seizure Predisposition in the Genetically Epilepsy-Prone Rat. In: *Neurotransmitters and Epilepsy III,* (J. Engel, R. G. Fariello, K. G. Lloyd, P. L. Morselli, and G. Nistico, eds.) Raven, New York (in press), 1986.

Jobe, P. C., Woods, T. W., and Dailey, J. W.: Proconvulsant and Anticonvulsant Effects of Tricyclic Antidepressants in Genetically Epilepsy-Prone Rats. In: *Advance in Epileptology: XVth Epilepsy International Symposium* (J. Roger Porter, ed.) Raven, New York, 1984.

Johnson, D. D. and Tuchek, J. M.: The Epileptic Chickens. In: *Neurotransmitters and Epilepsy* (P. C. Jobe and H. E. Laird, II, eds.) Humana, Clifton, New Jersey (in press), 1986.

Jones, G. L. and Woodbury, D. M.: Anticonvulsant structure activity relationships: Historical development and probable causes of failure. *Drug. Dev. Res.* **2:** 333–355, 1982.

Krall, R. L., Penry, J. K., Kupferberg, H. J., and Swinyard, E. A.: Antiepileptic drug development. I. History and a program for progress. *Epilepsia* **19:** 393–408, 1978a.

Krall, R. L., Penry, J.. K., White, B. G., and Kupferberg, H. J.: Antiepileptic drug development. II. Anticonvulsant drug screening. *Epilepsia* **19:** 409–428, 1978b.

Kresh, M. J., Shaywitz, B. A., Shaywitz, S. B., Anderson, G. M., Leckman, J. L., and Cohen, D.: Neurotransmitters and Human Epilepsy. In: *Neurotransmitters and Epilepsy* (P. C. Jobe and H. E. Laird, II, eds.) Humana, Clifton, New Jersey (in press), 1986.

Krnjevic, K.: Loss of Synaptic Inhibition as a Cause of Hippocampal Seizures. In: *Physiology and Pharmacology of Epileptogenic Phenomena* (M. R. Klee, H. D. Lux, and E-J. Speckmann, eds.) Raven, New York, 1982.

Kuriyama, K., Shuji, I., Chihiro, N., and Seitaro, O.: Distribution and Function of Taurine in Nervous Tissues: An Introductory Review. In: *Sulfur Amino Acids: Biochemical and Clinical Aspects* (K. Kuriyama, R. J. Huxtable, and H. Iwata, eds.) A. R. Liss, New York, 1983.

Laird, H. E. and Huxtable, R.: Taurine and Audiogenic Epilepsy. In: *Taurine and Neurological Disorders* (A. Barbeau and R. J. Huxtable, eds.) Raven, New York, 1978.

Laird, H. E., II and Jobe, P. C.: The Genetically Epilepsy-Prone Rat. In: *Neurotransmitters and Epilepsy* (P. C. Jobe and H. E. Laird, II, eds.) Humana, Clifton, New Jersey (in press), 1986.

Lomax, P., Lee, R. J., and Olsen, R. W.: The Spontaneously Epileptic Mongolian Gerbil. In: *Neurotransmitters and Epilepsy* (P. C. Jobe and H. E. Laird, II, eds.) Humana, Clifton, New Jersey (in press), 1986.

Loscher, W.: Influence of pharmacological manipulation of inhibitory and excitatory neurotransmitter systems on seizure behavior in the mongolian gerbil. *J. Pharmacol. Exp. Ther.* **233:** 204–213, 1985.

Loscher, W. and Meldrum, B. S.: Evaluation of anticonvulsant drugs in genetic animal models of epilepsy. *Fed. Proc.* **43:**(2) 276–284, 1984.

Loscher, W., Frey, H.-H., Reiche, R., and Schultz, D.: High anticonvulsant potency of γ-aminobutyric acid (GABA)mimetic drugs in gerbils with genetically determined epilepsy. *J. Pharmacol. Exp. Ther.* **226:** 839–844, 1983.

Masserano, J. M., Takimoto, G. S., and Winer, N.: Electroconvulsive shock increases tyrosine hydroxylase activity in the brain and adrenal gland of the rat. *Science* **214:** 662–664, 1981.

McNamara, J. O., Bonhaus, D. W., Crain, B. J., Gellman, R. L., and Shin, C.: Biochemical and Pharmacologic Studies of Neurotransmitters in the Kindling Model. In: *Neurotransmitters and Epilepsy* (P. C. Jobe and H. E. Laird, II, eds.) Humana, Clifton, New Jersey (in press), 1986.

Meldrum, B. S., Millan, M. H., and Faingold, C. L.: Seizure induction by focal infusion of exitant amino acids into the inferior colliculus and superior colliculus. *Soc. Neurosci. Abst.* **11:** 2, 1317, 1985.

Nyquist-Battie, C., Lints, C. E., and Stearn, D.: Midbrain alpha-adrenoreceptors in audiogenic seizure susceptible mice. *Life Sci* (in press), 1986.

Ojemann, L. M., Friel, P. N., Trejo, W. J., and Dudley, D. L.: Effect of doxepin on seizure frequence in depressed epileptic patients. *Neurology* **33:** 646–648, 1983.

Peterson, G. M., Ribak, C. E., and Oertel, W. H.: Differences in the hippocampal GABAergic system between seizure-sensitive and seizure-resistant gerbils. *Anat. Rec.* **208:** 137A, 1984.

Piredda, S., Lim, C. R., and Gale, K.: Intracerebral site of convulsant action of bicuculline. *Life Sci.* **36:** 1295–1298, 1985.

Prince, D. A.: Mechanisms of Epileptogenesis in Brain-Slice Model Systems. In: *Epilepsy* (A. A. Ward, Jr., J. K. Penry, and D. Purpura, eds.) Raven, New York, 1983.

Roberts, R. and Ribak, C. E.: Immunocytochemical evidence for GABAergic abnormalities in the genetically epilepsy-prone rat. *Life Sci.* (in press), 1986.

Schildkraut, J. J. and Draskoczy, P. R.: Effects of Electroconvulsive Shock on Norepinephrine Turnover and Metabolism: Basic and Clinical Studies. In: *Psychobiology of Convulsive Therapy* (M. Fink, S. Kety, J.McGaugh, and T. A. Williams, eds.) John Wiley, New York, 1974.

Schuvee-Moreau, J., Lepot, M., Brotchi, J., Gerebtzoff, M. A., and Dresse, A.: Action of phenytoin, ethosuximide and of the carbidopa-L-dopa association in semi-chronic cobalt-induced epilepsy in the rat. *Arch. Int. Pharmacodyn.* **230:** 92–99, 1977.

Traub, R. D., Wong, R. K. S., and Miles, R.: In Vitro Models of Epilepsy. In: *Neurotransmitters and Epilepsy* (P. C. Jobe and H. E. Laird, II, eds.) Humana, Clifton, New Jersey (in press), 1986.

Trottier, S., Chauvel, P., Dedek, J., and Gay, M.: Alterations of the cortical noradrenergic system in chronic cobalt epileptogenic foci in the rat: A histofluorescent and biochemical study. *Neuroscience* **6:** 1069–1080, 1981.

Wong, R. K. S.: Postsynaptic Potentiation Mechanism in the Hippocampal Pyramidal Cells. In: *Physiology and Pharmacology of Epileptogenic Phenomena* (M. R. Klee, H. D. Lux, and E.-J. Speckman, eds.) Raven, New York, 1982.

Index

Absence (petit mal) seizures, 10, 62, 63, 98, 99, 181, 202, 207, 242, 322

Acetazolamide, 291

Acetylcholine (ACh), 5, 6, 22, 29, 65, 75, 79, 80, 88, 120–123, 146–148, 165, 171, 195, 198, 206, 208, 215, 217, 222, 224, 225, 228, 229, 235, 237, 240, 241, 243, 244, 246, 248–250, 285–287, 302, 305, 308–310, 352

Acetylcholinesterase (AChE), 22, 80, 198, 217, 228, 233, 237, 246, 249, 250

Adrenergic
  agonists (*see* Noradrenergic agonists/antagonists)
  neurotransmission (*see* Noradrenergic)
  receptor (*see* Noradrenergic receptors)
  system, 126

Afterdischarge, 116, 117, 121, 124, 125, 132–135, 142, 167, 176–180, 244

Akinetic seizures, 181

Aldrin, 224

Allylglycine, 222, 224, 235, 240, 248, 249

Alpha-methyl-*m*-tyrosine (*see also* Noradrenergic and Dopaminergic), 66

Alpha-methyl-*p*-tyrosine (AMPT) (*see also* Noradrenergic and Dopaminergic), 26, 125, 288–290, 309

Alpha-receptors (*see also* Noradrenergic receptors), 73–76, 125, 126, 195, 354

Alumina model of epilepsy, 192–195, 203, 208, 220

Amino acid transmitters (*see also* Excitatory amino acid agonists/ antagonists and Inhibitory amino acid transmitters), 4, 29, 64, 80, 88, 333

Aminooxyacetic acid (AOAA), 25, 142, 199, 205, 208, 251, 299–301

Amitriptyline (*see also* Anti depressants), 63, 126

Amphetamine, 28, 290, 294, 331

Amygdala, 116, 117, 119, 121, 122, 124, 127–135, 137, 138, 140, 141, 143, 144, 225, 286, 305, 324, 325, 332

Amygdaloid kindling, 117, 122, 124, 131, 134, 137, 138, 140, 143, 146, 303

Anoxic brain damage, 329

Anterior neocortex, 124

Anterior sigmoid gyrus, 196

Anticholinergic drugs, 287

Anticonvulsants, 3, 21, 23, 24, 26, 28, 33, 42–45, 49–53, 62, 63, 68, 75, 78, 81, 84, 96–99, 106–108, 110, 117, 132, 133, 139, 141, 142, 195, 199, 200, 202, 203, 205, 206, 208, 221, 226, 234, 246, 249, 250, 282, 286, 287, 290, 291, 297, 301, 302, 304, 306, 308, 310, 328, 331–334, 340–342, 348, 359, 360

BASKETVILLE®

The Autobiography of

Frank G. Wilson

VANTAGE PRESS
New York • Los Angeles

Contents

BASKETVILLE®

1.

Stricken by an Unknown Disease

I GOT MY YEARLY PHYSICAL checkup from Dr. Burtiss of Brattleboro in December 1987. I visit our retail basket stores on the way to Venice, Florida, and work at the Florida store through March, so I take care of this first. Dr. Burtiss forgot to give me a flu shot, which is recommended in Florida for people over sixty. My wife, Connie, and I planned to take a side trip to Mexico to buy baskets for the stores and also to find some nice Indian baskets for my Putney, Vermont, basket museum. When I got to Venice, I visited a drive-in clinic for the shot.

We traveled to Mexico and had a good trip and purchased some nice Indian baskets made in the interior of Mexico at Oaxaca. Six weeks later, when I got back from Mexico, while attending a one-lunger engine show in Zolfo Springs, Florida, I started to come down with the sickness. The illness started when I woke at daylight at the Zolfo Springs show. While I was walking around the asphalt pavement at the show, my legs felt sort of funny. I can't really explain the feeling. As I am used to walking in the forest, I moved to the grassy section of the show area. After a few hours of walking at the affair, I stooped down to get a better look at some information on a gas engine I was thinking of purchasing for my museum. My body felt completely disjointed, and I fell to the ground. There were quite a few people around, and no one paid any attention to me. They probably thought I was drunk! I got up after a while and still did not realize what was going on. I continued at the show for an hour or so, and I told myself that something had to be wrong with me. From this point, it was two days before I was admitted to the Venice Hospital and three days before I became paralyzed.

This was March 4, 1988, and it took about four days for the doctors to figure out what I had. I was diagnosed as having Gullain-Barré syndrome, caused by the flu shot. They told me that it could have happened any of the times I received the flu shots. In my case, I was given a swine flu shot in grade school, as the whole class got one because of the swine flu scare. Also, while I was in the U.S. Navy we got all kinds of shots and I am sure the flu shot was included. I received flu shots every fall after I reached the age of sixty-two.

I believe it was the last flu shot I received in Florida that did me in, and I ended up in the Venice, Florida, Hospital.

I had to use a catheter all the time until I completed my rehabilitation. The nurses had to use a hydraulic lift for a time to get me in and out of bed for therapy.

All of the doctors and nurses at the Venice hospital did a fine job, but I have little recollection of what went on. Finding that it would be a long illness, after four weeks my family decided to fly me home to the Brattleboro Memorial Hospital. We got home in a Lear jet in two and a half hours and landed at the Keene, New Hampshire, airport from where I was taken to Brattleboro by ambulance.

Honestly, I remember little of how hard and well the doctors and nurses helped me in Brattleboro also. I was paralyzed from my legs to my toes, and my arms and hands would not function. It was a big deal when I could just wiggle a toe. The Brattleboro staff got me up to a position where they felt I could be moved to the Farnham Rehabilitation Center in Keene, New Hampshire. In Keene I was able to get out of bed and mix with the occupational and physical therapists and become familiar with all the nurses and doctors. Much of my Keene rehabilitation time I can remember, so I got a better idea of what I was going through with the terrible illness. I can now better put a name on and give thanks to those helpful people in Keene who directed me back to health, but hesitate to do so in fairness to all those before who did the job they did in the times during my illness that I cannot remember.

At this time, six months later, I can do just about anything I want to do. There is still a lot more muscle and nerve repair to be done and a lot I have to work on every day.

Yes, in Florida I certainly was very sick. My whole body was like a great big charley horse. I wanted to die and get out of my misery as the syndrome crept up my body from my toes to my fingers. Once I was paralyzed, the sharp pains ended and I decided to stay around.

The way I figure it is, I have led a wonderful life. I have had a lot of fun in my work and leisure and have been just about everywhere in the world, so it was just my turn to get ill. Earlier I felt that there were few people who had such a full and happy life. I have always been restless and energetic and have had more than my share of varied experiences. I have always been inquisitive and learned why and how other people lived, played, and worked, and I have always been anxious to know what was on the other side of the mountain.

I learned the bigger a person in government or business is, the easier he or she is to know, get along with, and learn from. I've always believed that one should go to the top when seeking help or advice and tell things like they are so he doesn't have to remember the lies.

As I came from a poor but happy family of thirteen, I know what it is like to be poor, and I have a good idea of how it feels to be rich. It must be hard on those who don't know otherwise—the ones who are filthy rich and never had to meet a payroll or have never been cold or hungry. There is evidence in our society of these bleeding hearts and do-gooders with complete idle time and plenty of cash who are tearing down American society as our forefathers envisioned it. These people have missed a lot, and certainly the point is that it is not everything to be born with a silver spoon in your mouth.

After being paralyzed and out of pain, I decided that with a little help and luck I would like to live. There are many things of interest to me that are not finished yet, and I have always gotten a kick out of life. After the first few days of the sickness until now, there has been little pain. There have been continued aches, but I learned to stand them. The worst ones were when the Gullain Barré syndrome left my body and the muscles and nerves started to come back. It is uncomfortable but bearable. Actually, I felt better when I was paralyzed than I did when I was coming out of the paralysis. In the long haul I expect to regain better than

90 percent of my health when this syndrome has left all of me.

My mind would like to do great things, but I still tire easily. Be assured I am very happy with my situation and I am in a good frame of mind.

While at the Farnum Center, I read the book *No Laughing Matter,* by Joseph Heller and Speed Vogel. It is about Heller's bout with the same illness as I, Gullain Barré syndrome. It is hard for me to understand how Heller and Speed could write a whole book, and a good book, about Heller's catching and overcoming Gullain-Barré syndrome.

Also while convalescing, I read *The Art of the Deal,* by Donald Trump. Part of my therapy included using my hands and fingers to combat the loss of strength and mobility caused by the illness. My friends and therapist suggested I write this autobiography using the hospital computer. I thought that if Heller and Speed could write a whole book on one sickness and Donald Trump could write on how to make a deal, I ought to be able to write about my sixty-nine years' worth of life experience most interestingly, because I wrote the book on how to manufacture, import, and sell baskets, and how to farm and became chairman of the Commerce Committee in the Vermont legislature, among other varied experiences. When I say "I wrote the book," I mean that no one before me ever manufactured, wholesaled, retailed, and put together a national sales organization to sell baskets and buckets in volume. The buck stopped with me. I had no one to show me the way to go.

As I built my business I found it easier to do business by simplifying the business basics. You just have to stop and think of it in three simple words: know-how, capital, and sales. With these three words you can make it go. First you must have all the available know-how and really know the business as well as anyone in the same type of enterprise. Second, you have to be sure you have the money to start and have good relationships with a banker, and third, sales—you must be able to sell the merchandise or idea. Absolutely nothing happens until somebody sells something. I have found that any of each of the three basics is as important as the others.

If you make something, a basket, automobile, or anything that increases in value, you are adding to the value of the things in America. If you mine for coal, oil, or other minerals and put trees into the woods and haul lumber, you add some value to America. Trading money back and forth, as in the personal exchange of money from giving someone a haircut, waitress work, and some other professional services, adds little to the value of the country. When you spend money between people with no added value it does a little to merit wealth for the country.

As you read this autobiography, bear in mind that everything I did was a big deal to me considering my finances and twelfth-grade education. What I have to tell is how a poor lad with luck and persistence, bound and determined to succeed, progressed during hard times. You cannot compare me to the individuals of this period such as Donald Trump, Lee Iacocca, and Armand Hammer. Many people have done more business or real estate, dealing in one hour or one day, than I have done in all my life. When I write about my collecting of Early American baskets, artifacts, farm and horse equipment, cast-iron farm automobiles, I am writing of a lifetime search for old things I am fond of and which took me thousands of hours during my lifetime to collect. The whole bunch of stuff may not be worth as much as a nice oil painting done in a short time by a good artist, but they mean a lot to me.

This autobiography is a contribution, making a record in a small way or part, to posterity as I found and lived it.

How It All Began

MY NAME IS FRANK GORDON WILSON, and I live on Kimball Hill in Putney, Vermont. I founded Basketville. The main office is in Putney.

The first Wilson, Ezekrial, came to Dummerston, Vermont, from Rehoboth, Massachusetts, in the 1700s, traveling up the Connecticut River and turning up the West River in Brattleboro. He went ashore on the east side of the river and chose his home at the site that is now Bolster's camp. I suspect he chose this site because it was out of sight of the West River and out of sight of the Indians who used the river for passage. Ezekrial came to Putney because it became too crowded in Massachusetts. The site he chose is what is now the hunting camp of the Robert Bolsters of Dummerston.

Ezekrial's son, Fairing (1771–1842), chose to build just east of his father, about one thousand yards down a gully. Fairing's son, Sanford (1841–97), located between his father Fairing's home, halfway down toward the covered bridge over the West River.

My Grandfather Charles's home in Dummerston, Vermont, is at the dead-end road at the top of the hill, now the Van de Water homestead.

My father and his brother Arthur owned the Wilson Brothers' Garage on Flat Street in Brattleboro. They had a nice business as pioneers in selling and repairing autos and parts. Unfortunately, the garage burned down and they lost everything and were without insurance. My father took a job as an auto mechanic in Connecticut while he decided his next move. I seldom admit it because I consider myself to be a Vermonter, but because my parents were out of Vermont temporarily, I was born in Warehouse Point,

Great-grandfather Sanford Wilson.

Connecticut, on August 2, 1919. My grandchildren should be the eighth generation to be born in this area of Vermont—that is, there are Ezekrial, Fairing, Sanford, Charles, Cassius (my father), myself, my sons, and my grandchildren.

Though my father's oldest brother, Arthur, was the smallest, Grandpa Wilson said he was the strongest and the best scrapper in the family. He was supposed to like to get into scraps, and he always came out of the fight standing tall. Grandma told of a fight that Art was in at a local dance when a group ganged up on him and he beat them all.

Uncle Arthur was always good to us kids and enjoyed taking

Author's father (right hand on fender) and his partner, Arthur, on his left (leaning against tire).

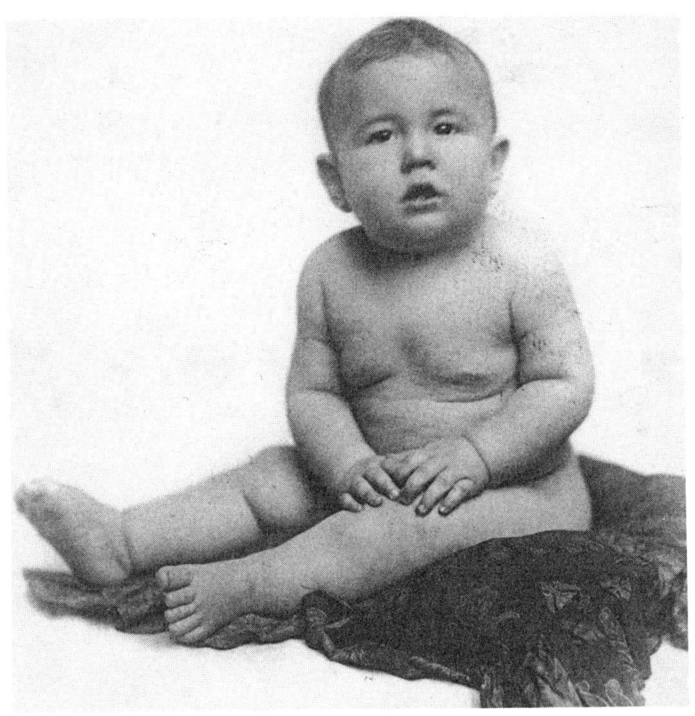

The author, Frank G. Wilson

us to his vacation farm in Guilford, Vermont, in one of his open-top garage cars.

In those days, there were little hills on the back roadways and he would speed up the car at the start of a bump and coast over the hills. It made our stomachs feel like they were coming out of our chests, but it was fun.

Arthur had a wild horse in the pasture, and I, as a kid, went into the pasture once and the horse ran for me. As I tried to avoid him by going through a pole gate, he turned around and kicked me in the head. I never went to see a doctor, and I still carry the scar. Maybe that kick in the head was my big problem!

During the evenings in the summer at the farm, in the big field beside the house there was a pond and a million fireflies that lit up the field like stars in the sky. We would catch a glass jar of flies and use it as a flashlight.

This is the same farm where Arthur entertained his garage customers and served them hard cider or liquor if they wanted it. One day I did a favor for one of his customers and the man tried to give me a tip. I never understood why I should get paid for doing a favor for someone. I thanked him anyway but did not take the money. When the man left, Art took me aside and told me that he never wanted to hear of my refusing a gift. He said that the man would not have offered it if he did not want me to take it and I probably made him feel bad by refusing it.

One event at Uncle Art's that I will never forget is one Fourth of July Party that my family attended. Art had all his friends from Guilford there. Firecrackers, Roman candles, twinklers, the whole bit, were demonstrated, but the crowning performance was when my father and Art blew the big wood cider barrels into the sky! I could tell they had done this before as they seemed to know what they were doing, as they used a carbon substance of some sort for powder. They put the powder into the barrel through a hole and saturated it with water and plugged the hole. After a while there was a chemical reaction that caused the barrel to explode with a bang and it shot up into the air like a rocket. All the visitors and us kids sat on the porch in the dark of night and enjoyed the performance. To top the evening off, the shingles on the wood roof of the house caught fire from the Roman candles and it caused some excitement to put the fire out.

The Wilson cemetery, which is difficult to find as it is in the forest and has huge trees growing among the monuments, is located between Ezekrial's and Fairing's homesite and has about sixty people buried in it. It is located to the right of a very steep gully. There is a stone marked for Fairing, who died September 2, 1842, at the age of seventy-two, but there is no stone marking where Ezekrial is buried. Most of the monuments are plain head-size stones with no identification. Other stones are for Betsy Wilson, who died March 17, 1853, Abel, who died March 16, 1854, William, son of Wheaton and Sally, who died July 31, 1832, and Alonzo, son of Wheaton and Sally, who died October 13, 1828. The Fairing home is now gone, but there remains a nice cellar hole. The original barnyard with big stone enclosures that was used to keep the farm animals in and the wild animals out is still showing.

Ezekrial was a captain in the Revolutionary War and was one

of the original members of the West Dummerston Baptist Church. He was married in 1776 to Sarah Turner and had ten children and forty-five grandchildren.

Our family history tells us that Sarah Wilson went down to the nearest blacksmith shop in Massachusetts by horseback, by herself, to get some nails. In early days, nails were made of iron, one at a time, by blacksmiths. The early homes were made of logs and were temporarily built until you had time and materials to make a better home. When you progressed upward, you would burn down the previous domain in order to save the nails in the ashes when they cooled off. Nails were a really necessary valuable jewel.

Sarah got lost on her way back home and gave the horse free rein. It freely took her home.

The family cut and burned down all the forest they could to clear the land for cultivation. The ashes were saved and made into lye potash cakes in big iron kettles for export to England. Our settlers burned all the big beautiful trees they could for lye, from which they got their biggest income. History tells us that a person walking in the Connecticut River Valley walked all day, much like in the dark of night, because there were such huge, tall forests and the sunlight hardly penetrated the passage.

What a difference there would be in all of America if our settlers left some of the nicest trees for seed trees rather than cutting down everything. More recently, in the Northeast I know of no forest still with virgin trees. Can you imagine if the loggers who cut all the nice trees in all the Northeast in the 1800s through to recent times had left just one tree every acre rather than rape the forests? How wonderfully nature would have replenished our lands and forests, and what a sight we would behold.

This generation had an entirely new look at the hills than I did as a child, because the early settlers cleared lots of land for cultivation and especially because they burned up the mountains. In my early days many of the tops of the highlands were bare with open hills, and you were able to see much more of the hills and the wildlife and cows and horses and other animals as you toured the area. Nowadays, the scenery is not as pretty as it was then and driving along the highways you get a feeling of being closed in by the vegetation.

Going into the forest to pick out a Christmas tree was a tradition my immediate family used to practice when my kids were growing up. It was supposed to be a fun deal, but it never turned out that way for me, not because I am fussy about a tree but because my family could never seem to agree on the proper tree. Spruce trees in the forest are all slender and don't have the look of Christmas trees that you see for sale on the market, which are nice and bushy. In the woods, when they saw a tree that appeared satisfactory, it would turn out to be two or three trees growing together, which gave it the appearance of being much larger than it actually was. Believe me, there are not many nice wild spruce Christmas trees in the forest. I am glad I don't have to venture into the forest in search of the perfect Christmas tree anymore.

Sanford was the justice of the peace, clerk of the West Dummerston Baptist Church, and a life insurance agent, and he farmed three hundred acres. He had six children. My grandfather, Charles Wilson, lived with his father, Sanford, until he married Annie Holton.

Part of this autobiography includes a summary of when I purchased some of my properties and a family tree I had Mrs. Samuel Bunker (Marjory) get together for me in 1957. I had copies made of the tree, the lineage of Cassius Irving Wilson, my father, and I gave copies of it to members of my family as a Christmas present. If you refer to this, you will also get a lot of information about the history of the family of my grandmother, Mrs. Charles (Annie Holton) Wilson. This history goes back to 1611 and is a very famous part of the family heritage.

On my grandmother's side, Annie Holton, a neighbor of Grandfather Charles, lived nearby at the homesite on the current property of Ellsworth Bunker. In those days of little movability, most men found their future wives nearby.

In the Holton history, William Holton came to America on the ship *Francis* in 1634 and was an original proprietor of Hartford, Connecticut. He moved to North Hampton, Massachusetts, where he earned great trust and respect.

Something of importance that needs to be a part of this autobiography and because it happened nearby about eight miles away in Westminster is the story of Chileman Houghton, born 1731 in Lancaster, Massachusetts, whose nephew, Daniel, lived in Dummerston. He is one of my grandmother Anne Holton Wil-

Grandfather Charles H. Wilson.

son's kin. Daniel was mortally wounded along with William French, and they were the first to lose their lives in the Revolutionary War at the Westminster, Vermont, massacre.

My wife, Connie Roby Wilson, also has a very Early American family history. Her people were out of New Hampshire, which is another interesting story to tell at another time.

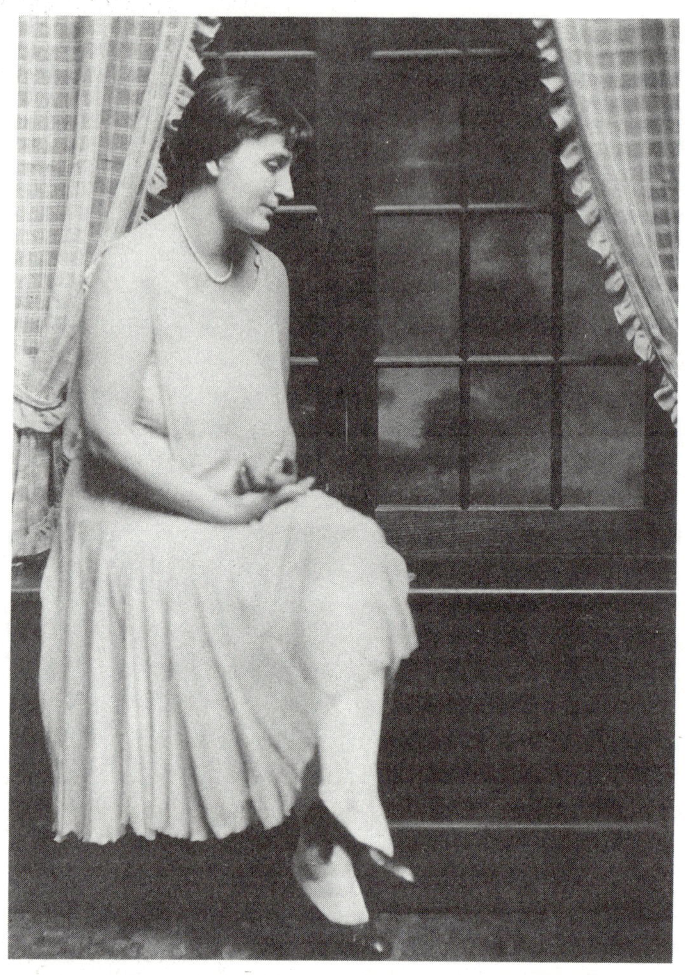

My mother, Naomi Alice Darden Wilson.

Robert Roby, Connie's brother, and I went to the 4H camp Waubanong in West Townshend, Vermont, and it was there that I met and started dating Connie Roby. My dad had an old pickup truck that I used in my courting, and we often joked about it as we pushed it to get it started. We figured we pushed it as far as we ever drove it.

Connie and I were married December 3, 1939, and the first of our sons, James, was born November 8, 1942. Our wedding took place in Connie's parents' living room with my father and a few other relatives in attendance. We borrowed my father's car for our honeymoon and got as far as Northhampton, Massachusetts, the first night and rented a room in a fleabag of a hotel on the main street of Northhampton. The next day we drove to Stamford, Connecticut, and parked the car, as I was scared of driving into New York City. We took the train into the city and the subway to Avenue M in Brooklyn and stayed overnight with my Aunt Emma and Uncle Ernie Parent. We returned and stayed two nights at the Roger Smith Hotel. On the way home, we stopped in Greenfield and spent every last cent of our funds on groceries.

My mother, Naomi, bore thirteen children, of whom I was number two; Frederick, myself, Sanford, Barbara, Naomi, Cassius, Betty Anne, Charles, George, Suzanne, Ruth, Barry, and Simone. Three of us, Frederick, Betty Anne, and George, are gone.

My dad, Cassius, built his home at 70 Oak Grove Avenue in 1928. He did much of the work himself, but he did hire out some

My father, Cassius Irving Wilson

My parents, Cassius and Naomi.

of the professional work, such the electrical, plumbing, and fancy wood trim like windows and doors. The house was sold in 1937 still unfinished. He worked on the house just enough so he could move into it, and then little by little he worked on it while he lived there. He always held two jobs, one as a salesman for Manley Brothers Garage and doing the erecting of highway billboards.

Dad worked hard to support the family, and the community recognized his efforts. Every day a Guilford milk farmer would leave a dozen quart bottles of milk for us kids on the front porch

16

of the Brattleboro home. He was N. W. Drury Dairy, a Guilford farmer. I know when we left we owed Drury Dairy a bit of money, and I am not sure if Drury ever got paid in full.

Dad had a vegetable cellar where he kept a crock each of his salt pork and sauerkraut and the vegetables that he purchased and raised in a potato bin. Once in a while he raised pigs, but mostly we had chickens. A peddler used to come by wagon frequently, and my mother would buy corn hominy from a small wooden barrel out of which he ladled the hominy.

Our home in Brattleboro had three bedrooms, and with nine of us at home at the time and our parents using the master bedroom, seven of us had the run of two small bedrooms, each with a double bed. It was common for us to sleep four to a bed, and there was no separation of males and females. It was best to sleep head to toe, criss-crossed, and I have no memory of having any trouble going right to sleep. The big problem was having to get up in the morning. It seems like we could have slept forever.

The choice for those getting up first was the pick of the wardrobe! We knew that if we wanted to look the best in the house we had to get up early. Fighting over the clothes was routine, and as one grew out of sizes that fit him, the next smallest child would inherit the secondhand stuff. I wore the most inexpensive heavy work shoes and cheap clothes all during my childhood. I did not have a low dress shoe until I went to the seventh grade and was able to buy my own. Friends like my father's barber, Mr. Orie Labert, who had one boy and knew my dad had a lot of kids, for years would send lots of clothes to us. The oldest and biggest in my family, my brother, Fred, was the one who most often got the new clothes. He always dressed the best of us and was the one I had to fight with to get suitable attire.

My earliest work in town in the season when we did not live at the summer camp in West Brattleboro when I worked on the Carpenter farm was when I worked setting up bowling pins at the Masonic and Odd Fellows halls in Brattleboro in the evenings. At the end of the night I would have to walk home about one and one-half miles. Having been paid, I almost always stopped at the downstairs of the Billings Hotel Restaurant to play the nickel slot machines, which actually paid off the winners with cash. To get down there you had to walk down a scary flight of open grate stairs directly over a brook.

I enjoyed stopping here and did it often. It was music to hear the machine cough up the nickels one at a time. Some of the smart kids had a trick of opening up the glass cover of the machine to make it pay off illegally. We would hover over the machine to block the manager's view of the action.

When we tired of Billings, we would stop at Ed's Diner on Canal Street and work on his slot machines to get a bite to eat. We never took much cash home with us!

All the businesspeople in Brattleboro were always nice to the kids. To this day I have never seen any evidence of drugs, but I must admit, we did sneak some drinking of 3 Point 2 beer in those days and at the country club once we stole some drinks of hard liquor, but we never got into trouble. Once during my grade school years we did try smoking the ripe corn silk from the neighbor's garden.

During this time, in the daytime, I caddied for the more affluent citizens of Brattleboro at the local golf course. My father told me at that time to always rest when circumstances allowed, and it was at this time that I first found it helpful to rest whenever possible. It was then I learned you have to take control of your own health.

We had a lot of kids to play with on Oak Grove Avenue, and our house was most frequently full of the neighbor's kids. Dad set up a full-sized professional pool table, a Ping-Pong table, and a real five-cent one-armed bandit, which kept us busy down in the cellar. The one-armed bandit had a hole in the bottom of it so when you put the nickel in the slot and pulled the arm down you could catch the nickel and use it again.

On nice days all the neighbors would play baseball or go swimming. In the winter we would skate on the lily pond nearby, where we would build big wood fires to keep warm. In the early evening darkness, under the streetlights, we played Duck on a Rock in the road at Oak Grove Avenue. Dad and Mother and the neighbors always knew where us kids were, and that is why they kept us in toys.

My father took me hunting and fishing often. I used to hunt deer, squirrels, and partridge by myself when I was big enough to carry a gun.

My father took sick once with appendicitis and could not work for a couple of months. This was the worst time my family had making ends meet. I was set, with my job setting up bowling pins at the Masonic temple in Brattleboro. I was aware that Dad was having trouble accumulating groceries during his sickness, so I took the money made from the bowling pins and went to the grocery store looking for something to buy. I wanted to get the most for my money, and I brought home to my mother a huge front shoulder of ham. I had never seen my mother so over-whelmed as she was when she saw this meat, and I was hugged and kissed more fondly than ever before.

One day during this same period, I was with my father and he was buying a piece of steak at the A & P grocery store in Brattleboro. At the same time, my aunt, Ruth Jones, was buying fish and implied that our family should not buy expensive meat such as the piece Dad was buying but should consider inexpensive fish like she was getting. The Jones family had plenty of money, and with Aunt Ruth's sister, my aunt Emma Farrington, she probably has as much money as anyone in Brattleboro, but they both avoided us all they could. My cousins, the Jones boys, Carter and Owen, were in the same school as us when we went to Brattleboro High School, but they never recognized us or spoke to us. They considered themselves part of the elite and did not associate with those from "across the tracks." We still don't hear from the family and don't have any idea what they are up to or where they live.

While my family lived at 70 Oak Grove there was a Mr. Nolan who delivered mail to our house by horse and buggy. His horse was well trained, and Mr. Nolan delivered mail and instructed the horse to follow him down the street while he went door to door along the path. The horse would stop whenever Mr. Nolan wanted to replenish the mail or make whatever move the wagon had to make. All Mr. Nolan had to do was tell the horse what to do, and he would do it. Mr. Nolan became older and tired and ready for retirement, and it was somehow found that he was taking to drink-ing on the job. One day the horse pulled into the barn, with Mr. Nolan fast asleep in the buggy drunk after the route had been attended and finished.

My uncle, Arthur Wilson, kept a barrel of cider way down on the ground floor cellar of his home, buried in the ground to keep it hidden. Aunt Bertha Wilson, his wife, happened down in the cellar one afternoon and Mr. Nolan, her mailman, was lying on the ground sucking hard cider through the hose. You can bet that Mr. Nolan lost his source of booze and Uncle Arthur had to find a new hiding spot for his hard cider.

While in the Green Street School in the seventh and eighth grades and high school, when we could get together, my buddy was Harold Carlson. We both played football and were on the track team. He was a nice big, gentle Swede and, like me, addicted to hunting and fishing. We did a lot of hunting in the woods around Brattleboro and ice fishing in the West River near the Brattleboro Retreat. When the fish are biting, ice fishing is one of my favorite sports.

We would put up twenty or thirty tips, scattered around the meadow, and sometimes the fish would bite so fast we had little time to put up the whole bunch of tips. The most fun was when the ice was in good smooth condition and we would race by skates to the red flags (minnow lures) that flew up when the fish were biting. We would hurry over on our skates and gently pull the line before the fish swam away. The secret was to let the fish have enough to swallow the line and when we thought it was in its throat, we would yank the line and hope the fish was caught on the end. We had to hope that we had chipped a hole in the ice big enough for the fish to get through!

A secret few people know, which helps catching fish through the ice, is chopping your holes in the ice in a line just as the sun goes from east to west. As the sun progresses during the day from hole to hole, the sunlight attracts the schools of fish to your lines.

Harold and I also did a lot of hunting in our early days. A good way for us to get experience shooting was to go to the town dump. In the thirties, dumps were all open and not covered as they are now. At the dumps, almost always we would see rats running all over the place and we used our .22 rifles for shooting them. We would stop shooting only when we ran out of shells, for which we had saved all our money. We bought all we could afford. We not only shot the rats, but shot those interesting targets, glass jars, bottles, and cans that were spread all over the dump. We would also throw bottles and cans in the air for each other. I got so that I hardly missed a missile flying through the air with my trusty .22 rifle.

My wife, Constance Roby Wilson.

Harold was a good student and a smart one, too, but a little slow to learn, and he had to spend a lot of time on his homework. He went along to college at the University of Vermont, and I have not seen him in years. He became an important engineer in the General Electric Company in Schenectady, New York. I think of him often and remember the times we had.

After taking apprenticeship of basketry with my father at the Sidney Gage Basketry in North Westminster (Gageville), Vermont, from 1937 to 1941, I bought the West River Basket Company in Putney, Vermont, and founded Basketville in 1941.

My father had little money. However, he gave me the first week's payroll of perhaps three hundred dollars.

While I was working with my father, I married Constance Roby and we lived in the little white house directly across the road from my dad's big house. Connie worked for Eddie Kent, a carpenter, as his office manager. Before we started raising a family, she used her income to buy the household utilities such as the refrigerator, washing machine, et cetera. My pay at the Sidney Gage and Company was the house rent, use of the company truck after hours, and a little cash if I could find it.

Because there were so many of my family living in the area, I could see that there was not much of a future for my father and me if he was to support all of us, so having just married, I decided that it would be best if I moved on to become independent. I had three years of apprenticeship under super basket makers, and I felt I could improve some of the business principles learned at the family business.

3.

Starting My Own Business

WHEN IN 1941 IN PUTNEY I started my new company, which I bought from Dwight H. Smith, I sold all my products back to him. The first day into production, the hardware and basket nails were missing. Mr. Smith told me he'd sold me the business, not the hardware! I did a lot of business with Mr. Smith over the years, but I never forgave him for this, as he knew I had little money. I sold all my products to Mr. Smith, as he had a truck and sold the baskets to the little roadside stands scattered around the area. Smith picked up the week's production on Friday and paid for it by check after the banks were closed. Most often his check would bounce on Monday mornings because of insufficient funds. He only had one check that would bounce, and he always found enough money to cover it in a day or two or three. I had ten or twenty checks to cover when his bounced, and I had to chase all over to pick up my bad checks. It was a time-consuming job for quite a long time. The banks knew what I was doing but tolerated it because they knew me and they knew that I was trying to give jobs to people as both jobs and money were scarce and there were a lot of people with no jobs who wanted to work. We knew which stores would cash our paychecks under these circumstances, and most of them were cooperative. This went on for a few years, until I prospered and knew better or learned how to use the bank's money. The "float" is when my checks are not cashed at the banks for cash. Not all were cashed at the same time, but they were all out and sometimes not used for a few months, thus were floating freely.

After Dwight Smith sold me the basket shop in Putney, he stayed around and tended his little grocery and country store. Every evening anyone with nothing better to do was invited to sit down and visit with him. He was an innovative man who knew a

The Putney D. H. Smith Garage, later the Basketville wooden bucket factory.

little about everything that a small-town man needed to know. He had owned and run the old Putney basket shop, and he was a good blacksmith and mechanic. He could handle trucks and he sold cars, farm machinery and fertilizer.

One night after work, he invited me to see his secret perpetual-motion machine. We went up to the top of his store to a secluded work area at the peak under the rafters. His perpetual-motion machine was made in a wooden frame perhaps four feet long and two feet high. What he did was fill up, on a slightly inclined spot, about fifteen steel balls about an inch round. The balls would fill up a vertical conveyer with fourteen balls, and he had it geared so one ball would escape the bottom and one ball would come to the top to get ready for its return down the conveyor. The falling ball activated the machine. Of course, the machine did run a long time before it stopped. However, as I had learned in school, filling up the machine with steel balls in itself used energy so it was not really a perpetual-motion machine. He was very proud of this machine, though, and thought he had invented a great machine that would make him rich and famous.

Everyone was good to me when I first opened my business. Even Mr. Stack of the Green Mountain Power Company "forgot" to turn on the electric power meter for a few months, which helped us get started in the beginning. As I look back on it, I think that he knew what he was doing. Over forty-eight years, since we have prospered, we have probably put somewhere to the tune of $3 million into the power company by buying power and staying in business.

I got into trouble early on with one bank, as they thought they were lending me too much money. The bank officers found out we were partying in the bank with the loan officer. We would share good liquor, and the loan officer would keep me in loans. I didn't see anything wrong with it, but they called in my loan of a few thousand dollars, so I had to change banks.

For a-year after I went into business for myself, my father let Connie and me stay in the little white house in Gageville. I commuted to Putney in a 1937 Chevrolet sedan. Then Connie and I moved into a miserable upstairs apartment in Putney for a while. A little house came on the market for sale nearby, and a friend, Elbert Martin, was the realtor. Mr. Martin, a good Republican, was a former secretary to Teddy Roosevelt and had been a state representative to the Vermont Assembly for many years and was a man who shared confidences with me. I walked into the house and looked upstairs and into the cellar and bought the house for seventy-five hundred dollars. Connie and I moved into the little house and raised our family.

The water system in our first Putney home came across the main street from the hill in front of the house. We had a lot of trouble with it and were completely out of water a few winters. I lugged water for our household from our neighbor's animal trough across the street. We shared the water with animals, with the hay and barn waste still mixed with the water when I dipped it from the water trough. This was at the barn of our good neighbor Robert Austin. Later, the house we are currently living in, next door to the little white house, the home of the Bryant Jones family, came on the market. Jack Bitner was in charge of the sale, and he gave us a good deal, selling it for $13,500. I went to the bank to try to borrow the money. The banker was sort of a straight, pompous man, and I explained to him that I had a big family and a little house and that the house next door was for sale and was exactly what I needed. He said, "You screwed yourself out of room," but he loaned me the money. It has been a delightful house to live in. Since I purchased the home, I have accumulated the land from the house all the way down Kimball Hill right to all the houses to Sackets Brook. It goes across the brook to the basket works, across Route 5, through the Basketville store to the height of the land at the rear of the store, back across Route 5, south of the day-care area, back to Sackets Brook, and all the way to the north to Sand Hill Road. I have always been crazy over real estate and have kept myself poor paying interest and taxes. I figure they ain't making no more land.

Walter Hendricks, a new neighbor, moved next door to us. Walter founded Marlboro College and Windham College in Putney. He was a well-respected educator. Once I told him how I regretted that I only went through high school and realized that because I did not receive a high education it interfered with my personal contacts and business and it was too late to go back to school because of my family and business responsibilities.

"Well, Frank," Walter said, "if you carefully read the *New*

York Times every day, it is as good as a college education." From then on I read the *New York Times* and also the local *Brattleboro Reformer* and the *Wall Street Journal.* In my opinion, he was right, and I still learn something every day by reading newspapers. I certainly cannot expect to get an honorable doctorate with my twelfth-grade schooling, but I have done well with the education that I have. In addition, when I can get the Canadian papers I keep up to date and learn from the Canadians, as they give me more accurate accounts of news than the U.S. papers do. My policy is, if you never stop learning and have good luck, which is half the battle, you should prosper. If I had a choice of going into business with a partner with a lot of money or one with a lot of luck, I would choose the lucky one!

Also, at this point I would like to offer my opinion on who to ask if you want to get a job done quickly—ask a busy person to do it.

4.

The Lure of Real Estate

IN THE FORTIES, I DID logging and lumbering for the war effort along with basket and bucket manufacturing. I purchased and logged sixty acres of East Putney forest east of the Hill Farm, which I purchased twenty years later for thirty thousand dollars (115 1/2 acres) and is now owned by my son, Steven Wilson. It takes about thirty years for the partly cut forest to grow big enough to warrant another harvest, so my father insisted I get rid of the property, as it was not worth keeping and paying taxes on it while waiting for the trees to grow. I gave this property to Marvin Temple, who was a teamster and drove horses for us. This is the last time I gave away any land. These sixty acres that I gave away recently sold for eighty thousand dollars. My sons and I (and the banks) have title to our homes, businesses, stores, farms, forests, warehouses, whatever. We seem to be more comfortable with ownership. To put it a subtler way, we can pee on the land and not get thrown off.

In the area of this Hill Farm, I know of a secret spot where there are thousands of lady slippers growing wild. Shortly after the snow goes away, I make it a point to visit these flowers. They are very hard to find, and I understand the soil has to be a specific type to allow them to grow. I also understand that few people have ever been able to transplant lady slippers.

I also know of a nice spot at the Putney River Farm where there are quite a few nice mayflowers (arbutus) growing. In both cases, I try to be careful who I show the arbutus and lady slippers to. They are both protected from picking by law, but all people don't know it. The arbutus smell real nice, and at times in Vermont you will see kids selling them beside the roads.

Also at the River Farm there is one native chestnut tree still living, which produces nuts in the fall. But seeing it lacks another tree to cross-pollinate the flowers, the nuts were sterile and incapable of reproducing. Twenty years ago I could see huge chestnut trees in the area, but most have died now and are left in the forest decaying. Some of the logs were as much as four feet through when they succumbed to the blight. Most nice big old New England barn frames were made of chestnut or black walnut, which also grew in Vermont.

I brag to people when I tell them that I purchased a thousand-acre ranch for twenty-three thousand dollars. I never tell them that Marvin Temple got his land for nothing an acre.

The real estate that is now the Putney retail store came on the market in 1956. At the time, the property was being used for the converting of paper to tissues such as napkins and toilet paper. It was a partnership. A carload of paper was rejected by a customer and was never written on the company's books. Lo and behold, one of the partners must have taken the money and forgotten to replace it in their joint account, so it was time to dissolve the partnership. A New York factor was factoring all the paper company's accounts receivable so they had to recover the best they could by taking the properties from the paper company. (A factor is a firm that collects all of your accounts receivable monies for a percentage of the bill and risks not getting paid for the merchandise. The factory pays the manufacturer for the merchandise as soon as it is shipped and billed, so that relieves the manufacturer from having to wait to get paid.)

I thought because the basket shop was always out of room and this property was right across the road, I ought to try to buy it. Bear in mind that in 1956 I was quite naive about business and proper dress, et cetera. However, I looked in the town records and found that a factoring company on the fortieth floor of the Empire State Building in New York City owned the property. I put on my best clothes (I didn't know I should do business in New York in a suit), including a plaid shirt, took the train to New York City, and stood out in front of the Empire State Building. I looked up toward the sky at the fortieth floor in the clouds and wondered what the hell I was doing there. I got my courage up and found the elevator to the factor office. I walked into the office over a rug that seemed six inches deep and told the receptionist what I was there for, and then I went into the office to see the man in charge. He could see right off what I was and could tell I was not familiar with modern business. I was really over my head in this kind of matter. Anyhow, when I made the really nice deal he laughed like hell and told me, "Son, it does not make a bit of difference what I sell it to you for because it is covered for any loss to us because we are insured for any losses like this in our company." I think this deal, around $10,500, was one of the best deals I ever made. This store is worth more than that; it would go for about $1 million if I were crazy enough to sell it.

This Basketville property was formerly Parker's Garage, run by Roy Parker of Westminster West. He ran an automobile service station here. In those days before the Eighteenth Amendment was enacted, the directed railroad and highway route from Quebec, Canada, where liquor was legal to make, ran to New York City and west along Route 5. The Putney Parker Garage was on the main highway route for the illegal booze to flow. Many people made their living running liquor, and the Putney Garage was a main link in the liquor enterprise. When the big cars loaded with liquor came racing down Route 5, the big main door of the garage was opened and the loads of liquor went through the garage, out the back door, and down the steep bank, where the bootleggers would hide out from authorities. One enterprising farm in town had a large barn, and whenever a load of booze came roaring into town in hot pursuit from the law, the farmer would step in. He had his barn prepared so that a big load of loose hay overhead in the barn was released and it would come down and completely cover the car full of liquor, hiding it from the chase.

Doc Prouty, our friendly Putney neighbor and veterinarian, was a good man to know if you needed liquor. In his younger days, he was said to have a good side job changing numbers on "hot" cars. One really nice lady, Minnie Coombs, was known to help you find a drink in the old days if you needed help.

John Kazmajack came to Putney in 1940, a year before I purchased the West River Basket Company in 1941. He was a master papermaker with a lifetime knowledge of paper production in New Hampshire mills. He purchased the almost valueless antiquated Robertson Company Mill, right next door to the Basketry, from the selectmen working for the interests of the town

The original West River basket crew at Williamsville, Vermont. Photo by Porter Thayer.

of Putney. John was called John Smith by us because most of us could not pronounce his last name.

John borrowed money from a friend and with his son-in-law, Frank Potash, undertook the challenge of getting the papermaking facility to a position of making tissue paper. Few people have worked as hard as John and Frank to start a business.

It was a trial of sweat and total personal sacrifice to finally get a roll of paper off the machine, which took a couple of years. John's two daughters, Mrs. Gertrude Potash and Shirley, teamed with John and Frank all the way to their dream's fruition. As John got closer to getting a sheet of paper off the machine, everything he had or could borrow was exhausted and it looked like the end of his dream.

Directly across the street from his mill was a country store owned and operated by Mr. S. L. Davis. Of course every day Cy Davis could see what was going on nearby and sensed the critical situation that John was in. It is a credit to New England neighborly generosity that Mr. Davis loaned John a little money to finish this paper venture. It is also a credit to John Smith that he paid off all of his lenders with interest as soon as he got on his feet.

John ran the mill all his lifetime, for over thirty years, and it was and still is a very profitable paper mill. In his later years, John took great pleasure in playing the horses, and when he lost, over time, a million dollars of his own money gambling, it was thought he deserved it, but it did not hurt anyone else or his business.

John liked to play the nickel slot machines at John Montville's restaurant for hours. Another business in town that John frequented for this was Dan Corey's, a small grocery and meat business in an area that is now part of the Basketville parking lot.

Cy Davis ran a really well-stocked country store where you could buy anything. He even had a scale outside where you could get, for a price, a whole load of weighed hay, coal, or whatever. Cy had groceries, hardware, medicine, clothes, shoes, and gasoline, anything a small-town resident might need. You only had to travel to the city to purchase the few things he did not stock. He was known to be very generous with credit.

Cy's warehouse was located on the Putney-Dummerston town boundary, and it is said that for the many years he was in business that he never paid a personal property tax because Putney thought he paid taxes in Dummerston and Dummerston thought they were paid in Putney! He made sure that a majority of the store inventory was in the warehouse on the day the tax was counted and figured.

The Chickering boys in New Hampshire, just the other side of the river, played a similar game. On tax-counting day in Putney the Chickering cows were found grazing across the river in New Hampshire, and on tax day in New Hampshire they were moved to Vermont—all of which was legal!

I have many memories of peculiar people in the area from over the years. At the time when there was a dirt road in front of our little white house in Putney, a neighbor up the road a piece often trained a racehorse with a high-wheeled sulky, trotting quickly along the highway. He was Arthur Lovell, a middle-aged man who I suppose sold racehorses for a living. Whenever I talked with him, his voice whistled while he spoke. I don't know whatever happened to Arthur, but he did have a reputation for being lazy.

His method of feeding hay to his horses in the winter was to rake it into long rolls and leave the rolls of hay outside so the cold weather would freeze them up so all he had to do was draw the hay into the barn and it was eaten by the horses as it thawed out.

My friend Benny Pollard, who went to grade school with me, went into the junk business in Brattleboro. During his later years, I visited him. He was a person who complained about everything all the time. He complained that he paid too much for the junk he had to buy and got too little for it when he sold it. I asked him why he prospered so well, and he replied, "I don't really know. I buy for one dollar and sell for two dollars. I mark it up 1 percent, but it seems to come out all right."

Another person who is known by everyone in Putney is Gordy Rhodes. Gordy is a regular visitor downtown and along the highways, as he is the person who took on the job of retrieving all the bottles and cans in order to return them to the stores. He does a fine job with regular hours and is fun to be around.

Gordy worked for many years at the Putney Paper Company. He didn't amount to much, but they kept him on for over twenty years out of the goodness of their hearts. He is slightly retarded, but he is certainly the happiest man in town, with absolutely no worries. Even with his state of mind and some physical problems, I bet he will live to be one hundred. He likes to get into our big boxes of waste that are ready for the dump, and I kick him out every time I see him doing it, as he cuts open all the rubbish bags and the wind ends up scattering the trash all about the yard. Gordy is persistent and when I'm out of town or when he knows I'm not around, he goes right back to opening up the bags and boxes, looking for the goodies he collects.

Gordy is quite popular with the employees of the store and the office and strikes up conversations with all he comes in contact with. His biggest problem is that the retired Putney pastor keeps cutting into his territory and he feels he doesn't need any competition or help because he does a fine job. Gordy says he is going to kick him and punch him in the nose if he catches him in his territory. Gordy's only problem is that his competition knows Gordy keeps pretty regular hours on the roadsides and works Gordy's territory in his off hours.

All in all, Gordy is a heck of a guy and known and loved by

all in Putney. The town of Putney would certainly not be the same without Gordy Rhodes.

Another memory I have is of Roger Fellows, the son of an old family in town, who got married about forty years ago to a new lady in town whose profession was that of midwife. Roger got sick not long after their marriage began and passed away. His wife inherited all his goods and property and took Roger's body home and buried him in the backyard!

Jenny Mellon, along with her daughter, Betsy, ran a really nice, clean, neat grocery store on Kimball Hill, and later Caroll Brown would get together a group to open a cooperative grocery store. D. H. Smith ran a small grocery and hardware store, later taken over by Sam Howard. A group of us started a Putney Credit Union, which acted as the only bank for many years, until the Bellows Falls Trust Company came to town. This credit union is one of the best things ever to happen in Putney, and it was engineered by the Robert Rosegrants.

Good luck was also with us when we bought the Hill Farm. We purchased this property from the Cassidy brothers, Ed and Bud. Doc Prouty, our local veterinarian, who I never knew to practice his profession (if he was still allowed a license), was our nearby jack-of-all-trades. He could fix most anything from a gun to a machine, if he felt in the mood. We would see him first for some uncommon problem. What I remember him mostly for is his preaching to us not to consume any products made from dairy products: "Do not drink milk or eat ice cream or anything with milk in it. Milk is the cause of much sickness and is deadly. Cow's milk," Doc said, "is for calves—not humans."

My negotiating with Ed and Bud Cassidy is a good example of my persistence. Remember, Ed and Bud owned a lot of land in Putney, so one farm (the farm my son Steve purchased) was not that big a deal for the Cassidys. Many times I visited them but they would not break down and sell this farm to me. The evening when I got the okay, Ed and Bud were at the Cassidy family farm on Cassidy Hill in Putney, looking at and listening to that modern contraption, the television. As I negotiated, they did not take their eyes from the TV. I would venture a question, and usually it would be quite a while before either of them said anything. Then one or the other would speak, while still looking at the TV, making a remark like, "What do you think about it?" to the other brother, and of course the other would take a long time to answer the question—with another question. I presume they were mostly interested in the TV program, but I finally got loads of information. They were driving me nuts with the long length of time being used to talk about buying the farm, but we seemed to be finally getting a meeting of the minds, even though they used the very minimum of words in their discussion. But it did happen. I bought the farm. The Cassidys did exactly as they said they would, and I am sure they are pleased at how the farm is cared for and is used gently and productively.

The Cassidy family are all great people. Ed has another small piece of land near the farm that a friend of mine would like to buy. Ed told me he would sell it to him but he didn't want to be bothered by him when he stayed too long at his favorite watering hole and had to drive home, sometimes with questionable driving abilities. "If I sold it to him," Ed said, "I would lose its credibility."

This is the Putney Trapper's Farm, the Sid Bailey place at East Putney, Vermont. Sid Bailey was a man who left us with many memories of a most unusual personality. I only knew of him when I tried to buy the mature trees from his forest when I was actively buying forests for woods for the war effort and the manufacturing of baskets, buckets, and furniture. Sid, of course, was foremost a trapper of fur, and his work included long treks into the woods to tend to his traps. When I inspected the forest with him, I was in top physical shape, but I found it very difficult to keep up with his smooth, swift, long-legged passage through the woods. I just about had to run to keep up with him. He was the sort of man who, for example, was one of the first to own a newfangled gasoline-operated one-lunger engine, a six-horse-power early American Abaneque gasoline engine that had been invented and made in Westminster Station, Vermont, and Sid used this machine in his idle time. The engine pulled his horses to his neighbors, whereby he, with his Abaneque engine, would saw the woodpile to the length everyone used for heating and cooking at that time. This method of sawing to the desired stove or fireplace length by machine sure beat the previous method, when it had to be cut up by manpower with a buck saw or hand-held ax.

Sid showed a cruel, unnecessary insistence and demand on

the team of horses as they just stood in place while the Abaneque engine fired away with the circular saw he used to saw the wood doing the work. He made the horses lean into the collars and rigidly stand at attention. If at any time the horses showed any desire to relax while in harness, he would club them to attention. This same unusual man when he had trouble starting his new engine at a neighbor's farm backed his engine to the bank of the deep Connecticut River and let it roll into the river, where it probably still is today.

Sid had a trapping shed in his front yard where he kept his trapping paraphernalia and had a skinning section with hot water and stretching boards, et cetera, set up. He kept his furs ready for market upstairs in one of the bedrooms where he kept the furs neat with a door padlocked for security reasons. His speciality was trapping red and silver fox. In order to capture a fox you have to remember that they are very foxy. It was my opinion, you really had to be good to catch a fox. Well, I will tell you how Sid caught a lot of foxes, which is a good way to make money if you know how to do it.

First of all, it has to be done in the cold weather when the fox pelt becomes prime with long, warm fur. Second, the law won't let you trap them until fall, when the weather gets cold. Sid, however, started trapping long before open season. What he did was make a deal with a chicken-processing factory in Bellows Falls, Vermont, and for free carried off much of the waste part of the chickens. He loaded this into his open truck and scattered it about all over Windham County where he knew foxes hung out. He kept the foxes well fed for a month or so before the trapping season opened. The first day of trapping season, he set out all his traps amongst the bait he had been feeding the fox. The second day of trapping season was his most productive day. However, all season he kept up his free chicken dinners for the foxes until he had caught them all.

One evening Sid heard a padding of footsteps and other factors that led him to believe someone was in the process of stealing his furs. He quietly tiptoed out the front door with his trusty rifle and shot the person to death while he was coming toward him across the porch. The person was his wife, who had been coming back from the privy shed connected to the house. He satisfied the law that he had shot her by accident.

This Hill Farm is now owned by my son Steven and his wife, Polly, who live in it with their son, Jonathon, and daughter, Melissa. They have done a lot to the 120 acres, modernizing the house and adding an outdoor swimming pool and modern maple sugar house. The whole acreage is in production. The forest is in the highest stage of growth, and there are about five thousand Christmas trees ready for sale. Steven has one thousand rock maple trees planted for future maple sugar production. This is a real country gentleman's farm and a really happy environment. The East Putney Brook runs down the whole farm and is pure enough to drink from. They have a true artisan water system to the house, and there are barns, stone walls, vegetable gardens, flowers, and blueberries.

My son Gregory and his wife, Annette, have their own farm of over one hundred acres on the River Road in East Putney. They live there with their daughters, Leah and Anna. They purchased it from Mrs. Nancy Watson in 1982. It was the Floyd Rice farm. It has a nice brick main house, and they are gradually improving the home, grounds, barn, and forest and this is becoming a really nice and happy showplace. It is funny how the family got to own this farm on River Road also.

Years ago, Floyd Rice, with his son, Ed, and daughter, Marjorie, owned and lived there and Floyd did much of his own logging with his really nice horses. I had an eye on the farm and told him that if he ever decided to sell it, I would buy it. He told me he would give me first refusal on it, and I let it go at that. The next thing I knew, he had sold it. I felt really bad about it, but he did what he had to do. He had a buyer (Nancy Watson) with the money so he sold it. This taught me early on not to procrastinate. It is forgotten now, but probably the real reason I did not get the property then was that I felt I could not swing the deal aggressively.

Everyone should know that most big American fortunes are made because of the increasing value of real estate. We later had the chance to buy this land from Nancy for quite a bit more than we could have purchased it from Floyd Rice. Nancy Watson dropped by a while after she sold it to us and saw that we had fixed up the real beautifully large barn and told me, "Frank, if I knew what you were going to do to the farm, I would have married you!"

5.

Life on the Farm

TO ME, MY GRANDFATHER, CHARLES, on his Valley View Farm, seemed a very gentle man. I first used to hitchhike from our home in Brattleboro at 70 Oak Grove Avenue to his Dummerston farm. I would usually get a ride and get let off on the main road at the covered bridge. From there I would walk two miles to the farm of two thousand acres. Sometimes it would be after dark when I arrived. I think I was in seventh or eighth grade, eleven or twelve years old, at the time. When walking in the dark of night on a narrow dirt country road, the proper way to walk and tell where you are going is to look up above the road and walk where the moonlight is showing like a lane through the sky and trees. One night while walking to my grandfather's, I walked right into a deer in the road. It was a really frightening experience. My grandfather told me about a time in the same spot when another Wilson, my Uncle Arthur, shot a deer and thought it was dead. He laid down his rifle and got out his knife for dressing, and as he got close to the deer it got up and started to run away. He jumped on the deer's back, held onto the deer's antlers, and rode the deer across the road for several hundred feet before it lost its strength and died.

My grandfather Charles was a breeder of Ayeshire cattle and Southdown sheep and had six hundred sugar bushes and five hundred tree apple bushes when he farmed what is now the Van de Water property in West Dummerston. He also had a few peach trees. I used to help him harvest the peaches.

Grandpa Charles needed a wooden stick about three feet long to put on his shirts and coats because he had arthritis in his body. He could put one arm in a sleeve but had to use the stick to manipulate his other sleeve to a position whereby he could enter his arm into the other sleeve.

Aunt Edith Holton, who was later Grandpa's housekeeper, never went to bed. She just leaned back into a chair and slept in whatever clothes she had on. I never knew the reason she never went to bed, but it may have been because she had a big goiter the size of a grapefruit on her neck and it might have caused her some discomfort when she lay down.

Aunt Edith had sold the family farm, the Anne Holton Farm, which used to be the tavern on the main road to Boston on Putney West Hill (now owned by Shannahan and Dyer), earlier to Dr. Bugbee. When she partially retired, Aunt Edith purchased a little home just north of the basket shop on Route 5 to live in, and when she retired from teaching, she sold it to Fred Howard. She had no money when she came to keep house for my grandfather, Charles, on his family farm, now the Dummerston, Van de Water, place. However, she owned a beautiful black horse she called Black Beauty, which had the run of the whole of Grandpa's farm. I never knew it to do any work and it was infrequently used for horseback riding, as it was very wild and untrained for any useful purpose. When Grandpa Wilson sold the place, Black Beauty was getting old and Aunt Edith had "him put out of his misery" rather than let him go to a new, untested owner.

Charles Frank Wilson (1851–1933) is buried at the Putney Cemetery on Westminster West Road at the Holton-Wilson plots.

I remember seeing him take his bath in a large galvanized round tub in the kitchen. Aunt Edith helped him bathe. He used a long-handled brush to scrub his back. Grandpa was very arthritic and his hands were quite gnarled. He liked to play the fiddle and harmonica and did so frequently.

I used to sleep in a room upstairs, east of the large kitchen. The room was reached by a little stairway at the back. I used a "thunder mug" for a toilet in the bedroom for urinating, and I used to empty it out the north window onto a wood shingle roof. You can imagine what it smelled like, but Grandpa never complained to me.

There was a nice three-holer outhouse in the open unheated shed connected to the house, and by going through one room you could make use of it without going outside.

Great-grandmother Anne Joslyn Holton, Grandmother Annie Wilson, and sister Minnie Holton Wilder, with son, Richard.

I can just barely remember Grandma Ann Holton Wilson (1861–1930). I was twelve years old, but I remember her dying in Grandpa's arms in the living room overlooking the West River. It was her sister Edith, the retired old-maid schoolteacher who never married, who came to keep house for Grandpa. Aunt Edith always said her scholars were her children. When Grandpa sold the Valley View Farm to Van de Water around 1931 and moved

The hired hands at Great-grandmother Annie Holton's farm in Putney, Vermont.

to Stewart Place in Brattleboro, Aunt Edith went with him. I happened to be at his Brattleboro home when he died. I recall Dr. Klain trying to revive him at the time.

On the lawn of the West Dummerston house there was a nice, deep well with great cold water. The well had a great long log sweep that let down a bucket to gather the water in. I don't remember if they had running water in the house, but there must have been a gravity spring in the east. I used to help Grandpa do the chores: cut the hay and corn into the silo with the help of a big one-lunger gas engine. I used to help him when he had a big white horse to gather wood and get out logs. He showed me how to chain the logs so the horse would pull them. Grandpa was old and moved slowly, so I was able to scurry around and hitch the chain and he would rein the horse, and when he located the log, I would scramble to unhitch the chain so Grandpa hardly had to bend down.

Grandpa Wilson also raised a team of oxen that were an important and necessary work team needed when we had something heavy to get done. Our forefathers used them to get out logs, move heavy stones from fields, plow, and clear land of stumps. Early settlers and farmers did not have the physical strength to function without the oxen. Because of having a team,

I learned something few people understand; that is how to make an ox yoke.

An ox yoke is that heavy piece of wood that goes over the neck of two oxen and has big wooden bows that go down their necks, around their chin in a U-turn, and back up to the heavy neck plank, which has to be strong enough to pull heavy loads when the load is hitched to this yoke by heavy chain. The ox yoke should be of one piece, and for mature oxen the wood is usually of hard rock maple and around six inches wide and eight or ten inches deep and runs the whole length of the necks of the two oxen hitched together, which is about six feet. In order to keep this big piece of wood strong and because it is guaranteed to crack if you don't know how to dry the wood properly and it weighs, green from log, about seventy-five to one hundred pounds, somewhere back in history a secret was handed down generation to generation, and my grandfather handed it down to me. The best and only way it can be done is to place this piece of wood under a pile of cow manure for a couple of years. This keeps the wood moist all over, and the manure acts as a water separator as it blots and draws out the moisture evenly. Now you know how to cure an ox yoke or any large piece of hardwood, and you are probably only one in 10 or 20 million people who know this secret.

Grandpa's main house was quite nice. It was removed from the hill north and east and relocated to its current position with the help of the neighbors at a house-moving bee with the use of something like twenty-three pair of oxen, two barrels of hard cider, and a small cask of cider brandy. The kitchen, which was not that fancy but quite large, was built on later.

The keg of cider brandy was kept handy on the doorway of the house while they moved it down the hill. It took two days to move it because the heavy iron moving chains kept breaking and they had to be repaired by a blacksmith the evening of the first moving day.

The farm was the Willis Ray property, and the house was originally located on the property about a mile north of the current site. Grandpa moved it so the location would command a great southern view of the West River, the railroad bridge route, and the valley. I wanted to find out why none of the Charles Wilson family carried on the farm, so I asked the youngest son, my Uncle Rick, why no one took it over. Uncle Rick told me, "I had seven cows which I paid twenty-five dollars each for. I bred them and sold them the next year for twenty-five dollars with calf. . . ."

According to Aunt Edith, who wrote to her nieces and nephews, the old Wilson homestead was purchased without a dollar in hand. I quote here from a letter she wrote to her family discussing the purchase.

My Dear Nephews and Nieces,

I would like to talk to you about how your father and mother bought the old Wilson homestead without a dollar in cash on hand as a starter.

Your mother, as you know, had a very gracious manner and made friends wherever she was all her lifetime.

Oliver Blood, Putney's richest man in his day, promised her that when she got married, he would help her and her husband to buy a farm. When the Willis Ray farm was offered for sale at a low price, they came up and, taking me with them, went up to see Oliver.

Annie said she had come for him to redeem his promise to her to help them buy a farm. After some persuasion by your father, he finally agreed to buy the place for $2000 and give them a bond for the deed. They had to pay 6% interest and $200 a year principal, both exorbitant conditions.

Your father sold his driving horse for $150 and Annie had $150 that came from her grandmother Joslyn and $50 for a cow that Mother gave her as a calf and also she took with her a young cow that later died.

That was all the capital they had to stock and tool the farm and furnish the house except what their mothers gave to them.

Your father borrowed $200 from Aunt Sybil somebody and they worked like slaves to carry on.

Charles used to come over the hills through deep snowdrifts on foot to pay the interest. He cut wood to draw to town. Ormando Norcross used to come up very early to start the oxen with a load to Brattleboro and would get there about the time Charles did with the horses after doing the "chores." They got up at unearthly hours in the morning. They boarded the choppers and Annie used to bake big Johnny cakes, 2 inches thick, for breakfast and made pies, doughnuts and bread every other day.

With the children coming along so fast and other expenses, it

Sometimes in the fall at Grandpa's Valley View Farm he would have a harvest festival. When I was about twelve years old, he had one and he invited all his friends and neighbors. He got together some musicians and we danced into the night and he played the fiddle and harmonica with whoever came to the party and could join in the band. The highlight of the evening was the red-ear deal. Grandpa wanted to get his corn husked, so what he did was mix some corn with red ears in the huge pile of corn. Anyone who husked a red ear could kiss anyone he wanted at the husking bee. There was plenty of hard cider and cider brandy, and it was a rollicking affair. Grandpa Wilson made kind of a fool of himself around the pretty girls. It was a great fun affair with lots of help and food.

Frederick Van de Water wrote a book, *A Home in the Country*, about his purchase of the farm and the people around the farm. Later he also had a husking bee, to which he invited all his friends.

He talked with Uncle Richard Wilson about Grandpa's party and tried to duplicate it in its entirety. Van dé Water bought the farm from Charles for six thousand dollars, and he sold it to the current owner, Ostroff, for twenty-five thousand dollars in 1980. I would say the farm is now valued at $250,000.

Another deal Grandpa Charles had was a toll road to the top of the mountain. It was about two miles from the farmhouse, and he kept the road maintained. There was a great view at the top of the mountain, and you could look off a long ways on a clear day. The toll road did not bring much business. Grandpa could tell the weather from the house by looking up to the hills where the sheep were pasturing. If the fog was still on the mountain—no change in the weather. If the fog was rising—good weather on the way.

Grandpa Wilson was happy to live on the farm in peace and tranquillity. He did not hanker to do big business. He just wanted to happily work the farm, get the property paid for, and enjoy his family and friends. He thought everyone was honest and lived accordingly.

Just above the farmhouse where there was a ledge the Wilson boys had a merry-go-round to play with. It was not fancy, but it went round and round for fun. Also they had an air glider that glided from the top of the hill downgrade for a mile or so.

As you came out of the big S curve and headed up the steep hill toward the farmhouse, you could see the barn. The barn had a sign in large white letters the whole length of the peak that read "Valley View Farm." I remember my father, Cassius, a master mechanic, whom everybody called Cash, going into the S curve below the house, and he could not get the old car he was driving to go. I was a baby in the car and was crying because he was beating the poor old car up with the crank.

When my father was fifteen and a crack mechanic and driver, he was hired as a chauffeur by Major Houghton of Putney, who lived at what is known as the Philip Chase Home. Mr. Houghton was a man of means and intelligence. He was the contractor in charge of building the railroad that ran down the Connecticut River into Hinsdale, New Hampshire. He was paid for building the railroad whatever he spent and a percentage of commission of the total cost. Most of the laborers were Italian emigrants, and they were good workers and knew how to build nice stone over-passes and underpasses. The workers lived in caves or holes dug into hills beside the railroad and moved as the railroad progressed. One of their choice foods was robins, and they ate all the birds they could catch.

On payday the workers lined up at the Hinsdale post office and sent their wages home to their families in Italy. They were paid in cash from the automobile.

All the people my father chauffeured with were armed with pistols for security. Dad told me about two of the workers who were involved in some animosity. Remember, the workers were at the height of strength and dexterity. One of these men ran down Main Street toward his adversary, swung his body down on the ground onto his hands, flipped upward with his legs and heavy shoes, and struck the poor fellow in the chin and killed him.

Interestingly, my father told me while we were riding on the New Hampshire side of the iron railroad bridge from Vernon that years ago, while the railroad was being built, the workmen lost an iron drilling rod that had a big, expensive real diamond on the end of it. The riverbed here was very muddy, and they never found that piece of machinery with the big cutting diamond. As far as we know, it is still there.

When Major Houghton finished the railroad, he wanted my father to go with him to South America, where he had to build another railroad, but my dad's parents would not let him go.

6.

Blackberry and Dandelion Wine

GRANDPA TAUGHT ME HOW TO split out ash to make arrows and to bend and shoot a bow and arrow. He said that for many years the Indians went through his property on the way to and from Canada. He and my father claimed to see a catamount go through and the leap was over thirty feet when they measured the tracks in the snow.

While I was quite young, when I was hunting with my father on the last day of the season my father shot a nice buck in the apple orchard. My dad would not dress it out, and we dragged it down to the house, where Grandpa Charles dressed it for him.

There was a big crank telephone on the wall at Grandpa's house. It was a party line, and everyone listened in to whatever was being said. You took it off the line to listen to hear if anyone was using the line before you cranked out. In those days there were few secrets between neighbors. Those were the days when neighbors helped neighbors. The telephone was the great communicator, and in case of emergency all the neighbors pitched in. They would drop everything in case of fire, tragedy, death, et cetera and help each other. If it was an emergency, you would ask someone to get off the telephone. Most marriages were confined to the small areas traversed by horse and buggy, so many of the marriages were discussed over the party line.

Grandpa had a lot of visitors, and he would stop everything to entertain them. He really enjoyed himself, and I think he seldom let work interfere with the fun.

At the sale of his farm Grandpa had an auction and sold everything to the highest bidder. It would be really nice if I had some of his things now. I was at the auction as a child and not yet able to think about the future and how nice it would be to have some of the family goodies.

One day in later years, my son Steven told me he did not have, or own, anything my father, Cash, had had and he felt badly about it. It occurred to me that at one time my father had a 303 Savage sport rifle—his favorite possession. He had let a friend, Whitey, come and live in his big Gageville house as a favor, and the house was all furnished, including his personal stuff like this rifle. Dad in later years spent the cold part of the winter in my little house in Florida, and in his absence from Gageville his friend Whitey sold many of Dad's possessions. I remembered Whitey sold the rifle to Russ Howard, a gun dealer in Dummerston, and I went to see old Russ and asked him if he remembered whom he'd sold the rifle to. Russ looked in his records and told me who had bought the gun. I called the man, who still had the gun, and told him that if he would let me have my dad's rifle, I would would buy him any gun he wanted. He accepted my offer and I gave the gun to Steven, which pleased both Steve and myself a great deal. Now Steven has a rifle he is fond of—a gun used by his grandfather.

This same gun has other history to it. Ray Pratt, the county game warden, who lived in Newfane, had taken this gun from a person who broke the law, probably by jacking deer. As it became state property, Dad was able to buy this nice rifle from the game warden.

When Grandpa Wilson sold the farm to the Van de Waters and moved to Brattleboro, I used to spend a lot of time at my grandmother Naomi Darden's little farm in West Townshend. She was the mother of my mother, also named Naomi. Grandma Darden left her husband after some scandal in Atlanta, Georgia, and she came first to Brattleboro, with all of her children: May, Rodney, Emma, and Arthur. A few years later, she bought the little farm in West Townshend. To get to the farm, if I had the money I would take the West River Railroad from Brattleboro, and the train would stop to let me off and to let me on for the trip home, even if I was the only customer. It left me off right at the wooden covered Scott Bridge. All I had to do was walk through

Grandmother Darden's little farm in West Townshend, Vermont.

the bridge straight ahead about one mile to the first house on the right. The bridge is closed now, and the farm is part of the Federal West Townshend Flood Dam System. I could hail the train at the bridge for the return ride home to Brattleboro.

Arthur Darden was the only single person left at the West Townshend home. At the time, he lived there with Charles Houle, who was also a boarder and roomer at my grandmother's when she lived in Brattleboro. Mr. Houle became a good friend and lived at Grandma's home until she died. He acted pretty much as the man of the house all this time and is the one I had a lot of fun with and worked with around the farm. Uncle Arthur during this period was attending the University of Vermont in Burlington and went along to graduate with a degree in engineering. Arthur first graduated from LeLand and Grey Seminary in Townshend. He used to walk to school from the farm, a distance of four or five miles, sometimes in cold, snowy weather, but he always made it.

Arthur joined the ROTC at college and was a crack shot on the rifle team. The team was so good they beat the army team from West Point Academy. During summer breaks, the riflemen were allowed to take home the .22 caliber long rifle cartridges so they could keep in training off campus. Art brought home a gallon of cartridges, and he would sit on the porch and practice in the sunshine. His practice included setting up in the dirt road a hundred feet from the porch wooden matches in a row for targets. He could hit the sulfur tip of the matches, which would ignite into flame. Grandma tried hard to get him into West Point to become a professional soldier. Senator Aiken had the privilege of assigning him to the West Point Academy, and Art passed all the necessary conditions but lost out because he was too small.

Later Art became an important man in electrical steam and atomic generation in the New York City area. Incidentally, he married the sister, Hazel, of my good friend Ernest Parent, who married Art's sister Emma; in other words, they married sisters and brothers.

Grandma Darden raised a few brown Swiss cattle. She got up to three mother cows, and they were just like members of the family. Grandma only had a little pasture, and the cows spent most of their time grazing up and down the highway. Grandma Darden did not often have a bull around the farm, and when she had to have a cow bred we used to halter the cow and lead her down to the farm right by the Scott Bridge. There the farmer had a bull we could use.

Grandma's farm was very much self-sustaining. She would barter with the owner of the in-town grocery stores, trading butter and eggs and wild berries for flour, sugar, kerosene, and whatever she could not get from the farm. If Grandma Darden happened to run out of or need something, she would barter with a near neighbor, as the nearest store was ten miles away, in Townshend.

All my family ate and raised lots of potatoes. Grandma, like the rest, always peeled the skins off the potatoes, which were traditionally not eaten, and mixed them in with the pig food. Nowdays, of course, potato skins are an expensive, popular food. She also threw the smaller pig potatoes into the rations for the hogs, and today the smaller potatoes boiled, skins and all, are a delicacy.

Grandma would feed the pigs from the dishwater, which had food leftovers in it mixed with whatever else she could scrounge to feed them. Charley, the boarder, kept the home with a lot of

My mother, Naomi Wilson, my grandmother, Naomi Darden, my brother Fred (in foreground) and I.

trout and venison year round. He also kept the farm in good wood, as they burned wood year round for heat and cooking. At night only the kitchen and living room were heated. The bedrooms would be cold, so we would take soapstone heating blocks to bed with us. The stones, about two inches thick and a foot square, were heated in the oven, and we tucked them into the foot of the bed to make it comfortable and warm to sleep.

Grandma smoked her pork in a big wooden barrel with oak and hickory wood, which became one of my jobs. To do this we had to cut one head out of the large wooden barrel, drill holes, and wire all the meat we could to the head left on the barrel. We would get a good fire going on the ground, and when we got a good bed of coals, we would cover them with corncobs. When the corncobs were smoking well, we would roll the barrel with the meat upright over the cobs. The barrel would be tight, and the smoke would penetrate the meat and cure it. Before smoking, Grandma would rub the meat all over with some kind of liquid seasoning paste.

Grandma also always had plenty of pies in the pantry, and

she always had oodles of food around. She made her own lard and venison mincemeat, and we always had fried pork rinds left over to chew on. Often she had fried raised doughnuts, fresh from the oil kettle. The raised doughnuts were so good we would eat them dipped in sugar until we got bellyaches!

In the fall at Grandma Darden's, my job was to gather the butternuts from the many trees that grow in West Townshend. The soft, sticky butternuts would stain our hands heavily when we harvested them. We would spread them all over the floor in the attic for drying, and once in a while we would move them around to help them dry completely. The really hard work was done in the late fall after the nuts had dried and the husks had fallen off and the meat had to be removed from the shell. After a while, our hands would get cut from the sharp protrusions on the nuts. In order the separate the nuts, we had to place them on a heavy piece of iron and hit very hard with a hammer. If hit correctly, they would spread in half and the meat would have to be pried from the shell. Mostly we had to be satisfied with small pieces of nut meat, as it often crumbled upon removal. We might crack butternuts for four or five hours and only end up with a third of a cup of nut meat. It was a wonderful delicacy to eat. What an experience when we mixed these with ice cream, cake, or cookies!

I have never heard of anywhere where you could buy butternuts in the marketplace. First of all, there are few and scattered butternut trees in New England and they don't bear nuts every year. Second, you could not afford to pay for the long time it takes to get the nuts ready to eat. What I can tell you is that if you never have tasted them and ever get a chance to eat some, you would be wise to do so.

There were acres and acres of wild blackberries on the hills around West Townshend, and Charlie and I had little trouble picking thirty or forty quarts of them in a short time. When we went to Brattleboro by Model A coupe on Monday mornings in the early fall, Charley to work and me to school, we could sell the fresh wild blackberries easily to the hotel and restaurants downtown. A neighbor in Dummerston, Charley Taft, made and bootlegged blackberry and dandelion wine and hard apple cider. He was caught and put into Newfane Jail, but they let him go home nights to tend his cows.

Grandma would visit us at Oak Grove Avenue in Brattleboro once in a while, when she could bum a ride or when my mother would get her in the Essex Terraplane sedan. Grandma was an avid reader, and she would scrounge around all the time for used magazines. I can just see her now lying in her featherbed propped up, reading the *Saturday Evening Post,* in the unheated bedroom. She also had an old Edison Gramophone player and had Caruso records and other classics. She had a battery-operated Atwater Kent radio and used earphones to hear with. She was from a rich old Georgia family, but she never had any money and I don't know why not.

Just before Arthur left for college, in Burlington, Vermont, he discovered someone left his brand-new leather mittens on top of the radio's battery and acid had eaten holes in them. Because of the money crunch, this was real bad news for Art.

One way Art was lucky. When he was attending college, Grandma Darden had an all-metal mailing suitcase into which Art mailed his dirty clothes home right to her door so she could wash them and mail them back from her roadside mailbox to him at the University of Vermont in Burlington.

When the sucker fish ran in the West River, Grandma served sucker cakes. Being from Georgia, she did a lot of southern cooking. When in Brattleboro, Grandma also visited her two sisters, Mrs. Glen Jones and Mrs. Emma Farrington. Grandma's relatives in Atlanta were Zastrows. My brother, George Zastrow Wilson, was named for Uncle George Zastrow. The only time I saw the Zastrows was at my brother George's funeral. When the Zastrows passed away, they left a million dollars to the Audubon Bird Society. My aunt, Emma Farrington, said the birds needed it. As far as I know, any consequence of money went to the Joneses and Farringtons.

George Zastrow was reported to have a lot of money. They were supposed to have more cash in the Atlanta banks than anyone. Aunt Emma said they were "eccentric" and when they went from one room in the house to another they would lock themselves in. They bought their groceries in large quantities, and Aunt Emma said, for example, they would buy a whole hank of bananas or a bushel of fruit and eat those that were overripe first. But after a while they had to eat *all* overripe ones, so they could never enjoy the best ones.

(From l. to r.) Emma Carter Farrington, Ruth Carter Jones, Naomi Carter Darden, and Amelia Carter Zastrow.

One vivid memory I have of my Grandmother Darden was one day when she was leaving for home from Brattleboro with two little piglets in a gunny sack, all a-squeal.

At the West Townshend farm, the back house was built off the kitchen at the same level at the back of the shed. It was a cold, unheated, very drafty four-holer, and it was at least twenty feet to the bottom. She used old Sears Roebuck catalogs for toilet paper. She kept a big picture of Pres. Franklin Roosevelt to look at while you were on the seat. I don't remember if Grandma Darden was fond of FDR or if she disliked him!

Charley Houle often had some alcohol mash working somewhere. Grandma would not allow alcohol around, so when Charlie knew she was to be away he made alcohol. The house water came from a spring about a mile away to the west on a steep hill through a lead pipeline, but the line was often broken down. Anyway, the water flowed constantly to a wooden fifty-gallon barrel in the kitchen. Charley set up the still on the kitchen stove, and the cool water in the barrel acted as a cooler and condenser for the still and the alcohol would condense into a container. When Charlie thought he had enough, he would run it through a second time. At that point, to get the impurities out of the drink, he would filter it lengthways through a loaf of Grandmother's homemade bread and it was fit and ready to drink. It was said that drinking this would cause your toes to curl. At the Joneses', the farm nearby, they also ran a still. They operated it from a stove under the bridge and used the brook for the condenser.

Another poor family, with a lot of kids, took their free welfare groceries all in raisins only. They were the Kaisers and Mr. Kaiser was famous for selling raisin whiskey to the hunters.

The Kaiser family had some pretty daughters, and one of the younger ones married Martin Jones, their nearest neighbor. Martin was middle-aged at the time. Martin told me once that the farthest he had ever been from West Townshend was Brattleboro.

At the Jones farm, directly off the house, was a long meadow reaching the forest. Many a deer was shot with jacking lights, mostly at night from the doorstep. Another specialty of the Jones family was maple sap whiskey and beer. I never had any and don't know the process.

When Grandmother Darden died, Charley Houle, her boarder, went to live with Mrs. Jones until she passed away.

Everyone hunted venison and I have been involved in venison hunting. One neighbor, Herb Jones, earned his room and board keeping Mr. Rawson's boardinghouse in venison.

I had a lot of fun at Grandmother Darden's. I bummed a ride from Brattleboro to West Townshend many times or went by train if I had the money, and occasionally some of the neighbors would ask me to get liquor for them at the Brattleboro liquor store, "the only legal liquor dispensery at the time," on the next trip up. Even though I was very young, I did not have any trouble getting a fifth through older friends. They wanted the cheap green river whiskey—I did not drink the stuff!

One day I was with Herb Jones getting venison for the boardinghouse when he shot a deer. We were in an area where the trees and brush were thick, and the deer was in an open space about one hundred feet away. Herb held his rifle, hands high about the brush, pulled the trigger, and dropped the deer in its tracks.

Charley Houle's brother, Tib Houle, a wandering gypsy who did not work, used to visit at Grandma's, usually in the winter, as he was passing through with no place to stay. He was a crack shot with a rifle and was helpful in keeping venison in the house.

My family was always interested in hunting and fishing. Hunting or fishing for food must be a strong inherited Wilson trait. Guns were our fun and we contributed to our food on the table with them and were told to never waste food or shoot more than we could use. Before we could have a gun of our own, we followed in the woods in back of our father while he explained the use and care of our shotgun or rifle. He told us that we should believe a gun was always loaded whether it really was or not and for no reason at all should we ever let our weapons point toward a person. Even walking around or handling a weapon in the house that you knew darn well was harmless, you never let the weapon point at anyone anytime or anywhere. It was always in the back of my mind to be extra careful with a weapon.

We also learned the habits of the game—where the deer, bear, rabbit, partridge, or whatever was to be found, their eating and sleeping practice, and their wild habits. We were careful to stalk the game into the wind, as many wild creatures have a great sense of smell and excellent eyesight and hearing capabilities. With time, we learned to see any little thing that moved in the woods. By habit and experience our eyes became accustomed so we would see things that were out of place. A deer's ear would twitch; a white tail would tip up; a bush or grass or tree branch would move when it should not. We kept alert, intently looking for signs that were out of place in the environment. We learned what the wildlife liked to eat and would carefully look for game that hung around the area where there were nuts and wild fruit trees and berries. My father told us it was best to keep stopping in the woods, not walking too far at one time, and to use all of our experiences and skills to get quietly up to whatever we were hunting. According to him, not only should we be careful so we would not step on a branch and make a noise that would scare the game, but we had to be careful not to make other noises that they could detect. If we didn't make any noise, we had the best chance of getting near the game. In addition, we had to be clean of body and our clothes had to be cleaned with water only, because we did not want to have a scent while hunting.

As an afterthought, when I was in the navy, even though I never saw combat, the training that my father gave me about hunting came in handy. I can tell you that if I was ever in a foxhole in battle, I would want to be in there with someone who could hunt and shoot. Someone with experience in the woods who knew how to survive because of this experience would be my choice.

In the dead of winter in West Townshend, in the deep snow up to our waists on the hill in back of Grandma's farm, in the same area we picked wild blackberries and shot squirrel in the oak trees, was the winter deer yard. On this rather small hill, up to one hundred deer spent the winter. They would eat the twigs of any trees their necks could reach, and together they pawed the snow to reach the ground, where they ate the undergrowth and the nuts or whatever was on the ground. All the deer were forever hungry if the snow cover remained for three months or more. They became very weak and the weakest ones died or became prey to the bobcats or dogs. We could get up really close to the deer as they had hard work to travel in the deep snow. The dogs could scamper easily on top of the snow crust and attack the deer and kill them. They killed the deer for fun, as they did not need the meat. It is bred into many dogs that they have to kill and learn to hunt to survive, and it is natural for them to take advantage of this tragic situation. Most dog owners let their dogs run all year in rural areas.

Grandma knew her little pet bulldog, Sparkey, would not hurt the deer, and I found it hard to tell her what I saw her pet dog was up to. While studying and enjoying watching this herd one sunny day, I heard barking in the distance and suspected what was going on. I went to the scene of the noise and found that Sparkey had joined other dogs in killing and chewing on the deer they were attacking. The snow was bright red all over against the pure white snow in the killing fields. Mangled heads, legs, and bodies were all about. The memory of this scene will never leave me. The deer had little or no way to protect themselves from this savage situation. It was hard to convince Grandma that her little dog was involved in this affair, but I convinced her when I showed her Sparkey had fresh blood all over his face and chest when I dragged him home from the slaughter.

Grandma Darden always had a little rock garden and a delightful flower area at her little farm in West Townshend. There were a lot of rock ledges in the yard, and she planted all kinds of flowers all around the ledges throughout the yard. The only problem was, she had a lot of work to do, like weeding and lugging in stones and loam and fertilizer. This is what she used her company for. I got in the work detail and spent a lot of time in the garden when I would rather be doing something else, like fishing.

The favorite outdoor game around the farm was croquet. The neighbors were also into the game, and there were contests between us.

Travels With My Family

BECAUSE WE WERE SUCH A big family, Dad had to use all of his wits to keep us in food, shelter, and clothes. There was a Dunham's Shoe Store near Manley Garage on Main Street in Brattleboro, and Dad knew the clerks very well. He would bring us kids into the store when our shoes got bad. Sometimes they got so bad we had to put paper in them to cover the holes in the soles. Dad would tell the clerks how the shoes did not stand up as well as they ought to, and we always got new pairs of shoes. The Dunham Shoe company certainly was generous to a fault.

My father, Cassius, worked at two jobs in Brattleboro most of his life. His night and weekend job was the erection of billboards along the highways. After dark he used the spotlight attached to his truck to see his work. Once a month he pasted new advertisements upon the big metal backed highway boards. He did about one hundred a month, and he got a dollar for each one he pasted up. I helped him a lot. We would go to work before daybreak at the C. E. Bradley Corporation on Flat Street in Brattleboro, where we used the steam from their boilers to heat up ten gallons of paste.

In the early morning before daylight, we would stop at the Dutch Bake Shop in Brattleboro, owned by Carl Vaetch, to buy day-old bakery products. My dad would buy all they had, as it was sold as pig food and cheap and they wanted to get rid of it before they opened for business in the morning.

Our big family lived like kings because of this bakery. My dad had an oil-burner kitchen range, and every night before going to bed he would make a huge double boiler of oatmeal so we would be able to have it in the morning with jelly doughnuts, cookies, or whatever else was in the bakery's collection. We would give the surplus to our pigs and chickens when we had them, otherwise

Oakland roadster (on left) about to be driven by my father, Cassius I. Wilson.

to the neighbor's hogs. Dad bought the coarse oatmeal at Farmer's Grain Store in big fifty-pound bags, and we really ate well. I still enjoy doughnuts and oatmeal. I have eaten enough oatmeal in my lifetime to fill a big room. My favorite breakfast now is regular oats cooked for eight minutes in a microwave, mixed generously with raisins, sweetened with maple syrup, and topped with milk.

We would paste up as many billboards as we could until 9:00 A.M., when my father would have to go to his regular day job selling Hudson and Essex cars and Federal trucks for Manley Brothers on High Street. When we got behind in our billboard work, we would work into the night after supper. My job was to help scrape off the old advertisements, and I would roll the big new posters into a long, round roll. My dad, standing on the ground, would slide the poster up to position and with a big long-handled brush unroll the poster and paste it in position.

My father was a great salesman, and he won a contest for selling the most cars. He was awarded a trip to Detroit, where Amelia Earhart was helping to introduce the new Essex Terraplane. He drove back to Brattleboro in the first new Essex Terraplane sedan. Then the Hudson Essex Company had a contest

My father, Cassius I. Wilson, with the Terraplane he won.

CASSIUS WILSON'S
LETTER WINS CAR

Manley Bros. Salesman One of Eight Among 2,000 Contestants to Be Given Essex Terraplanes.

Cassius Wilson, salesman at Manley Bros. garage, was one of the eight winners in a contest conducted among 2,000 Hudson-Essex salesmen in this country, and has received one of the eight Essex Terraplanes offered as prizes by the Hudson Motor Co. Mr. Wilson recently won a salesman's contest conducted by the Manley garage and as a reward he was sent to Detroit by the garage. There he was entertained, with 2,000 salesmen, by the Hudson company and was taken through their factories. After the visit 2,000 Terraplanes were driven away by the salesmen and all wrote letters about their trip home, regarding the performance of the new car, to the company. Eight of these letters were prize winners and the winners were given the car they drove away from the factory.

The judges in the contest were Thomas G. Wade of the Curtis Publishing Co., Claire White of Automobile Topics, Robert Ross of the Detroit Times, and Athol Denham of the Chilton Class Journal. Only three winners were in the East. Salesmen in the following cities received the other seven cars: Shelton, Conn., Milford, N. J., Columbus, O., Kewanee, Ill., Sharon Springs, Kan., Aberdeen, S. D., Fresno, Cal., Winnipeg, Man. Mr. Wilson's prize car has been used as a demonstrator at the Manley garage and has been ridden in by many local residents.

asking people to tell them why the Terraplane was such a great car and my father won the contest. The prize was a new Terraplane sedan, which he gave to my mother, and she drove it for many years.

My father sold a lot of cars and was Manley Brothers's best salesperson. John Manley gave him a free hand to trade as he saw best. Dad told me of how he started out one day with a New Hudson auto and traded it down four times and hitchhiked a ride home. In those days he also took in trade maple syrup, cows, pigs, and whatever he could trade for and the company could make a profit. The help of Manley Brothers garage had a room where there was a poker game going on whenever things were slow. Also, in one section of the garage there was a still where you could make alcohol. One of the faucets in the washroom ran clear alcohol. I don't know if this alcohol was for in-house use or if they ran a bootlegging business.

My dad, also known as Cash Wilson, did not hang around the Manley garage very much. He beat the bushes for sales. His special route was out of town up the West River Valley to old and potential customers. His best customers, other than those in town, were the farmers or the people in the many sawmills and the loggers who were scattered all over the boonies. Much of his sales pitch and no doubt his great fun was bantering with his customers. They all seemed happy to see him and delighted to get up to date with the news and gossip. He did not seem to try to sell the cars or trucks. If they wanted or needed one, they would tell him to get one for them.

When any of us kids were not in school, Dad would take us on the sales trips, and I never missed the opportunity to go, as motoring in the country was a great treat. In addition, we were able to share lunch with him, which he purchased in the old country combination grocery and hardware stores. The fare he always had at lunchtime was sardines, local cheese, salt crackers, peanut butter, and fig bars, washed down generously with C. H. Eddy soda pop or Moxie.

My dad may have been the strongest person I ever met, as he could hold up an automobile so you could place blocks under it! He bought a frame that was on a top floor of a car parts store in Boston to use to repair an early automobile. He asked for help in taking it down the five flights of stairs, and the parts people told him that getting it out was his problem. Dad took a hold of one end and dragged it down the stairs and exited from the premises. While dragging the car frame down the wooden stairs, Dad popped off all the lips of the stair treads.

John Manley was a very successful businessman and a gambler. They tell of him driving by an auction where they were auctioning off a hydroelectric plant. He stopped the car and bid on and purchased the plant, even though he knew nothing about the electric business. At one time his garage had a lot of extra automobile paint of all colors and he wanted to paint his house. They mixed the paint all together, and the house color turned out as gray.

When my brother Fred had a chance to buy a used square-back 1936 Essex sedan, he wanted me to be a partner in the deal. At the time I was caddying at the Brattleboro Country Golf Club and had saved a little money. I thought it would be nice to be part-owner of a vehicle and to come and go independently as long as my older brother had the time to chauffeur, as he had the driver's license. I never got to have a ride in the car because soon after we purchased it, Fred totaled it into a tree on the road to our West Brattleboro summer camp. He failed to negotiate a corner, but no one was hurt.

This was the period when I added caddying to my profession and learned how the *other* people lived. It was a new lesson to learn that some people paid real money to walk around the grass. It was a time to see what money could do to add to a person's pleasures. It was fun to talk with the Brattleboro merchants, doctors, dentists, manufacturers, and other elite of the community. Over time, some of the golfers asked especially for my services to carry their clubs, find lost balls, and help select the proper clubs for shooting. The caddy master would take me up front of the waiting caddy line to work for my golfers.

We caddies were at the bottom of the society pecking order. The golfers enjoyed the facilities of the club house dining room and bar, and we caddies sat on boxes out of sight in back of the pro shop waiting to be called and hoping we wouldn't go home broke. During this period, I thought rich people were the ones who were fat.

My father always liked big cars with lots of gadgets. This obsession must have been inherited by me, as later in life when I had some "wild" money, I started to pick up big used Chrysler Imperial cars for my museum. In 1988 I had thirty-five Chrysler Imperial cars. I bought those Imperials that had good bodies. As long as the body was good and the price was right and the car was not in too bad shape otherwise, I would buy it. I have an Imperial for every year they were made from 1957 through when they stopped making the big Imperials in 1975, not including 1958 and 1959, which I am still searching for. I seldom paid over fifteen hundred dollars for an Imperial, and the most I ever paid was thirty-five hundred dollars for a 1961 sedan.

When I was twelve years old, my mother and dad and all us kids went on summer vacation to Quebec, Canada, in a big Willys Knight open sedan, a big, long, beautiful black car with a big rubber ball horn we kids blew all the way up and down. We sure had a lot of fun. My mother had a pint bottle of Canadian booze hid in her stocking that she smuggled out when we crossed the border back into Vermont. Stranger still, my folks did little drinking of alcohol. Because other alcohol was prohibited, when my parents had something they wanted to celebrate they drank Lydia Perkins tonic, which does have some alcohol in it.

The summer I was fourteen years old, I got a job as water boy for the Brattleboro Water Department. At the time, my family had a shack like summer camp my dad built in West Brattleboro. This was in the Brattleboro watershed, which fed water to the town of Brattleboro from Sunset Lake, where they built a pipeline near our camp. This was my introduction to how to work in the government. My job was to carry drinking water to the workmen laying the iron pipeline.

I have always enjoyed traveling. As long as I am moving, I seem to be satisfied. The best way to see the countryside, other than walking, is from a pickup truck or a recreational vehicle. Something that sits higher in the road than an automobile is the best. It is best for me to try to see every place or thing of interest as I travel. That is, I keep an eye out for wildlife, interesting people, and the beauty of the land. Every state or country has something of beauty and interest in it for me. I lose most all navigational instincts as I try to grasp all the interests of my surroundings. As I travel, I do not recall north or south, and if I stop, I sometimes don't know where to go to get back on the highway. It is hard for me to understand seeing a person not enjoying the ride but reading a book or newspaper while traveling in interesting country. It is like reading a book while making love. Because of my attention to the delights of touring rather than knowing exactly where I am, increasingly I get lost. It does not bother me to get lost. Some of my best adventurers have happened when I have been lost!

8

Different Ways of Making a Living

DURING THE SUMMER OF MY water-boy job with the Brattleboro Water System, I went on a B&M excursion train to New London, Connecticut. That was in the days when the railroad liked to increase the revenue with excursions to the ocean, and in reverse we used to get a lot of trains from New York, New Haven, and Hartford full of people who came to our area for winter sports, especially skiing. As this was one of the first times I was ever away from home, I bought from Nick Lasasser, a fellow worker, a used pair of stylish pure wool white dress pants. They were quite reasonably priced, and only had a few mothholes in them. My mother cleaned the pants with a kerosene solvent the night before the excursion. I put them on and realized they were damp and smelly but did not realize what was about to happen to me. I broke out in a rash and was very uncomfortably irritated and there was nothing I could do about it, as I was stuck on the train. Upon arrival in New London, I went to the Red Cross tent on the beach for help. They did relieve me, but I was stuck wearing my damp pants until I got home later in the evening.

Before the Water Department job, I worked for a couple of summers for the Carpenter family, Frank, George, and Mary, who had a farm near our summer camp there in West Brattleboro. I helped with all the farm chores, and I learned how to drive oxen and horses, mow with a hand scythe, rake with a one-horse-drawn rake winnroller, and load and unload hay with a pitchfork. This was before the invention of a hay bailer. To store loose, dusty hay to the height of the barn, they drove a long, sharp fork contraption into the load of loose hay. A big rope was attached to the fork and led to a pair of oxen who would pull the fork of hay via a pulley to the highest bay in the barn, where it was dumped. Then the man in the bay would pitch it by fork to the next man in the bay and so on until it got to the highest point. My job, as a wiry, fresh lad, was at the highest point. I had to tread it down and jam it into the corners and eaves to save room. Looking back on this treading job, I think it was harmful to my lungs because it was very hot and dusty in the airless barn and in between the wagon loads I would go out into the fresh air and my nose would run with a lot of heavy phlegm. This would go on for a few days of hard treading, and I think it permanently harmed my lungs.

I took my meals at the Carpenters', and I ate quite well. They all ate a lot of homemade doughnuts fried in lard, and at every meal, including breakfast, they would draw hard cider for drinking. They drank the best hard cider (two years old).

I helped them pick the apples to make the cider. What they did was grind the apples upstairs in the barn, worms and all! (This was before the use of pesticides, when most apples were wormy.) The pulp would go down through the floor into a press about eight feet square. First we would put about a foot of straw in the press and then a foot of apple pulp and more straw and apple pulp to perhaps eight feet high. Then there were two huge wooden jack presses. We would keep turning the jack screws down as long as the apple juice would keep floating out to a big wooden catch barrel. The left over apple pulp, mixed with straw, was fed to the animals. The cows and horses ate it like it was candy.

The Carpenter brothers made a lot of hard cider, and they had a lot of company who liked to drink the intoxicating stuff. I imagine they used to sell it because I can't imagine them giving away so much hard cider. They must have been bootleggers. The nice, cool cellar must have held forty or fifty wooden barrels of fifty-two gallon casks.

I recently spoke to Herbert Ingalls of Brattleboro, and he told me that once he was at the Carpenters' buying two-year-old hard cider for a friend. The price was twenty-five cents a gallon, and he counted forty-four casks in the cellar.

The Carpenters' principle income was from beef, which they sold by the pound, and as they did not have a scale, they sold it by a measure. They would tape measure the body of the beef cow with a cloth tape that had a scale on it that indicated how much the critter should weigh. When the beef buyers showed up, it was a real fun affair to watch the haggling and bantering as they negotiated the value of the animal.

One memory I have of the days I used to work for the Carpenters was when I had a bad scare one summer day when I was driving their best pair of huge Durham oxen off a steep hill. When you drive a pair of oxen, you drive them from the front and direct them with a whip to their bodies as you travel along. The hay load coming down the very steep hill was heavily packed, unbalanced on the rear, and it raised the oxen in their yokes by their heads and front feet, with their hind feet up in the air above my head. The wagon and oxen were flying, gaining speed down the hill and I was scurrying for my life to get out of the way. We made it okay, though!

Earlier one of the oxen wandered into the area upstairs in the barn where the apples were ground up for cider. The ox was so heavy that it went through the floor and hung in the air by the beams and planks. It was necessary to get a sling under the ox, and we had to use pulleys to pull it out of the hole. It was a very hair-raising experience.

The Carpenter brothers were bachelors. The old maid, Mary, was alive when I first worked there. After I left, the Carpenters sold out to Bob Gannet and moved to a smaller farm in Halifax. They really did not want to sell, and I think they were unfairly induced into signing a sales agreement. There was some uncomfortable litigation. They had no near relations, and I do not know who finally got their estate.

Down the road a piece lived Doc Gage, an old man who was retired from veterinary work. Doc Gage had a terrible reputation of being a hell-raiser all his life and living a lot of sin. In his waning years he spent a good deal of time reading the Bible because he thought it could get him into heaven.

The Carpenters told me their father raised a lot of turkeys and when the turkeys were of size to sell they would drive them down the main road to the Boston market. This was before the automobile.

The Carpenter brothers used oxen for most of their work instead of horses. One reason was that beef is always salable and in New England horse meat is not too marketable. I saw evidence of the economics of this while one of their best pair of oxen was gathering maple sap in the extra-deep hip-high snow at the rock maple sugar bush. While turning in the deep snow, one of the oxen broke its neck. If this was a horse, it would have been a complete loss. In this case, the ox was dressed off and sold for meat to the public at the butcher shop.

The Carpenters' open fields were where my family harvested the wild strawberries in the summer. The berries were in abundance all over the farm. My sisters especially were adept in picking the berries and often showed us the proper way to do it. The wild berries are very small, "pea-sized," and it would take you all day to produce a quart one at a time. We discovered that the best way to harvest them was to kneel down on the ground and pick them stem and all. This way we could fill a half-bushel basket with a multitude of stems, that held all the berries. When we got home, we would comfortably pick the little berries from the stem. In most cases, mostly all the berries were dead ripe and those that were not ripe we would leave on the stem and cast the stem and unwanted berries and leaves away. If you have never eaten dead ripe, juicy, wild strawberry shortcake, you ain't lived yet!

It is interesting to note the difference in health standards now from when I was a lad at the West Brattleboro summer camp. In those days, we used to get a metal can of milk at a neighboring farm. There were thousands of flies around the milk house, and it was common to have flies floating around in the milk jug. We paid little attention to the flies and just skimmed them off the top. Also, when I worked during my high-school days during the school season washing dishes at the Hotel Brooks in Brattleboro, they had a large refrigerator in the kitchen area and it was solid black all over with frozen dead flies. I lived in the hotel during this time, and when I left my dirty clothes around they were sure to end up with cockroaches under them. The owner of the hotel, Sandy Daniels, took me under his wings and helped me grow up properly. I am indebted to Mr. Sandy Daniels for his help and advice.

My first job at the hotel was that of dishwasher. I ate some of the nicer morsels that came back into the kitchen from the tables uneaten. Later I graduated to busboy and cleaned up the tables and learned to carry overhead by tray the dishes to the dishwasher. I would do any other jobs the waitresses wanted done. For this I got a percentage of the girls' tips. We learned a trick with the hotel beer dispensers—those long-handled spigots under air pressure where you filled the glasses and pitchers hitched to the cold beer barrels just out of sight. We learned when the bartender closed the restaurant and shut off the air machine that pumped the beer, any beer that was still in the pipes, was still available to capture under the pressure from the pumps. We had plenty of free beer to drink after the crew shut up for the night and went home!

As I was the youngest person working at the Hotel Brooks in the thirties, the men and ladies all were extra nice to me and I really learned a lot about life with their help. Mimie Towle, a senior waitress, was always available to help me mature. A very small, gentle Danny, quite an old man, a hotel worker in the next room, was one I spent a lot of time with. He had a lot of stories to tell and was crazy over boxing and its latest developments. His radio was turned on all the time, and when boxing was the event, it was his most happy experience. Any sport kept him occupied, and radio was his best friend.

9.

In the Navy

When I graduated from Brattleboro High School in 1937, I thought that because jobs and money were scarce, I would join the U.S. Navy. I bummed a ride to Springfield, Massachusetts, had my interview and physical, and was accepted for duty. When I left the navy recruiting office, it was late in the day and I was very hungry. I had five cents in money, so I checked Woolworth's Five-and-Ten-cent Store for some food. The most I could get for the nickel was five cents' worth of Spanish peanuts. I satisfied my hunger and hitched a ride home.

The navy needed my parents' signatures in order for me to get in the service, and both my father and mother refused to sign the release papers. My father told me to hang in there and assured me he was going to see that I did better. What he had in mind was to keep all us eleven kids busy in the basket business he was buying.

In 1944, I was drafted into the U.S. Navy. I have little to say of it, as I did not get overseas, but I did all that was asked of me and more. I learned fire control—the firing of all guns aboard a ship. I hated the whole time I was in the navy, and worst was being away from my wife, Connie, and sons, James and Frank. From what I saw of the navy, I do not understand how we won the war.

I took the navy aptitude test. Here again, because of my varied experiences in having run my own business, I knew a little about most everything. The navy asked basic business questions as to your knowledge, and if you could keep business records and run a typewriter, take shorthand, et cetera. I could do all these things. They gave you simple math questions and asked if I could tell

*Author Frank G. Wilson, high school
graduation, 1937.*

what the picture was, showing a hammer, pipewrench, et cetera,
if I could drive a truck, et cetera. I had been exposed to all those
things, so I got the highest test score in my company. Because of
this, they put me in the office, sitting around all day doing office
work in the personnel department. Every day I got softer and
softer and lost much of my good physical condition. To top it off,
eventually, because my math test score was so good, they sent me
to fire-control school, which requires high technical mathematics
and the algebra-type figuring needed to fire the guns and cannons
on board a ship. What they did not know was that I was lousy in
algebra and geometry and disliked both! So I went to fire-control
school and disliked it and went nuts learning to fire the guns. I
hated every minute of it, and it was hard to keep up with the class.
Fortunately, the war ended and I was discharged.

While I was going to fire-control school I got a job while off
base on liberty at the Hunt's Catsup Factory because my navy pay

Frank G. Wilson, U.S. Navy.

was not enough to keep me in beer and cigarettes. Because I had gotten soft at my navy job, I looked stronger than I was at that point. At the Hunt's factory in the Oakland, California, area, they gave me the job at the end of the packaging of catsup where there was a big mechanized metal roller conveyor. My job was to pick up the heavy boxes filled with catsup bottles from the conveyer and pack them high on pallets so the forklift could take them to storage. The conveyer ran at a speed where you had little time between pallets to rest because if you stopped, the catsup cartons would spill all over the cement floor. Remember, as I was no longer as physically strong as I had once been, I barely made it through the shift. I never went back to work there and never went to get my pay.

I had to shut down my business to get into the service. During the war period, before being drafted, I produced large quantities of lumber for the war effort. I had the lumber trucked to Connecticut, and they made big reels to hold wire from the lumber. I figured I produced enough lumber during the war to build a bridge from Vermont to New York City. It was really silly for them to draft me into the service. It was the county draft board that was in charge of deciding who went into the service and who did not get summoned to go to war. A lot of guys who did nothing to help the war never got drafted. I think my neighbor in Westminster, Mr. Ranney, a dairy farmer and the closest neighbor to me who was on the draft board, had to be the one who finalized the deal. I have no animosity about this phase, but I believe that Mr. Ranney had little knowledge of the problems of the world war and business activity toward the war effort outside of his little dairy business. Someone had to do this job, and who would want to send a person to war who had a good chance of getting killed. Mr. Ranney never met me, and as far as I know, he probably did not know of my situation and my contribution to the war effort.

The last thing I did before leaving for the navy was visit Napoleon Martin at his Putney diner. John Rice, a nice big, young neighbor, was in there at the counter. When I got back from the navy, John was sitting on the same stool.

Once I was sitting on a stool eating in the diner talking with Napoleon about what he would do if someone held him up in a robbery, of which there had been one recently. Almost instantly, right in front of him, from under the counter he pulled out a big long-barreled revolver and stuck it in my belly.

The first few days I was in the navy I used to get up at daybreak, before reveille, and run the track to assure myself that I was ready for combat. You have to know that during this period part of my civilian business was the cutting of trees in the forest, bringing logs to the Basketville mill, sawing the logs, and delivering wood to Connecticut for use in wire reels for the war effort. I personally spent most of my time doing this very heavy work, and I was in perfect physical condition. No one ever did, nor I think no one ever could beat me in physical combat. I strongly believed that if I was dropped off in enemy territory I could win the war all by myself. I was ready to fight. I did not want to waste any time!

Before being drafted, during the war I purchased the trees in the forest around Basketville from farmers. We had a crew of choppers, teamsters, and truck drivers for the logs and timber, and we had the necessary saws, axes, trucks, skidders, horses, and equipment to do a good lumbering business. The lumber was then processed at the basket works. It is important to remember that I pitched in personally to work on any of the processing, and I was in perfect physical condition.

When working at the Gage Basketry, my job was to keep the logs moving into the sawmill from a long skidway where the logs came off the trucks from the forest. I liked the job. I could roll hundreds of logs a day all alone with baseball-sized cant hook with long-wooden handles, which would curl from a fulcrum bolt held around the log. The trick to rolling the logs, I found, was to throw the hook around one side of a log so the sharp point of the hook would embed itself into the log. I would bear down on the hook from the end of the handle and roll the log along. It was a fascinating tool, and with a little skill, I was able to roll a log that weighed as much as a ton.

At the Putney Sawmill operation, I also worked the cant hook as much as I could. I learned to use the hook like a real professional, and not only could I roll the logs down along level or uphill; I could move the log either way endwise. I could roll a huge log all by myself from the ground all the way up to the level of a truck body. I think I could eat breakfast with it! It is a very clever trick to learn to use this tool, and few people can master it. I should write a book on how to use a cant hook!

I added to my lumber enterprise when I purchased another pair of horses from Mr. Butinsky of Vernon, Vermont. They were a young, beautifully matched pair of Belgian draft horses of prime value. Before the purchase I watched them work and saw they were well trained. Mr. Butinski made sure they were shown doing light work, pulling a wagon with little weight on it, and the team of horses performed magnificently. I purchased the team and put them to work on the Gorham lot in Westminster West. My teamster, Art Gifford, put them to work scooting out big loads of logs to the truck skidway. One of the horses kept falling down under the pressure of the heavy load, and soon we learned that the horse had a stifle, a disjointed bone and muscle problem. He was worthless and would never be able to do heavy work. Seeing as I was gypped badly, I rushed right down to Vernon to get my money back from Mr. Butinsky, and when he saw me coming he knew right off why I was there. He ran into the barn and barricaded himself inside, and I could not reach him. I am glad I could not get to him because I was so furious that I could have gotten into a lot of trouble. I took my licking, learned a new lesson, and went about my business.

When I got drafted into the service and closed down my sawmill and basket and bucket business, I was shipped to Sampson Naval Station near Geneva, New York, for my basic training. From Sampson I went to Bainbridge, Maryland, where I attended fire-control school, and from fire-control school I was shipped to Oakland, California, where I awaited assignment to the USS *Massachusetts*. While I was in California, the war ended, and I was not going to be sailing if I could help it! My neighbor, D. H. Smith, wrote to our U.S. senator, George D. Aiken, and explained to him the situation and that he thought it best for Putney and society if, now that the war was over, I be allowed to come home and reopen my business and get the folks working again. Senator Aiken advised Mr. Smith to write to some navy person of importance (I don't recall his name) and explain my history and to enclose a can of good maple syrup and that the senator would brief that person about the situation. Mr. Smith did as he was told, and a short time later, I was called to headquarters and informed that I was to be discharged and to present myself back there the next day with my seabag and personal belongings for discharge.

I arrived at the headquarters office the first thing in the morning and presented my papers to the man in charge and was told to sit down while my papers were processed. I already had my "lame duck" patch sewn on my uniform, so I sat down. I sat in the office all day, and the papers just sat on the officer's desk.

At quitting time, 5:00 P.M., the officer signed the papers and threw them on the floor and went home.

It did not take me long to pick up the discharge papers, get out of there, and head for the railroad terminal, where I got a ticket to Vermont. I was alone on this trip and free as a bird with my papers. I was an honest-to-goodness civilian again. I had a little time on my hands before the train arrived and wanted to celebrate, but the area was dry and no liquor could be had for the public. I got on the train, of course of light heart and pleased to be going home. The train stopped in Reno, Nevada, the next morning. I looked out the window and noticed that the train was parked right in front of an open liquor store. With speed and sharpness of intelligence that I didn't know I had, I ran down the steps to the liquor store and grabbed two quart bottles of the handy liquor right at the cash register. I paid for them and caught the train before I left the station. "Oh, boy," I said to myself, "I am all set for this party on the long, slow ride to Manchester, Vermont." Would you believe that I had grabbed two bottles of Southern Comfort, a sweet concoction that I disliked. But for lack of better I started to drink the damned stuff. To make a long story short, I got sick on this liquor and was sick all the way to Manchester.

10.

Back to Civilian Life

WHEN I GOT TO MANCHESTER, it was the first day of deer season. I started to hitchhike up the mountain and a lot of shooting was going on around me. This was a fun day that I always looked forward to as a civilian, the first day of deer season. Finally I got a ride and got into Brattleboro, the city nearest to my home in Putney. I forget how I got to Putney, but I was happy as a lark to get home. No one knew I was coming. People ask me what the first thing I did when I got home was, and I tell them that I put down my suitcase!

During this period, five of us brothers were in the navy and we all came home. My wife, Connie Roby Wilson, lost two brothers to the war. She lost her brother Bertrand, who was serving near Munda Airport in the South Pacific, and she lost her brother Robert, who went down with the submarine *Tullibee* in the South Pacific when it was sunk by one of its own torpedoes as it came around and blew up their submarine.

I was visiting my brother Fred in the navy in New London, Connecticut, which was chucker-block full of sailors during the war. The sailors on leave away from home filled the city as they went walking and wandering around with little to do. The liquor and beer distributors and all the public dispensers in New London could not satisfy the sailor's need and purchases, and all the public dispensers would run out of liquor early. As you went down the main streets, you could see temporary beer retailers along the way with big galvanized tubs filled with ice and beer bottles. They all seemed to run out of beer about the same time in the early evening, and hell was to pay. That is when I joined the Elks Club, as they had a better purchasing system and never ran out of beer or whiskey. My life membership card is number eighty-nine, and

I have been a member for forty-three years. In all the time since I have been out of the navy I have paid my dues properly and I have been in the Elks Club less than twenty-five times in my lifetime.

Whenever there is a big congregation of men off duty on leave in town, you will always find those who have set up a business of prostitution. It has always been my honest conviction that prostitution should be government-controlled and legalized. If I was in the position to vote on its merits, I would certainly vote for legal prostitution. It is not proper, in this case, to be against human nature as long as it is mutually agreed upon. Personally, I have done little sinning. One event I will always remember, though I do not recall where I was on military leave, was when I saw a slow-moving line of about fifty sailors at a building. Sitting on the ground, standing, and reading comic books, they were waiting their turn for action. Vividly I recall seeing a girl whom I thought was the most beautiful lady I had ever seen appear for a bit at the site. Believe me, I never participated in these events, but I wonder if I would be in better health and a better citizen if I had let nature control naturally!

I know I have been to the top of the hill, but I am fairly sure that I have never been to the moon. I have learned from experience that all good, successful businessmen have a great sexual drive. One of the few things that will stop the negotiating of an important deal is to have a pretty girl appear somewhere in the vicinity of the meeting. The talk will stop automatically and from where one is looking, all the men know what thoughts are going on in his mind. After the pause in the gentleman's subtle way, one would say something like, "Jesus, take a look at that." It is one way to take the seriousness out of a deal and to see the nicest gift God ever gave to mankind.

A fond memory I have is of a good friend and neighbor, Dick Phelps, who worked for the state of Vermont road department. Dick was a plain middle-aged man, typical of a plain Vermont-raised farmer, a nice guy with a minimal education. One summer day a nice new convertible with four nice pretty young girls in it hailed him while he was working on Route 5 in Putney. They asked Dick if they were on the right way to Keene, New Hampshire. Dick said "No, you go back the way you came and eight miles down the road at the Howard Johnson's restaurant you turn left toward Keene where you see the intercourse in the road." At that remark, things turned hilarious, and Dick still doesn't know the problem that caused the girls to turn around and drive away.

11.

Rebuilding My Business

AFTER THE WAR, I WENT too fast to try and regain my lost business and got into financial difficulty. I had to go into Chapter 11 bankruptcy to get back on my feet. At this time, Jerry Lippe of Leipzig and Lippe in New York City at 1166 Broadway sent me checks for all my production of baskets. I must have done business with Leipzig and Lippe for thirty years. I visited the Leipzig and Lippe New York office and showroom once a month. The new samples and deals and problems were a constant duty, and I had to keep updated to keep ahead of the competition. There were few big stores, catalog houses, houseware retailers, or gift shops of any consequence all over the country who did not buy from us.

Jerry Lippe handled the big accounts like Sears Roebuck, Montgomery Ward, and S&H Green Stamps personally. His customers had a lot of faith in his presentation. Actually, the smart buyers would let Jerry make out the order himself. Jerry would choose the items, decide how many to buy, spell out the details of the deal, and hand the buyer the pen to sign the contract.

Leipzig and Lippe also sold for other manufacturers and importers. They sold Sears Roebuck an imported round willow clothes basket that was imported by R. Greenspan of Brooklyn, New York. The clothes basket retailed for one dollar, and Sears and Roebuck sold so many it was a chore to find room in the warehouse in Brooklyn to handle the shipments. There were piles or mountains of clothes baskets in the warehouse and some outside in the street. I was aghast to see the workmen actually treading on top of the basket piles in order to move about the area.

Bill Leipzig, Jerry's partner, introduced the fitted picnic basket that was the most popular basket we made. We sold hundreds of thousands of picnic baskets fitted with knives, forks, spoons, cups, and dishes. The next most popular basket we make is our pie and cake basket. This basket has a partition that separates whatever you have on the bottom layer, such as a frosted cake, from whatever is on the top layer. The pie and cake basket is popular in Vermont. The basket is taken to club affairs, church suppers, et cetera, and the owner's name is printed on it so she can claim it back.

When we worked at Leipzig and Lippe's showroom at 1166 Broadway, New York City, we all went for lunch at the Black Crow Restaurant, since gone out of business. Many of the merchants of the area used this as sort of a social club also. One thing that went on there all the time at lunch was poker, using regular paper money in place of playing cards. The merchants could take out their cash and mix their bills equally, shuffle well, and deal out the bills. The winner was the one who had the better poker cards using the numbers on the bills.

My dad, at Sidney, Gage & Company in the forties, sold much of his department-store business through a sales organization by the name of Hamilton Cornwall in New York City. Hamilton was also a crack salesperson who sold a lot of baskets. He wanted everyone to know he was in the sales game, and one of his selling methods was to have a huge black man drive him around the country on his selling trips. The man would drive his car and carry his big showcase samples into the buying offices to show the products he had for sale. Another unusual method he had was to carry a picture on the shipping label of all the merchandise that he sold. Of course, anyone would know that the only people who saw the shipping labels were the transportation companies and the receiving clerks, who had no reason to need Hamilton in the first place. I asked Hamilton Cornwall why he did the unusual things he did, and he replied, "You are talking about it, aren't you? Everyone mentions these things to me. I don't care what anybody says about me, good or bad, but don't stop talking about me. Don't forget me." As I thought about it, I guess the things he did to get attention and be remembered paid off for him.

In order for me to get to New York and lose as little time as possible at the factory, I had to take the B&M Railroad Bootlegger out of Canada. I would pick it up in Putney and travel to Pennsylvania Station in New York. I got on the train in Brattleboro at 2:30 A.M. It was never crowded and I would sleep crossways on the seat and I was ready to go to work at nine o'clock in the morning when I awoke at Pennsylvania Station. At four thirty in the afternoon, after work, I would take the Montrealer from Grand Central Station to Putney. At Grand Central Station I would always visit their famous oyster bar, where I would pleasantly fill up on oyster stew and free popcorn, sausage, and crackers and cheese, generously washed down with William Penn whiskey mixed with ginger ale. I would get on the train in good shape. I only missed getting off in Putney one time by oversleeping on the train.

Lippe and Leipzig were really a fine outfit and always did exactly as they said they would. The only trouble was, I could not make any money doing business with them. They just kept me afloat. They must have made a few million dollars selling my stuff because they marked it up from 15 to 25 percent above cost. All they needed was a pencil to take orders, and they sold a lot of other merchandise at the same time. In the meantime, I began selling to a few other accounts and started my retail operations. When I finally weaned myself from them, I had to give them commissions for six months for all orders of any accounts they ever sold to, and they did not take any orders for me or service my accounts for these six months. It was a tough time. While in the office of Leipzig and Lippe, I had a friend named Ken Steele who was bookkeeper there. Ken told me, "You see that man working over there, named Harry Bach? If you ever need an accountant, he is a top-notch man." He said I should definitely try to hire Harry.

Harry Bach was apprenticing for Lambert, a big accountant firm, at the time. Later, when increasing business required an accountant, I asked Harry if he would come to Vermont and work for me. He consented and came to work one day a month in November 1955, more than thirty-five years ago. His wife, Evelyn, who has a master's degree, taught economics in the New York City school system. She too is a very competent accountant in her own right. During the summers, Evelyn came to Vermont with

Harry to help out with the records. After a while, the Bachs spent two days at Putney, and now that Evelyn has been retired from teaching since 1957, both she and Harry spend three days in Putney each month. When the Bachs visit Putney, they stay in our home and are just like members of the family. We enjoy their company and are grateful for their many years of important contribution to our business.

My family is also grateful to our lawyers, the Barbers, who, with their office girls in Brattleboro, were able to do our personal and legal work for us. Earlier Frank E. Barber did all the work for my father and all members of the clan, and Frank's son, Elliott F. Barber, has done my legal work since the first day we went into business. Elliott retired recently and Tom Costello, an attorney from the Barber office, is our attorney now.

When I quit Leipzig and Lippe as our sales agents, I personally spent all the time I could acting as sales manager, hiring a national sales organization, and assuming all the duties needed to merchandise the Basketville line. Also, personally I traveled widely and reserved the big accounts like Sears Roebuck, Montgomery Ward, S & H Green Stamps, and others for house accounts (buyers who, because I serviced them, were able to save the sales commissions, which ran from 5 to 15 percent of the value of the sale). During this period, I met some of the biggest men in the business. This is when I learned that the bigger the person, the easier to do business. I also learned that some people are "too busy working to make money."

Part of my duty was to attend trade shows and show samples of our products and explain our offerings. The important shows were in January and July in Chicago and at least twice a year in New York City. Smaller shows, scattered all over America, were worked in when we thought they were opportunities. We also sold a lot to the big food packers and mail-order houses. One of our biggest customers was Figis of Marshfield, Wisconsin, one of my house accounts who used to buy baskets and wooden pails in full trailer loads.

When I first went to the shows, we did everything as inexpensively as possible. For example, in Chicago, before we were able to get into the big shows like the Navy Pier or McCormick Place Exhibition Hall, we used to show at the Hotel Morrison, a satellite of the main shows in town. The shows in these rooms were for us beginners. It was a place for us to familiarize ourselves with the job of meeting the buyers and selling. Later we were able to be in the main arenas. We still have these positions at important shows, and because of the longevity of our participation in the shows the positions are coveted and worth a lot of money and our customers and sales agents expect us to be there.

During the early shows I would take an assistant, the latest creations, and a bale of catalogs and drive nonstop to Chicago. At the hotel, we would sleep in the showroom in the off-show hours and during the show we would put the beds in the bathroom out of sight.

In all the time we spent in Chicago, those in the show who wanted to party and haunt the bars and nightclubs and meet wild women went to Cicero, a Chicago suburb. All you had to do was tell the taxi driver or doorman whatever you wanted to do, and Cicero would accommodate you, no matter what. In those days in Chicago, and I presume in this area, it is still the same, where anything and everything goes on. It is a very safe place for strangers who want to raise hell away from home. It is safe because it is run by the Mafia.

There was no danger of getting in trouble in Cicero. The mob wanted the area to be wide open to the conventioneers, and if anyone in town bothered a visitor or picked your pocket or whatever, the boys would see to it that whoever did you wrong was no longer seen in Cicero. During trade shows in Chicago we would fill up every motel in town, and you could tell that a trade show was on when most of the windows at the Hilton Hotel had lights on in the darkness. The whole area would really rock.

I have always felt that Basketville had the best merchandise at the best prices and the best service in the industry. If anybody needed merchandise such as we made and did not buy from us, they were wrong, and no doubt about it. I can't understand how anyone could sell merchandise other than ours when they knew darn well we could not be beaten. That is why we prospered and kept in business. We know we are best, and it is easy to sell under these conditions. Good buyers know if it is bought right, it is half-sold.

12

Baskets International

IN 1956, I STARTED the importation of baskets. It was quite a simple decision. How it happened was once our good neighbor, former governor George Aiken, then the U.S. senator in Washington, visited me at the Putney Retail Store and I asked him how I could compete with the overseas competitors when I was paying thirty-five cents an hour and they paid their help twenty-five cents a day. George Aiken told me, "If you can't beat them, join them," and this I did. All I did was write to the U.S. ambassadors at the various basket-exporting countries asking for information on baskets exporters. We started buying by mail and bringing in a few baskets from here and there, and it is still happening. I have traveled the world over several times, buying baskets in many countries.

We have visited many countries, and not all were peaceful. While I was staying at the Atlantic Hotel in Beirut, Lebanon, in May 1967, it was an armed city. Lots of tanks were rumbling around the city, and we had the feeling that something like a war was about to happen. At that point in time, you could say, "Stay home and see America first!"

Everyone in Lebanon was concerned about the Israel-Jordan business, and I predicted some action there soon. We got a kick out of shaking hands with the people as they called us the "last American tourists" and told us they would see us after the war. The government of Syria had passed out arms to all households, and the Lebanese Congress was screaming for the same. The Gulf of Agaba, international waters that are only two miles wide, was then mined by Jordan. The whole thing boiled down to the fact that some administration would start something of great importance to the peace of the area.

From my visit to Beirut I will say that some people are crooks, as for two days running I went to the Atlantic Hotel desk and bought a newspaper that cost twenty-five piasters and no effort was made to give me change for my one hundred piasters that I gave them. So I sat in front of the desk and read the paper, and when I was done I asked for my change and they said, "Oh, yes," and gave me fifty piasters back. In other words, they would have liked to have kept the whole one hundred, but seeing as I asked for change, they gave me fifty back, which still left me twenty-five short.

I bought a lot of baskets from Hong Kong. I was in Hong Kong when Nixon took off the embargo on the exporting of Chinese merchandise, so I went into China with my good Hong Kong friend and factory owner Mr. Mock. I went into Kuang-chou (Canton) and bought heavily. Among other things, I bought the Kuang-chou sample room. Some of the samples I put into my personal basket museum. We sent out a group from the Hong Kong factory to buy all the nice farmers' baskets at the retail stores in Hong Kong and Kowloon. The Hong Kong retail stores sold only the bamboo rice baskets. A rice basket on a China farm in the field is like a pickup truck in the USA—they are woven tight enough so a kernel of rice won't fall out of them while you are harvesting or storing rice. They are sturdy enough to take abuse in the carrying and planting in the fields. In the home, they are down-sized and used for storing food. Rice baskets, like all baskets, enhance the circulation of air, which helps prevent mildew and rot. These are real heavily built but lightweight for easily handling. The baskets are in sizes of about eight inches by eight inches to about a bushel and a half. At first I put together eighteen hundred sets of ten pieces of the best bamboo work baskets ever made. We purchased the whole lot of the Hong Kong store inventories at all of the basket stores in Hong Kong and Kowloon. All the baskets were of the same high quality. Interestingly, some of the baskets in the back rows were very dusty and old and could have been one hundred years old or more when I bought them. As the baskets sold, the Chinese basket merchants used to buy more and put the new stock in front and the ones in the back rows just stayed there. I sold them in Putney at fair prices, the biggest

bushel and a half baskets for $12.50 each retail. All the baskets
now should be collector's items of museum quality. Baskets have
always appreciated in value. What happens, and not only in China
but in Europe, is they become little by little of poorer quality. For
example, when baskets like the rice baskets were reordered, they
would tell us that it would take them hundreds of work years to
make four thousand sets of those baskets—however, if we were
to take off this and lighten up that, they would take our order.
Every time around, they would cheapen the quality a bit more.
Now, though the baskets are still quite nice, it is hard to find really
good baskets. Also, overseas the young people do not want to
make baskets because it is hard work. As the old basket makers
around the world die off, there is no one left who knows how to
make nice baskets and basket furniture at all, like the quality I
used to buy in 1967, twenty-one years ago.

While I was in Hong Kong, the Red Guard young people of
China were all around Hong Kong, making a lot of noise and
demonstrating Mao's doctrine. It was scary living at the Hong
Kong Hilton and trying to sleep with all the noise going on. We
could look out the window and see thousands of these Chinese
demonstrating all over town. As Hong Kong is British, the young
Red Guard never bothered the tourists—only people of Chinese
origin.

In May 1967, while I was staying at the Hong Kong Hilton,
the city was full of Red Guard youngsters hooting and hollering
all over town. Thousands of hooligans and demonstrators were
shouting communist slogans and throwing stones. Wild mobs took
over the streets, and the China newspapers claimed that nine
hundred police struggled to break up the hostile mobs. The Hotel
Hilton, which was where we were staying, was the scene of a day-
long battle of the police with the communist-led rioters on May
22, 1967. Groups of teenagers carrying red books and chanting
Mao songs tried several times to break the police bloc to our hotel
but were unsuccessful. Most of the Hong Kong businesses came
to a standstill, and the consolation to us, as American citizens, was
that there were always U.S. warships in evidence visiting in the
Hong Kong harbor, close enough to shore so I could have swam
to them if I needed to.

We were friendly with David Young, who was a newsman with UPI in Hong Kong. We told him we were going into the mainland to buy at the Kuang-chou fair. At the time, the Red Guard was still active in China and David neglected to tell us he would like fresh pictures of the uprising in China, which was still going on. We saw a great deal of demonstrating and marching in China, and when we got back to Hong Kong David asked us for some of the latest photographs and we were very sorry that we could not accommodate him, as it had never occured to us that news was his business. We never thought that we could have been able to give him the latest on China, and we felt badly about it.

My son James and I went around the world in 1959 looking for an experienced basket designer and weaver of bamboo, rattancore, willow, et cetera, but did not have any luck in finding one. On a later trip, I was in Kowloon doing business with a Mr. Chan of the Kowloon Basket Company and he told me of a Mr. Sek Tim Chan who lived in Kowloon and was the world's greatest basket designer. Sek Tim had designed many of the new styles already on the China market. Connie and I visited Mr. Chan at his government apartment in Kowloon, where he lived in a room about twelve feet by eighteen feet with his wife, mother, grandmother, and four children. He consented to come to Putney to design for Basketville. Sek Tim designed for us for twelve years, and his wife and some of the children worked for us off and on. He has helped us and the basket industry very much. His children also are very intelligent and are good and prosperous citizens. It is hard to imagine how well the family got along in the United States. One of the girls works for the FBI now. The family came from a very humble beginning, dating back to the People's Republic of China, where Sek Tim originated.

Sek Tim was sent to China, expenses paid, by Basketville to visit his friends after a few years in Putney. Sek Tim has tales of a very interesting life to tell. He smuggled himself in and out of Hong Kong several times during World War II to see relatives and friends and for some other reasons he never told. He was in Hong Kong when the Japanese were there and had to swim the China Sea to Macao once to escape from the mainland. Politically, he was a follower of Chiang Kai-Sheck. When Sek Tim Chan visited his China from Putney, he always brought gifts, like a TV set, to some old lady, a nonrelative, whom he was very fond of and somehow had saved his life during the war.

Sek Tim Chan retired in 1986 and at this writing is visiting overseas. Certainly he must have quite a lot of savings and he collects social security, so he is able to live quite well. I know he is enjoying himself wherever he is! Sek Tim is very fond of brandy, gambling, and playing the horses, and he also loves the ladies! I have learned that he has purchased a house in his China hometown, where he has taken his mother, who must be nearly one hundred years old, home to die amongst her friends. This is a very great gesture for Sek Tim to do for his mother. I am very fond of the Chinese, as they have a very close family arrangement.

In 1961, I started my own national sales organization and I registered the name Basketville. Basketville is still going strong twenty-seven years later. Now both my sons, Steven and Gregory, are running the business and I am gradually retiring. Steven is in charge of the manufacturing, warehousing, forests, and all matters other than retail. Gregory is responsible for the retail operations and matters concerning retailing situations. Steve handles the wholesale operations, and they work together to solve problems such as the large Basketville factory in China, where they stay in touch with all important problems needing both of their imput.

One of the big deals I made in the fifties concerning imported Maderia willow baskets was when I made a contract with the American Sugar Company to furnish a basket and fill it with an assortment of Domino sugar packages in Putney. We packed 16,500 baskets of assorted sugars and shipped them all by Parcel Post to the stockholders and friends of the sugar company. At the time, with the postal income from the mailing of this order, Putney earned the right to become a first-class post office and the postal service built a nice new post office here.

For the new operations in China the Putney, Vermont, basketry, saws the ash logs, shaves out the standards and filling, cuts the handles, and gets them ready for assembling. All of the basket parts are put into big containers and shipped via water to Hong

Kong. From Boston, the containers are taken inland to China. The very labor-intensive baskets, designed especially for China assembly, are put together in China and shipped complete back to the Brattleboro Warehouse in these containers.

Putney makes all the baskets they can and concentrates on the larger simple baskets, like market and clothes type. It is hard to find help to ensure production in Vermont.

In 1988 my son Steven has his farm on the Cassidy place on East Putney Brook. My son Gregory has his farm on River Road, which he bought from Nancy Watson. In 1966 I purchased the Canney and Boyd farm on the Great Meadow in East Putney. The best way to get to it is to go from Putney village about five miles north on Route 5 and east down Fort Hill Road, and the house and barn are at the intersection of River Road immediately on the left. This was purchased for the company not as a personal property, but I consider it my farm. This property consists of 235 acres. There are one hundred acres of the best river crop-growing land bordering the Connecticut River, the most northerly part of the Great Meadows, with maybe one half-mile frontage on the Connecticut River. This is the land the Putney Fort was located upon in Indian times. The other 135 acres is on the house's west side of the road, which has about 25 acres of pasture, and the rest is in a carefully attended forest. I tried to raise Charolais cattle here for a few years, but I lost my shirt and had to give it up.

Jack Williams now rents this farm and gets milk from a big herd of Holstein cattle. The house is rented to three tenants. I use the barn to store some of my museum stuff. There was quite a lot of work done on the main barn a couple of years ago, and it is in great shape. It has a complete slate roof and is a really nice barn. Several years ago, the rest of the Great Meadows came on the market because the owner, Mr. Bazin, was killed on the meadow as he was unloading corn from a silage pit. His part of the meadow I think was 335 acres or so. If I bought it, I would own around 435 acres of the best cropland one could buy. So I went to the bank and borrowed the money to purchase it. Mrs. Bazin, her lawyer, and my lawyer were in Mrs. Bazin's house waiting for me to close the deal, but my wife Connie was needed to cosign the note with me. Connie refused to cosign the note, saying we owed too much money already. I went to the Bazin house and told them the sale was off and went out of there in a hurry with my tail between my legs. It was perhaps the best deal I could ever have made, and my wife still says she would not cosign for it now under similar circumstances.

13.

Some Notes on Hunting

WHENEVER I TAKE A LONG hunting trip I like to keep a detailed diary of what we did and what went on. I have included in this story some of the accounts of such a trip. The following is an account of a hunting trip that we took in 1953:

We plan to take a moose hunt this October to the Perabonca River section in Quebec, Canada. Since August, we have been busy accumulating gear and anticipating this event.

The Hudson Bay Co. of Pointe-Bleue, on Lake St. John, are to be our outfitters and we are to have three Montmagy Indians as guides. My partner is Dexter (Cub) Kathan of Putney, Vermont, considered by many to be one of the best deer hunters in our section. Both Cub and I are new to this exciting adventure of moose hunting.

The day has come and we are off to Canada—Saturday, October 3, 1953 at 7:30 in the morning. The season in Quebec opens October 9 and we want plenty of time to get to our hunting grounds in time. Cub is to use his trusty .35 Remington Automatic with 200 grain bullets and open sights. I am using a worked-over Gov't 30-06 Springfield with 150 grain bullets and Weaver 2 1/2 power scope.

Sunday we drove through the Lauration Park, which is just above Quebec and extends nearly to Chicoutimi. Most of the way this is an exceptionally fine tar highway and the country is most beautiful, very wild with no houses and few signs of civilization.

Arrived at the Hudson Bay Co. at 9:30 P.M. Saw Mr. Cooter, who is the manager there, and he has all arrangements made so we can start into the bush Tuesday, October 6.

Monday, we met our Indian guides, Clement and Gerard Simeon, two brothers age 37 and 31, respectively, and William Balim, age 58. I must say, these three Indians are gentlemen. They are neat and clean, strong and clever both in the forest and on the water.

Everything is set for tomorrow. We assemble at the Hudson Bay Co. at 8:00 A.M.

Had supper at dark on the Peribonca River. Don't know what time it is as no one in our party has a time piece. Am writing this under the light from a Coleman lantern. Left the Hudson Bay Co. this morning about 9:00. A Hudson Bay truck took us five plus 3 canoes, grub for twenty days and gear about 60 miles to the point of our launching on the Peribonca River about noon. With the help of an 8 H.P. Johnson Motor, we have come about 20 miles upriver to this spot, where we set up our tents.

We have two tents at all times. One for Cub and I and the other for the 3 guides. There is one large 22 foot canoe on which the outboard motor is attached, one 16 foot and one 14. The canoes are floated single file with the large one with motor up front towing the middle sized one and the small one being towed in the rear in tandom.

Wednesday, we broke camp and went upriver all day in rain and snow. The water is very rough and it is cold, wet and uncomfortable for us all. The wind stopped by mid-afternoon but the weather did not clear up until later in the evening. Clement said we will arrive at the hunting grounds tomorrow evening.

October 8, 1953—Thursday—brought with it a clear, nice day and the sun is trying to come out. About noon we left the Peribonca and went up the Manouane River. The last five miles was up rapids which are very swift and dangerous.

Now we are at the spot where we will establish our base camp. This is as far north as we will go. Tomorrow the season starts and we are anxious for the morning. William is to stay at camp and the rest of us are to start the hunt.

The following is how I remember this hunt when I look back on it in 1988.

This eventful trip was our hunt to the wilds of Lake Saint John, Ontario. Dexter Kathan, Sr., and I rode to Lake Saint John at Mistassini, Quebec, in 1953. At the time, the Hudson Bay Company had an outpost to trade with the trappers at Mistassini and also ran a guide service. We contracted for a complete service for the trip, looking to get bull moose or bear. The Hudson Bay Company furnished everything but the guns and sleeping bags. The company trucked us east to the Peribonca River, which flows into Lake Saint John. We chugged up the river with one big cargo canoe and two small canoes for overland portaging. We were three days in the river in wind and rain, with dangerous currents and waves. At points on the river we would hide and store gasoline and food for our return trip. About seventy-five miles north, we branched off to a smaller stream in the area of Peribonca Lake, which is the farthest point we traveled. We made main camp about ten miles up this stream and left the cargo canoe. The river was very steep and dangerous, wide, and fast, with shallow areas. The water was ice cold, and if you crashed in the water you could only survive a short time. The danger of our situation was that the outboard motor had to work hard going upstream to progress. If for any reason the canoe lost upward mobility and backed downward, the motor would hit a boulder in the water and, because the rear canoe motor holder board is quite fragile, the whole back of the canoe would rip off as if being opened by a can opener. There was nothing I could do as I just sat there while the Indians maneuvered the craft. We were all day in the worst part of the river and could look behind us for a half-mile and see that there was at least a fifty-foot drop in the elevation of the mad water. I was so scared and, with nothing to do, lay down in the bottom of the canoe so I could not see the problems and left my life in the hands of the Indian guides.

I am a good swimmer, but it is doubtful that anyone could survive the shock of the cold water. The Indians could not swim at all.

Kathan got his moose the first day, and a few days later our guides agreed to meet at a certain spot and build a fire for a lunch of hot tea and sandwiches. Dexter (Cub) Kathan got there first, taking his rifle with him, and they set up for lunch a hundred yards from the canoe. When we got there, for some reason I left my rifle in the canoe while eating lunch, and while eating I looked behind Kathan and this big bull moose was walking along the hill. I did not say anything but quickly picked up Cub's rifle and ran toward the moose. I stopped and fired twice at him as Kathan was running beside me. He wrestled the rifle from me and did some shooting with it also. The moose walked to the pond nearby and waded out about two hundred yards and died and disappeared in the water headdown because his horns were so heavy. As he was wading through the pond a lot of blood was following him, and interestingly enough, there were many hungry trout following the bloodstream.

The water is as thick as ink in this area, and you are unable to see anything underwater. We happened to have a bear trap with us with three long eight-inch hooks that are used to carry the trap chain and drag along and slow down a bear if it gets caught in the trap. Remember, the moose was two hundred yards from shore. We used this trap to drag the pond to try and hook the moose with the hooks if we were lucky enough to locate him. After most of the day, we thought we had him hooked, but at the bottom of the lake there were a lot of old logs, which impeded the progress of towing him to shore. In order to drag him to shore, we hitched the canoe by rope to the chain of the bear hook, which was hooked to the moose. The Indians paddled as hard as they could but could not move the moose (we hoped it was the moose) a short way. Our next move was to hitch the leather carrying straps the Indians used to carry the canoes. We hooked about one hundred feet of strap to the back of the canoe, and we also had about one hundred feet of heavy rope that we tied to the straps. We happened to have a lightweight roll of string in camp, and we hitched this to the heavier rope. So from the trap to the canoe, with the Indians paddling like hell and Kathan and I on the shore pulling the string and rope connected to the canoe, we saw the bull moose break water near the shore. We pulled the moose to shore, dressed it, and cut off its head with an ax at the lower neck for a trophy. The bull moose horns were fifty-one and three-quarters of an inch wide. We should have left him in the pond for the big fish! It was two miles through two small lakes to the main camp, and his twelve hundred pounds of everything was not worth the effort.

Author and his moose.

Here is some of the original account of another trip that I wrote about in a diary while in the bush in 1954.

September 23, 1954

A typical trip to Canada!

Trip to Lac de Saultes in Quebec, Canada. Started negotiation with Hudson Bay Co. in July. Three of us to go. My dad, Cassius, my brother, Junior, and myself. Decided to fly in from Lac Saint-Jean this year and stay in bush nine days. At the last minute Junior decided not to go. My good neighbor, Bob Austin, consented to go with us. He found he could go at 8 P.M. last night and we were headed north the next morning from Putney at 7:00 A.M. Pretty fast work. Mr. Cooter, of H.B.C., is going to try and get at least one of our last year's guides to go with us, either Clement or Gerard. In Cooter's last letter, dated Sept. 20, he said we would fly in Sat., Sept. 25, in afternoon. The guides will go earlier by plane. So there will be 2 flights in and 2 flights out. At a cost of $70.00 a trip it will cost $280.00 approx. I took my 222 Remington and 30-06. Dad took his 303 Savage and Jr.'s 300 automatic. Bob Austin took his 250-3000 Savage.

We expect to hunt bear especially and Dad wants to get a bull moose. Went thru Norton's Mills customs at 2:30.

At Quebec, speedometer 40,636.

Friday, September 24, 1954

Arrived in Quebec last night at 8:30 P.M. Stayed at the Chateau Frontenac. A very nice hotel. Had dinner at a restaurant in the square. Bob and I had a drink at the Frontenac cocktail room. Went to bed about 11:00. Got up this morning about 8:00. Went to the Parliamentary Building for a topographical map of Cherobougameu and Mistassini Counties. They had none. Did get a geological map of the whole of Quebec, which is not so good for us. Paid $7.95 for five forty-ounce bottles of 4 Roses whiskey. Arrived in Roberval about 5:00. Checked into Hotel Lake St. John. Saw Mr. Cooter at H.B. Co. Bought our licenses. Met our head guide, Jimmy, and his son. Neither Clement nor Gerard, our favorite Hudson Bay guides, can go on this hunt.

Sept. 25

Met Mr. John Carpenter of Tunbridge, Vermont. He wants us to write to him and tell him what luck we had and how much it cost. Make a mental note never to hunt with this guy. Seems to be a crab. Went to H. Bay after breakfast. We were to fly in today but the weather would not let us. Hung around until about 3:00. Then they called saying we could go. Loaded the gear into the truck. By the time we were loaded the airway called again and said the weather was not good. Took the canoes and gear to St. Felician anyway by truck. We drove up in the car with Jimmy and they said they would call us in the morning. Bob and I shot a crow on the way back to hotel at Roberval. Went up to Hotel Maison Blanch and had a bottle of beer then went to a restaurant and had supper. To bed about 9:30. Up at 7:00 and breakfast at 7:30. Mr. Carpenter had breakfast with us again. Airline called up while eating breakfast and said to come right up. When we got there weather had closed in again. Finally the guides went out at 11:00 in a Beaver plane. We left in Husky plane about 2:15 and arrived at Lac de Saultes in 50 minutes. The guides finished erecting camp and tents about 3:00. Dad started to fish standing up in canoe and fell into lake. Got all soaked. He said water is warm. Started to fill magazine of my rifle. Discovered magazine does not work. Doc Prouty has cut too much wood around the magazine and the shells are too deep and won't come out. Lugged the damn gun all way here and it is no good. Bob worked on it 2 hours to try to get it so it will work. My luck!

Sept. 27

Got up early. Breakfast at 7:00. I went north off lake. Bob went up hill near camp. Dad went farther up north and Dad and his guide got their bear about 10:00. An old male bear with quite a few teeth missing. Probably weighed 300 dressed. Henri dragged it out all by himself about 1/4 mile. Did not skin it. It is whole. Saw a cow moose cross the lake this morning about a mile down the stream.

Sept. 28, Tuesday

Up about 7:00, Henri and I went south. Stalked all day but saw no game. Bob and Dad saw none either. Dad's guide, Richard, caught a good batch of trout for supper. Was in good country west of camp today. I am going to get something tomorrow!!? The blueberries are dead ripe and millions of them are all over the area. We skinned the bear this morning.

Sept. 29, Wednesday

Didn't shoot any game today. Neither did Dad or Bob Austin. Saw 3 partagarin on our stalk west. Henri, my guide, threw his ax and got one. When we got to where we could see the river Ashuapmuchuan (Indian River of Many Moose), saw two bear on the other side. No canoe so could not chase them. Later saw a moose in the other side, still no canoe. Caught three fish at camp and had another good trout supper. Dad heard a moose today but did not see him. Dad feels good today. Punched a hole in my boot. Bob is fixing it now. Sun-up and warm—not too cold. Have seen no ice or snow yet.

Sept. 30, Thursday

Has been a strong wind all night. Today is overcast and foggy. Left the Coleman lantern on all night to keep off the dampness. Caught 9 speckled trout after breakfast, caught 17 more trout for dinner in small brook near camp. Went moose hunting about 2:00 till dark. Was on a hot moose trail but came to a lake and lost him. Saw another canoe with 2 Frenchmen on lake. They said they have a big moose they got last night.

October 1, Friday

Been a dull, wet day till after 1:00 P.M. None of us saw any game today. I did not sleep well last night and am kind of tired and dizzy so came into camp early, about 4:00. Had the rest of the speckled trout for supper. Had dinner in the woods today as we did yesterday.

Slept good last night. Started hunting about 8:00. Went south down the lake then east at the outlet. Had Jr.'s 300 automatic, the spare rifle, about 1:00 saw a big bear about 1 1/2 miles on the opposite ridge. Ran 1/2 mile to the stream, which is an outlet of Lac De Saultes. Could not get across because too deep and swift. Went up the stream to a narrow section so we could wade out to where there was only about 15 feet of deep, fast water. Henri found a spruce log laying on the ground about 18 feet long and laid it to another rock so we could walk across the pole. Then Henri, with his hatchet, cut a small tree for use as a pole to balance ourselves across the stream on the spruce log to the other side. Then we ran a mile and intercepted this big black bear about 100 feet away. I took careful aim and pulled the trigger. All I got was a click. At this last click, the bear took off out of sight. Of course I was greatly disappointed. I looked at the cartridge and there was a small dent in the primer. Am of the opinion the firing pin was rusted and later realized that it was greased in, as I sat right down and shot at a tree and the damn thing went off. Later in the day, about 6:00, Henri called a moose at a small lake in the general vicinity. Saw the bushes part here and there till he got to about 200 feet from me. Was waiting for him to show his head when someone shot three times, which sounded nearby. My bull moose took off, crashing in the forest like hell and I did not get a shot at him.

Later I learned that Bob was nearby and did the shooting. He came upon his bull about 200 feet off. He shot and the bull backed up like he was going to sit down. His guide, Jimmy, said, "OK, don't shoot" and then the moose took off running and Bob fired twice more. They trailed the bull moose perhaps a half a mile and he was bleeding badly. As it was nearly dark they have decided to go out in the morning and find him. I think they will.

Oct. 3, Sunday

Bob Austin did not get his moose. He and Jimmy tracked him from after breakfast till 1:00 P.M. The moose bled all the way for approx. 5 miles then went into a lake and they could not find any evidence where the moose came out of the lake. Dad and I did not see anything today. Dad has caught nothing either. Jimmy said his tribe chief is 102 yrs. old. One of his tribe lived to 125 years. Some of the old Indian in the bush have never left the bush or seen an automobile.

October 4, Monday

Got up early at 4:45 and was hunting about one hour before daylight. Called on east side of lake until dawn, then went into bush in east and called some more. Henri thought he had a bull coming but after a couple of hours gave it up when he did not show. Went into camp about 10:30 in the rain and wind. The plane is supposed to pick us up at 1:00 but there is fog now at 2:00 all the way down. Played cards before and after dinner, also all of us shaved before dinner. Cleaned up a little about 3:30. Around 5:00, two planes showed up but did not land because it was too foggy. Showed up again in a half hour but still could not land because the water was too rough. Another plane showed up later but now it is 7:30 and after supper so guess we are here for the night. Played cards again tonight till about 9:30.

October 5, Tuesday

Very windy last night. The tent billowed and shook all the time. Ran out of eggs so had ham, bacon, coffee and trout for breakfast. At 9:00 still a bit windy and sun may come out if we are lucky. Cannot hunt or anything else, just wait for the plane. Henri is very scared of flying. He shakes every time he sees one. Especially yesterday when they were shook around in the wind. Peter Crux, aviator in Norseman, picked us up about 10:00. He radioed in and had to go back and take one canoe. Left two canoes and tents etc. of H.B. Arrived back in St. Felician 12:30. Arrived in Pointe-Bleue about 1:30. Left Pointe-Bleue at 2:00 P.M. Arrived Putney about 6:00 A.M. Sixteen hours. 461 miles from St. Felician to Putney.

The flying charge was $380.00, which is $100.00 more than

Mr. Cooter told us. Peter Crux, the aviator, said he was caught in a hurricane last night with wind at 80 miles per hour. He landed on a lake ten miles from our lake and slept in the plane overnight and was lucky to get down. His motor stopped just before he hit the water, the wind presumably being stronger than the motor at that time. In his 14 years flying, he said it was his worst experience.

Here I would like to share with you accounts from the journal I kept while on this trip in 1954.

Note—What a hunter needs to hunt Quebec in October

1 air mattress
1 good sleeping bag
1 pillow
1 pair leather top rubber boots
1 pr. heavy pants
1 pr. light pants
2 pr. heavy stockings
2 pr. heavy cotton underwear
3 heavy cotton shirts
1 wool hat
1 pr. suspenders
1 belt
1 rifle and 60 cartridges (spare rifle and cartridges a good idea)
1 cover all plastic parker
map of territory
canoe seat
rope
blanket
hunting knife
compass, flashlight, extra batteries and bulb
shaving utensils and small mirror, smokes
boot dryer, 2 pr. gloves, camera and film
toothbrush and paste
1 fish pole, pliers, playing cards
1 Coleman gas lantern with 2 mantles with gas
medicine—aspirin, iodine, support bandage, adhesive tape,
    cold tablets, 1 qt. whiskey, safety pins

Resources—

	Cash in Advance	HB	$100.00
	Frank in Cash		423.00
	Dad		415.00
	Bob Austin		437.00
	Grand Total—		$1375.00

Car's Speedometer reading at Bellows Falls 40,272

White River, Vermont— Thursday—

	coffee, breakfast,	.50
	cigarettes, batteries, iodine, cards and film	8.00
	cigars, cigarettes	8.00
	Bob's ammunition	9.00
	Dad's ammunition	6.00
Littleton, New Hampshire—	1 qt. William Penn	4.10
	hat and gloves—Frank	1.98
	jelly donuts	.60
	hat and gloves—Bob	2.86
Lancaster, New Hampshire—	1 plastic parker—Frank	5.59
		46.49
Stewartville, New Hampshire—	Dinner	4.15
	flashlight and bulbs	.36
Quebec—	Supper in Quebec	9.50
	lunch—liquid	2.00
	breakfast	4.00
	Hotel Room—Hotel Chateau Fronternac	19.00
	5 bottles of 40 ounces 4 Roses	35.85
	11 gallons Canadian gas	5.26
Sept. 24, 1954		
	Supper	4.00
	toothbrush	.30
	Hotel Room	7.00
	Breakfast	2.00
	Dad's moose license	101.00

<table>
<tr><td>My bear license</td><td>10.50</td></tr>
<tr><td>Bob's bear license</td><td>10.50</td></tr>
</table>

Sept. 25, 1954

<table>
<tr><td>Dinner</td><td>4.35</td></tr>
<tr><td>Hudson Bay Co.</td><td>453.00</td></tr>
<tr><td>(makes $553.00 in all)</td><td></td></tr>
<tr><td>fishing tackle</td><td>2.00</td></tr>
<tr><td>beer</td><td>4.50</td></tr>
<tr><td>supper</td><td>3.00</td></tr>
<tr><td></td><td>682.27</td></tr>
</table>

Sept. 26, 1954

<table>
<tr><td>Hotel and breakfast</td><td>9.50</td></tr>
<tr><td>dinner</td><td>2.20</td></tr>
<tr><td>rent Coleman lantern</td><td>2.00</td></tr>
<tr><td>Map of Chibougamo County HB</td><td>.25</td></tr>
<tr><td></td><td>14.95</td></tr>
</table>

<table>
<tr><td>Plane by cash</td><td>290.00</td></tr>
<tr><td>owe plane</td><td>91.40</td></tr>
<tr><td>beer, cigars, cards</td><td>4.05</td></tr>
<tr><td>jar of jam</td><td>2.75</td></tr>
<tr><td>gas</td><td>3.43</td></tr>
<tr><td>dinner</td><td>5.05</td></tr>
<tr><td>2 lunches</td><td>1.25</td></tr>
<tr><td></td><td>397.93</td></tr>
</table>

In different groups, for seven or eight years we went to James Bay in Canada for blue geese, and these were the most fun hunting trips of all. We went by auto to Cochrane, Ontario, where we took the Ontario Northern train called the Blue Goose to Moosonee, which is on the shores of James Bay. It is about a day's trip by the train. From Moosonee, which means "the end of the rail" in Cree Indian language, we canoed toward Hudson Bay to land and camp at the adjoining big Canadian wildlife preserve. We choose to camp just outside the sanctuary.

The Blue Goose of the Ontario North Railroad is a fascinating

About to board the "Blue Goose" train to Moosonee at James Bay.

train to take. It runs 140 miles out of Cochrane, Ontario, and is a dead-end road to Moosonee on the edge of James Bay. The Blue Goose lays over a day at the dead end in Moosonee, and it takes the return trip the following day. The train also services the great Ontario hydroelectric system, which captures the runoff water of the great area of James and Hudson bay headwaters at the Moose River. The people of Moosonee depend upon its services, as there is no highway into town. The train will stop anywhere you want, and it will let hunters and fishermen off at their favorite spots. The railroad will let you use the empty boxcars parked on the side of the tracks for camping. The train stops at the Indian tepees and shacks along the way to take customers and let someone off and on. At the time there were only dirt streets in Moosonee, a fairly modern Hudson Bay store, and a few scattered small general stores. The squaws would clean and dress your geese if you would let them keep the down feathers, which they use for pillow stuffing. Indians traveling into town to visit usually brought the host a generous piece of moose meat. If a girlfriend accepts the gift of meat, he is most welcome.

The Indians have a clever way of calling a flight of geese right out of the sky. I have seen the Indians call a flock of blue geese right down to the Moosonee railroad station.

Our friend and hunting partner, Earl, became friendly with a railway counter girl, Shelia, on a trip north, and the next morning at the lodge where we stayed Shelia appeared at breakfast. "Gee," said Earl, "if I had known she stayed here I could have been schlepping with Shelia."

At Moosonee we hired local Indian guides and canoes to take us out to the area of the game preserve, where we set up tents. When the time of day and weather is right, there are ducks and geese all over, but it is hard to get them in range. I have shot so much up there that I have had to quit because the concussion of the shotgun makes my fingers and hands so sore it is not worth the pain to fire the weapon. If you go to James Bay after blue geese here is a tip: Don't bring any decoys, wood or plastic, fancy or otherwise. Bring several packs of Kleenex tissues with you, as the Indian guides throw them into the wind and from the air the Kleenex, as they scatter on the ground, look like a big flock of white geese decoys. The Indians call the flying blue geese into the scattered Kleenex tissue, which look like geese on the ground.

You have to be very careful in this area and should have a guide, as the tide gets high and dangerous. It is easy to wake up surrounded by water, and it is hard to walk, as the mud sucks your feet down and impedes your travel.

One year we took a whole case of twelve-gauge #2 shot and used it up. Some of us had some spares in our packs, and we shot those, too, and did not get one goose. We had to send the guide back to Moosonee for ammunition.

The blue geese would come down from all over Canada to the preserve, where they do not eat. The geese fly long distances on empty stomachs and get ready for the flight down south to the Dakotas, Florida, Mexico, or wherever they fly nonstop. We were lucky, being at James Bay, to see their southern embarkation twice.

On the day geese fly south the morning wind has to be just right and swiftly moving south. A small group goes up to investigate the situation and you can almost hear them talking to one another. When it is just right the geese start south. Of course they fly overhead just out of gun reach.

All day long they come out of the preserve and make great noises talking to each other. Thousands and thousands, all day long they fly. The sky is darkened with tens of thousands of geese on their way to their winter southern residence.

I would take my children out of school so they could join me in experiences like this. My son Steven takes his son, Jonathan, on similar trips, too. We do not believe in "letting school interfere with a child's education."

During the trips, the guides cook the geese in camp on an open fire of wood in boiling water with vegetables in a big cast-iron black kettle. Their webbed feet are left on while cooking. It is great stew to eat in camp. While they cook you can see the boiling feet circulating around the top of the iron pot. Geese are very tough to eat. Some say you get the rock in the pot good and hot and throw away the goose and eat the rock.

We experienced a sight seldom seen by anyone at a little stream that flowed into James Bay. The stream was only 2 or 3 feet wide and full of slim, hungry brook trout. The trout were so hungry that they ate each other.

Cub Kathan was a very good sportsman. He was into all kinds of sports—baseball, pool, you name it—but his true love was hunting.

Cub Kathan wanted to go into the Canadian bush with the Cree Indians to learn if he could get some trapping tips and ideas. Fur trappers as a rule have many secrets, and they do not choose to share this knowledge too freely. Cub made contact and went into northeastern Quebec to live with the Indians. First he learned that in trapping the beaver, the Indians do not use traps at all. They merely break open the beaver dams and the beaver just stay there dazed and they hit the beaver in the head with a hammer. Cub also learned how they hunted the moose. Moose hunting, as generally known, is a very learned, scientific sport, and all hunters have their favorite theories of stalking and methods of getting their trophies. The Indians have their own style, as Cub discovered. The moose spend a lot of time eating water lilies in ponds and lakes, heads underwater as they forage. They move across water easily because their skin has hair that is like a porcupine quill, hollow and buoyant, and they circulate on water smoothly.

The Cree know how fast they travel and the usual habits a moose performs in water. The way the Indians hunt the moose is quite simple. They sit on the bank of the lake with a good view of the area and their canoe nearby. The Indians know at just what point they can jump in the canoe and paddle fast and overtake the moose in the water. When they get up to it, they hit it in the head with a hammer!

Cub and I visited at the Cree Indian camp of one of our guides. The family lived in tepees and came out of the bush in the summer to replenish their food from the Hudson Bay Company, where they get salt, sugar, flour, tin cans of butter, shotgun and rifle shells, and whatever they need that they can't get in the woods.

The Indians also smoke the game like deer, bear, and moose, and we watched as they smoked bear meat in a miniature tentlike smokehouse. There were some things resembling hot dogs hanging in the smokehouse, and when we questioned what they were the Indians explained that it was blueberry season and the bears ate a lot of blueberries. The things hanging were the upper intestine of the bears, full of berries that had been masticated and when smoked become a really delectable food for the Cree.

When cool weather sets in, the whole family moves into the bush, where they trap for fur five months of the winter. They sell the furs in the spring to the Hudson Bay Company. Hudson Bay also sells nice wool blankets. The best blankets have three stripes, which means they are worth three beavers. Two stripes mean medium quality, and one stripe is for the lower grade. The Indians have to give two geese to the Hudson Bay Company for one shotgun shell because the Indians must shoot more than one goose to buy one shell. At the time the Hudson Bay Company, the oldest company in Canada, was owned by the English and in order for the manager to have a good job in England, he had to spend three years in the wilds of Canada.

While hunting the moose with the Cree Indians on one trip, I saw my first and only big gray woods wolf. He was a really nice, big specimen. The day was rainy and everything was quiet, and we were able to stalk in the forest with no noise at all, as the rainwater softened the whole environment. This wolf must have smelled us and possibly thought we were a newborn moose, which makes nice eating for wolves. Like a ghost, he came running right straight for me, appearing about one hundred feet away. Straight on he came at full speed up to about fifty feet away, when he recognized us. I had my rifle up by then and took my first shot as he turned, running a quarter-circle around me. Twice more I shot at him as he went past me at full speed. I'm sorry to say that was the last I saw of him and the end of this story.

14.

Dexter "Cub" Kathan, an Unforgettable Character

ON THE WAY HOME from one of our trips in Quebec, after roughing it in the bush for ten days, we decided to go first-class and stay at a classy Chateau Frontenac in Quebec City. We thought we ought to live high on the hog after the tough time in the wilds. We showered and put on the best outdoor clothes we had brought and went into the fancy restaurant the Beaver Room downstairs, but they would not let us in without jackets. However, after a bit, the manager found us jackets and let us in.

Cub was a real American fellow of the same caliber as the early American independents. He would rather work in the outdoors but could always get a job in a paper mill. He was a top machine tender in papermaking and did not have to belong to a workers' union to have someone do his talking and negotiating for him. The owner of the Ashulet, New Hampshire, Paper Company asked Cub to run his paper machine as his crew was out on strike. This is the sort of thing Cub liked to do. The papermakers' union thought they would scare Cub and make him give up his scabbing. With all the strikers out of work, they had nothing to do but harass him and his son Billy, who worked with him. The strikers waylaid Cub on his trips from Hinsdale to his nearby home in Putney by bumping him with their cars and trying to get him to wreck. The last straw for Cub was when they nearly succeeded in bumping his car into the Connecticut River from the iron bridge in Brattleboro. Well, Cub had a big 1932 four-door Willys Knight sedan, and he took the back window out for his son, Billy. He drove the Willys and showed his favorite 35 Remington Automatic Rifle when the union boys came around. Billy sat in back with a twelve-gauge shotgun loaded with buckshot. Everyone knew Cub would use his power if things got out of control again,

Dexter "Cub" Kathan, Sr., a master frontiersman who should have been an Indian.

so the strikers did not try any more tricks. When the union people got hungry they went back to work and Cub went back home to Putney. The paper mill gave him an unexpected present of one thousand dollars for his trouble.

As Cub ran the mill, the paper machine kept getting into trouble as the sheets of paper traveling over the machine rolls kept getting torn and no one could ascertain the problem. Finally Cub discovered the strikers had punched a small nail hole in the roof over the machines and drops of tar were falling onto the moving paper, causing much trouble. The extra tar on the roof around the hole would drip once in a while, especially when it was a warm day. It caused havoc to the production.

The Dupont Powder Company initiated a contest and whoever shot the most black crows would win a big prize. Cub wanted to get a live owl, which crows hate, so the crows would fly and peck at the owl wherever they saw one. Cub went to where owls were known to be and put a mirror near a small spring trap. He knew the owls would look at themselves in the mirror, and he set up the trap so it would trigger his trap as the owl looked at the reflection. He caught an owl! He would keep the owl in a cloth bran sack so the owl could not see when not in use. When Cub hunted the black crows, usually down by the Connecticut River, he would tie ten feet of rope to the owl's leg and take him out of the bran sack. Cub would hide nearby and the black crows would go crazy swooping and pecking at the owl. As the crows came in, Cub would shoot them by the hundreds.

Cub worked days chauffeuring and doing yard work for Mrs. Andrews, a rich retired widow in town. This left him plenty of time while the contest ran for shooting crows. He must have shot thousands of crows and counted them by the barrels. He thought for sure he would win the prize, but he didn't. Later he learned that the winner did not play fair. The winner put out corn for the crows as bait near a group of trees. When a lot of crows got in the grain they would touch off a stick of dynamite and hundreds would fall down out of the trees. Either way, the Dupont Company made money selling Cub their shotgun shells and dynamite for the winner.

Cub worked in the basket shop when he had nothing better to do.

Cub bought a small piece of land on Putney Mountain from a Putney lady, as this was one of his favorite hunting grounds. Later, for some reason, the town wanted to get back the property, on which Cub had built a very modest camp. John Maley, the senior selectman, thought he had a great idea to pay off Cub with a great pile of one-dollar bills. Cub was furious that he was thought to be of such a simple mind and read the riot act to Maley. Cub did later let the town buy the land with a check. However, in the meantime, the town wanted to get the camp off the land it was on and pull it to some land that the town owned nearby. Cub got wind of it, and he took up the floorboards in the middle of the cabin, dug a big hole in the ground, poured cement in the hole, and embedded a heavy chain around the main carrying timber and nailed the floorboards back up so no one would know what he had done. The town sent up a light truck and crew and could not pull it away and gave up on it. When Cub settled with the town, they had to disassemble the camp piece by piece.

All of Cub Kathan's boys always got their legal limit of deer every year. The Kathan family hunted together as a group and had certain rules on how to hunt together. Also, they had certain patterns and rules when they were together as a group and also when they allowed others, like myself, to hunt with them in the woods using certain routines. For example, in a hunt of, say, ten people, there might be five sitters at predetermined stands and five pushers with predetermined routes, all to change to other certain duties at prearranged instructions. In larger or smaller groups you were told what your contributions to the hunt were, and if someone outside the family got their game it was okay, but any game the Kathans got was distributed so all had one, no matter who shot it.

Remember, "blood is thicker than water." This particular year, I hunted with Cub mostly every day of the season and I had not gotten my deer and neither had Bud Kathan, who was unable to hunt that day. Cub and I had hunted hard all day in the Chester, Vermont, area and we went home early with no luck. The next day Cub's son Bill told me that Bud had gotten his deer in Chester just over the wall out of sight of where I had parked my car. Of course Cub had kept Bud's record of always getting his buck, but do you call this fair to poor old me, not sharing the deer with me or at least going to all that jazz to deceive me?

Bob Kathan tells me of a year when all the kids in the family had their limit of deer and Cub told his wife, Gladys, she had to claim one. "Gee, Cub," she said, "I never did any hunting and don't even know how to shoot a gun."

The Kathan boys tell me of a time when Cub shook up his whole family when they lived on Route 5 in the most northerly part of town. It seems like there was a tough bunch of men in Brattleboro that Cub knew to be robbers. Cub thought they had followed him home and suspected they were going to rob him of his furs, and he went to bed with this on his mind. In the night, he was awakened because there was a car in the yard and someone was knocking at the door and hollering for someone to open up in the name of the law. The reason was that the sheriff and deputies were looking for mink stolen from the West Brattleboro Mink Ranch and had suspected the trunk of Cub's car, as it had the rank smell of mink in it. They knew Cub's interest in fur and thought Cub could be the robber they were looking for. Truth is, Cub has been to Portsmouth, New Hampshire, and brought home a bunch of lobsters, oysters, and clams, which smelled rank like mink, and there was evidence of fresh stain in the trunk of the car. Cub opened the upstairs window and shouted down, "Get the hell out of here, or I'll shoot the whole bunch of you!" Then he cranked a shell into his pump shotgun, which vibrated a scary clang in the house and terrified the family. He shot a charge in the air. Fortunately, about the same time Sheriff Pat O'Keefe, whom Cub knows well, was able to show he was the law and no one was hurt.

Cub is no longer with us, as he sat on a sharp stump while hunting and it penetrated far into his body. He could not overcome its effects as the poison went through his body and killed him. I miss Cub. He was a lot of fun to hunt with, and I did learn much of hunting and the ways of wildlife and habits of the wild birds and creatures from him.

Our good friends and neighbors the Ellis family, old inhabitants of Putney, were involved in all phases of our basket manufacturing and sales. The patriarch of the clan, old Jim Ellis, was an outdoorsman—hunter, trapper, et cetera—who enjoyed talking about trapping and his hunting experiences. At times he explained to me some of his experiences hunting, fishing, and fur trapping. He also gave some of his trapping secrets to Floyd Ellis to carry on. He told me one of his favorite productive trapping spots was in the upper west part of my River Farm, which is a spot where I also have seen many kinds of wild creatures, such as deer, partridges, fox, loon, and, a surprise one day, a family of coon dogs. Anyhow, one fall day I happened to be at the uppermost part of this open long meadow and I saw Jim a half-mile away visiting his trap line. It may have been the last or among the last times he was in the area. Jim could see little at that time, and he walked slowly and rested often as he progressed up through this area. I did not let him know I was there, and he positively could not see me anyway. However, I feel Jim knew this may have been his last trip and he was slowly savoring his personal memories.

15.

Basketville

BASKETVILLE ALWAYS TRIED TO BE fair with wages to its people. We know all our help's names and families and consider us all in the same fix, trying to make a living. About 1950 we hired a stranger to work for us, and after a while he started to talk up a union in the shop. I learned about it and asked him into the office and asked him how much money would satisfy him to get the hell out of town. He said three hundred dollars cash. I figured it was worth it!

It is supposed to be illegal for a company to shut down the business if it is related to the question of becoming unionized. You have got to be kidding if you think you can't have union troubles as a reason to shut down. I promise you I would not stay in the manufacturing business if my business was subservient to the help. I don't have to manufacture baskets and never made much money at it anyway.

We hired Sam Roth, an elderly friend of a friend, to work with us in Putney, and he wanted to know if we were doing him a favor by giving him a job or if he was doing us a favor by working for us—a good, unusual question. He thought he was doing us a favor by employing him. However, I think that after he had been with us awhile, he could say that after we had educated him about the job he was doing us a favor. Still, it was a good question and a hard one for me to answer. Both employer and employee should be satisfied to work together.

I had a similar situation when we ran a nice basket shop in Milo, Maine. We had a great crew and ran the basket operation with no problems for eight years. The help were real workers and made better-quality baskets than we make in Putney. We used to have whoever was on hand run the boiler, and the manager was

Basketville retail store, Franklin, North Carolina.

The Putney "town tavern."

The Putney town hermit, outside the old Post Office.

The original "Putney" West River Basket Company and an uninvited guest.

Parker's Garage, now the site of the Basketville flagship operations and retail store.

The original basket store in Putney, Vermont.

responsible for the nice, big boiler, which is one reason in the first place that we bought the big American Thread Company in Milo. We got along fine for eight years, but one day the Maine Department of Labor informed us that we were running our business illegally. They said we needed licensed boiler engineers on hand all the time to run the boilers. We did not need high-pressure steam to run the steam boxes and in the winter little pressure was needed for heat, so hiring expensive engineers seemed ridiculous and expensive. However, in order to keep going and make our future plans, we had to hire them. Our manager learned that some of the engineers did not know as much as he did and that he had to show them how to do certain things. In our opinion, we ran a very safe steam operation, and as it progressed, we realized this little basket business could not afford the union engineers. All that some of the licensed engineers wanted to do was sit in a chair and look at the steam gauge. It was below their dignity to throw wood in the furnace. According to them, that was the job for registered firemen, not engineers. It was not that expensive in warm summer months, but in the winter with below freezing weather in order to keep the basket shop from freezing it was important to have a little steam pressure in the vessel twenty-four hours a day. Paying at least four engineers for 168 hours of

A Basketville factory and retail operation, Milo, Maine.

122

work a week to keep our thirty people making baskets was simply too big a burden for Basketville. The buck stopped with me when I told them, "Shut her down." We still feel terrible because we had a great crew who made beautiful baskets and Maine needed and still needs badly work and products made from Maine forests for rural Maine residents' benefits.

Getting back to the Carpenter Brothers of West Brattleboro, where I worked for a few summers, they never gave me any wages during the summer. At the end of the summer, they would add up all the stuff from their farm that I might have taken to Dad's camp nearby, such as apples, vegetables, et cetera, and took it out of my wages.

One year I got paid for the whole summer's work probably less than one hundred dollars at the end of the summer. I went home and invited my mother to go with me and some friends to Island Park, which was just across the bridge toward Hinsdale, New Hampshire, from Brattleboro. (Later the park and island went out in a flood.) There was a big traveling carnival in town with Ferris wheels, merry-go-rounds, girly shows, et cetera. When we got to the carnival, not realizing that the carnies were mostly crooks and pickpockets, I opened my wallet to show my friends and schoolmates how much money I was lugging around. I took my mother to the gambling wheel to try to win a prize, and when I reached for my wallet I realized that it had been stolen. I went to the sheriff, Pat O'Keefe, and town officer, Mr. McKinnon, who were in the area. Pat told me that after the carnival was over we would probably find my wallet in the bushes, empty. Since then I have carried my wallet in my front side pocket and I am very conscious of it, sort of hoping that some pickpocket would try to steal it and I could get even.

Besides making splint baskets at Basketville in Putney, Vermont, much of the time since 1941 we have also made and still do make wooden white pine buckets, pails, and tubs. Now we make an average of seven hundred or eight hundred baskets a day and a hundred wooden buckets.

Heat and steam for bending rims and handles are an important part of basket manufacturing. Boilers have been a big problem for me in my operations.

When I first opened the West River Basketshop in Putney, there was a good sixty-horsepower horizontal boiler in use. This gave me quite a bit of trouble because the water from Sackets Brook was conducive to rusting out boilers. I made a mistake when the original boiler got to leaking. I should have welded it and kept using it, but I put in an upright steamer of about twenty-five horsepower. This was too small and it was hard to keep enough pressure or enough steam to heat and maintain the steam for bending the basket rims and handles.

After my original shop burned down, I inherited the boiler from my father's mill, which had also burned down. This was a good boiler and worked very satisfactorily, like the original boiler in Putney did.

However, the state of Vermont, due to pressure from certain townspeople, condemned this new operation because the wood smoke from the stack was too dark. The activists in town made

Basketville, Putney, Vermont.

The first Basketville retail operation in Putney, Vermont.

it impossible to continue using this system. Little old ladies from the church and the pastor's wife would sneak and prowl the mill yard looking for trouble and telephone the state officials in Montpelier if the smokestack seemed too dark for satisfaction of the law.

As I was trying to be cooperative, my move was to buy a new oil-fired boiler that emitted little smoke, and we built a new boiler room to accommodate it. I spent forty thousand dollars for this boiler, hoping to satisfy everybody.

We found this boiler needed thirty gallons of oil an hour to run, not counting the electricity needed to keep it going. At over $1 a gallon, $30 an hour, or $720 a day, it was prohibitive. We ran the boiler for a few days and then shut it down, and it now sits there rusting away.

We had to go back to running the old boiler. The fuel for it was free, as we burned waste wood. It needed no electricity and we only needed part-time help to throw in the wood.

Eventually, my sons invented a system of burning waste fuel in a new eighty-thousand-dollar boiler system that burns wood with the right color smoke emission and satisfies the law and the activists. With little labor needed to keep it going it is economical.

Another thing that caused us considerable agony, time, and money in the late years was that a Mrs. St. John purchased as an investment the property next door on the bank overlooking the basket shop, up a steep bank, not more than seventy-five feet from the basket-shop building. It was strictly a business deal according to her, and she purchased it to rent to a day-care center and rent apartments for the teachers. It is hard for me to believe that she did not know that there was noise, smoke, and smells from production from our business all the time. Now we are frequently harassed by many of our new neighbors who are crowded into the apartments. They report us to the government and are a burden to us. Can you imagine putting a school practically on top of us? As if that is not enough, now there is a church adjoining the day-care center.

We originally purchased the property across the street at Putney for storage and manufacture. We also opened a part of it for the sale of baskets we made, and eventually the whole building was used for the retail of our splint baskets and wooden buckets. Now, of course, the retail is a most important operation, and we have nine stores scattered down the Atlantic coast, all the way to Venice, Florida.

We opened the first retail store in Putney, Vermont, in 1956. Shortly after, I realized we would have to supplement imported baskets to the store variety because customers asked for baskets that cannot be made at the Putney basketry, where we use native ash and oak logs for the construction of splint baskets. The industry of splint baskets in Putney came about because we have such wonderful oak and ash trees in the area. Soon the customers were asking for baskets made of different materials. We really wanted to sell only our own local baskets, but it seemed silly after a while not to accommodate their requests. We began to buy from Import Basket Jobbers in small quantities. After a bit, we realized

The Manchester "Sunderland" Vermont, Basketville retail operation.

An early picture of our Venice, Florida, retail facility.

the baskets and other merchandise were becoming very important to our volume of sales.

Our second retail store was opened on Route 7 in Sunderland, Vermont, in 1957, giving us continuous operation for thirty-one years. We purchased the property from a Mr. Schultz, fifteen acres, with a frontage on the Battenkill River of one-half-mile. It was bought for eleven thousand dollars, including the barn on the property, which houses the retail operation. This is a great store, and the building has been improved. The property was recently appraised for almost six hundred thousand dollars.

We bought the third store, the Florida operation, in 1957 also, because in those days anyone with any money went to Florida for the winter. This was before the ski business in Vermont became successful and drew people to the area to buy. There were few people and little buying of baskets in wintertime in Vermont in 1967. We have almost ten acres on Route 41, the Tamiami Trail, in Venice, Florida. The store was put together in three purchases and cost twenty-five thousand dollars in all for the land. This type of land sold empty in the Venice area would now cost around $1 million.

We built a modest house in back of the store for seventy-five hundred dollars for the manager who moved from Vermont. Now the new managers have their own home and are Florida residents and we use the house when we are working in the winter in Venice.

Originally we shipped, by railroad, a carload of our baskets to Venice because we wanted to keep our Putney factory busy in the winter and of course keeping the help busy was the reason we went there in the first place. We did not prosper for three years and we tried to sell the store, but fortunately, no one bought it. Now the store is one of our best and most profitable operations.

Mrs. Iris Woolcott and Mrs. Philip Chase, both from Putney, encouraged us to set up the store in Venice. Mrs. Chase's college roommate, Mrs. Robert Baynard of Florida, and Mr. Baynard, her husband, owned the only bank in Venice, the Venice Nokomis Bank. He owned practically all the land in the county, including miles on the Gulf of Mexico. You have to remember that this was when land in Florida was worth little or nothing. This was before the invention and common use of the air conditioner! Bob Baynard paid $.25 an acre for this land, and it cost him $2.50 to buy what is now the Basketville land. Since we went to Venice, there has been a whole city built up around us. Bob Baynard's bank built the store, paved the parking lot, and loaned us capital to get started—all this because our good Putney friend and neighbors told the Baynards to give us a break. We believe this Venice store and all of our Basketville stores are assets to their communities. We think we could charge a fee for the privilege of visiting our basket stores, and we think you would have fun and enjoy seeing what we have to offer.

When my son James returned home from the Vietnam War, he visited his flat work furniture factory (the James Wilson Company), which he had started as a Basketville subsidiary on Birge Street in Brattleboro. He decided that the factory was going well and had plenty of work and a good foreman, Dick Sanborn, who had been operating it as well if not better than when James started it. He decided there was little for him to do there. We traveled to Northboro, Massachusetts, and purchased the Gothic Company, which used to be the woolen mill, 120,000 feet of space, on the Assabet River. James got the shop going and made a lot of furniture. He later decided to put a retail Basketville store in on the top floor. He set up the store and got it ready for operation. He married Janet Flanders, who clerked at the store, about the same time.

James decided to have a grand opening when everything at the retail operation was to his satisfaction. He hired a hot air balloon to try to make the event interesting and make it an exciting grand opening. As they inflated the balloon, it raised up and laid against the electrical wires overhead and because the balloon was of a material that acted as a conductor, my son, standing on the ground holding the ropes, was electrocuted. This was a hard tragedy for our family to overcome, and we will never be the same because of it. Fortunately, we have a grandson, James, Jr., though James never got to see him.

I must report that we had also lost a son at age fifteen, years earlier. Frank Wilson, Jr., was riding with a Putney friend in Milford, Connecticut, and the car he was riding in was hit at an intersection and Frank was killed. Fortunately, I still have two sons, Steven and Gregory, who are taking over the business and each have given us two grandchildren. We are thankful for our sons, who have given us great pleasure, and our daughters-in-law, who have blessed us with fine grandchildren.

The complete list of the current Basketville stores is now Putney and Manchester, Vermont; Lake George, New York; Cape Cod and Sturbridge, Massachusetts; Paradise, Pennsylvania; Williamsburg, Virginia; Myrtle Beach, South Carolina; and Venice, Florida.

My dear and helpful friend and ex-Putney neighbor Reed Alvord told me it grieved him when we bought and opened a store in Paradise, Pennsylvania, instead of the next town, Intercourse, Pennsylvania.

Slowly, as we can afford it, we constantly are on the lookout for new sites for Basketville stores. Soon we will be buying land somewhere for another store and are pleased to hear from those who think they have a perfect spot that we should buy.

16.

The Art of Dealing with Oriental Merchants

I CONTINUED TO STUDY THE overseas market and went overseas seeking new baskets. I went to Nassau in the Bahamas, which is a big market for straw goods, and made a deal with a lady who worked at the Nassau Chamber of Commerce. She agreed to get together straw goods in the Bahamas, add on for her profit, and forward the goods to us in Putney. I left her one thousand dollars in a bank checkbook to get her started and went home and waited for delivery of the first samples. When nothing came, I started writing letters and trying to get in touch with her to find out what was going on; very simply, I had been cheated. I never heard from her again. I guess it was a good lesson, as I never did business that way again.

My wife, Connie, and Florida co-managers, Bob and Jeanette Austin, went with me to Haiti, and we realized the best way to buy in Haiti was to buy what we saw and then ship it to the states ourselves. We became acquainted with an American citizen who lived in Port-au-Prince, and we learned that all the business was conducted at the Iron Market, a government-owned building where there were at least two hundred vendors in small booths. My American friend went to the bank with us, and we changed a lot of U.S. money into Haitian bills. We had enough money to fill up a large shoulder bag, and we set off to the marketplace. Soon we made a deal with a vendor and bought his whole store stock with cash from our bushel of money. We would skip a few because the price was too high, but we purchased the entire inventory from quite a few vendors. By this time some of the businesses where we did not buy because we thought the prices were too high were hounding us, as they had realized we were buying heavily. This was a lot of fun, and we purchased a lot of stuff for our retail stores. We had to pack the merchandise outside the Iron Market on the ground. The Haitians make a huge palm-and-sisal basket about a twenty-bushel size, and we filled quite a few of these up with smaller stuff and shipped the filled baskets without further packing. It was quite a problem finding cartons to put the purchases into for transporting by air to the United States. We hired a truck and trucked the whole bit to the airport. We sent the stuff to Miami, Florida, and we hired a truck to get it back home to our Florida operation. We sold most of the stuff at the Venice retail store. We went to Haiti in 1975 and 1976 for more.

I like to meet the general public of Haiti, just as we do in all areas where we do business. The population is as poor as you can get and still survive. The whole of Haiti is becoming much like a desert. In order to cook, the people search the forests for scraps of twigs and wood to fuel the fires, and here and there a whole tree is lost. I predict that unless something unforeseen happens, soon Haiti will not have one tree left.

When we visited Haiti, Papa Doc Duvalier was running the country and it was a safe place because they needed the income from visitors and catered to them. We learned that if Haitians bothered visitors the government would systematically abduct and shoot them. We enjoyed the people and no one ever bothered us.

We did a lot of business in Hong Kong with the Fookloon Company. They were among the first we imported from, dating back to 1957. We had a shipment at least once a month for many years, and I visited Hong Kong many times. A very fine owner of Fookloon, a basket manufacturer, Mr. Mock, made a lot of stuff for us. In the fifties and sixties we sold oodles of handbags Mr. Mock made for us at his Hong Kong factory, along with the rattancore baskets we brought into the Vermont operation for distribution to our other stores and to wholesalers. He had an associate, Mr. Henry Lee, later on, who also took care of our account.

We went to China with Mr. Henry Lee in 1960, and in 1987 my sons, with the help of Mr. Lee, opened the Basketville factory

Hong Kong factory owner, my son Gregory (with hands in pocket) and Henry Lee, Basketville's representative.

in Sheki, China. They are doing a lot of assembly of splint baskets there. Lately, a young Chinese from Hong Kong, Mr. Steven Wang, is stationed in Putney and is working on Chinese and Japanese situations for Basketville along with other business. At the current time, China is our largest supplier of overseas baskets, wicker furniture, and basket-type merchandise such as woven rugs of sisal, straw, et cetera.

In the late fifties, I was in Hong Kong when Nixon removed the embargo on China-made merchandise. I was one of the first American buyers to go to Kuang-chou, China, the year after I bought the rice baskets. Among other things, I bought the Kuang-chou sample basket room, and one each of these samples are in my basket museum in Putney.

We went to Manila one time, trying to buy some of their basket-type items, but had no luck finding a supplier. The Philippinos make terrible baskets. They are inexpensive, but we do not like to sell the junk. The baskets really look better than they are.

We have purchased Jamaican hats, and this is about all they make that we can sell. Interestingly, the Jamaican (Panama) hats are hand-woven underwater.

We went to the city of Taipei in Taiwan or Formosa for bamboo-type baskets early on in our importation. They are famous for their bread and bun baskets. Most of the baskets are made at the place of residency. While visiting some of these basket places, we unexpectedly found the finished baskets all over the yards drying in the sun. What we did not expect was to see chickens and ducks all about and sometimes on the baskets roaming around. The economic health of Taiwan has made it impossible to compete now. We do not buy there anymore. It was a fun trip, especially riding the little, narrow railroad in tiny cars with tiny engines all the length of Formosa from Taipei to Tainan.

Whenever we fly to the Far East we always spend a couple of days in Hawaii because of the jet lag and from the problem of habit. It is best to rest awhile here before the hustle and bustle of negotiating and buying baskets in the Orient. When haggling with the Chinese over the deal and prices, we are competing with a whole roomful of Chinese who are talking in a language we do not understand. We know we are getting the shaft, but we get worn out and tired in a hurry and hope we make the right decisions so we can get away and do other things. Coming home from the Orient, we also plan to spend time to rest in Hawaii before the final plane home.

We like to buy, for our personal use and the museum, antique ceramics and coins in Hong Kong and Kuang-chow. The Chinese government had one store only in Kuang-chow, and the merchandise was stamped and authenticated with wax stamps. There were a lot of nice antiques in the sixties left to buy in Hong Kong in the Cat Street area. We used to buy what I thought was a lot, spending up to three thousand dollars in U.S. money on each trip. We thought we were real big spenders until we saw the Japanese

"Travel No Closer," Republic of China.

China's Chairman Mao.

Looking for merchandise in Bangkok, Thailand.

buying on one trip in Hong Kong. The Japanese I believe bought all the really nice stuff of Chinese origin, as these were the times when the Japanese started to become rich and were able to get into Chinese antiques first and heavily. What I saw were several Japanese buyers who hired carriers with big boxes in the retail stores to lug out their purchases as they bought and filled up boxes and carried off a mountain of antiques.

Over the years we did pick up a few nice things for our home from Hong Kong. We got quite a bit of nice ceramic pieces and cloisonné on one trip. I asked Mr. Choy, a very learned old Chinaman, to visit the antique stores in China and help me pick out one nice piece for my Putney home. Mr. Choy was the right-hand man, controller, and adviser to Mr. Chan, the owner of Kowloon, the big company I bought from in Hong Kong. I purchased, with Mr. Choy's help and advice, an unusually large pair of lion dogs from the Ming Dynasty for three thousand dollars, as I remember it. I saw identical lion dogs at a New York City antique shop recently priced at ninety thousand dollars.

A typical Hong Kong Street. The area on the right is the Kowloon Basket Company, a major supplier of Chinese baskets.

Mr. Mock, the author, Mrs. Wilson, and Mr. Chow outside the Hong Kong Handbag Factory.

My sons, Steve and Greg, were partying with the Chinese people who negotiated the factory deal in Sheki, China. The Chinese were getting pretty high on liquor, and it was becoming evident that some of them were no longer able to socialize. As they all were conversing in Chinese, my sons did not know what they were talking about. Steve leaned over to Greg and said that it looked like whoever could remain the most sober must be the boss communist. The Chinese manager nearby, speaking in perfect English, replied, "That is right, Steven."

Connie and I happened to be buying at the Kuang-chow fair, which they have every fall. All the Chinese merchandise, minerals, spices, equipment, rugs, machinery, and crafts, can be purchased. Everything is offered every year at this fair, which had been going on as long as anyone can remember. We were invited to the closing banquet of the fair, and there must have been four or five thousand people there. All the merchants, important politicians, and army leaders attend the banquet each year. There had to be up to twenty great varieties of food courses to eat and enjoy and thousands of bottles of Chinese beer and one-hundred-proof Mao Tai liquors, and plenty of "bottoms up toasts" were offered. This has to be one of the best and most unusual events anyone could ever attend. We were two of the few who never learned to use chopsticks, and they did have cutlery for our use. At this time in Kuang-chow, you just did not sit down and buy. Everything took time to go through the chain of command for verification. We became acquainted with buyers for the McCormick Spice Company, buyers who spend millions of dollars for spices and especially need cinnamon, which the Chinese are important suppliers of. The American McCormick buyers were sweating out their contracts and were not sure if their orders and offers would be accepted. At the last minute at the banquet, with their bags packed to go home, they were able to learn with relief that they were going to get the spices. The Chinese government is very honorable and does exactly as it promises. You will get your order as negotiated. It may take a few years to get all you ordered, but you will get it at the prices at which it was contracted.

Connie and I made arrangements to be in Hong Kong to visit with our son Jim, who was serving with the army in the Vietnam War. The army allowed a rest and recreation for the recruits who

The author out front of the Quangchow Fair Building.

served one-year duty in the conflict. We joined him on his R&R at the Hong Kong Hilton, where we liked to stay. We had a good reunion, and he and I reminisced about the war and other things, recording it into an Oriental recording machine I had purchased in Hong Kong. The only problem was that the recorder ran under an electrical system unfamiliar to our American one. As I did nothing about the tape until we had lost Jim in the accident in the U.S. later, we realized this was the only time we had recorded him. But we had lost much of the conversation and are not able to hear his voice and experiences on the recording in any useful manner.

We went to Barbados, as they make a nice little shopping basket out of palm tree leaves. They are supposedly looking to get work for their craftspeople and are subsidized by the U.S.

taxpayers. I chose one of the styles and gave them a sample order for one thousand pieces. It was a considerable job for me to negotiate the order, and they never did anything to fill the contract. I wrote them a few letters asking how they were doing with the sample order, and they never answered the mail. Apparently, they really do not want to work. They just talk a big deal to keep on the U.S. dole.

17.

Around the World in Search of Baskets

WHEREVER I GO, the forests are of special interest to me. In Barbados, the Dutch carried on a big business making rum and in the late 1800s there was a big rebellion of black working people and the Dutch lost much of their help, who were needed for the operation. To replace the hand labor, the rum makers brought in steam engines to do the work. Before the steam engines, the island of Barbados was covered with beautiful mahogany trees. With the exception of just a few in town and along a few streets, the vast forests are gone and grown up to weeds and brush. They cut down all the mahogany trees and fueled the engines with the wood.

One of our first and best countries to buy baskets from is Madeira, which is a Portuguese possession off the coast of Africa. We fly into Lisbon and take a local flight into Madeira. The runway on the island is very small, and it is scary going in there.

Madeira is where they make very nice willow baskets, and basket making is an important business to them. They even grow willow trees very much like American farmers raise apple trees in orchards. Willow is important to them, and they take good care of their trees to ensure they have the willow branches they need to make nice strong, dainty baskets. In the spring, the natives cut all the branches off so all that is left is the right-foot trunk and the base of the branches, which now is shaped like a big ice cream cone. They take the harvested branches to the stream nearby and soak them in warm vats beside the stream. They take a handful and strike them against the hard ground, knocking off the bark. Then the branches are sorted by size and ready to assemble into baskets. The willow is of the highest quality. The closest country

Native Barbados basket workers bringing in the palm leaves.

to have such fine willow is Poland. The Chinese make a lot of willow baskets, but the willow is of poorer quality. The quality of baskets from Madeira is good and constant and not subject to cheapening for the sake of price, as other countries' baskets are. Most countries do not have a younger generation learning and continuing to manufacture nicely crafted baskets. It would be hard or impossible to find anyone in China, for example, who can make the beautiful rattancore or wicker furniture that was last made about 1900. Basketry is hard work, and the new generation has little pride or know-how to make fine baskets. The current generation would rather work in shops making things like radios, televisions, et cetera. With the exception of Madeira, soon you will not be able to buy really nice baskets anymore, as basket making is becoming a lost art.

In Madeira, the people make a lot of nice lace and fancy soft goods. Madeira is one of our favorite places to visit. They have nice hotels and great inexpensive restaurants, and the weather is always great for us. The people are friendly and Madeira has nice shops and pleasant and agreeable citizens. We always rent a small

Two of many Mount Pleasant, South Carolina, basket makers.

Fish baskets and the author's wife on the island of Madiera.

car and travel and explore the little island. Everybody should visit and enjoy Madeira if it is at all possible.

We were also able to visit Macao, which is also a Portuguese possession, located not far from Hong Kong. Mr. Mock, the Hong Kong factory owner, took us by boat to their bun festival, an important holiday there in which they honor their parents. They make platforms much like a big tree and place pastry buns all over the trees, and at a schedule time there is a rush whereby everyone climbs and takes the buns from the tree. There are a lot of bun trees, and this carnival is a colorful festival with a very enjoyable fireworks celebration.

Yugoslavia is one of the first European countries from which I imported baskets. The business was done through the government, and the office is in Zagreb. There was not much incentive for the people to sell the baskets, and they seemed to care less whether we purchased or not. We did visit and import from Yugoslavia for a long time, but eventually they just faded away. The main reason is the quality got so bad the merchandise was hard to sell, so it was not much fun to do business in Yugoslavia.

As in most of the places we visited looking to buy baskets, toilet paper seems to be as bad in Zagreb as it is in India. The toilet paper is rugged, with slivers, and we forgot to bring our corncobs, and with diarrhea, which we get in those countries, it was quite uncomfortable.

We visited Warsaw, Poland, one year but could not find anyone who would sell to us. They claimed they had enough importers and did not need another customer. Poland also did business through the government only. They have good willow and they also make quite a lot of baskets out of maize. We bought a lot of Polish baskets from Skalny Baskets of Rochester, New York, a leading importer of Polish basketry until about 1980, when they sold out. I do not know where you can buy good Polish baskets at fair prices now.

China is the place to buy about anything in baskets at this time. They furnish all kinds of basket crafts made of willow, straw, sisal, maize, bamboo, rattancore, fern, grape, or twigs; about anything you want in our industry you can get in China if you look hard enough for it. Recently, for example, we sent to China from

Putney a forty-foot container of ready to assemble basket stock. The same day we received a container from China holding thirty thousand baskets of premium quality. We have been sending about one container a month to China and receiving one container from them.

We explored Bangkok, Thailand, for basket-related products but could not find any. However, Bangkok is a nice place to explore and if you like silk you had better go there.

We tried India, where they make some nice basket-type products, but the prices were too high. We could not understand how a country with so much unemployment and so many workers has to charge so much for their crafts, but if you want to buy brass you should go to India, as they seem to make and export lots of it.

We have been to Mexico a couple of times and I have bought from them regularly. My impression of Mexican basket craftwork in most cases is that it is of poor quality. If you look around hard enough, you can find a few good baskets, but it is a hard place

Bangkok, Thailand, native transporting merchandise.

to ship out of. I would like to buy from the Oaxaca Indians of Mexico, as they make a nice basket, but there is no way to get them together to ship the baskets to the United States.

On one trip to Mexico at the time we were running our Green Mountain Woodenware Company in Brattleboro, we wanted to try to buy Mexican manufactured glass, which has the best prices in the industry. We went to the main office near Mexico City and tried to give them our business. We wanted to buy full carloads of glass, which can come from Mexico by railroad all the way to Brattleboro. We thought they would welcome us with open arms, but it seems that they did not want to rock the boat. Certain people wanted all the sales income, and unless we agreed to pay certain people commissions for doing nothing for us, we could not buy the glass. This is typical in most countries where governments run the business communities. There is no incentive for the ordinary worker to try to improve the management and improve or expand efficiency. There is little reason for anybody to do anything, as there is no reward. If you are lucky enough to be part of government, you are assured an income, legal and otherwise, but you are at a dead end.

18.

Recollections

MY FIRST EXPERIENCE WITH WOOD was in high school in Brattleboro, when I took a very limited shop course taught by Mr. Ralph Burgess. I graduated from the high school in 1937; while there I played football on the first Brattleboro team, in 1935–36. I never played in a losing game.

I also threw the discus on the track team, and Brattleboro always won. I broke the state record for the discus throw, but I only kept it for a few minutes, because a track member from Rutland threw his farther than I had. My brother Sammy later broke that state record.

My father purchased a steam-operated sawmill in Jamaica, Vermont, while I was in grade school, and he sawed a lot of logs, mostly custom sawing for others. My job at the mill was to run the steam engine that ran the mill. It was very exciting work. I learned I could keep the steam pressure up with the waste green slabs even if they were covered with ice. The exhaust from the steam engine was piped into the smokestack so every time the steam pistol exhausted, it would make a loud pressure blast as it blasted into the smokestack. This pressure, like a fan, would excite the wood fire to an inferno. The whole mill was run by flat belts from pulley to pulley, and the steam engine pulley was at least seven or eight feet in diameter. The pulley had to be thirty inches wide and accommodate a leather belt the full width. This job was a good experience. There was a lot of noise from the steam and the squealing of the belt and the screaming of the big board saw crawling through the logs. The oil that lubricated the steam piston smelled sweet in the mill. One day a twelve-foot board edging got caught on the back of an edging saw and, like an arrow, sailed through the mill through the back wall and out the wall. If there had been anyone in its way, he would have been a goner!

My father hid the payroll one payday, and when he went to pay the help, the money turned up missing. I don't remember how he got the pay together again, because we were always poor and lived from hand to mouth.

Apparently, my father always had a longing to be around wood, because he ran and owned the Jamaica mill, the Basket Shop at North Westminster on the Saxtons River, and also for a time a mill in Marlow, New Hampshire. He and his brothers also had a sawmill way back when he lived on his father's farm in West Dummerston, Vermont.

In the thirties, my father had a chance to buy the basketry called Sidney Gage and Company. This firm was founded by William P. Gage in 1842 and had a reputation for making the best baskets in the industry. Dad borrowed the money to buy the company from a good friend, Winn Hosley of South Londonderry, Vermont. I expect one of the reasons Dad purchased the mill was because he had a built-in labor force. At the time there were ten of us brothers and sisters at home, and he put us to work.

I have fond memories of Winn Hosley, who was one of my dad's closest friends. He traded horses and ran a sawmill in South Londonderry, Vermont, for a living. He was a very successful businessman. We use a lot of logs in our business, and we have to keep a good inventory on hand. Our log piles are rolled off trucks and piled haphazardly in the log piles however they came off the truck. Not Winn Hosley—he made the log scaler file the logs exactly in neat piles with the ends of the logs in perfect symmetry. He smoked cigarettes all day long, lighting a new one from the old one before he threw it away. My father did all the mechanical work on Hosley's personal car, and Winn was very particular. One fall he asked Dad to get his car, a little Star automobile, into perfect condition, as he had a bet that his little Star could go farther up Winhall Mountain in high gear than one of his rich neighbors' cars, which was a big new Buick. When Dad got there to drive up the hill, the whole town was out and all were betting on the contest. Dad won the hill climbing, and he and Winn Hosley made a big chunk of money. The secret of the win is that Dad let most of the air out of the Star and the tires hugged the gravel nicely. The big, powerful Buick, with full air in the tires, bounced and rocked on the hill, which lessened the distance it could travel.

I learned the basket business from the best splint basket craftsman ever. He was William LeClair, over eighty years old, from Saxtons River, Vermont. I apprenticed the crafting of fine baskets from Bill LeClair, so I consider myself a master basket maker. Because of my lifetime association with baskets and basket manufacture worldwide and because I am familiar with basket making and all phases of basket merchandising and know what baskets are made of, I consider myself to be the most well-versed basket man around.

Bill LeClair, of the Gage mill, purchased all of his groceries at the Corner Store—Newton's nearby. He said he squared up his store account every Christmas and Newton always gave him a cigar because they were even.

The Gage mill was all water-powered, and it had a dam on the Saxtons River that was built up to the maximum height of the water with a clever assembly of logs. The top deck was decked over with heavy plank and gradually sloped back upriver for a couple hundred feet. It had twin waterwheels. In its main bearings there was a coal-black wood that lasted for a long time for bearings as long as it was kept wet.

The mill had a weird feeling when it was activated. You cranked up the gates to let the water into the turbines. Everything was hitched together by counter shafts and pullies all over the basketry upstairs and down. When activated everything was silent, but the whole factory works was under full power. Every half-minute or so the silence was broken by slap of the main belt where the long, wide, main leather belt was connected. It was connected here by a thicker lacing, which would go slap when the lacing hit the end pully. There was silence as the mill machinery got up to speed, and as the machines all started to work there was a large assortment of different noises, but there was no evidence of how they were powered.

I used to go to the mill an hour early to oil the line shafting, and I would have to crawl on my belly to reach some of the oil boxes at the lower level of the mill. I would guess there was at least six hundred feet of shafting and at least one hundred pullies.

This included an elevator that ran to all four floors. This operation also had to have a sixty-horse D. M. Dillon boiler, which was needed to heat the mill and the drying forms and kiln. Also, much of the steam was needed for use at the bending boxes, where steam was necessary for the bending of the rim handles and bottom yoke reinforcements. We would have to go to the mill and stoke the boiler with wood for fire about half past five in the morning so that when the crew came on there was enough steam to make the baskets. At noontime we would refill the boiler with enough wood to keep us going all afternoon.

This great basketry operation burned down in May 1943. Dad tried to make baskets in a new shop nearby, but the facilities got worse and worse and never amounted to much. The finished baskets in stock, which burned with the old mill, were all of museum quality and were some of the last of the really nice splint baskets ever made. The good old-fashioned machinery and patterns were never duplicated.

When another water mill burned up at Saxtons River about the same time, I was watching it burn down and I heard one man say to another, "There used to be a dam here by the mill, but there sure ain't a mill here by a damn sight."

At the downriver area of the Gage mill there was a really nice spare steam engine made of lots of shiny brass, which could be used when needed. All of it was destroyed by the fire. I wish I had this beauty in my museum collection.

The unused mill dam washed out in March 1968 after a small hole appeared in the deck planking. Over time the water washed down the hole and undermined the whole base and let the dam wash out. I had noticed the hole for a long time but did not realize the hidden damage being done.

Picture a pretty junky factory of simple log foundation, cheap wood sideboards, tar-paper siding, up-and-down, irregular flooring, rickety open stairways and everything of the most inexpensive building materials, and you are seeing the basket factory that I purchased from Dwight H. Smith and Ernest Parker of Putney, Vermont, in March 1941. I purchased this for nothing down and one hundred dollars a month, and it included the basic machinery and boiler needed to make splint baskets. Today it will be hard to believe, but we had no bathroom! We built a portable two-holer bathroom outside that was used by both men and women. We would just move its position once in a while when it overloaded. The crew brought in food and their own drinks, as there was no drinking water. Any water needed for the boiler was pumped out of Sackets Pond. The roof sagged like a camel's back, and it leaked.

19.

Providing Employment for the Community

TODAY, WITH ALL THE GOVERNMENT laws and restrictions on businesses, it would be impossible for a person to enter a business such as I was able to do. Picture any young energetic person wanting to start a basket business today. Where would he get the money? Where would he learn all the things he would have to know to satisfy the bureaucrats? How would he have gotten the water inspected and the test done for purity? He'd have to have a bathroom; that's for sure. He'd have to educate the workers on how to use the tools so they wouldn't get hurt and sue him. He would have to collect monies for retirement into the Social Security system and carry many mandatory insurances. I never knew about this Social Security system until the collector came to me explaining the law and fining me for my ignorance. And don't forget that to start in business he would have to buy or make special basket machinery. He would certainly need logs, nails, and tools. Some have tried to start in the business, but one would have to be nuts to try to manufacture baskets and buckets to be competitive, because you wouldn't make any money anyway after you got the shop running. To do it right, even on a small scale, today you would need four or five hundred thousand dollars. How crazy can you get? To top off his problems, he would have to satisfy those do-gooders wanting to eliminate manufacturing in Vermont.

When I went into business, this was the start of the bureaucratic system of government we now live under. Before this, everyone took care of themselves and their family and neighbors. This was America, the land of the free. This was the American business revolution before Uncle Sam made laws governing all aspects of your life. This was before people found out that hard work never hurt anyone, "but why take a chance?" Be happy; don't worry; the government will take care of you! Gradually, the government and the new Vermont citizens are trying to eliminate all manufacturing in Vermont.

I ran this shop from 1941 until it burned down flat in 1959. At that time we were making about one thousand baskets a day. We weren't making much money, but we steadily improved the operation and kept growing. We always did and still do make baskets and buckets, and you probably find it hard to believe that we never really made much money manufacturing. We made a lot of money in the retail and importation of merchandise, not from the factories.

From my house in Putney I can look right downtown and see the basket factory. When my factory burned, I heard the town fire alarm ring; it awoke me, as all the fire alarms did. The fire was in the nighttime, when the shop was empty of help. Any time the fire alarm rings, I just go all to pieces and shake and tremble. To give you an idea of what people do, when I saw we were involved in a bad fire I ran down into the cellar of my house to look for my shoes. Who would ever keep their shoes in the cellar? I hunted quite a while before I came to my senses and went to the fire.

In the afternoon of the day of the fire, our set-up lady, Minnie Crawford, noticed the main electric line to the factory from the main switch was bright red. The electrical people were called in, and I expected the problem had been fixed. For some reason or other, this problem must have had something to do with the loss of the basket shop.

Interestingly, our neighbor Glen Davis, who lives right near the shop, rushed over and took pictures of the fire early on in its burning. Unfortunately, Glen did not have any film in the camera so the photographing was in vain. If Glen, who was Maj. Glen Davis of the Vermont police force, had film in his camera, we would have been able to determine what happened.

The basket shop I purchased was called the West River Basket

Company, and I used the same trade name until 1961, when I changed the name to Basketville. The West River Basket Company had been founded in Williamsville, Vermont, and was owned by a Richardson and Lazelle. The shop was right downriver about a mile from the bridge in the town of Williamsville. The basketry burned down and was moved to Putney. I expect the Putney shop was built in a hurry, as it was of such poor construction. I am not aware how Smith and Parker got to own it, but they sold it to me. I did get quite a few good craftspeople at the Putney shop who learned the basket craft from working at the Williamsville mill.

The first day I was in business, I inherited Charles Wade, Lake Styles, and Albert Hakey from the old owners. Charles Wade was my foreman for many years until his death. They were a good bunch to work with and good basket men.

Besides being a good man and worker, Charley had a good sense of humor. He once said to me, "Frank, I wish I knew where I was going to die."

"What do you mean, Charley?" I said, and he replied, "If I knew where I was going to die, then I wouldn't go there."

A similar story was told by a richer townsperson whom Charley asked why he didn't spend some of his money. "You only go around once, and I never met anyone who went around the second time," Charlie explained.

"Hell," his neighbor said, "if I can't take my money with me, then I ain't going to go."

Since Charley Wade, as foremen I have had Ernest Parent, Bill Graham, Bill Kissell, Paul Wade, Jr., Ronnie Simonds, and Robert Dunham, who is still with us. Bobby started with us while in school, and I expect he will retire from us. Harty Bryant worked quite a while as our splint cutter, and he was the best. Harty's brother, Warren, was our Hall Scott log truck driver.

Charles Wade and his family lived upstairs over D. H. Smith's Garage next door toward downtown. He had a large family, and most of them worked for me at one time or another, including his wife, Marge. Lawrence Wade worked with us until he went to work for and retired from the state police. Paul Wade worked with me for a long time and was foreman for a while. For many years, the basket-shop help fought all the fires in Putney. The fire station was nearby, and when the fire alarm blew, the firemen would leave the job and run to fight the fires. Paul Wade was the basket-shop foreman and the fire chief also, and it raised the devil with my costs and production fighting all the Putney town fires for free. I told Paul Wade he could either be fire chief or run the shop, and he chose to be chief.

Bill Graham was our factory foreman for a while and a good boss, too. He had help from his father in running production, as his dad was boss at the American Optical Company in Brattleboro. I did not want to loose Bill Graham, but he went along to be a state policeman and then became Windham County sheriff, a position he still holds.

Many of the people in Putney worked with us at one time or another. In the old days, before the child labor laws ruled that youngsters could not work for money, the school kids could work in the shop after school or on weekends. Before the law, many of the youngsters worked making baskets at their own time. I made sure the kids in Putney always had a jingle of money in their pockets for this. I wonder if the law did our children any good. I look around me and see kids who need an income and they are getting into a lot of trouble because of the idle time, little money, and nothing much to do.

Many of the people who worked with us went along and became very important people in our society. Minnie Crawford worked with us all of her working days. She was the set-up person and a good worker. Most of the Dunhams and Jewetts worked with us at one time or another. Bobby Dunham's mother retired from us, and his father, Gene, worked at the shop until he was eighty-four years old. Fred White, the undertaker, worked at the shop, too, when business was slow. We had three generations of Turners; Gib and Ernest, Sr., worked with us, and Ernest, Jr., is still with us. Jerome Turner, who was completely blind, wove for us. Ernest Turner, Sr., liked to hunt deer, and this hobby kept his family and neighbors in venison.

A lot of the members of the Sam Beam family worked with us, and the daughter, Etta Beam, is still here part-time.

Some of the people who worked with my father at the Sidney Gage Basket Shop ended up working in Putney when his shop burned down. Ernest Thomas from Peterboro Basket Company worked for us a long time, as did Roland Robittalle, a transient from Canada. We also had Howard Loomis, a veteran of the Spanish-American War, in our employ.

Our town clerk at the time, Ernest Parker, worked with us. He was once financially interested in the West River Basket Company. In off seasons some of the farmers and their wives worked with us, such as Pete Loomis, Sr., Ed Stockwell, George Stockwell, Carl Stockwell, his wife Esther, and many other members of the Stockwell family.

In the forties and fifties we had five or six camps where some of the employees lived, located in back of the basket shop. The camps were little better than tar-paper camps, much like the woodsmen and loggers used in the forests while they were logging. We did not charge any rent for them, and the tenants got their water from a spring near Sackets Brook. The Sam Beam family lived in one of the better ones for quite a few years. All the tenants hitched up to the basket shop for free electricity. There was plenty of waste shop wood for free fuel for heat and cooking, too. We had a couple of bachelors who were in one of the camps when we purchased the property. They worked elsewhere, but they had "squatter rights" and lived there for free, but we did not bother them. If someone wanted a camp and went to work for us, we would give them the lumber and tar paper to make their own home. Roland Robittalle was one of the tenants, as was Howard Loomis, the retired Spanish-American War soldier. Howard loved to play poker, and he had a lot of company playing poker with him. I also played some poker with Howard.

Phyllis Stromberg was the first office lady, and she worked for us before she married Roy Stromberg, who worked with us for a while also. We had a ten-by-ten building next to the shop, which we used as an office. About all we had in it was some writing paper and a checkbook. For the longest time Phyllis's chair was an Atlas Company tack box, a wooden basket nail box we always had plenty of.

Mrs. Phyllis Austin Graham, Bill Graham's wife, worked for us when she was single and continued on after she was married, remaining for a long time. I owe her a lot of thanks for helping me prosper during this period. Phyllis has started her own business, the Green Mountain Security Company located in Putney, which is a very well-run and successful enterprise. Her whole family worked for us at one time or another, and Phyllis's mother still works for us at the Florida Basketville operation, where Phyllis's brother, Robert, is co-manager. Shirley Austin, another valued associate, managed the Putney retail operation until she went to work for Margolin of Brattleboro, where she is still active.

Lona (Thurber) McGuire is our office manager now and has been for quite a few years. Lona runs a nice, fair tight ship and does a marvelous job. She has two lovely children whom she is raising by her own effort. Lona is fun to work with. I give her hell once in a while, and she thinks it is a great joke and tells me that I can be replaced.

Quite a few members of the office have come and gone in the past years. A few who come to mind are Lois Mech, Lou McCaffrey, Donna Houghton, Darcy Washburn, Pam Snow, Thelma DeOrsey, Ashley Breslend, and Barb Jones.

Shirley Ellis has been with us forever, it seems. She is mostly in sales and has great rapport with our many wholesale customers. Ken Ellis, Shirley's husband, is our over-the-road truck driver, and Gary Ellis, her son, does everything whenever needed. He has acted in all stages of the company, in computers, sales, and several managerial positions and wherever we have needed him.

Ellis Derrig, as well as his father, also worked for us for some time. Ellis went along and now owns a successful family-run construction company, and Peter Loomis owns a great paper-processing company. Roy Stromberg graduated to a very good position, establishing the Green Mountain Well Company.

The Howard families from Putney were with us off and on. Sam Howard, our next-door neighbor, downtown side, operated the D. H. Smith retail store and filling station successfully for many years. I was chosen to settle the Dwight Smith estate, and I had a hard time convincing Sam Howard, Sr., to buy the store. At the time, he worked for Jennie Mellon, who ran a nice, clean retail store on Putney Square. At the current time, Mountain Paul is running the store, having bought it from Sam Howard's sons. The store seems to be going to the dogs at this time and is no longer neat and clean, but he must be doing okay.

Sam always took care of balancing the cash register at his store, and he is the only one I ever knew who never counted his change. He said that his change never varied enough from day to day to bother counting it, so he only counted the paper money. Our Basketville cash register balancers in our many stores collectively spend at least ten hours a day counting the change in and out of the cash register. How much easier and smarter it would be if nobody had to count their change every day. Just count the bills and that is all you should have to do!

Minnie Coomes ran the restaurant downtown for many years, and we called it the Dirty Spoon. The businesspeople of Putney hung around the place about every morning, and I seldom missed a morning coffee break at Minnie's. Addis Robinson was there almost every day and had good stories to tell. Earl Stockwell, who was an owner of the Putney Paper Company nearby, was almost always at Minnie's, too. Earl and I, being the largest employers in Putney, would divide up the town workers to the best advantage of all. Earl sold out about 1984, and this kind of broke up the hanging around at Minnie's. I was very fond of Minnie Coomes. She was always pleasant and accommodating. Minnie used to say that in her shop millions and millions of dollars changed hands over coffee each day. Now the new businesspeople and others have been hanging around the Corner Country Store—Fairchild's the last few years.

Addis Robinson's wife, Harriet, worked at the basket shop for a long time. She is a really nice person and a real skilled and neat basket craftsperson.

Anita Coomes ran our office for a while and also worked in the office on other work at the same time. Anita is and has been voted Putney town clerk for quite a while since. Harris Coomes, her husband, worked for us until he got his electrician license and opened his own electrical business. Harris has also spent a lot of time promoting sports in town. All the kids and other residents of Putney should be thankful for all he has done, but I have never heard anyone thank him.

Frank Southard, our neighbor north of the shop (where there is now the Putney Day Care and Church), worked for us a long time, and his wife, Doris Southard, was our first and fine Putney retail store manager. Frank was a fine gentleman and a hard worker.

My aunt, Emma Parent, was the retail store manager the longest. Aunt Emma must have worked as store manager at least twenty years, until she and Uncle Ernie moved to Florida and went into retirement. Patricia Rogers is the current manager. She took over after Aunt Emma, and she must be going on her tenth or eleventh year as manager. Pat Rogers is a lot of fun to work with. She puts up a tough front, but she really has a big heart of gold.

The Bucky Clark family, Francis Mounsey, and Nelson Riendeau of Westminster West made baskets with us most of their working lives.

Valentine Danforth, a fine Basketville helper and strong log and lumber handler, worked with us. He was the clown of the town and always showed up at all the parties. He danced and clogged and kept everyone entertained with his stories and manners.

20.

Hunting Ain't What It Used to Be

UP UNTIL 1970, THERE WERE always plenty of deer around Windham County, Vermont. Most of us felt that like our forefathers, we could hunt game any time we needed meat. It was part of our inherited right and a part of our roots to have venison any time we wanted it. The problem was, of course, that it was against the law to shoot deer out of season, but we thought that the wild deer belonged to the townspeople and venison was shared freely with our neighbors. Many big venison parties were held, and when those outings were premeditated, we could count on a big crowd at the cookouts. Those friends, with whom I hunted and partied, had deer camps spread all over the county, and most had venison meat hanging up while at camp. Usually a couple of men would go into a camp a couple of days before hunting season opened to get the venison and to open and clean the camp and get in the firewood for the season.

Once I was out getting venison for camp meat up by Ed Dodd's (formerly Bill Reed's) camp with Earl, Frank Potash, Cliff Gilbert, and others. I was lying in the grass resting, and Frank Potash came by dragging a deer and threw the deer on me. The sharp, hard hooves of the deer hit my head and hurt me badly. I did not complain much and never did anything about it, but it was a long time before the bruises left me.

In the old days when hunting season started you might get thirty men in the camp for the opening day. Beds and mattresses would fill up the whole floor, and some of the campers' wives would send over large orders of supplementary food like baked beans, meat casseroles, homemade bread, and pastries, among other goodies. The men would kick in funds to buy things like sugar, tea, milk, store bread, pastries, cookies and candy, et cetera. Some stayed at camp the whole deer season or a week or whatever time they got from work. Some of the party would bring their musical instruments, and we would have real cowboy music hoe-downs.

Earl Stockwell was mostly there playing his guitar and singing "The Wabash Cannonball," and Raymond Stockwell showed up sometimes. Leon Bushey from Brattleboro was a favorite playing his violin, and sometimes friends from other camps would join the band.

There was always a poker game going on in camp, sometimes lasting late into the evening. There was always plenty of beer and liquor for those who wanted it, but later in the season the booze would dry up and someone would have to go into town on a beer run.

During deer season, the basket shop would stay in production, but it ran with a help shortage. On the first day of deer season forget it. Nothing much got done.

Before cooking venison, you should be warned, it should hang a day or two in order for the carcass to cool. I have known some of us to get in a hurry to eat the heart and liver and if they have not cooled, they will make you sick.

In camp, we would not eat the take-home trophy buck we shot. We cooked and ate the does or small buck with small antlers.

Out of season, the cookouts were mostly a family affair. We would gather at someone's camp or in other favorite places like Happy Valley in Westminster or Acton Hill in West Townshend. In these affairs, as some of them did not care for venison, all kinds of food was served: beef steak, hot dogs, corn baked in the husks (in season), and marshmallows for the kids, browned over the fire.

A friend in Putney, Curtiss Tuff, was hired to cater some of our parties, and he put on a good feed. He is especially good at cooking chicken and ribs and whole pigs. Curtiss came to town to pick apples part-time at the Darrow Orchards in Putney, and while he was here he learned to cater to parties and now enjoys a nice growing business in an expanding outdoor barbecue restaurant.

Sometimes a farmer, like Bob Cassidy in East Putney or Hugh Houghton's friend Charley Hitchcock, in Westminster West, would open his house for a special party. We would slice the whole deer, from the neck to the lower legs, into half-inch slices. We kept the frying pan really hot and threw in a big slice of oleomargarine to cover the bottom of the pan. Then we would brown the meat until the top side of it started to show a raspberry color. We would spread a little more oleo on the top and turn it over in the pan for a short time. This was a one-course meal. We would eat the venison with slices of bread with beer or alcohol chasers.

Up until 1979, the legal Vermont deer had to have antlers at least three inches long. The venison parties got more frequent when the state of Vermont decided there were too many deer around and decided to have a doe season to thin out the herds. The local hunters decided to do the state a favor, so we accelerated the venison cookout parties. Needless to say, when the state of Vermont encouraged us to shoot the does as well as the bucks, the herds dwindled so badly there are no longer many deer or deer hunters out hunting in Windham County anymore. Before the shortage, we used to see thirty or forty deer in one day's hunt. We waited until we could see and get a nice buck, letting all the does and little bucks pass by.

We had deer drives and four or more hunters would spread out a little out of sight of each other and walk the woods, making noises to drive the deer toward other stationary hunters located in good positions, quietly situated out in front of the drivers. Near Fed Knapp's camp on Acton Hill in West Townshend, five of us hunters shot a buck each in one deer drive.

Charley Hitchcock of Westminster West was typical of a rural Vermonter before the invention of the television. This was the time when you made your own fun and especially wanted your friend and neighbors to join in the festivities. Charley lived in a nice, neat farmhouse. His principal income was from the sale of maple syrup, as he had a large sugar bush. Including his and his neighbors' acres he had five thousand taps. In the winter, he had living space at the sugarhouse and spent the sugar season on the job in the big bush when the sap was running. The sugar bush was on the back side of the hill, a couple of irregular miles away. Hugh Houghton and Clarence Reed, nearby neighbors, were constant visitors of Charley Hitchcock's. Anyone could come and visit and jest with and enjoy Charley anytime. His home was open for the comradeship and enjoyment for anyone who walked through the door. I spent a few evenings with Charley myself. Everyone joked, told stories, drank together (if that was their bag), and played the violin, piano, or whatever they were in the mood for. Looking back, I wish I had spent more time visiting Charley. Charley Hitchcock played the violin and piano, and he also had a tinny-looking jigger about the size of a marble placed over the violin bridge that made the violin warble. Friday night was the big night for the festivities, when everyone chipped in two dollars each to buy beer and beef steak. Charley's was the in place before World War II, when Billy Reeds went out of favor. After the war, George Stockwell's place in East Putney became a favorite meeting spot.

The trio Charley Hitchcock, Hugh Houghton, and Clarence Reed had a fun time searching for wild honey. In the old days, it was much like the sport of hunting game. You captured a honeybee and back-tracked it to its home, where of course there would be a supply of honey. Vermont laws say that when you find honey, which is always in the hollow of a tree, you post notice that you found the honey in the tree and intend to come back and get it. These three made a sport out of the process, which is another example of how they made their own fun. It took a lot of time, and sometime days or weeks to find exactly where the honey larder was hidden.

During hunting season, Charley Hitchcock and Clarence Reed were partners in hunting coons with their coon dog, Dixie. In 1933, they shot thirty-three coon for meat and fur.

Billy Reed lived on Putney Mountain at sort of a hidden farm that is now the home of Ed Dodd. Much like Charley Hitchcock's home, Billy's was open to all. Many a venison party was held there. The home of the Dewey Newcomb family in Westminster West was a favorite get-together place, as he was into sharing good hard cider with the boys.

You must realize that in those days—you will find it hard to believe—there were deer everywhere, like around the farms of Billy Reed and Charley Hitchcock. It was nothing to see thirty or forty deer if you were familiar with the natural habits of deer and knew where to look in the vicinity of their homes every day.

When the guys got tired of eating venison, they would dine on beef. The Kathans spent time at Billy's, Dexter, Jr., and Bobby Kathan especially. The Kathans would blow their saxophones as Earl and Ray Stockwell played the guitar. Leon Bushey was on the violin, and Addis Robinson or his brother John at times strummed the banjo. You never saw a sheet of music in the crowd, as everyone played by ear. They were great people and we had some great times.

Joe Gould, another popular man, lived in Putney and was friendly with Doc Prouty, the veterinarian. Mr. Gould was a big man and drove junky old cars on his job as the rural postman. Joe and Doc Prouty were coming down Putney Mountain once as I had my pickup parked and was looking at some deer beside the road. As the law required, I was looking to see if the deer had antlers before shooting them. Joe and the doc stopped and asked me why I didn't shoot at the deer, and I told them because I couldn't see any horns. Their response to that was, "You can't eat the horns."

Mr. Ellsworth Bunker, a famous man dedicated to helping the American cause and who served in many capacities for our elected presidents, was a neighbor who owned a big farm nearby on the West Hill in Dummerston. He was also active in local small-town affairs and open to talk and visit with everyone. He was especially a close, good friend of Harry Truman, the president, and did some hard jobs for him. One day, one of us townspeople asked Elsworth how Harry Truman decided a real hard decision and Ellsworth said he had asked Harry that question and he told him that when he had to make that real major decision he excused himself to wrestle with the problem and went upstairs and had a pee.

21.

A Bit of History

CAPT. JOHN KATHAN, IN 1752, was probably the first to settle in Putney. He set up his camp on the Connecticut River in the most northerly section, right on the Dummerston-Putney town line, where the college boat landing area is now located. This is part of the land Phil Chase, Bill Darrow, and I purchased with the hopes that the state would buy us out at our cost and have the whole area for a public recreation place. We talked the legislature into buying part of it, but they balked at taking it all, which was too bad for the people of Vermont. My contribution was to clean it up and get it ready for use. There was a log cabin chicken house right on the Putney-Dummerston boundary that was falling down and sort of unsightly. I cleaned it up and for this I take credit for bulldozing down and destroying Capt. John Kathan's original log home! Capt. John Kathan was the progenitor of the Putney-Dummerston Kathan clan, some of my best friends.

The Dexter Kathan family of Putney were the best all-around game hunters I have known. Dexter Kathan, Sr. and Jr., were always hunting or fishing at every opportunity. If you wanted to hunt with them, you had to be prepared to go into the forest before daylight and stay in the woods until after dark. While hunting with the Kathans I have shot pheasants, partridges, duck, geese, rabbits, deer, bear, and moose.

Also, Bob Gragen of Putney was a really professional hunter and fisherman and fun to be with in camp. Bob used to like to cook, and if anybody complained about the cooking and service, the complainer had to take over. Once someone complained that the meal tasted like sh—— but caught himself and said, "But it is good."

When I was fourteen years old and going to Canal Street School in Brattleboro, I shot a big (410 pounds, dressed weight) black bear while hunting with my father on Bear Hill in Jamaica, Vermont. I used my father's spare 401 automatic Winchester rifle, which was very heavy and had sights built up with wooden wedges. I was on the side of the mountain at a location on top of a big rock from where I could see a long ways off. I heard my father shooting and soon saw dirt kicking up in front of me as the bullets were landing nearby. I ducked in behind the rock and peeked from behind it. The bear was running slowly parallel to me about one hundred feet away. I opened fire with the 401 rifle, and the bear turned and headed for me. I shot off all of my clip of bullets but one, and I could hear him getting hit. The bear stopped sharp and stood up, looking at me and shaking his head, and turned a bit to walk off. I shot my last cartridge and blew off his lower head and mouth, and he dropped to the ground, unable to function. My dad showed up shortly and shot him through the heart to finish him off. For Dad's lifetime he always talked about the big bear that he got. Today I am still convinced I shot the bear! For those who never dragged a newly shot bear out of the woods, I want to tell you it is a big job. The freshly killed bear weighed 410 pounds hog-dressed, and we dragged him two miles out of the woods. The body was like jelly, and it wrapped itself around everything as we dragged it. Even a sapling no bigger than my finger made the carcass bend around it, causing it to be very hard to untangle and pull along.

Because our big family could use the money, Dad sold the bear to a rich flatlander from New York. He took pictures of the bear with himself standing beside it and pointing his rifle at it. He was to take the bear home and brag to his friends that he shot the bear. We hung the bear up in the Oak Grove Avenue home garage, and the hind feet of the bear hung squeezed to the garage rafters. He was so big that his head lay out flat on the floor.

Personally, I have not done any deer hunting since they passed the new laws allowing you to shoot any kind of deer. We have over three thousand acres of land, much of it in hunting country, and I figure I am now raising game instead of harvesting wildlife. It is my turn to help replenish the woods and ponds and rivers with wildlife. There is a lot of full-growth woods on this land.

Our land is not posted, and I hope that I never have to put

up No Trespassing signs on the land. Unfortunately, more and more land is being posted because of the abuse people are causing to the land, like leaving the gates open, scattering rubbish, and shooting up targets like signs and trees.

In 1966 I went to Tamworth, Ontario, and bought a one-thousand-acre farm from an elderly man, Leo Fleming, through Jerry Storring, a realtor. It was a poor pasture type of property with a lot of rocks and some trees in the hedgerows. Originally I wanted to just have a nice big piece of quiet land and I built a nice little log cabin in the middle of the property and rented the land back to Mr. Fleming. The father, Leo, wanted to turn the property into money, as he thought that his heirs could not handle it. I purchased the land, one thousand acres, for twenty-three thousand dollars, a cost of twenty-three dollars per acre. At this price, nine square feet of land cost a penny. When Leo died, his son, Vince Fleming, rented the land for the same price, one dollar per month per head of cattle in the pasture. Vince told me that he made a lot of money on his cattle, but he did not do a good job, for me, of keeping up the property. He let the fences and barns deteriorate. One of the barns, which he used so the cows could get out of the weather, was so badly kept that manure got so high that the floor collapsed into the cellar. Vince would not fix the floor, so I kicked him out. That is when I started planting trees on the land for the next generation to enjoy. In thirty years or more they will start to mature. Since the one thousand acres, I have accumulated twenty-three hundred acres in all in Canada and all are covered with their original trees plus six hundred thousand more that I had planted. This has been good therapy for me, and I like to go up there and watch the trees grow. Newly planted trees take thirty-five to forty-five years to mature. I planted them so the next generation of people and wildlife can be assured of a nice environment.

Leo Fleming, whom I purchased land from, had run beef cattle on this farm all his life, and he had made a lot of money. Leo told me that once he bought a bag of fertilizer for the garden. That was it. His theory was that he pastured the land as heavy to cows as he could and never spent a dime to improve the property. With no outlay of money, his beef operation from the land generated easy profit.

One of the current problems in Vermont is that many of the new residents, especially those with money, are moving into Vermont and building their homes right on the spots of the residents' favorite deer runways. All of the newcomers build their homes in deer habitats that seem to be on the favorite deer runways. Now, instead of seeing deer passing through and grazing nearby, we have to look at the houses.

22.

Business and Politics

THE IMPORTANCE OF POLITICS CAME home to me when the government told me that I would have to take down all my highway outdoor advertising signs. At the time, Lyndon B. Johnson was president, and Lady Bird Johnson was the originator of the billboard bill. This was also about the period that former president Eisenhower was active in erecting interstate highways through America, and the highway route called for a bypass of our basket shop on Route 5 in Putney. This was the military highway system whereby our enemies could spread into America more quickly. The laws ended up outlawing all visible signs from all highways except a few in high-intensive industrial urban areas.

Before the billboard law, we were starting to enjoy a good tourist trade and had outdoor billboard signs advertising our business in strategic places along our highways. Because we were being bypassed by the new road system, I could not see how the bypassing traffic visiting Vermont would find our store and buy our baskets, buckets, and woodenware, which my family and many other townspeople depended upon to make their living. There is no way, now or then, that I know of to get the attention of strangers without direction signs along the highways. Many tourists are going through the Putney area perhaps for the only time in their lives, and if they don't pass our door, how can they find out that we have a nice store and unusual Vermont-made merchandise made from Vermont trees? We could advertise, like Marlboro Cigarettes, for example, in all the fifty state newspapers, on radio and TV stations, et cetera, in America, but obviously the volume of business we would have to generate to pay this bill would be impractical and impossible. So we are now isolated from this important helpful billboard advertising.

My friends and I also miss direction signs while traveling in strange areas, especially on the interstate systems. Without direction signs you don't know where you are or how to get where you want to be. Most people object to offensive signs and signs duplicated everywhere, like alcohol and cigarette ads. Few like signs like those explaining the value of Carter's Little Liver Pills. I do not object to the cute Burma Shave outdoor signs. This billboard law is an example of bad laws sometimes cooked up by the do-gooders.

The Putney environmentalists took the law into their own hands and had a great time chopping down my roadside Basketville signs. We would make new signs and put them back up, hoping they would not chop them down again, but it was too much fun and my friendly townspeople would get together and have a chopping-down party. One of our best signs, visible from Route 91, was a very expensive to keep up and legal, but we can no longer afford the expense of financing the fun for these people. I am pretty sure who they are, and I really feel sorry about their politics and policies.

The law has also enhanced the income of big advertising outdoor sign operators and big business corporations who have the money to own or rent all the signs the law allows. The law voted for and paid big tax money to the biggies for the loss of the signs that they took down. Small businesses, like us, which the signs were really for, did not get a nickel.

In the long haul, I believe, for selfish reasons, that the signs were good for us for reasons nobody thought about. We had been in the retail tourist business long enough before the laws so many people had visited our stores, and because we run an unusually nice, interesting operation, they come back whenever they are in the area, as they know how to find us, and they have told their friends how to find us. How is anyone, any stranger, any tourist, going to find a new enterprise if he doesn't know about it or how to get there? I made a lot of noise to the lawmakers trying to temporize concerning this law but could not sell my objection. Curiously, one of the do-gooders, an environmental evangelist who spent a lot of time helping Lady Bird Johnson, was Ted Riehle, who lives on a lovely island in Lake Champlain in northern

Vermont. He admits he would have never found his lonely island if he had not seen a highway sign advertising the island on a billboard while he was touring Vermont.

We have the best political system in the world. It may get off track once in a while, but we will fight with our lives to love and protect it.

I try to discourage my children from going into politics. Somebody has to do this, but no matter how perfect your contribution to your country, there will come a time when you are humiliated. Worse yet are the clever newspeople who turn around your perfectly honest, good ideas or voting records to make them appear exactly the opposite of what you said or wrote. Worse still, in my opinion, is to have an opponent (in my case my own minister) who thinks in his own mind God sent him down to represent the people and then gets the church and the neighbors to help him gain his goal.

My experience in the Vermont legislature for four years, 1969–72, was good. During this period, every vote taken would show that I was on the job. This was a very important responsibility, and I consider it the most important thing I ever did. I learned that in the legislature there are mostly good and honest people, properly elected and residents of Vermont. Unlike in many states, in Vermont the elected are truly working for all the people. As chairman of the Committee of Commerce, I never saw evidence of crookedness. The big problem was that there were no leaders. No more than ten of us really took the lead and ran the whole show. Most of those elected were followers.

Even though I was the owner of a small and struggling enterprise and had little time to campaign, the business took a backseat to the legislative duties and suffered accordingly.

In 1969 and 1970, as a freshman legislator, the activities of legislation seemed a natural and easy job for me to understand. I took to working in the legislature like a duck takes to water. My crowning experience was when, with only one term behind me, the legislative body chose me to be chairman of the Commerce Committee. This is the committee that heads up the laws of the Social Security system, insurance, banking, transportation, and all matters concerning the business activities.

BASKET-MAKERS — Following up an earlier photo showing the West River Basket Factory in Williamsville before it burned down in 1924, this is a picture of most of the men employed by the company shortly after the move to Putney in 1925. Front row: Loren Bryant; Charles Wade and Oril Clark. Back row: Charles Little, Ralph Austin, co-owner Clarence Lazelle, Ray Austin and Ray Willard. The other owner, Elwin Richardson, and another employe, Frank Smead, were absent at the time. The photo was sent to the Sunday Phoenix by Mrs. Charles Wade of Putney.

SUNDAY PHOENIX — FEBRUARY 12, 1967 — PAGE 9

The West River crew in Williamsville, Vermont.

I lost the next election to my church pastor in Putney, 984 to 751. He had begun to wonder if I really represented the voters of this district in the legislature. He did not run on a precise platform, preferring to state his views as biases and benefitting in part from two more years of voter dissatisfaction with me. Using his closeness to God, plenty of idle time in his profession, and new voting rights that we legislators gave to the mostly out-of-town residents and out-of-town students of Windham College, my pastor cross-filed and engineered a great campaign that unseated me. He organized voter registration for strangers and students

The author's last run for Vermont state representative.

at Windham College in Putney and put together a fleet of cars and buses to be sure they got to the polls. When I voted at the town hall, the room was filled with people that I had never seen before.

You may have some idea of the personal feeling I encountered when some of my church members, most of my neighbors, the college professors, and members of the Yellow Barn Music Association were openly working for the pastor, believing his rhetoric and his sermons, which obscured gritty realities. As a result, I no longer go to church.

At one time I knew nearly everyone in Putney, but now I

Chairmen of the Vermont House of Representatives in executive session with Governor Davis (extreme right). Author Frank G. Wilson is the fourth from the left in front.

only know the older inhabitants. We have lost our local New England identity as some of the new inhabitants have changed the character of the town.

The winner, my pastor, was out of the mainstream in the House of Representatives, and the local people lost representation. In the legislature, as a lawmaker, he did not get much respect. The legislature either ignored him or ridiculed him. It is probably fair to say that he did not make much of an impression at the State House in Montpelier.

I stand on my record for posterity to show who represented the people of Vermont, me or the pastor who lived in town for five years before being chosen as best to represent the people of the community.

23.

The World's Biggest Basket Museum

MOST OF MY SEMIRETIRED time is spent working on my museum. I have the world's largest collection of baskets. Wherever I go, I try to find interesting baskets for the museum.

Early American farm equipment and old cast-iron farm equipment are also collected for display. There are over three hundred cast-iron farm implement seats, all with a different name or design on them. I am trying to collect one each of the old one-lunger gasoline engines used by farmers during the period from the late 1800s to the invention of the mobile tractor. Every farmer had to have a gas engine, and I have 450 engines in my collection. The engines were used to saw wood, cut and grind grain, run milking machines and cream separators, pump water and oil wells, and run washing machines and butter churns. It is hard to believe these old Early American people could make and invent such wonderful things from strange metal and primitive tools.

I spend a lot of time touring the United States and Canada searching for suitable artifacts for the museum, and it is great fun for me. Besides touring the local New England areas, I spend time looking in northwestern Pennsylvania in the oil fields around Titusville, Oil City, and Bradford. Before electricity became common, gas engines were used to power the pumps that bring the crude oil and natural gas from the earth. When the electric motor became available, the gas engine went out of favor. The gas engine needs constant attention, so it is more practical to use electric motors for pumping. With the advent of the electric motor most of the nice gas engines just sat in the woods unattended. In time, most of the engine buildings in the oil fields fell in, exposing the engines to the weather. The engines thus rusted out and deteriorated. There are thousands of these big engines scattered all over the oil fields, and my job is to find them in as good condition as I can whereby they can be saved and repaired for placing in my museum.

To find the engines I go to the oil fields and search for people who can be of some help to me and ask them where the engines are and if I can buy them. I have met a lot of people chasing these engines, and it is fun for me to search all over the oil fields. Also, I go right to the top of the people at the oil company's offices, like Quaker State, Penzoil, and others, who may have used such engines of their own or might be able to tell me of others who have engines and might like to sell them. At a big office of Quaker State Oil Company, I was directed to an old-timer who used to be in charge of the production of oil in the fields. He had a nice white shirt and suit on. It was a beautiful, sunny day, and he had a perfect excuse to take me into the brush and swamps—dress clothes and shoes on—to enjoy the fresh air and pleasure of getting out of the office to show me the engines that were for sale.

Sometimes I can find some of these engines still running in the fields. The old engines make a large, loud noise from the exhaust system while they are running. A good way to find the engines that are still in top shape is to got to the oil-well areas in the still of the night, shut off the car, and listen for the thump of the engines pumping away at the oil.

No matter where I go, I try to get acquainted with the people I am doing business with, including those who might help me in my hobby of collecting and preserving early American artifacts and older Chrysler automobiles. Wherever I go there are cars, engines, forests—all kind of things I can buy if I am really alert. You can see American vintage in most unusual places. On the main streets of Canton, China, you can see one-lunger engines driving homemade vehicles as taxis. In South America lots of old cars and trucks are still in service, and in the Caribbean old steam and gas engines still run the sugar and rum factories. I keep a briefcase with me at all times with the names and telephone numbers of all the people I do business with and others I enjoy meeting who have common interests. Especially in the States and Canada, but all over the world, I make it a point to check in with interesting people when I am in their area.

I like to visit with people, and it is the best way to keep busy when you are out of town. I am not one to be away from home and standing still or hanging out in a bar room. Keep up to date with your interests, and when you have to travel about the world check your records and contact your friends. Keep up to date, learn with and from your friends and acquaintances, and if you work at it, you will find people you can visit and learn from everywhere.

In addition, Connie and I have the largest personal collection of Bennington, Vermont, pottery of anyone I have heard of. The Bennington Pottery Company was one of Vermont's first industries, but the pottery closed down because of financial difficulties in May 1858. We have about two hundred Bennington pieces, including the Indian maiden parian piece shown in their catalog. It is the only one in existence. Once we loaned a nice piece of unusual colored parian to the Bennington museum and several years later when we went to get it they begged us not to take it away from the people, so we left it there. The lesson we learned was that you should not loan out something that you enjoy having in your home. There just aren't any Bennington-made articles on the market today. If we happen to come across a piece to buy it is a great occasion.

My auto mechanics friend who owns and runs a modest repair garage in Quebec helped me find old one-lunger engines and cast-iron implement seats. It is about impossible to deal in Quebec, Canada, because of the language barrier. My friend can speak French and English very well and was helpful in trying to help me clean Quebec out of one-lunger engines. What I want to mention about him, however, is a motorcycle. He has a 1964 Knuckle-Head Harley Davidson motorcycle. This is a very rare and valuable cycle, and he told me that he would like to sell it but he cannot advertise it. He asked me to see if I could get him fifteen thousand dollars in Canadian money for it. He requested I not tell anyone where it was because a Canadian motorcycle club member would find it and steal it. I told him I would ask around for buyers, but I had to keep its location secret. He said that if they knew he had it, they would just come to his place and take it and there would be little he could do about it. He keeps the wheels in one shed and the frame in another to slow down potential motorcycle robbers if they come. So far I have not found a buyer, and it is hard to sell under these circumstances. If I can't help him sell it, I would like to buy it and put in my museum.

Many of the old things I have the people of future generations can see to get a better idea of the history of their forefathers' way of life. Most of the old Early American tools and machinery are no longer used and have gone to the junk dealers and are becoming increasingly harder to find. There are certain things of the past that ought to be maintained in order to know where we come from and help us know where we are going. The Smithsonian Institution in Washington, D.C., is the best example of saving and showing similar things.

I will explain about the Wild West of Putney. We became interested in Charalais beef cattle, which are a big exotic type of beef critter. We went to the Royal Winter Fair in Toronto in 1970 and bought our mother cow, Bamboulla, a full French Charalais who was a blue-ribbon fair winner and bred to a famous bull Encore for $10,500. We raised up to thirty-five Charalais, mostly of her own bloodline, before we quit. Because the calves were born so big and matured so large and because we sold them by the pound, we thought we were going to make a lot of money in the beef cow business. Because the calves were born so big, it was hard on the mother cows. We lost a lot of our newborn calves because of it. This also is hard on the mother, because she gives milk and with no calf to feed it is a chore to dry her off. We tried hard to make a profit, but because of this problem, we could not satisfy the IRS that we were trying to make a profit. The IRS rules say that if you do not show a profit over a period of time, you have a hobby, not a business. If you have other businesses that showed a profit, the IRS allows you to consider expenses as a new venture, so we enjoyed refund checks from the government while we were building up the herd. However, it came to a position where we were taking too long to show a profit, so we were not to expect any more refund checks from the IRS from the beef business.

The exotic mother cow, Bamboulla, cost us $10,500 and we

considered those few nice show-type offspring to be worth $10,000 each also in the exotic market if we could sell them as expensive exotic, big beef cows. The last year Harry Bach, our accountant, said we lost fifty thousand dollars in the beef business. I sold the whole herd for beef at thirty-five cents live weight per pound or some $10,000 cows for $350.00 each.

Among the other problems we encountered was getting the mother cows bred. If we missed too many heats in our cow's breeding cycle, the cow is just a burden and unprofitable, and many of them did not get bred at the right time. As I had very little time to work the cows personally all during this period when we were experiencing bad luck, we hired a herdsman, Dave Hoskeer, to take charge of the cows, and we thought he was doing a great job. We kept the cows in the summer breeding season at our River Farm pasture at the Great Meadows in Putney. At the house where Mr. and Mrs. Hoskeer lived, adjoining the pasture up the hill three-quarters of a mile, our neighbors, Mr. and Mrs. Eugene Boyd, had a house trailer. Dave Hoskeer spent a lot of time in the pasture checking the cows for breeding, and we figured he was doing a fine job, as they required a lot of time to catch them in heat ready for breeding. We later learned that Dave was not attending to the breeding of the cows; he was attending to Mrs. Boyd. The Boyds got divorced and Dave moved in with Mrs. Boyd when we let him go. Later they were married.

We worked hard at trying to do the right thing and make the cow business pay, but things just never worked out. I doubt if many ranches made much money on Charalais, and even with lots more luck now, I don't think it is a good business. To top it off, when I had to sell the market was flooded and beef was at an all-time low.

Epilogue

LOOKING BACK, IT SEEMS THAT those people who were successful in business and lived long enough did almost the same as I did, only, of course, in a much bigger way. They worked hard to build their businesses and were not able to spend as much time as they wanted with their families. When they got more independent and money was easier to come by, they had a house or houses in the country or at recreational resorts, raised animals of some sort (horses, beef, dairy, or whatever), liked to travel (depending on their money by jet, yacht, recreational vehicle, or whatever), collected, and even built museums, depending on the availability of money—paintings, jewels, artifacts, something to take up their withdrawal from business. Successful people have drive and they must keep busy. Many become interested in politics and worked at helping their people and country. Most tried to leave the country better than how they found it. One way in which most of them differ from me is that many have girl friends or have been married more than once.

If I had my life to do over again, I don't think I would change much if I could, but of course it can't be changed anyway. As I look back over the years, I see that everything that happened was meant to be. Everything I have done, everywhere I have visited, and every person I have met has fulfilled my life and made me what I am today.